Società Chimica Italiana
Divisione di Chimica Organica

SEMINARS IN ORGANIC SYNTHESIS

XXXVI "A. Corbella" Summer School

June 13-17, 2011
Palazzo Feltrinelli
Università degli Studi di Milano
Gargnano (BS)

ISBN 978-88-86208-68-0

SPONSORS

Con il patrocinio
del Comune di
Gargnano

Università degli
Studi
di Milano

GRACE

INDEX

THE INTERPLAY OF LIQUID CHROMATOGRAPHY, MASS SPECTROMETRY AND ORGANIC SYNTHESIS

ANYWHERE THERE'S CHEMISTRY

CRITICAL SURVEYS COVERING THE YEAR 2010

FOREWORD

This volume of the series "Seminars in Organic Synthesis" collects the lectures presented at the XXXVI edition of the "Attilio Corbella" Summer School on Organic Synthesis, held in its usual setting at Palazzo Feltrinelli in Gargnano, Italy, on June 13-17, 2011.

The School, now in its 36th edition, is promoted by the Organic Chemistry Division of the Italian Chemical Society with the aim of confronting PhD students and junior researchers from the Chemical Industry with the most recent advances in the field of synthetic organic chemistry and with synthetic chemistry applications in related areas.

This year, the Organizing Committee has selected the following four themes, covering advances in fundamental areas of organic synthesis, analytical methods and synthetic applications in (bio)material sciences:
- *Synthetic Chemistry and Biomaterials: Improving Nature with Chemistry*
- *Silicon and Phosphorous Compounds: Preparations and Synthetic Applications*
- *The Interplay of Liquid Chromatography, Mass Spectrometry and Organic Synthesis*
- *New Insights in Oxidation Chemistry*

Each theme has been covered by two prominent Italian scientists, with the editorial coordination of one member of the Organizing Committee. The resulting eight contributions represent as many chapters of this book.

Two additional topics have been covered by two selected industrial researchers who have animated the *Industrial Perspective* section of the School. Their contributions deal with *The Drug Discovery Process: What a Chemist Can Contribute* and with *The Discovery of Mandipropamid: a Potent and Selective Compound for the Control of Oomycetes Diseases.*

The first chapter of the book is a contribution from Benjamin List, the well-known pioneer in the field of organocatalysis, whom we are specially proud to host this year as our foreign guest speaker.

This volume also includes the usual critical surveys of the most significant literature reports (published in 2010) in the fields of:
- *Organocatalysis in Organic Synthesis*
- *Introduction and Transformation of Functional Groups*
- *Total Synthesis of Natural Products*
- *Biocatalysis in Organic Synthesis*

A seminar on the *Molecular Secrets of Gastronomy* concludes the program of the School with an overview of the (food) chemistry involved in a typical Italian menu.

In presenting this Edition of the "Corbella" School, it is a pleasure for me to thank all the Authors for their contributions and in particular for their extended presence in Gargnano: setting aside a whole week for our students is a gift which has become increasingly difficult to integrate in today's schedules and which we all fully appreciate. My thanks are due to the Organizing Committee for the selection of stimulating topics and for editing the speakers' contributions. I am also deeply grateful to Enrico Marcantoni and Gabriele Renzi (University of Camerino) who have coordinated the publication of this book with the usual passion and great patience and to Clelia Giannini, secretary of the Committee, for her active cooperation that actually made possible both this book and this Edition of the "Corbella" School. Also this year Pino Cascio has been the web-master of the School site http://www.corsoestivocorbella.unimi.it/ and I wish to thank him for the many additions and improvements which are making the site a valuable tool and a repository of the School presentations.

Finally, I would like to take this opportunity to acknowledge the University of Milan and the Town Council of Gargnano for their financial help. My sincere gratitude also goes to the following sponsors that contributed generously to the School: Aptuit-Verona, Biotage, Bracco, Bruker BioSpin, Carlo Erba, CEM, Grace, Indena, and Syngenta.

My final thought goes to the students of this Edition of the School: thank you for being here with your passion for Chemistry which makes all this work meaningful.

Anna Bernardi

Chairperson of the Organizing Committee

ORGANIZING COMMITTEE

ANNA BERNARDI	*Università di Milano*
ROSA LANZETTA	*Università di Napoli Federico II*
ENRICO MARCANTONI	*Università di Camerino*
PAOLO TECILLA	*Università di Trieste*
CLAUDIO VILLANI	*Università di Roma "La Sapienza"*
CLELIA GIANNINI	*Università di Milano*

Printed June 2011 by

ARTELI**TO** Industria Grafica

Camerino (MC)

Editors: Enrico Marcantoni

             Gabriele Renzi

Benjamin List
Professor of Organic Chemistry

Since 2005, Benjamin List has been a director at the Max-Planck-Institut für Kohlenforschung in Mülheim an der Ruhr (Germany). He obtained his Ph.D. in 1997 at the Johann-Wolfgang-Goethe-University in Frankfurt am Main. From 1997 until 1998 he conducted postdoctoral research at The Scripps Research Institute in La Jolla (USA) and became an assistant professor there in January 1999. In 2003 he joined the Max-Planck-Institut für Kohlenforschung in Mülheim. He has been an honorary professor at the University of Cologne since 2004.

Professor List's research focuses on organic synthesis and catalysis. He has contributed fundamental concepts to chemical synthesis including aminocatalysis, enamine catalysis, and asymmetric-counter-anion-directed catalysis (ACDC). In 2006 he introduced the concept of asymmetric counter-anion-directed catalysis (ACDC). This very general strategy for asymmetric synthesis has recently found widespread use in organocatalysis, transition metal catalysis, and Lewis acid catalysis.

The accomplishments of Ben List's group have been recognized with the Synthesis-Synlett Journal Award in 2000, the Carl-Duisberg-Memorial Award in 2003, the Degussa Prize for Chiral Chemistry, the Lieseberg Prize of the University of Heidelberg, and the "Dozentenstipendium" of the German Chemical Industry in 2004. He received the Novartis Young Investigator Award in 2005, the JSPS-Fellowship Award in 2006, the Award of the German Chemical Industry and the Astra Zeneca Research Award in Organic Chemistry in 2007, and was named a Thomson Reuters Citation Laureate in 2009. In addition he has held many appointments as visiting Professor and named lectureships.

Asymmetric Aminocatalysis

Benjamin List

Max-Planck-Institut für Kohlenforschung, Mülheim an der Ruhr (Germany), list@mpi-muelheim.mpg.de

We explore small-molecule amines as asymmetric catalysts for carbonyl transformations. "Asymmetric Aminocatalysis" functions by activating carbonyl compounds as iminium ions and enamines using a catalytic amount of an amine. Our strategy complements existing catalytic methods based on chiral acids and bases that typically involve metals. After our studies on the first asymmetric proline-catalyzed intermolecular aldol reaction in 2000, we realized that enamine catalysis has the potential to be a general strategy for the catalytic utilization of carbanion equivalents. Since then we have discovered several other transformations that can be catalyzed by primary or secondary amines. Currently our research aims towards the synthesis and identification of new catalyst structures to tackle further exciting and challenging problems and to discover new amine-catalyzed transformations.

In my presentation, I will describe the historical roots of aminocatalysis as well as its renaissance and more recent developments.

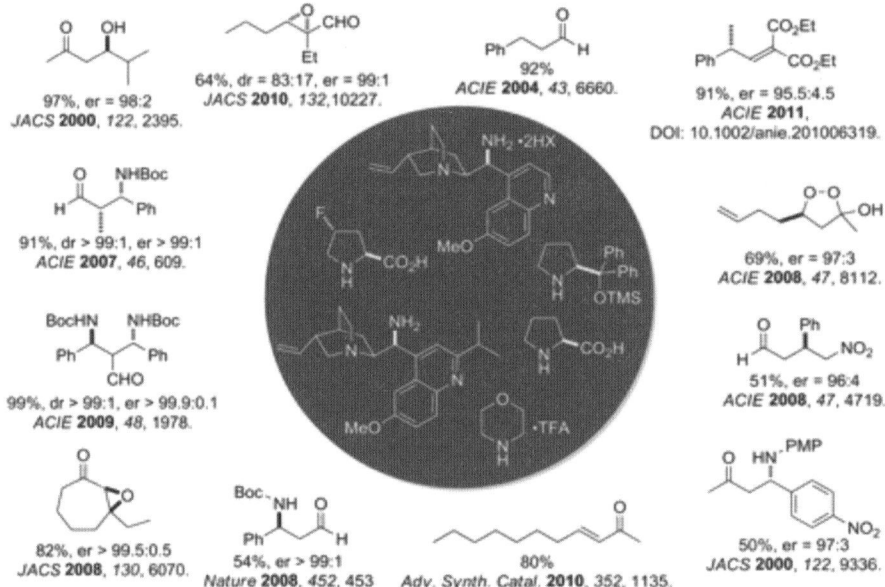

97%, er = 98:2
JACS **2000**, *122*, 2395.

64%, dr = 83:17, er = 99:1
JACS **2010**, *132*, 10227.

92%
ACIE **2004**, *43*, 6660.

91%, er = 95.5:4.5
ACIE **2011**,
DOI: 10.1002/anie.201006319.

91%, dr > 99:1, er > 99:1
ACIE **2007**, *46*, 609.

69%, er = 97:3
ACIE **2008**, *47*, 8112.

99%, dr > 99:1, er > 99.9:0.1
ACIE **2009**, *48*, 1978.

51%, er = 96:4
ACIE **2008**, *47*, 4719.

82%, er > 99.5:0.5
JACS **2008**, *130*, 6070.

54%, er > 99:1
Nature **2008**, *452*, 453

80%
Adv. Synth. Catal. **2010**, *352*, 1135.

50%, er = 97:3
JACS **2000**, *122*, 9336.

A Third Approach to Asymmetric Catalysis

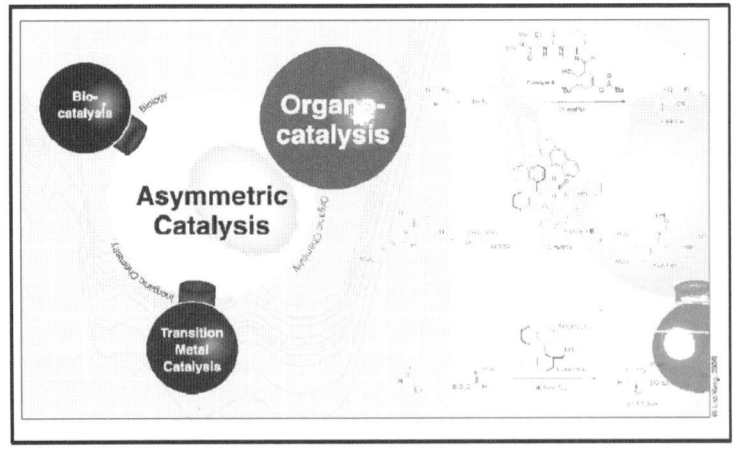

Benjamin List & Yung Woon Yang: The Organic Approach to Asymmetric Catalysis.
Science **2006**, *313* (5793), 1584-1586.

Historic Milestones

1912	**Bredig** uses Cinchona Alkaloids in first examples of non-enzymatic asymmetric catalysis	

ca. 10% ee

1920-40s	**Langenbeck** studies organocatalysis and writes book on "Organic Catalysis and Enzyme Action"	

1964	**Pracejus** achieves high enantioselectivities (>70% ee) using Cinchona alkaloid catalysts	

76% ee

1971	**Hajos** and **Wiechert** use Proline as a catalyst	

93% ee

Organocatalysis

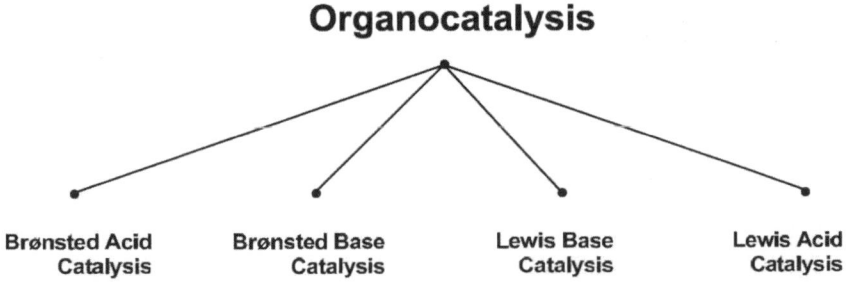

Brønsted Acid Catalysis **Brønsted Base Catalysis** **Lewis Base Catalysis** **Lewis Acid Catalysis**

Organocatalysis is the catalysis with small organic molecules that do not contain a metal as part of their active principle, and that function by donating or removing electrons or protons...

Aminocatalysis:
activation of carbonyls via
iminium ion and enamine
intermediates

Biological Aminocatalysis: Aldolases

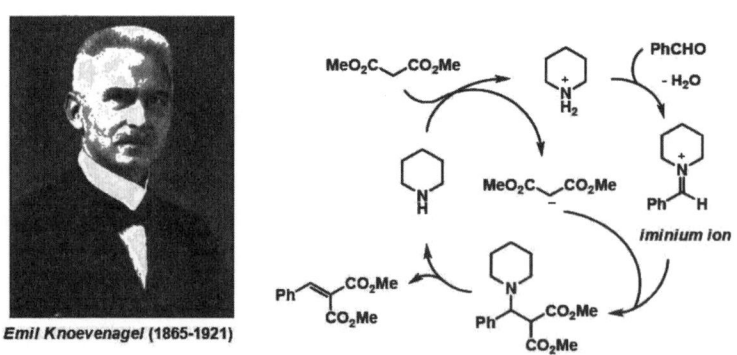

Emil Knoevenagel and The Roots of Organocatalysis

Emil Knoevenagel (1865-1921)

The Knoevenagel Condensation (1896)

Essay: *Angew. Chem. Int. Ed.* 2010, 1730 – 1734.

The Roots of Aminocatalysis: Kuhn

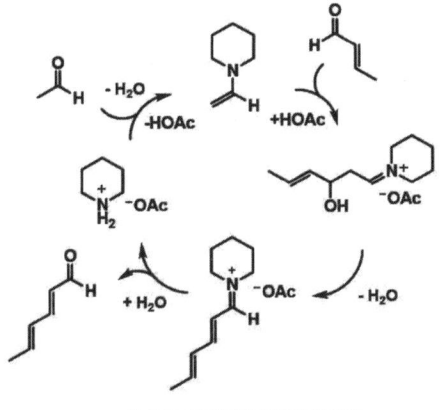

Aldehyde Aldolization (1936)

Richard Kuhn
(1900-1967)

The Roots of Aminocatalysis: Langenbeck

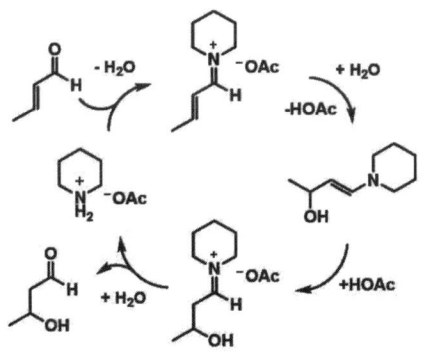

Crotonaldehyd Hydration (1937)

Wolfgang Langenbeck
(1899-1967)

The Synthesis of Steroids: Marker Degradation

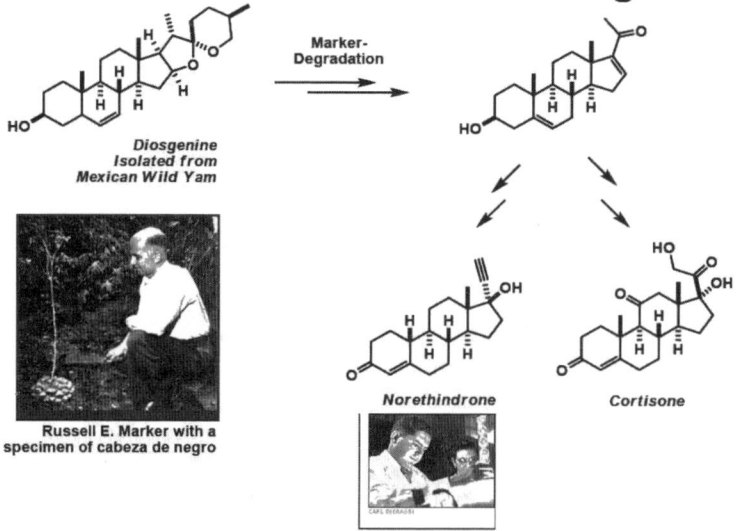

Marker-Degradation →

Diosgenine
Isolated from
Mexican Wild Yam

Russell E. Marker with a specimen of cabeza de negro

Norethindrone

Cortisone

Possible Industrial Total Synthesis of Steroids

Has been prepared
via resolution

Norethindrone

Cortisone

Asymmetric Aldolization
- - - - - - - - →

Hoffmann-La Roche (USA)
Schering (Berlin)

The Hajos-Parrish-Eder-Sauer-Wiechert Reaction

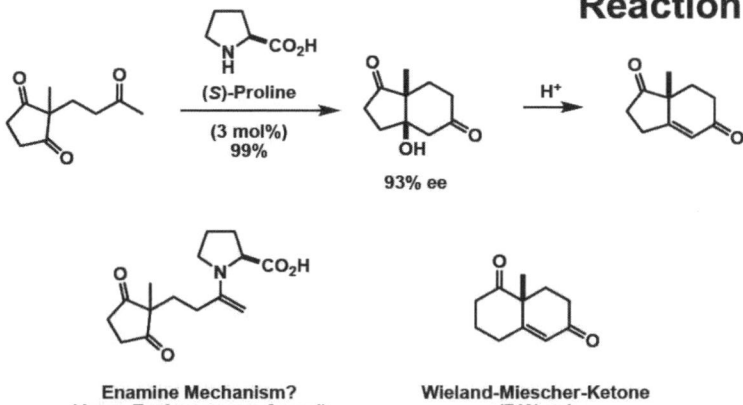

Enamine Mechanism?
(Jung, Eschenmoser, Agami)

Wieland-Miescher-Ketone
(71% ee)

Hajos, Parrish DE 2102623, July 29, 1971   Eder, Sauer, Wiechert DE 2014757, Oct. 7, 1971

The Proline-Catalyzed Direct Asymmetric Aldol Reaction

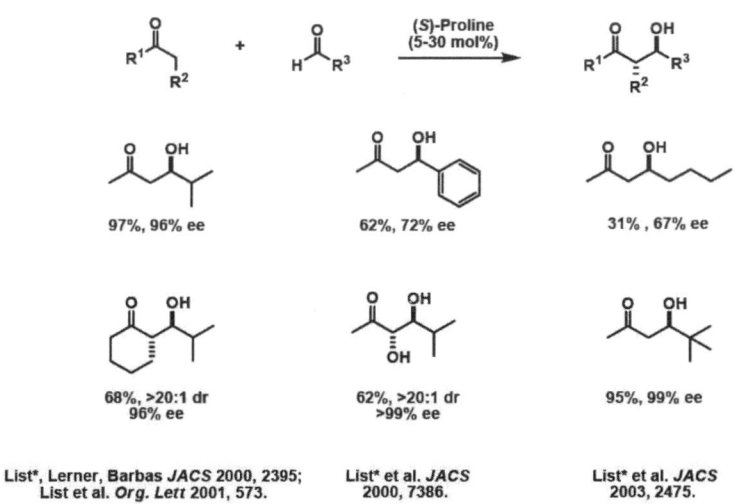

97%, 96% ee

62%, 72% ee

31% , 67% ee

68%, >20:1 dr
96% ee

62%, >20:1 dr
>99% ee

95%, 99% ee

List*, Lerner, Barbas *JACS* 2000, 2395;
List et al. *Org. Lett* 2001, 573.

List* et al. *JACS*
2000, 7386.

List* et al. *JACS*
2003, 2475.

The Proline-Catalyzed Direct Asymmetric Aldol Reaction: Mechanisms

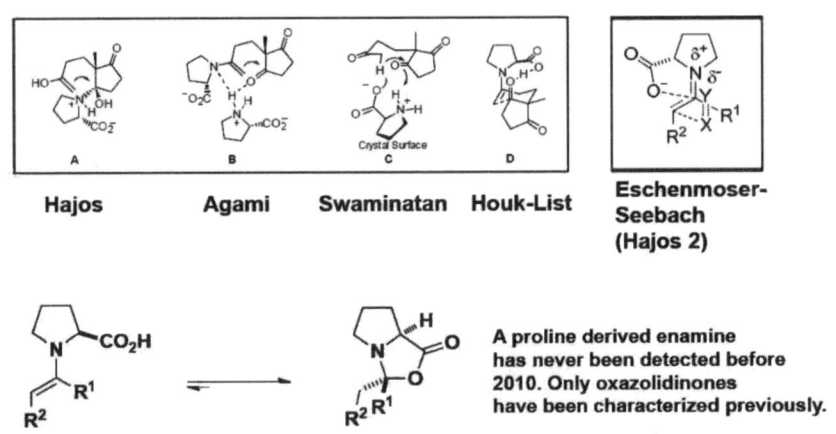

Hajos	Agami	Swaminatan	Houk-List	Eschenmoser-Seebach (Hajos 2)

A proline derived enamine has never been detected before 2010. Only oxazolidinones have been characterized previously.

Mechanistic Studies

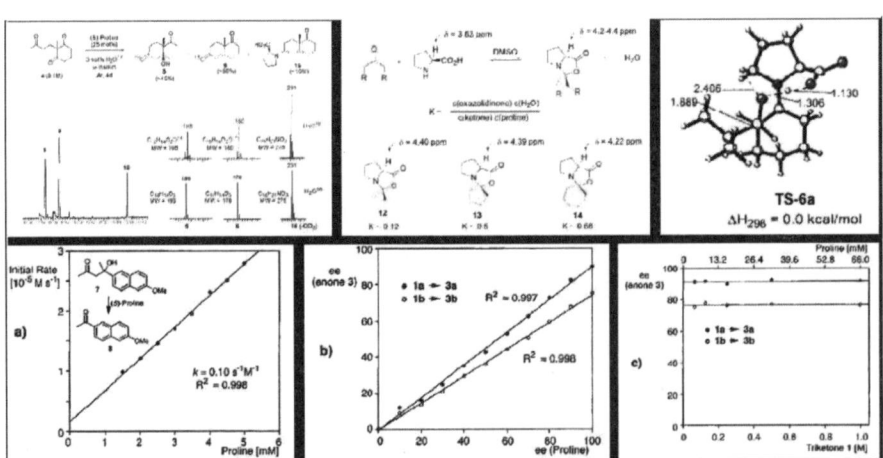

List et al. *PNAS* 2004, *101*, 5839. List & Houk et al. *JACS* 2003, 2475-2479. List et al. *JACS* 2003, 16-17.

The First Spectroscopic Characterization of a Proline Enamine

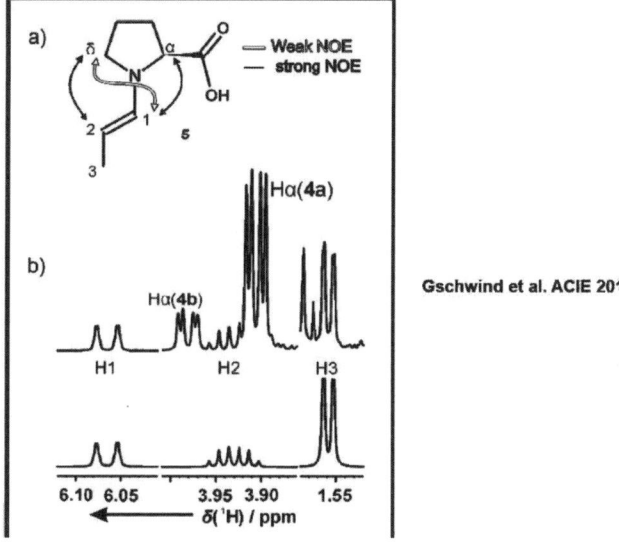

Gschwind et al. ACIE 2010

X-Ray Structures of Proline Enamines

With Bock and Lehmann, unpublished

X-Ray Structures of Proline Enamines

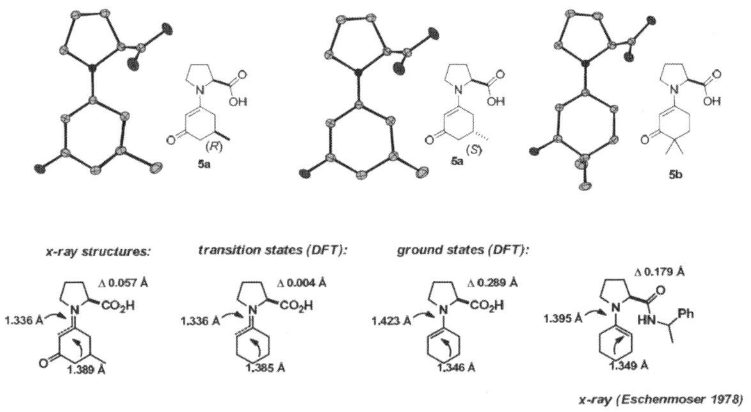

x-ray structures: transition states (DFT):   ground states (DFT):

x-ray (Eschenmoser 1978)

With Bock and Lehmann, unpublished

Proposed Catalytic Cycle

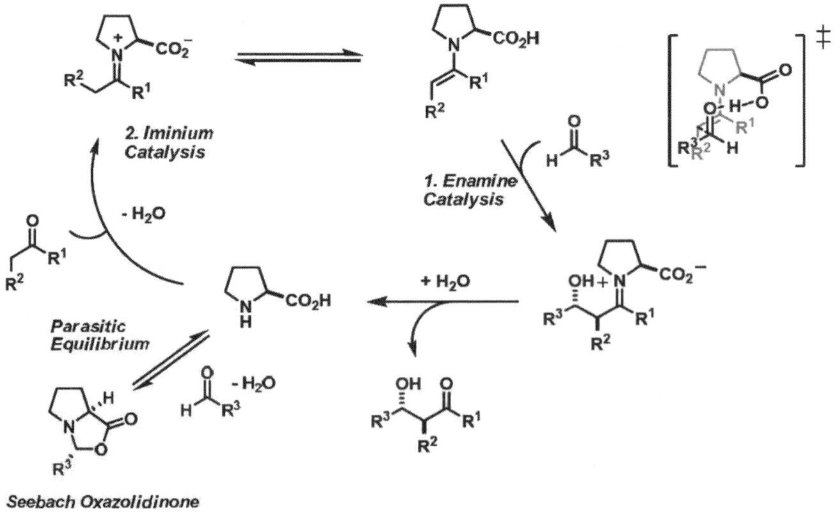

2. Iminium Catalysis

1. Enamine Catalysis

- H$_2$O

+ H$_2$O

Parasitic Equilibrium

- H$_2$O

Seebach Oxazolidinone

Enamine Catalysis

A Powerful Strategy for the Catalytic Generation and Use of Carbanion Equivalents

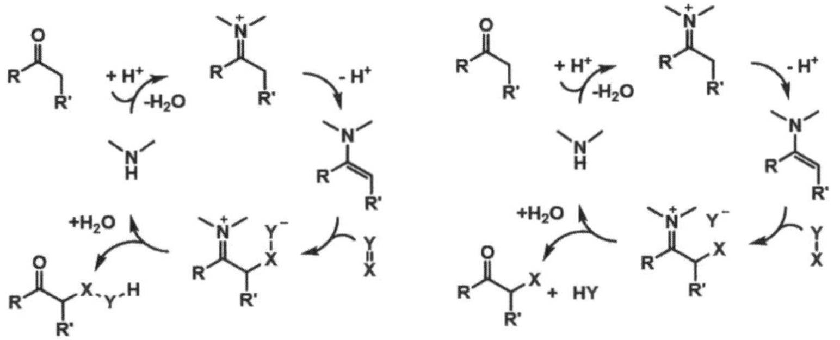

Nucleophilic Additions   *Nucleophilic Substitutions*

List *Acc. Chem. Res.* **2004**, *37*, 548-557.

Enamine Catalysis: Electrophiles (X=Y)

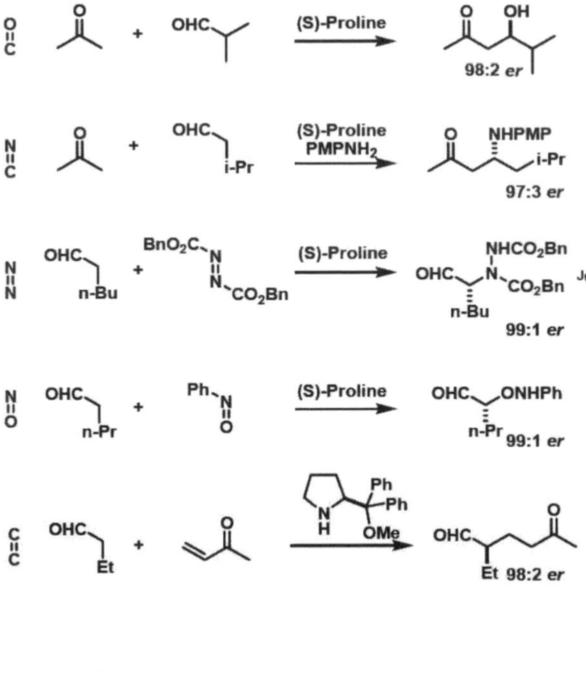

List*, Lerner, Barbas
JACS 2000, 2395-6.

98:2 *er*

List *JACS* 2000, 9336-7.

97:3 *er*

List *JACS* 2002, 5656-7.
Jørgensen et al. *Angew.* 2002, 1868

99:1 *er*

Zhong, MacMillan,
Hayashi 2003

99:1 *er*

Gellman 2005

98:2 *er*

Enamine Catalysis of Nucleophilic Substitution Reactions
Electrophiles (X-Y)

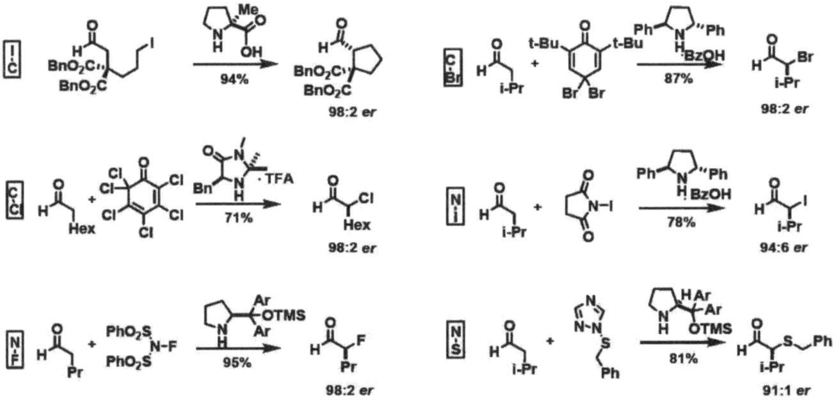

Alkylation: Vignola, N.; List, B. *J. Am. Chem. Soc.* 2004, *126*, 450-451. Chlorination: (a) Brochu, M. P.; Brown, S. P.; MacMillan, D. W. C. *J. Am. Chem. Soc.* 2004, *126*, 4108-4109. (b) Halland, N.; Braunton, A.; Bachmann, S.; Marigo, M.; Jørgensen, K. A. *J. Am. Chem. Soc.* 2004, *126*, 4790-4791. Fluorination: (c) Beeson, T. D.; MacMillan, D. W. C.; *J. Am. Chem. Soc.* 2005, *127*, 8826-8828. (d) Steiner, D. D.; Mase, N.; Barbas, C. F., III. *Angew. Chem. Int. Ed.* 2005, *44*, 3706-3710. (e) Marigo, M.; Fielenbach, D.; Braunton, A.; Kjærsgaard, A.; Jørgensen, K. A. *Angew. Chem. Int. Ed.* 2005, *44*, 3703-3706. (f) Enders, D.; Hüttl, M. R. M. *Synlett* 2005, 991-993. Bromination, Iodination: Bertelsen, S.; Halland, N.; Bachmann, S.; Marigo, M.; Braunton, A.; Jørgensen, K. A. *Chem. Comm.* 2005, 4821-4823. Sulfenylation: Marigo, M.; Wabnitz, T. C.; Fielenbach, D.; Jørgensen, K. A. *Angew. Chem., Int. Ed.* 2005, *44*, 794.

A Total Synthesis of Prelactone B

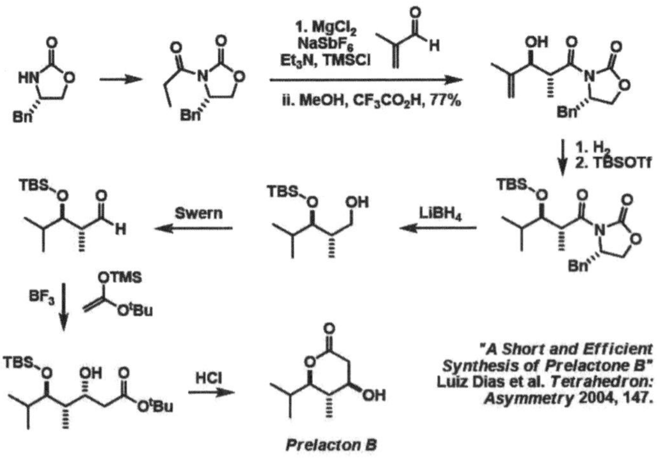

"*A Short and Efficient Synthesis of Prelactone B*"
Luiz Dias et al. *Tetrahedron: Asymmetry* 2004, 147.

Prelacton B

A Total Synthesis of Prelactone B

Proline-Catalyzed Aldol Reaction is ideally suited for Oligopropionate Synthesis

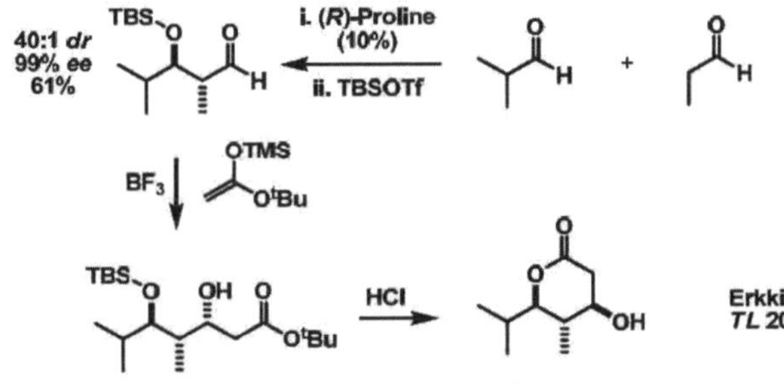

Erkkilä & Pihko
TL 2003, 7607.

Prelactone B

Enamine Catalysis in the Production of a Pharmaceutical

The proline-catalyzed intermolecular aldol reaction in the production of *Darunavir*

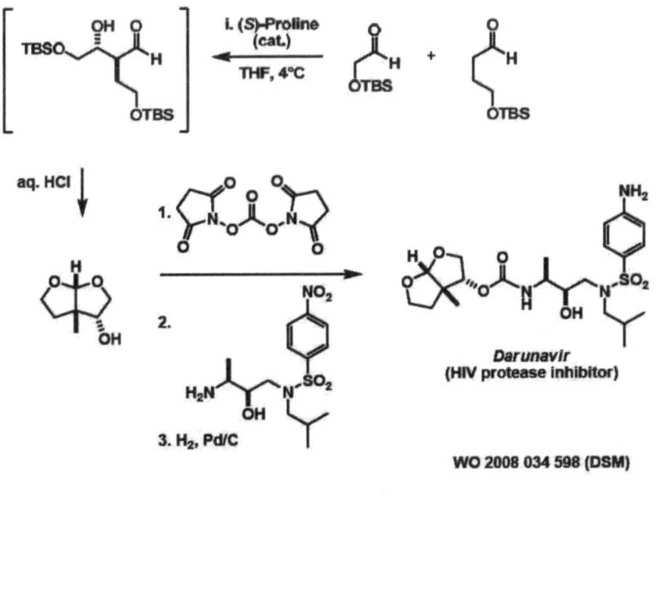

Darunavir
(HIV protease inhibitor)

WO 2008 034 598 (DSM)

Four Modes of Aldolization

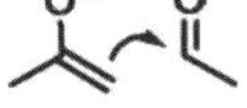

intermolecular

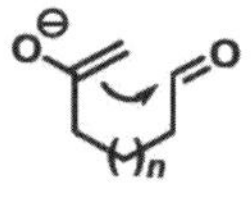

intramolecular 1:
enolendo

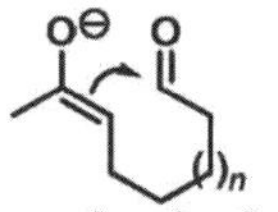

intramolecular 2:
enolexo

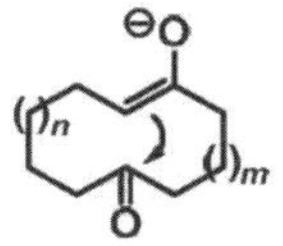

intramolecular 3:
transannular

Catalytic Asymmetric
Transannular Aldolization

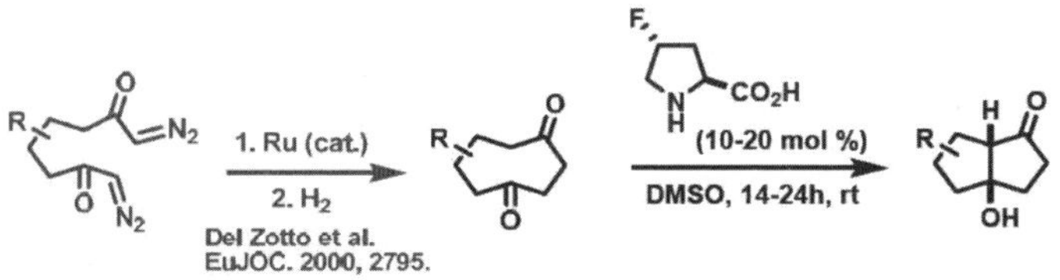

1. Ru (cat.)

2. H$_2$

Del Zotto et al.
EuJOC. 2000, 2795.

(10-20 mol %)

DMSO, 14-24h, rt

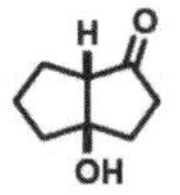

53%, 97:3 er

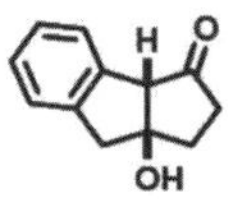

67%, 97:3 er

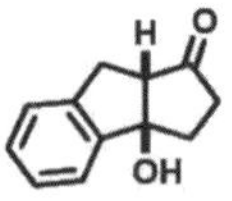

80% 97:3 er

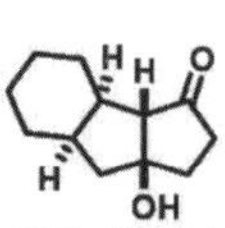

68%, 97:3 er

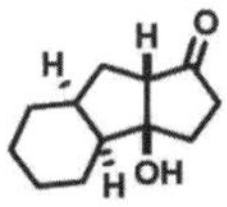

42%, 95:5 er (*KR*)

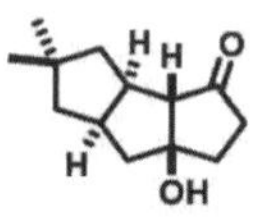

84% 98:2 er

Carley Chandler

Total Synthesis of (+)-Hirsutene

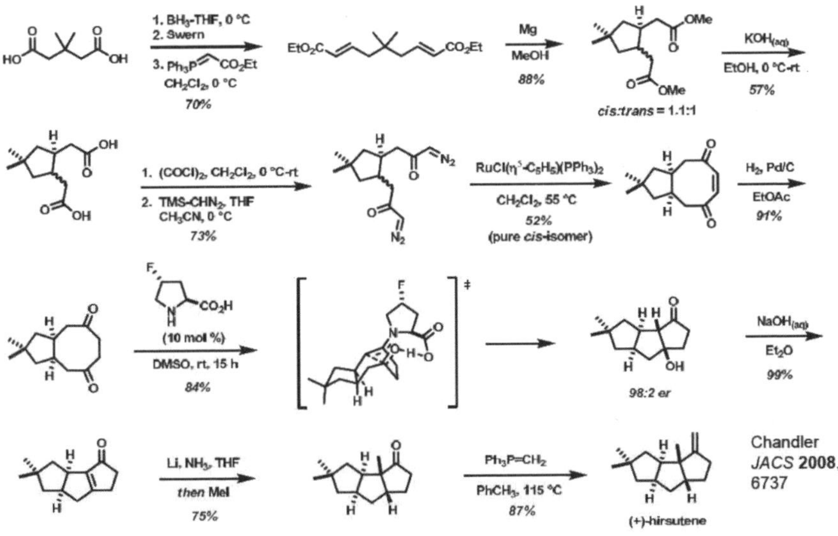

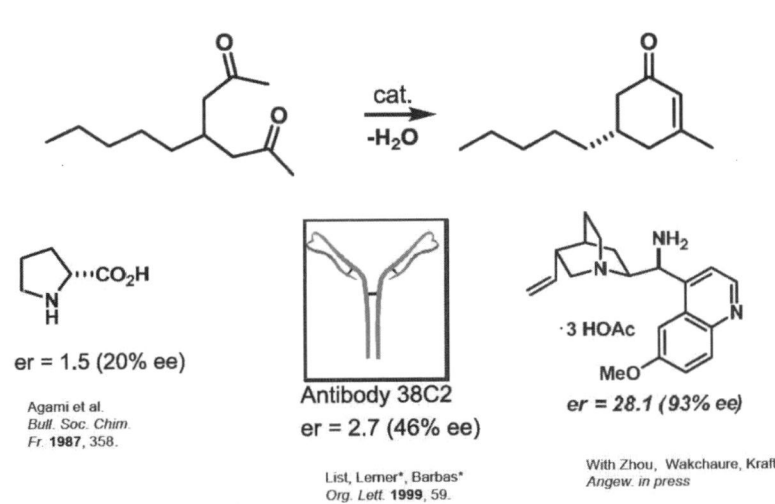

The Hirsutene synthesis scheme (Chandler *JACS* **2008**, 6737) with reagents:

1. BH₃-THF, 0 °C
2. Swern
3. Ph₃P=CHCO₂Et, CH₂Cl₂, 0 °C — 70%

Mg, MeOH — 88%; cis:trans = 1.1:1

KOH(aq), EtOH, 0 °C-rt — 57%

1. (COCl)₂, CH₂Cl₂, 0 °C-rt
2. TMS-CHN₂, THF, CH₃CN, 0 °C — 73%

RuCl(η⁵-C₅H₅)(PPh₃)₂, CH₂Cl₂, 55 °C — 52% (pure cis-isomer)

H₂, Pd/C, EtOAc — 91%

(10 mol %), DMSO, rt, 15 h — 84%

NaOH(aq), Et₂O — 99%; 98:2 er

Li, NH₃, THF then MeI — 75%

Ph₃P=CH₂, PhCH₃, 115 °C — 87%

(+)-hirsutene

Another Unsolved Aldol Problem: Diketone-Cyclodehydrations

cat. −H₂O

proline: er = 1.5 (20% ee)
Agami et al. Bull. Soc. Chim. Fr. **1987**, 358.

Antibody 38C2: er = 2.7 (46% ee)
List, Lerner*, Barbas* Org. Lett. **1999**, 59.

·3 HOAc, MeO: er = 28.1 (93% ee)
With Zhou, Wakchaure, Kraft Angew. in press

Enantiogroup-Differentiating Diketone-Aldolization

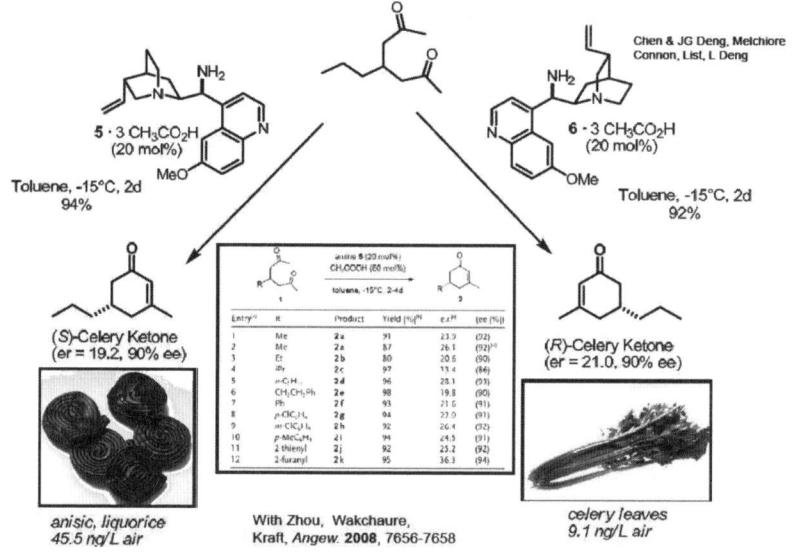

Chen & JG Deng, Melchiore
Connon, List, L Deng
5 · 3 CH_3CO_2H
(20 mol%)
Toluene, -15°C, 2d
94%
6 · 3 CH_3CO_2H
(20 mol%)
Toluene, -15°C, 2d
92%
(S)-Celery Ketone
(er = 19.2, 90% ee)
(R)-Celery Ketone
(er = 21.0, 90% ee)
amine 5 (20 mol%)
CH_3COOH (60 mol%)
toluene, -15°C, 2-4d
anisic, liquorice
45.5 ng/L air
celery leaves
9.1 ng/L air

Entry[a]	R	Product	Yield [%][b]	e.r.[c]	(ee [%])
1	Me	2a	91	23.9	(92)
2	Me	2a	87	26.1	(92)[d]
3	Et	2b	80	20.6	(90)
4	iPr	2c	97	13.4	(86)
5	n-C_7H_{15}	2d	96	28.1	(93)
6	CH_2CH_2Ph	2e	98	19.8	(90)
7	Ph	2f	93	21.6	(91)
8	p-ClC_6H_4	2g	94	22.0	(91)
9	m-ClC_6H_4	2h	92	26.4	(92)
10	p-MeC_6H_4	2i	94	24.5	(91)
11	2-thienyl	2j	92	25.2	(92)
12	2-furanyl	2k	95	36.3	(94)

With Zhou, Wakchaure, Kraft, *Angew.* **2008**, 7656-7658

Proline-Catalyzed Mannich-Reaction

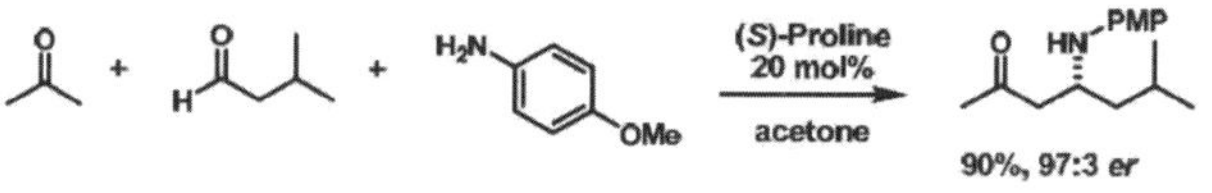

(S)-Proline
20 mol%
acetone
90%, 97:3 er

List et al. *JACS* **2000, 9336.**
***JACS* 2002, 827.**
***Synlett* 2003, 1903.**

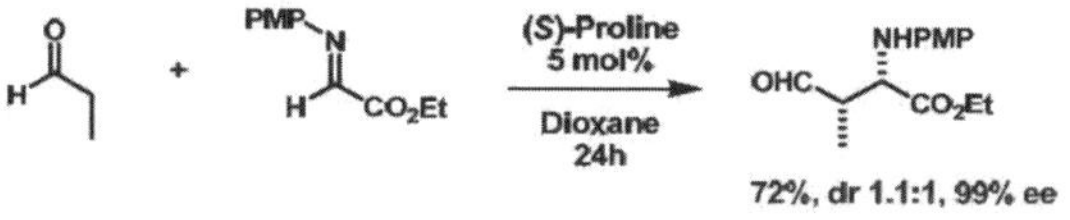

(S)-Proline
5 mol%
Dioxane
24h
72%, dr 1.1:1, 99% ee

Barbas et al.
***JACS* 2002, 1866.**

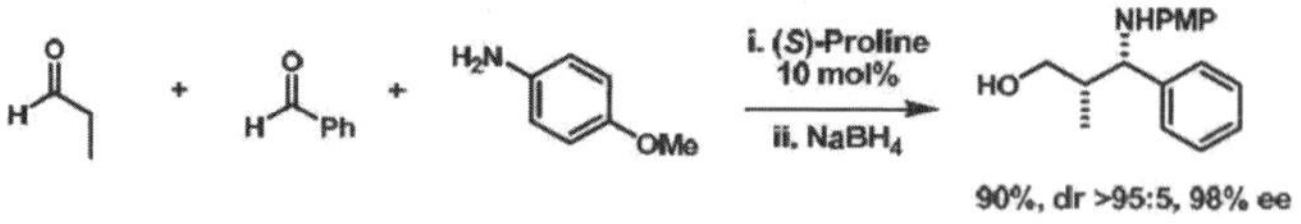

i. (S)-Proline
10 mol%
ii. $NaBH_4$
90%, dr >95:5, 98% ee

Hayashi et al.
***Angew. Int.* 2003, 3677.**

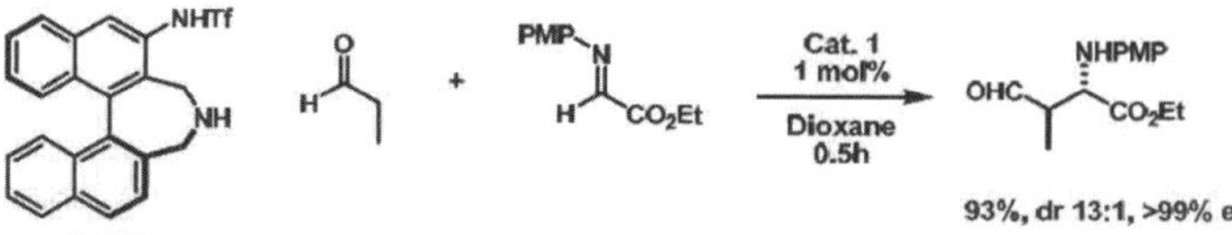

Cat. 1
1 mol%
Dioxane
0.5h
93%, dr 13:1, >99% ee

Cat. 1

Maruoka et al.
***JACS* 2005, 16408.**

A Challenge of the Proline-Catalyzed Mannich-Reaction: Removal of the PMP-Group

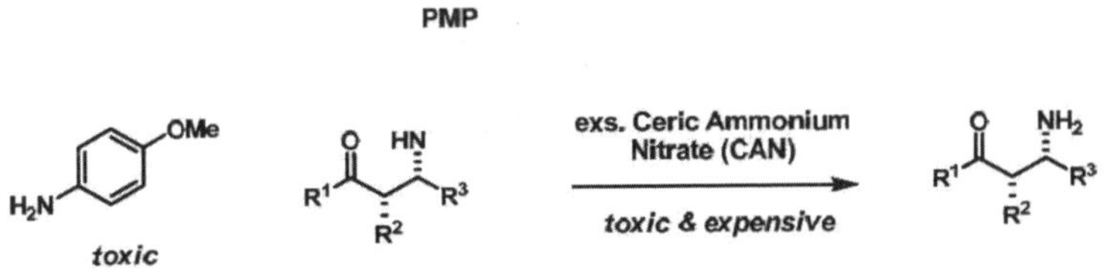

PMP
OMe
H_2N
toxic
O
HN
R^1
R^3
R^2
exs. Ceric Ammonium Nitrate (CAN)
toxic & expensive
O
NH_2
R^1
R^3
R^2

For Alternative Methods (without CAN): Rutjes & DSM TL 2006, 8109 & *ASC* 2007, 1332.

Attractive Alternative: N-BOC-Imines

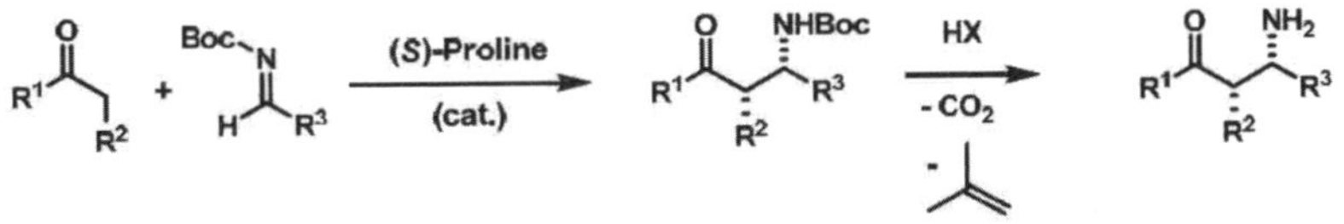

O
R^1
R^2
Boc
N
H
R^3
(S)-Proline
(cat.)
O
NHBoc
R^1
R^3
R^2
HX
$-CO_2$
O
NH_2
R^1
R^3
R^2

„You'll all be making beta-Amino Acids in the future"
D. Seebach, addressing pharmaceutical Chemists, Mainz, Germany 2004

Proline-Catalyzed Mannich Reaction with N-BOC-Imines

O
H
n-Bu
Boc
N
H
Ph
(S)-Proline
(20 mol%)
CH_3CN
O
NHBoc
H
Ph
n-Bu
i. $NaClO_2$, NaH_2PO_4
ii. TFA, CH_2Cl_2
O
$NH_2\cdot TFA$
HO
Ph
n-Bu
84% Yield, >99:1 er, >99:1 dr
Product precipitates from the reaction mixture!

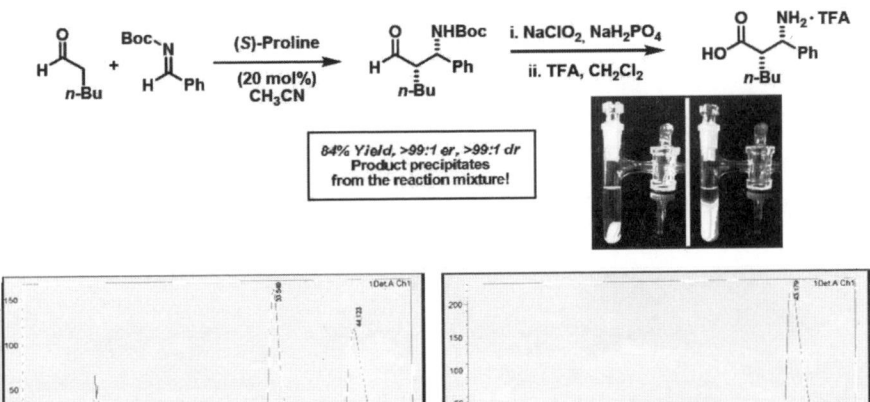

rac-Proline, rt

(*S*)-Proline, 0° C

The ideal chemical process is that which a one-armed operator can perform by pouring the reactants into a bath tub and collecting the product from the drain hole.

-Sir John Cornforth

Proline-Catalyzed Mannich Reaction with N-BOC-Imines

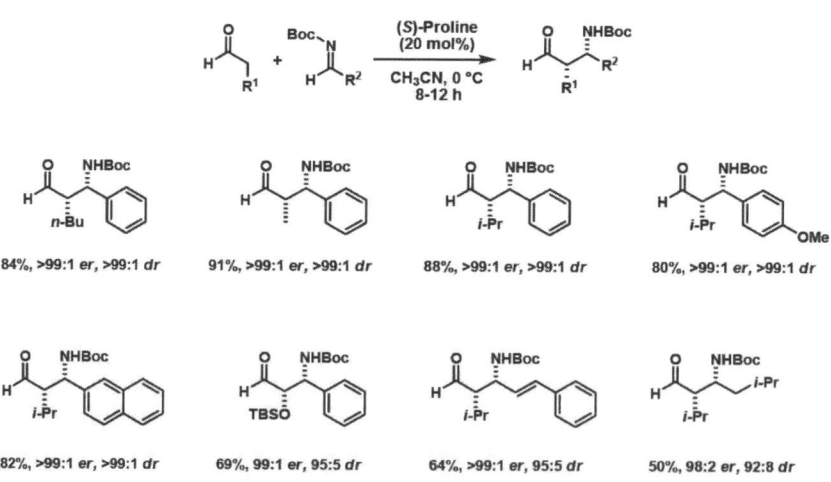

With Yang and Stadler *unpublished* & *Angew* 2007, 615, *Nature Protocols* 2007, 1937
For one example with a ketone, see: Enders et al. Synthesis 2006, 2155

Proline-Catalyzed Mannich Reaction with other Imines

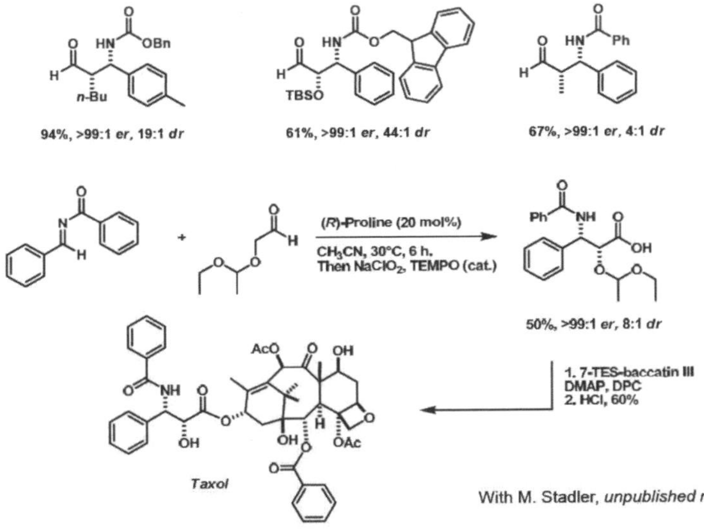

OBn
HN
n-Bu
94%, >99:1 er, 19:1 dr
HN
TBSO
61%, >99:1 er, 44:1 dr
HN
Ph
67%, >99:1 er, 4:1 dr
(R)-Proline (20 mol%)
CH3CN, 30°C, 6 h.
Then NaClO2, TEMPO (cat.)
Ph
NH
OH
50%, >99:1 er, 8:1 dr
1. 7-TES-baccatin III
DMAP, DPC
2. HCl, 60%
AcO
OH
NH
OAc
Taxol

With M. Stadler, *unpublished results*

Acetaldehyde: The Ultimate Challenge of Enamine Catalysis?

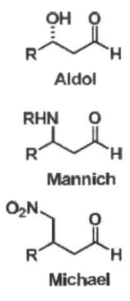

OH O
R
H
Aldol
RHN O
R
H
Mannich
O2N
R
H
Michael

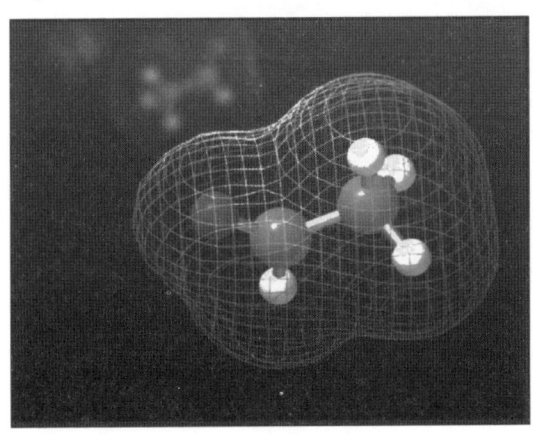

- Acetaldehyde is the "simplest" enolizable carbonyl compound
- Inexpensive, produced from ethylene via Wacker oxidation
- Would be an extremely useful 2-carbon nucleophile
- Untamable?

Proline-Catalyzed Mannich Reaction of Acetaldehyde

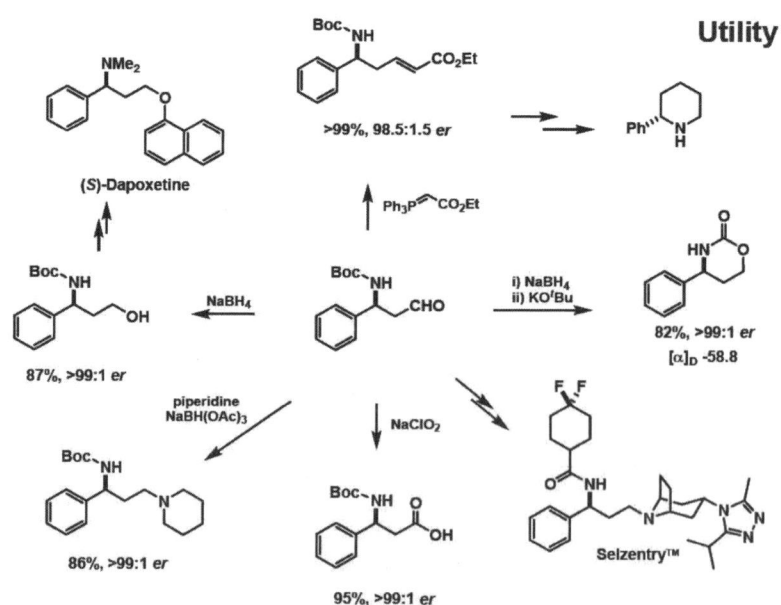

With Yang, Chandler, Stadler, Kampen *Nature* **2008**, *452*, 453-454.

With Yang, Chandler, Stadler, Kampen *Nature* **2008**, *452*, 453-454.

Organocatalytic Michael Reaction of Acetaldehyde

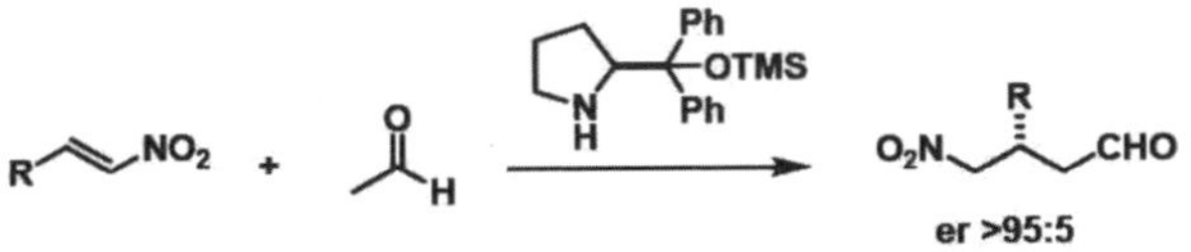

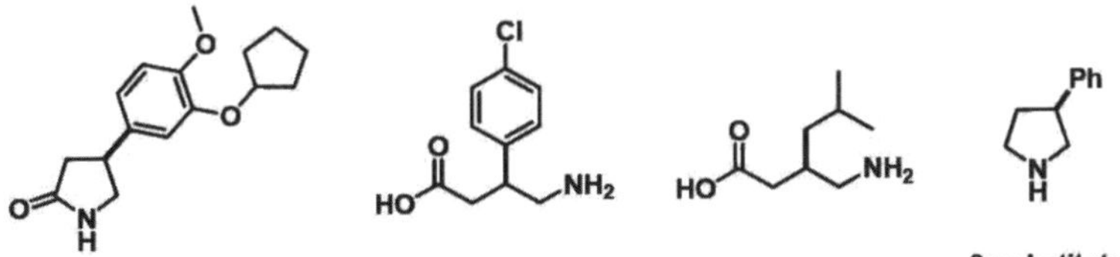

er >95:5

(S)-Rolipram

(S)-Baclofen

(R)-Pregabalin

3-substituted Pyrrolidines

With Garcia-Garcia, Ladepeche, Halder Angew. Chem. Int. Ed. 2008, 47, 4719.
Also see: Hayashi et al. Angew. Chem. Int. Ed. 2008, 47, 4722-4724.
Aldol: Hayashi et al. Angew. Chem. Int. Ed. 2008, 47, 2082-2084.

Enamine Catalysis Cascades of Acetaldehyde

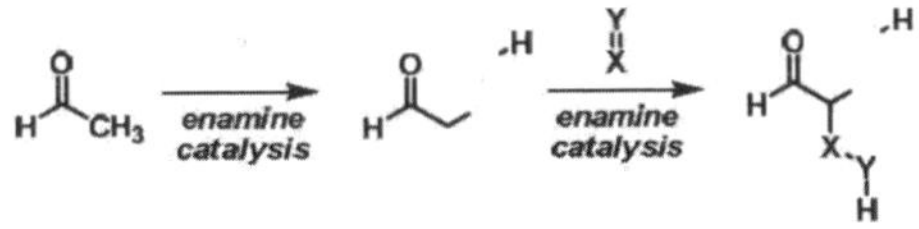

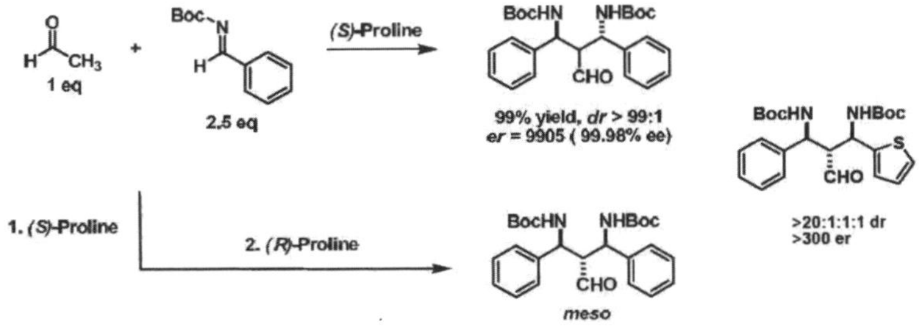

99% yield, dr > 99:1
er = 9905 (99.98% ee)

>20:1:1:1 dr
>300 er

meso

With Chandler, Galzerano, Michrowska ACIE 2009, 1978.

Proline Catalysis =
Enamine *and* Iminium Ion Catalysis

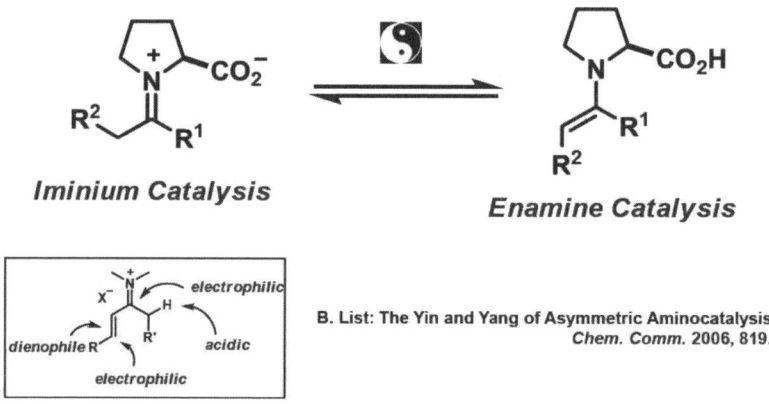

Iminium Catalysis

Enamine Catalysis

B. List: The Yin and Yang of Asymmetric Aminocatalysis
Chem. Comm. 2006, 819.

Iminium Catalysis: D. W. C. MacMillan

Ricciocarpin Synthesis via Reductive Michael Cascade

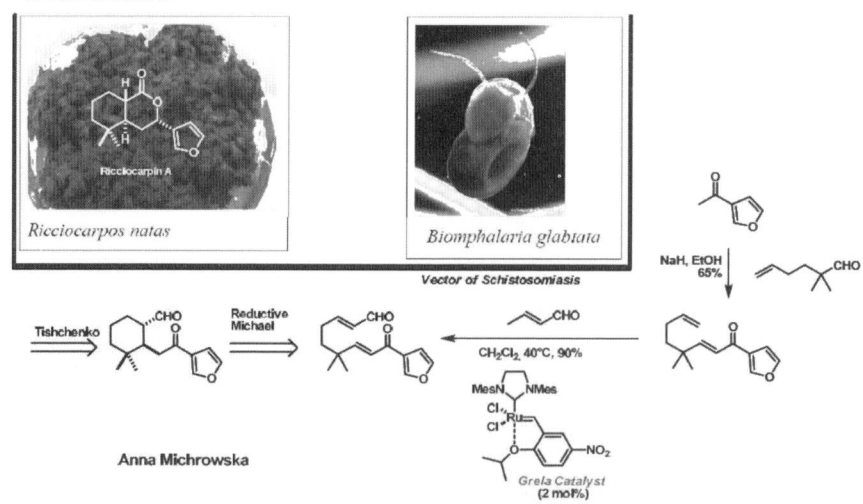

Ricciocarpos natas

Biomphalaria glabrata

Vector of Schistosomiasis

Anna Michrowska

Ricciocarpin Synthesis via Reductive Michael Cascade

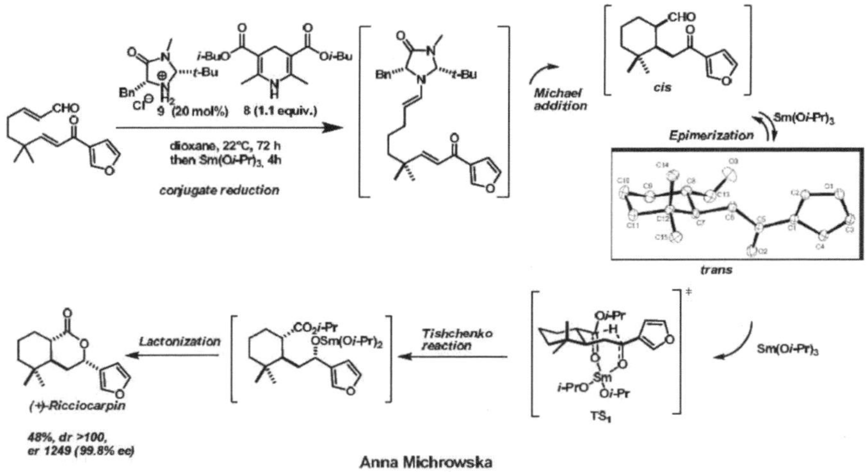

Anna Michrowska

Ricciocarpin Synthesis via Reductive Michael Cascade

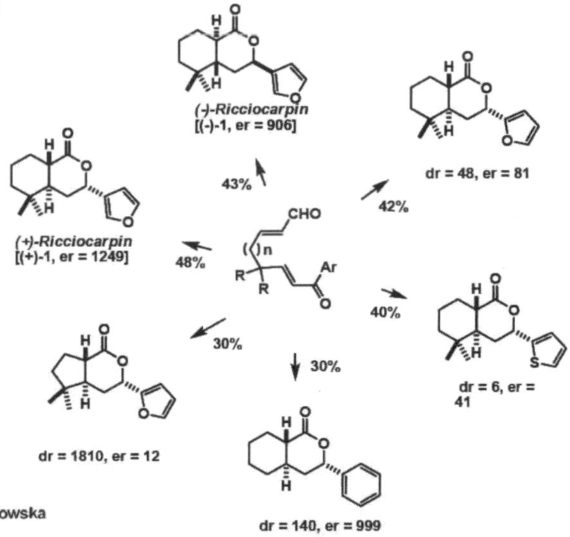

Anna Michrowska

Ricciocarpin Synthesis via Reductive Michael Cascade

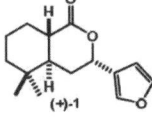

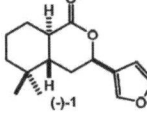

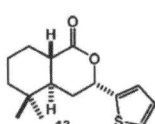

Compound	% Mortality of B. glabrata		
	11 µg/mL	22 µg/mL	44 µg/mL
(+)-1	0	0	90
(-)-1	0	0	0
13	90	100	100

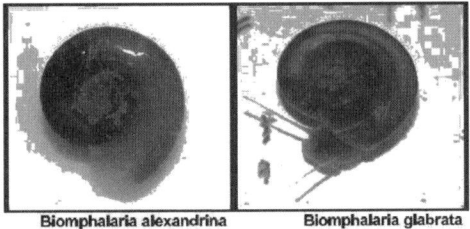

Biomphalaria alexandrina   Biomphalaria glabrata

With Michrowska *Nature Chemistry* 2009, 225.

Enones are Challanging Substrates for Iminium Catalysis:

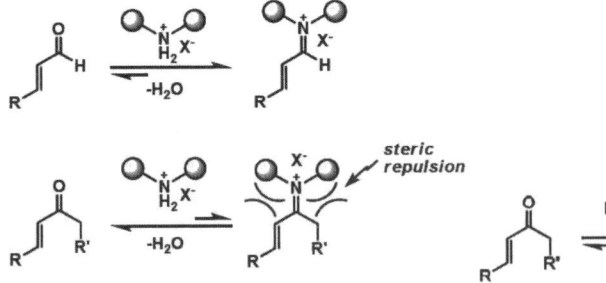

steric repulsion

secondary amines can activate
enals but typically fail with ketones...

...a possible solution: primary amines!

Iminium Catalysis of the Weitz –Scheffer Epoxidations of Enals *and Enones?*

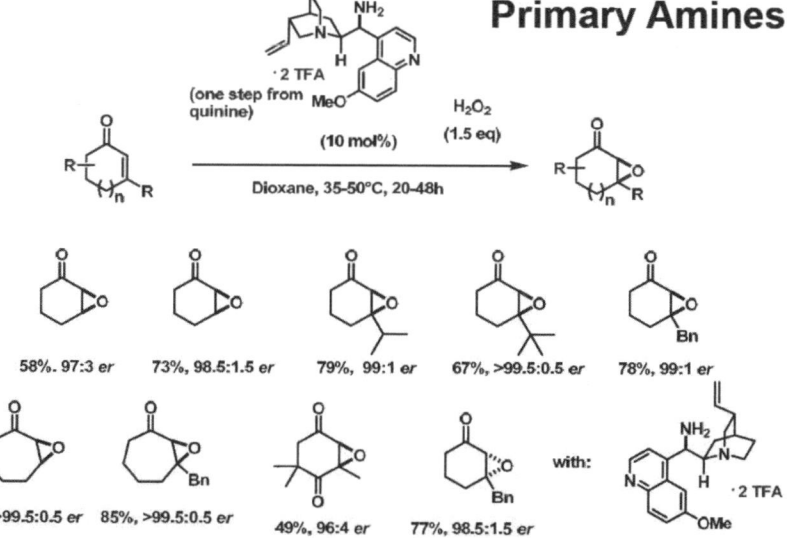

Ph⌒CHO + H₂O₂

Cat (10 mol%)

CH₂Cl₂, rt, 4 h
80%

93:7, 96% ee

Jørgensen *JACS* 2005, 6964

How about Enones?

Challenging substrates for catalytic asymmetric epoxidations:
Sharpless-, Julia-Colonna-, Jacobsen-, Shi- etc. *fail!*

Powerful Catalysts: Cinchona Derived Primary Amines

·2 TFA

(one step from quinine)

(10 mol%)

H₂O₂
(1.5 eq)

Dioxane, 35-50°C, 20-48h

58%. 97:3 *er*   73%, 98.5:1.5 *er*   79%, 99:1 *er*   67%, >99.5:0.5 *er*   78%, 99:1 *er*

62%, >99.5:0.5 *er*   85%, >99.5:0.5 *er*   49%, 96:4 *er*   77%, 98.5:1.5 *er*   with: ·2 TFA

With Wang & Reisinger *JACS* 2008, 6070-71.

Extension to Acyclic Ketones: Catalytic Asymmetric Hydroperoxidation

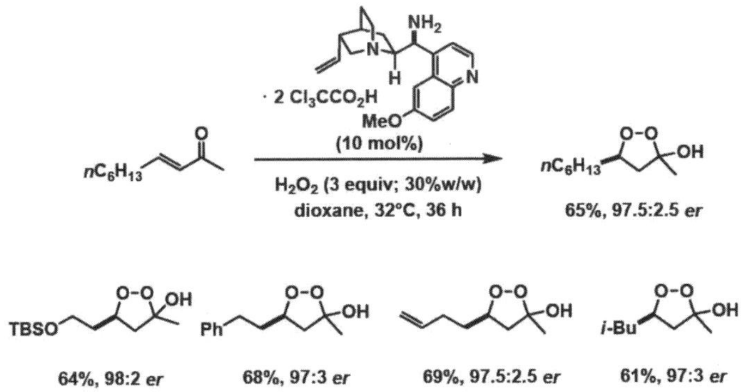

NH2
H
N
· 2 Cl3CCO2H
MeO
(10 mol%)
H2O2 (3 equiv; 30%w/w)
dioxane, 32°C, 36 h
nC6H13
O-O
OH
65%, 97.5:2.5 er
TBSO
Ph
i-Bu
64%, 98:2 er
68%, 97:3 er
69%, 97.5:2.5 er
61%, 97:3 er

With Reisinger & Wang *Angew. Int.* 2008, 8112.

Cinchona ACDC for the Catalytic Asymmetric Epoxidation of Branched Enals

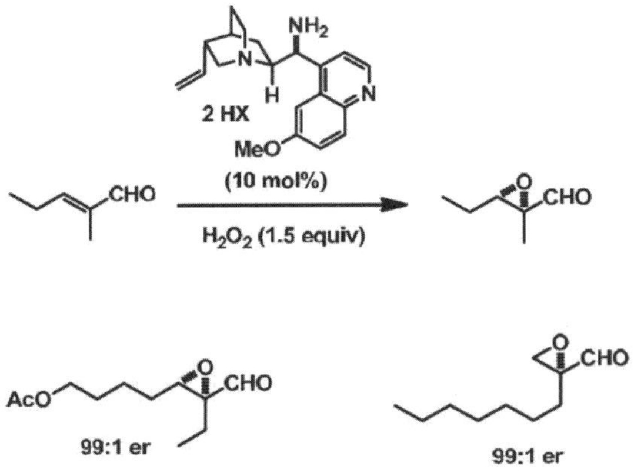

NH2
H
N
2 HX
MeO
(10 mol%)
H2O2 (1.5 equiv)
CHO
AcO
99:1 er
99:1 er

HX =
TFA 75:25 dr, 93:7 er
(*S*)-TRIP 74:26 dr, 36:64 er
(*R*)-TRIP 94:6 dr, >99:1 er

With Lifchits, JACS in press.

The Yin and Yang of Aminocatalysis

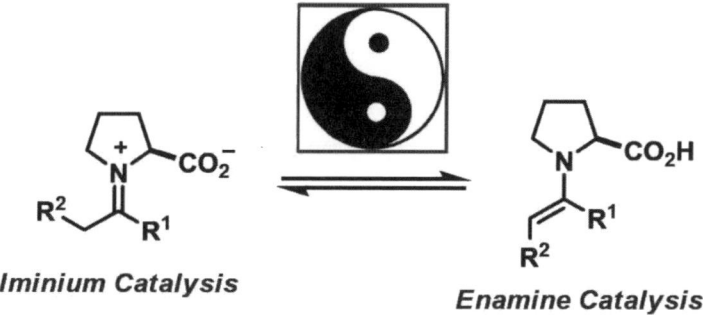

Iminium Catalysis   **Enamine Catalysis**

A Powerful and Versatile Approach to Asymmetric Catalysis

Not two but three!

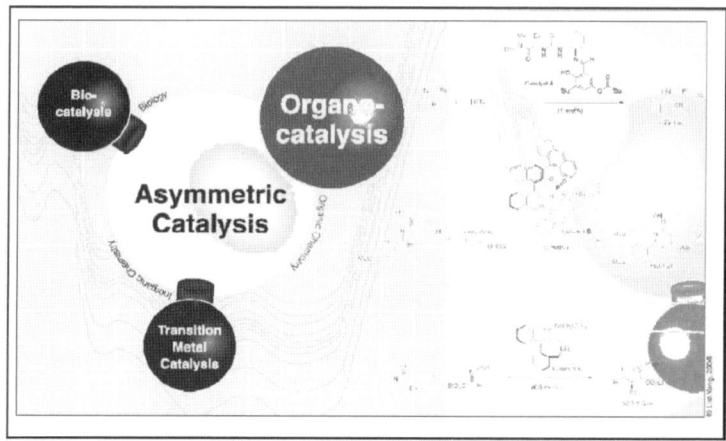

Benjamin List & Yung Woon Yang: The Organic Approach to Asymmetric Catalysis. *Science* **2006**, *313* (5793), 1584-1586.

Asymmetric Counteranion Directed Catalysis (ACDC)

Benjamin List

Max-Planck-Institut für Kohlenforschung, Mülheim an der Ruhr (Germany), list@mpi-muelheim.mpg.de

Most chemical reactions proceed via charged intermediates or transition states. Such "polar reactions" can be influenced by the counterion, especially if conducted in organic solvents, where ion pairs are inefficiently separated by the solvent. Although asymmetric catalytic transformations involving anionic intermediates with chiral, cationic catalysts have been realized, analogous versions of inverted polarity with reasonable enantioselectivity, despite attempts, only recently became a reality. In my lecture I will present the development of this concept, which is termed asymmetric counteranion-directed catalysis (ACDC) and illustrate its generality with examples from organocatalysis, transition metal catalysis, and Lewis acid catalysis.

Starting in 2006, we have discovered salts that consist of achiral secondary ammonium ions and a chiral binol-derived phosphate counteranion that catalyze conjugate reductions and epoxidations of α,β-unstaurated aldehydes with high enantioselectivity. These discoveries formed the basis of an extensive exploration of the ACDC-principle in catalysis. For example, we could show that primary ammonium ions can also be utilized in this context and activate α,β-unstaurated ketones in transfer hydrogenations and epoxidations. The high enenatioselectivities achieved in these epoxidations were entirely unprecedented in asymmetric catalysis before.

Subsequently, we have expanded the concept of asymmetric counteranion directed catalysis into the area of transition metal catalysis. While a distinction between an anionic ligand and a "true" counteranion cannot always easily be drawn in these systems, conceptually novel catalysts have been designed in the area of Pd, Mn, Ru, and Fe-catalysis.

Recently, we have advanced ACDC to yet another level. We have designed novel organic Lewis acid catalysts that generally activate simple aldehydes towards nucleophilic additions with silylated nucelophiles such as ketene acetals, phosphites, TMS-CN and similar reagents. Extremely powerful organocatalysts have emerged that begin to rival even the most efficient transition metal-based catalysts.

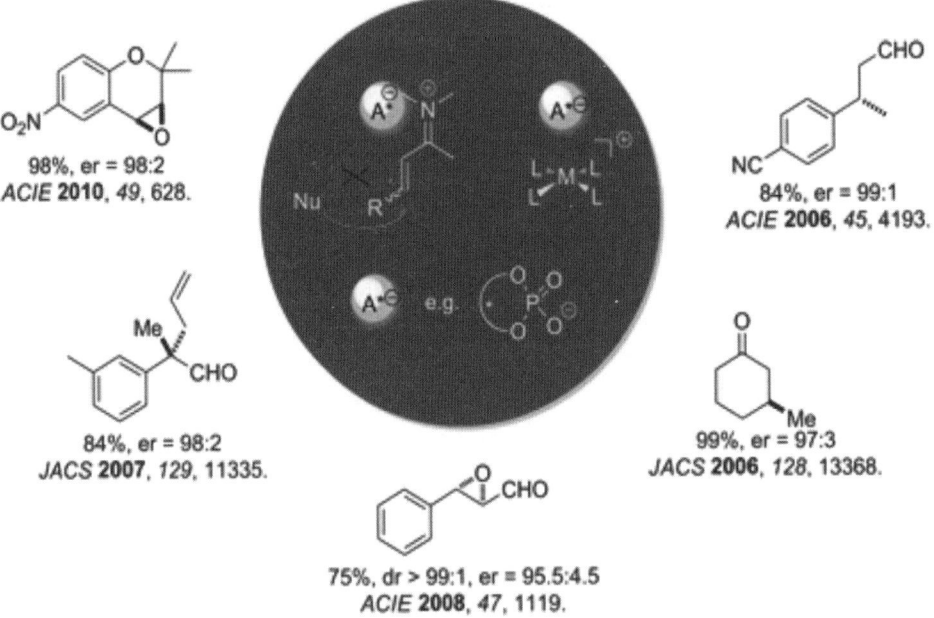

98%, er = 98:2
ACIE **2010**, *49*, 628.

84%, er = 99:1
ACIE **2006**, *45*, 4193.

84%, er = 98:2
JACS **2007**, *129*, 11335.

99%, er = 97:3
JACS **2006**, *128*, 13368.

75%, dr > 99:1, er = 95.5:4.5
ACIE **2008**, *47*, 1119.

Proline Catalysis Mechanism

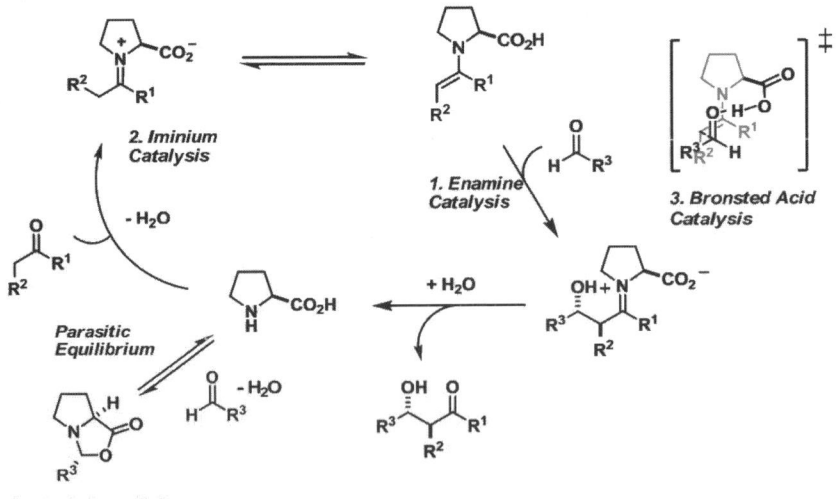

2. *Iminium Catalysis*

1. *Enamine Catalysis*

3. *Bronsted Acid Catalysis*

Parasitic Equilibrium

Seebach Oxazolidinone

Asymmetric Brønsted Acid Catalysis

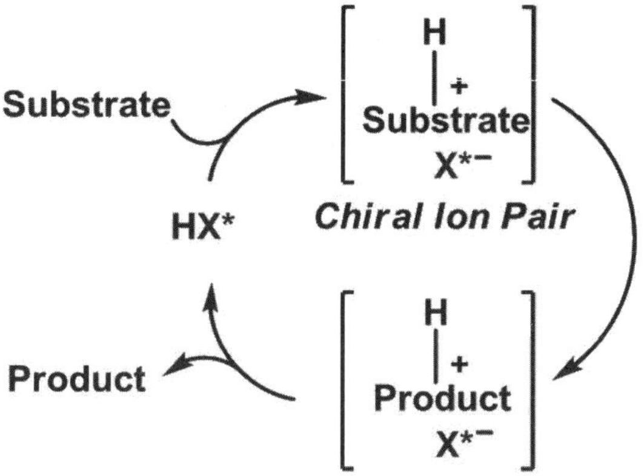

Substrate → $\begin{bmatrix} \text{H} \\ \mid + \\ \text{Substrate} \\ \text{X}^{*-} \end{bmatrix}$ *Chiral Ion Pair*

HX*

Product ← $\begin{bmatrix} \text{H} \\ \mid + \\ \text{Product} \\ \text{X}^{*-} \end{bmatrix}$

Asymmetric Brønsted Acid Catalysis

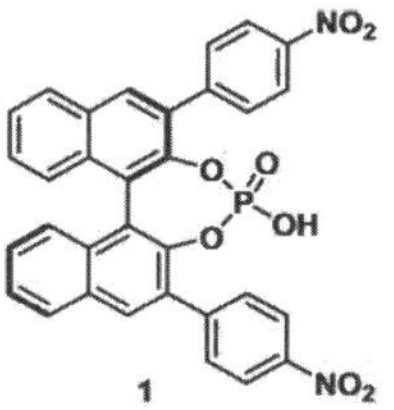

dr 87:13, 96% ee

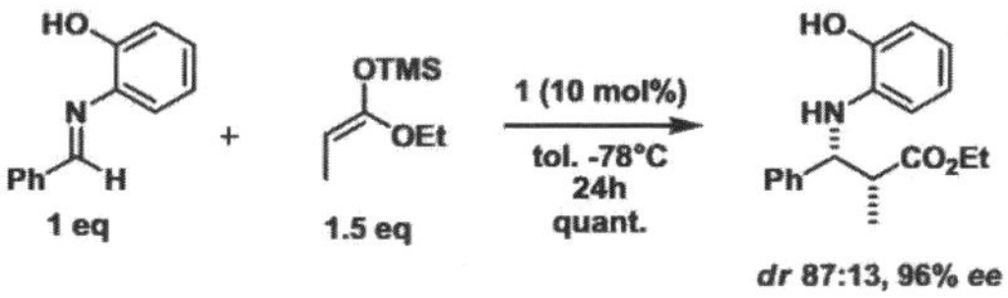

Akiyama et al. *Angew. Chem. Int. Ed.* 2004, 1566.

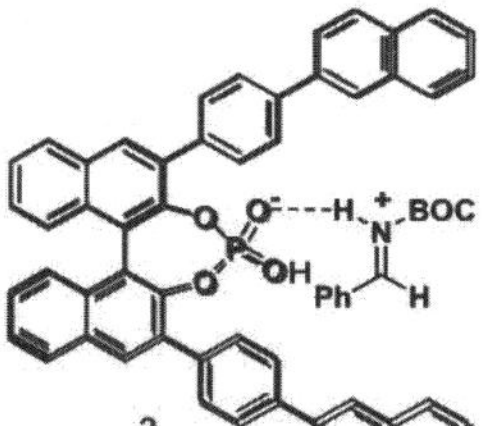

2
**Hydrogen-Bond
Assisted Ion Pair**

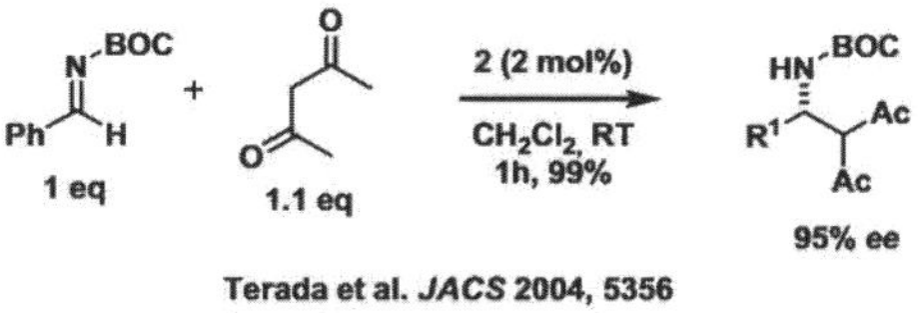

95% ee

Terada et al. *JACS* 2004, 5356

Catalytic Asymmetric
Pictet-Spengler Reaction

In Nature: *Pictet-Spenglerases*

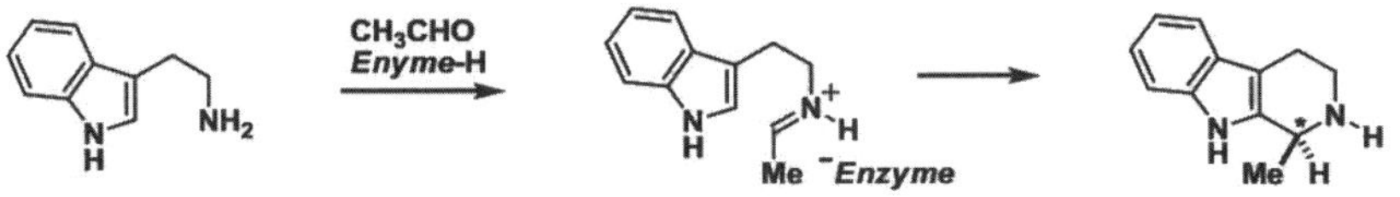

Catalytic Asymmetric
Pictet-Spengler Reaction

Our Studies: Bronsted Acid Catalysis

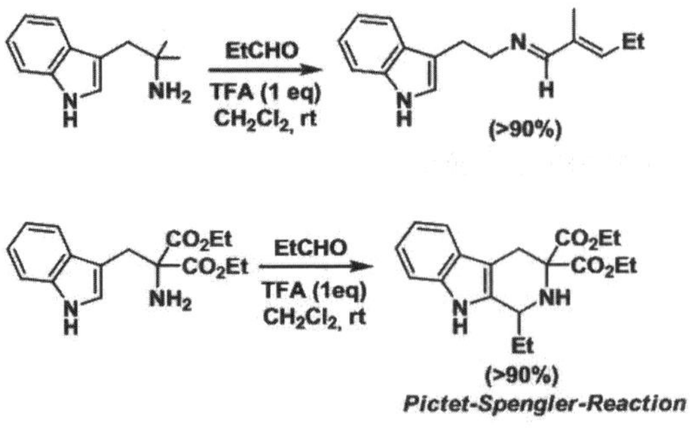

(>90%)

(>90%)
Pictet-Spengler-Reaction

Catalytic Asymmetric Pictet-Spengler Reaction

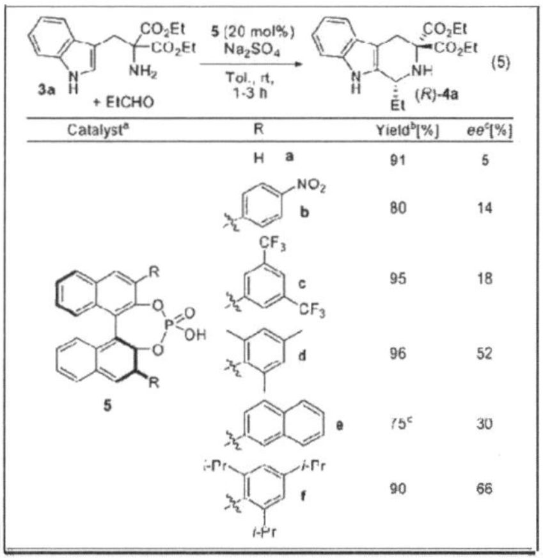

Catalyst[a]	R		Yield[b][%]	ee[c][%]
	H	a	91	5
	$\text{-}C_6H_4\text{-}NO_2$	b	80	14
	$\text{-}C_6H_3(CF_3)_2$	c	95	18
	mesityl	d	96	52
	naphthyl	e	75[c]	30
	$\text{-}C_6H_3(i\text{-}Pr)_3$	f	90	66

Catalytic Asymmetric Pictet-Spengler Reaction

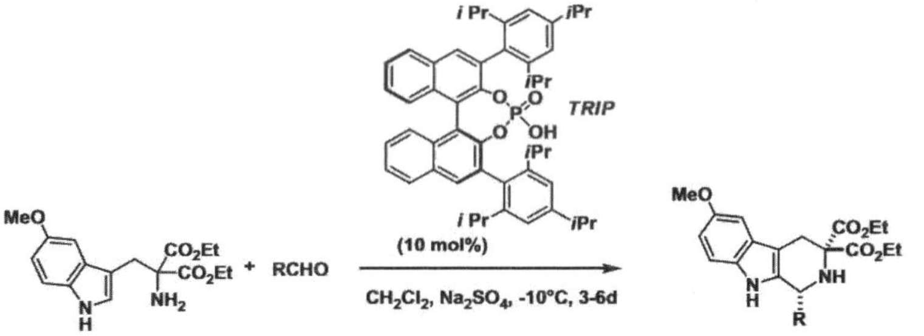

R	Yield	ee
Cy	64%	94%
Et	96%	90%
p-O₂NC₆H₄	98%	96%

With Jayasree and Abdul Seayad *JACS* 2006, 1086-1087.

43

Brønsted Acid-Catalyzed Reductive Amination

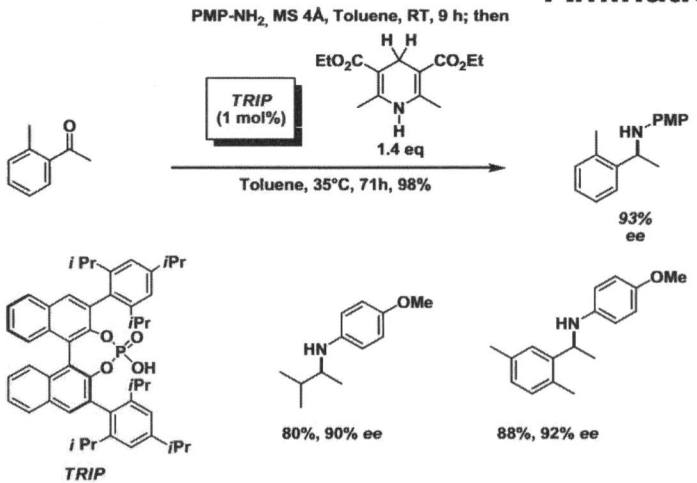

Brønsted Acid-Catalyzed Reductive Amination

Hoffmann, Seayad *Angew. Int.* 2005, 7424.

Brønsted Acid-Catalyzed Reductive Amination

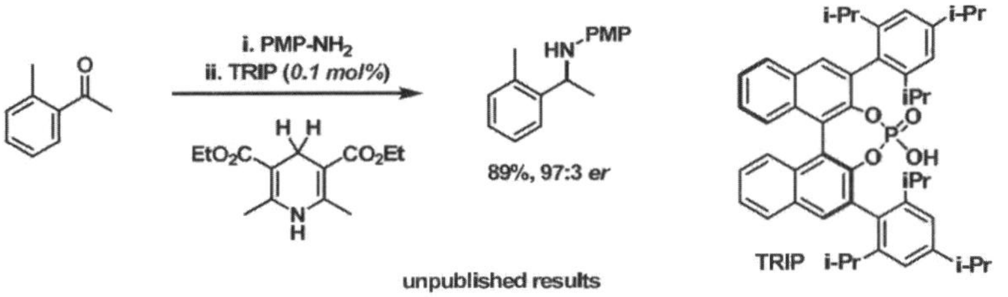

i. PMP-NH$_2$
ii. TRIP (0.1 mol%)
EtO$_2$C
CO$_2$Et
HN-PMP
89%, 97:3 er
i-Pr
iPr
OH
TRIP

unpublished results

Brønsted Acid-Catalyzed Reductive Amination

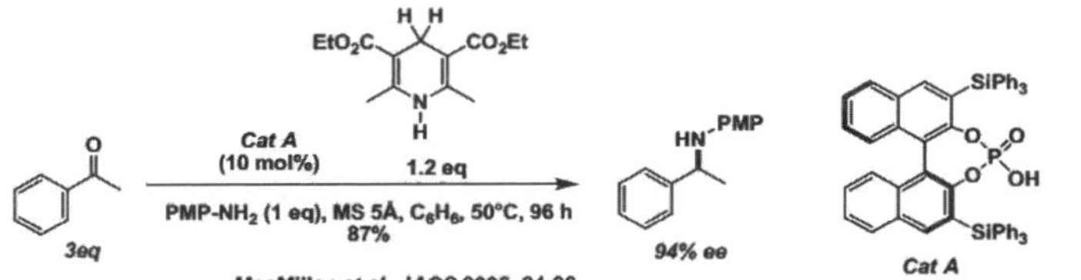

EtO$_2$C
CO$_2$Et
Cat A (10 mol%)
1.2 eq
PMP-NH$_2$ (1 eq), MS 5Å, C$_6$H$_6$, 50°C, 96 h
87%
3eq
HN-PMP
94% ee
SiPh$_3$
OH
Cat A

MacMillan et al. *JACS* 2006, 84-86.

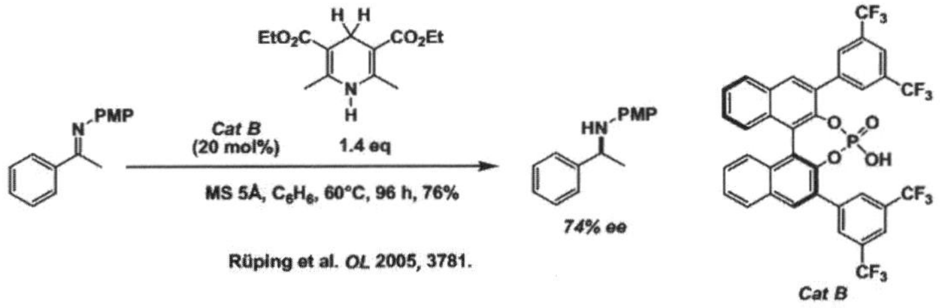

N-PMP
EtO$_2$C
CO$_2$Et
Cat B (20 mol%)
1.4 eq
MS 5Å, C$_6$H$_6$, 60°C, 96 h, 76%
HN-PMP
74% ee
CF$_3$
OH
Cat B

Rüping et al. *OL* 2005, 3781.

Brønsted Acid-Catalyzed Reductive Amination of Aldehydes

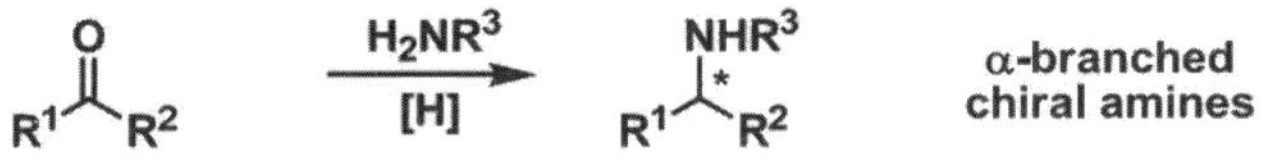

α-branched
chiral amines

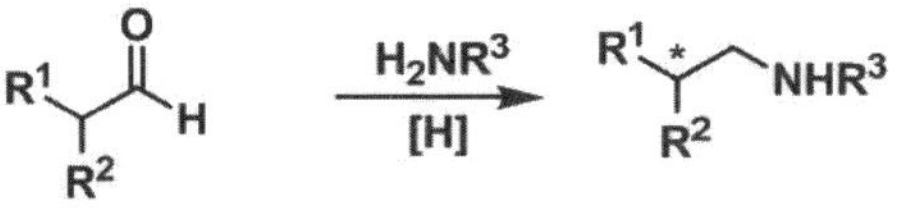

Brønsted Acid-Catalyzed Reductive Amination of Aldehydes

Dynamic Kinetic Resolution

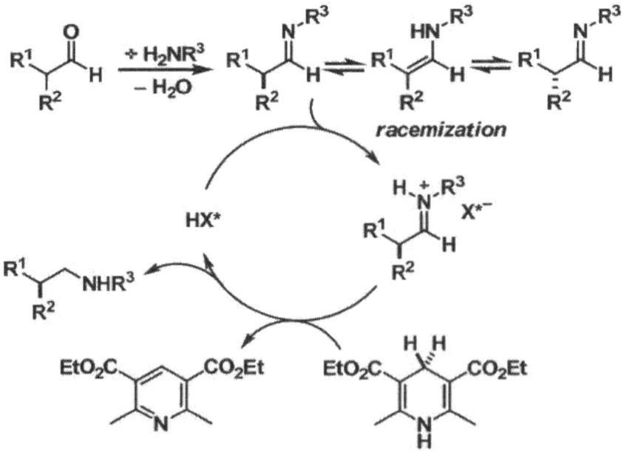

Brønsted Acid-Catalyzed Reductive Amination of Aldehydes

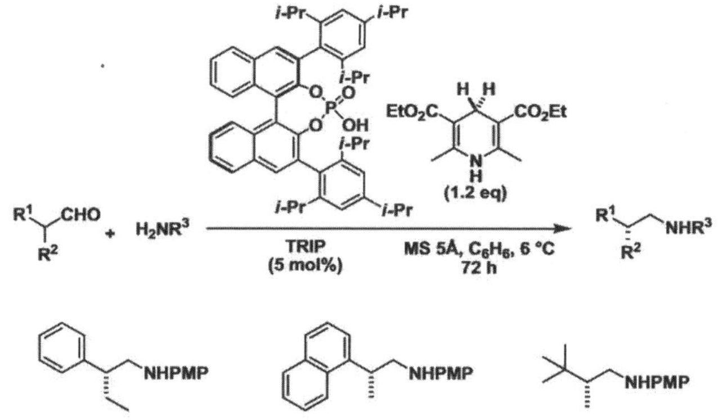

R^{1}R^{2}CHO + H$_2$NR^3 → R^{1}R^{2}CH-NHR^3

TRIP (5 mol%)
MS 5Å, C$_6$H$_6$, 6 °C
72 h

EtO$_2$C / CO$_2$Et, (1.2 eq)

92%, 99:1 er   88%, 99:1 er   77%, 90:10 er

With Sebastian Hoffmann, Marcello Nicoletti *JACS* 2006, 13074-13075.

Catalytic Asymmetric Aldehyde Hydrogenation

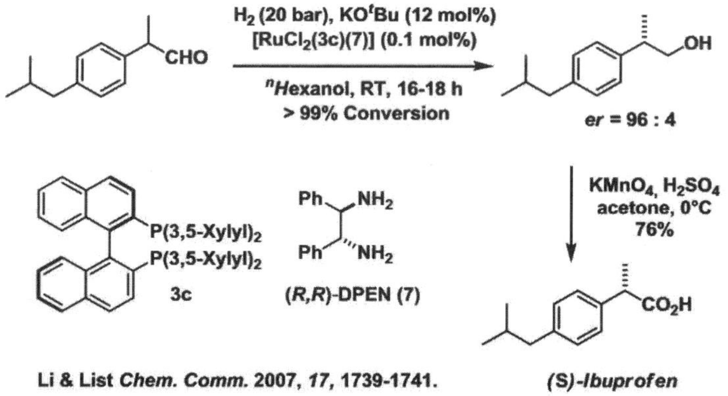

H$_2$ (20 bar), KO^{t}Bu (12 mol%)
[RuCl$_2$(3c)(7)] (0.1 mol%)

^{n}Hexanol, RT, 16-18 h
> 99% Conversion

er = 96 : 4

P(3,5-Xylyl)$_2$
P(3,5-Xylyl)$_2$
3c

Ph / NH$_2$
Ph / NH$_2$
(*R,R*)-DPEN (7)

KMnO$_4$, H$_2$SO$_4$
acetone, 0°C
76%

(*S*)-*Ibuprofen*

Li & List *Chem. Comm.* 2007, *17*, 1739-1741.

Brønsted Acid-Catalyzed Reductive Amination

DKR of ketones:

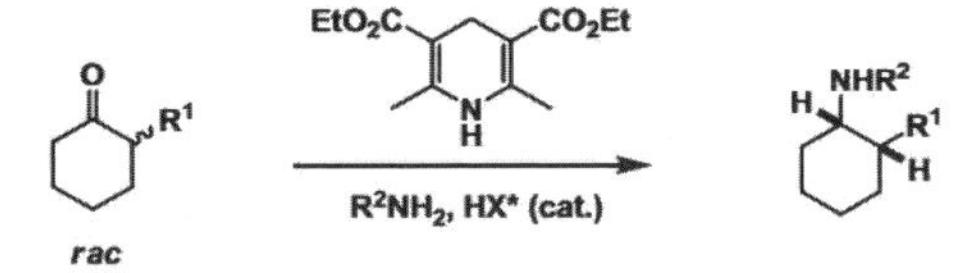

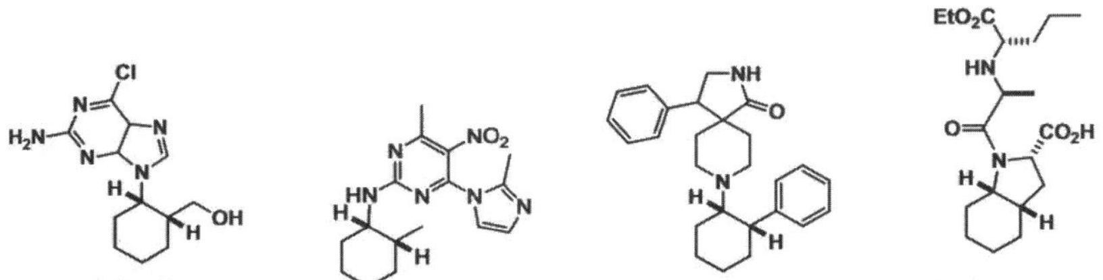

EtO$_2$C ~~~ CO$_2$Et

R^{2}NH$_2$, HX* (cat.)

NHR^2

rac

Anti-RSV anti-influenza

Cytomegalovirus inhibitor

GlyT1 inhibitor

Perindopril (ACE inhibitor)

Brønsted Acid-Catalyzed Reductive Amination

PMP-NH$_2$
TRIP (1 mol%)

EtO$_2$C ~~~ CO$_2$Et

PMP-NH

96%, 98:2 er
(10:1 dr)

Ph

"

PMP-NH
Ph

97:3 er

CO$_2$t-Bu

"

then KOt-Bu
THF, 0°C

PMP

78% of cis-isomer,
er 98:2 er

CAN, 72%

CO$_2$Et
HN

CO$_2$H

Perindopril

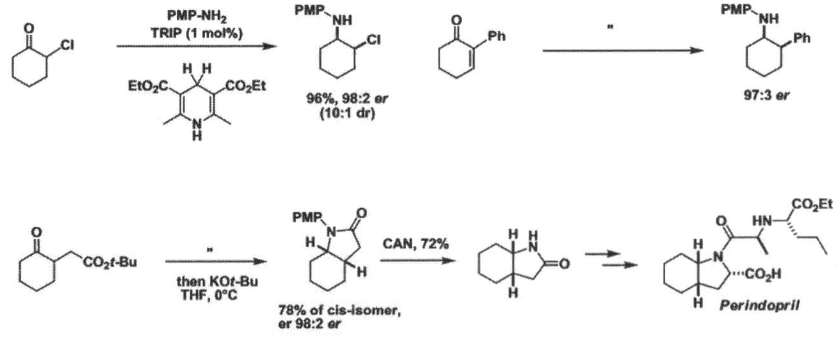

ACIE 2010

48

Brønsted Acid-Catalyzed Cascades

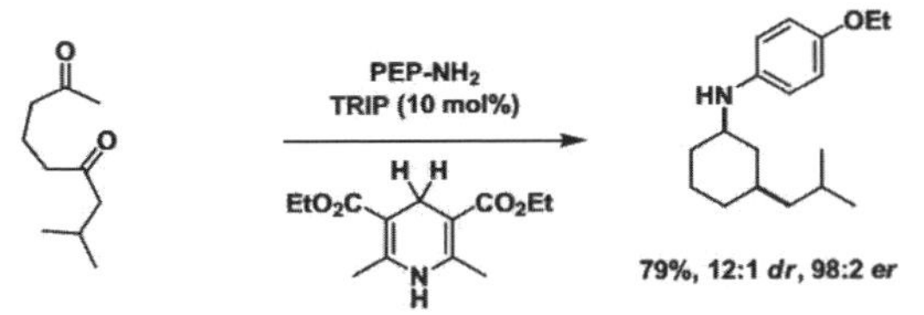

PEP-NH2
TRIP (10 mol%)

EtO2C, CO2Et

79%, 12:1 *dr*, 98:2 *er*

-via aldol, conjugate reduction, reductive amination cascade

With Jian Zhou
JACS 2007, 7498-7499.

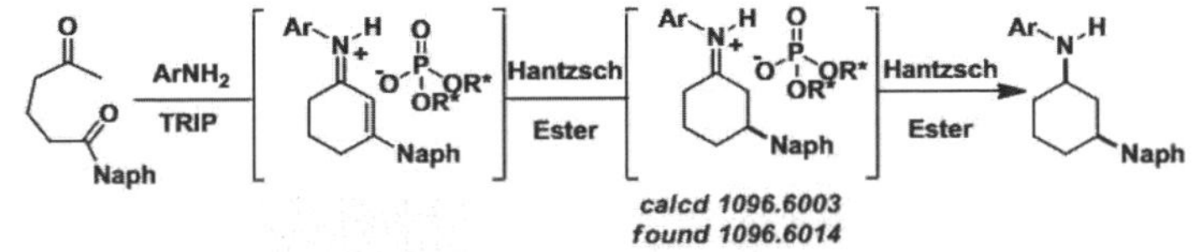

ArNH2

TRIP

Hantzsch Ester

Hantzsch Ester

calcd 1096.6003
found 1096.6014

-mechanism, *ACIE* 2010.

Also see: Adair et al. *Aldrichimica Acta* 2008 & Akiyama et al. *Synlett* 2006, 141.

Direct Catalytic Asymmetric Three-Component Kabachnik-Fields Reaction

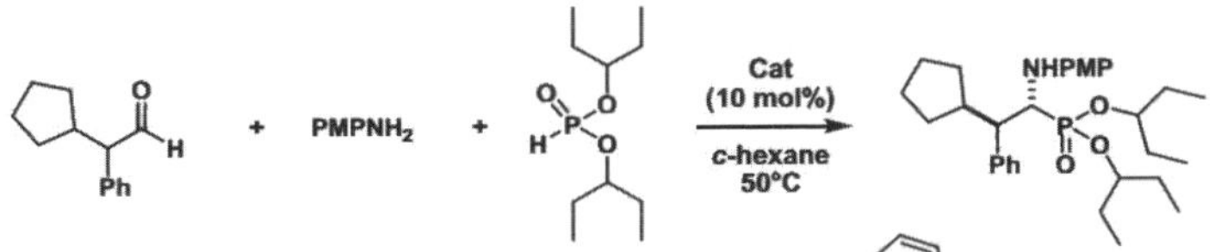

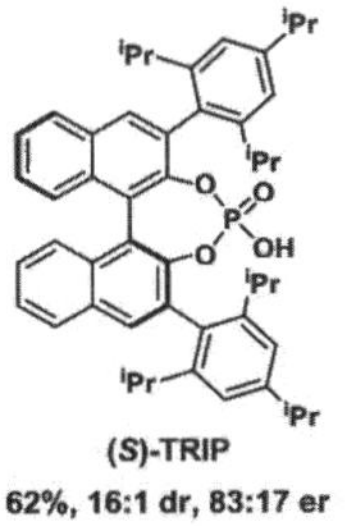

+ PMPNH2 +

Cat
(10 mol%)

c-hexane
50°C

NHPMP

(S)-TRIP
62%, 16:1 *dr*, 83:17 *er*

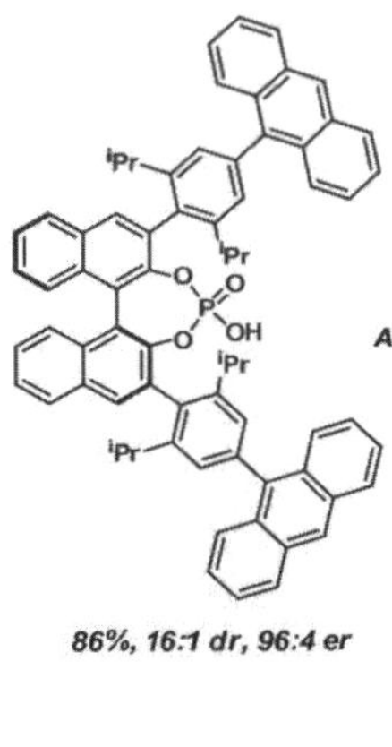

With X. Cheng
Angew. Int. 2008, 5079.

86%, 16:1 *dr*, 96:4 *er*

Direct Catalytic Asymmetric Aminalization of Aldehydes

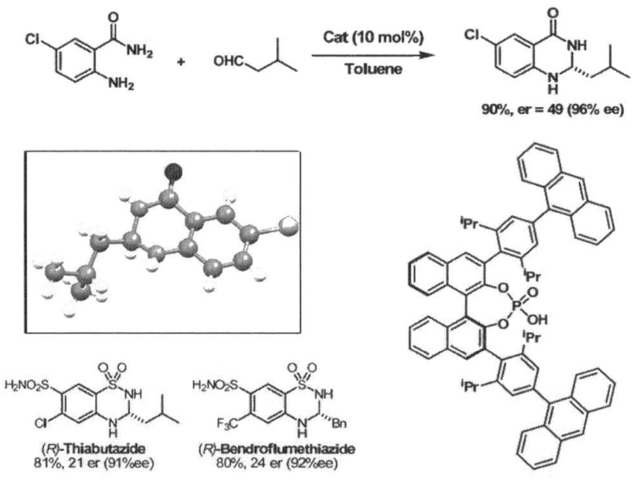
Cat (10 mol%)
Toluene
90%, er = 49 (96% ee)
(R)-Thiabutazide
81%, 21 er (91%ee)
(R)-Bendroflumethiazide
80%, 24 er (92%ee)

86%, 16:1 dr, 96:4 er

With X. Cheng
JACS 2008

A new Motif for Bifunctional Brønsted Acid/Base Organocatalysis

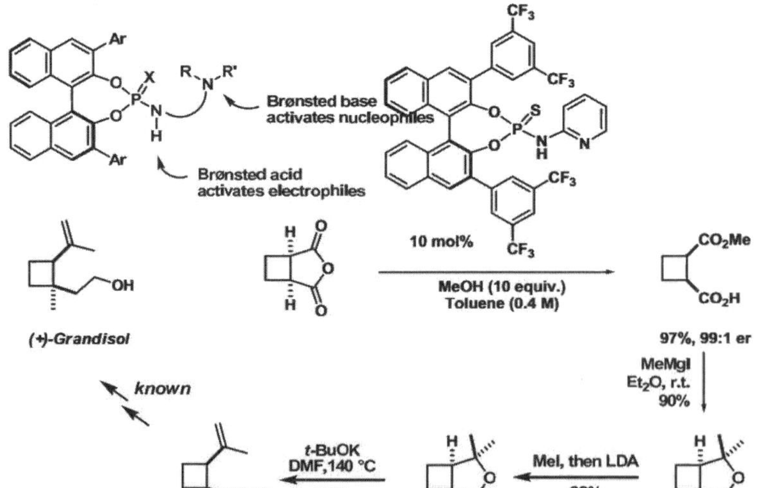
Brønsted base activates nucleophiles
Brønsted acid activates electrophiles
10 mol%
MeOH (10 equiv.)
Toluene (0.4 M)
97%, 99:1 er
(+)-Grandisol
MeMgI
Et2O, r.t.
90%
known
t-BuOK
DMF,140 °C
75%
MeI, then LDA
82%

With Vijay Wakchaure
ACIE 2010

Asymmetric Brønsted Acid Catalysis

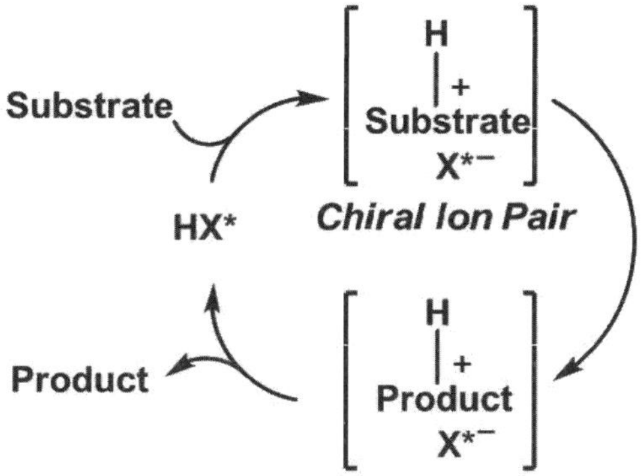

A General Concept for Asymmetric Catalysis

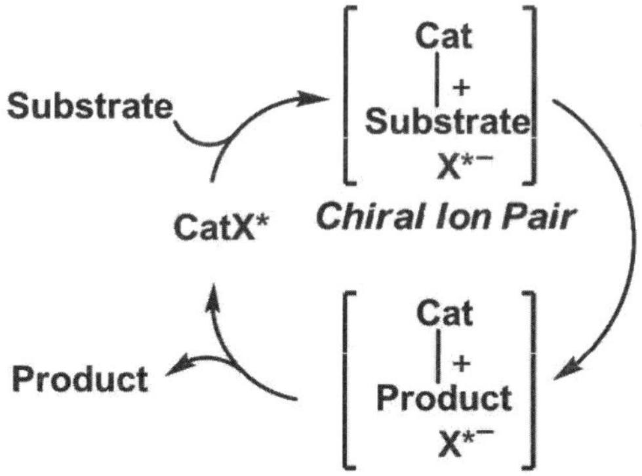

Asymmetric, Counteranion-Directed Catalysis (ACDC)
ACDC is a generalization of asymmetric Brønsted acid catalysis

Asymmetric, Counteranion-Directed Catalysis:
Previous Attempts

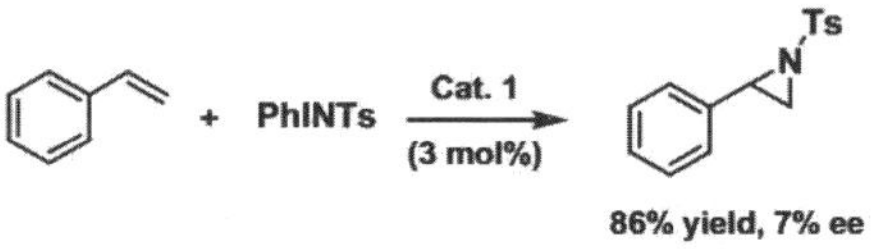

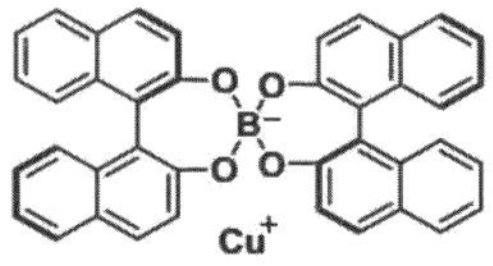

B. A. Arndtsen et al. *OL* 2000, *2*, 4165-4168.

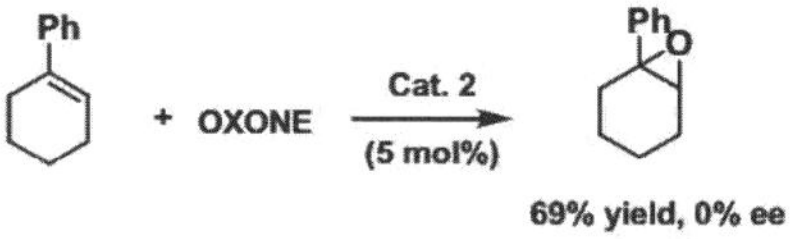

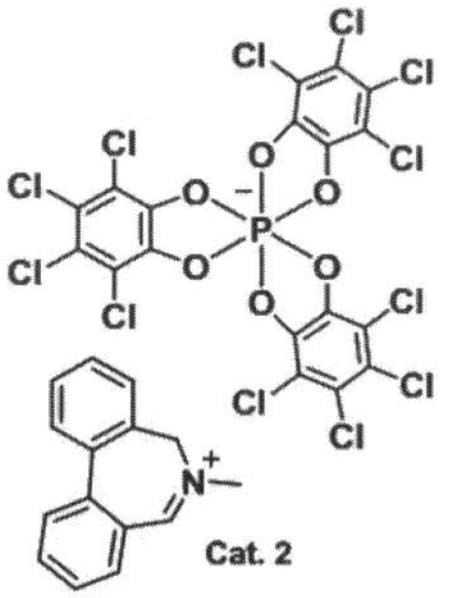

86% yield, 7% ee

Cat. 1

J. Lacour et al. *TL* 2002, *43*, 8257-8260.

69% yield, 0% ee

Cat. 2

Other attempts by A. Nelson and A. Pfaltz

Strong Counteranion Effects!

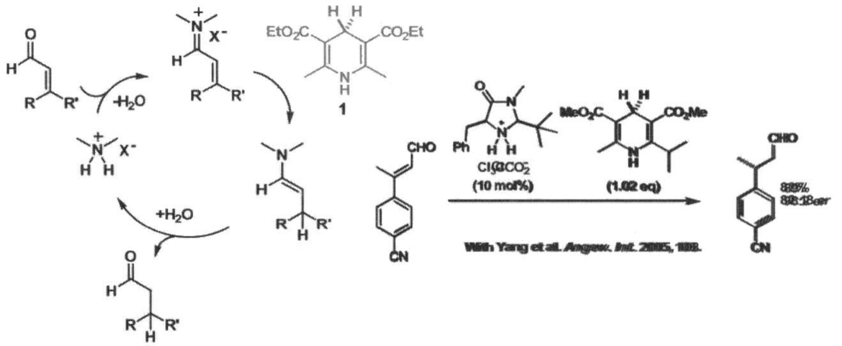

With Yang et al. *Angew. Int.* 2005, 108.

Asymmetric, Counteranion-Directed Catalysis: Initial Experiments

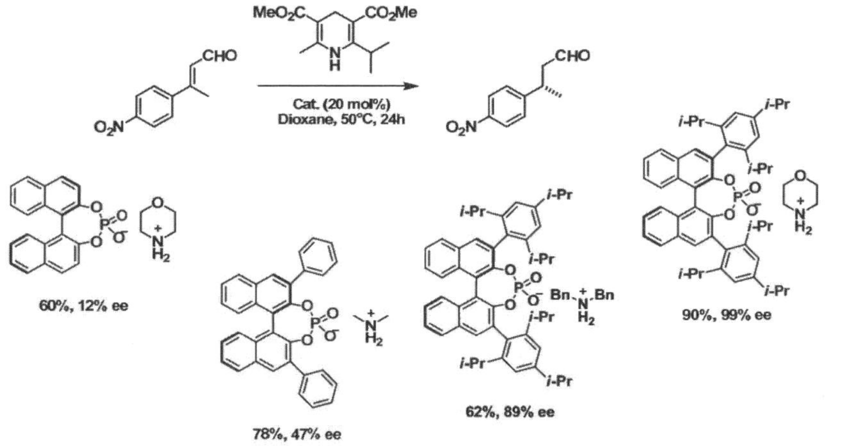

60%, 12% ee

78%, 47% ee

62%, 89% ee

90%, 99% ee

Asymmetric, Counteranion-Directed Catalysis: Enal Transfer Hydrogenation

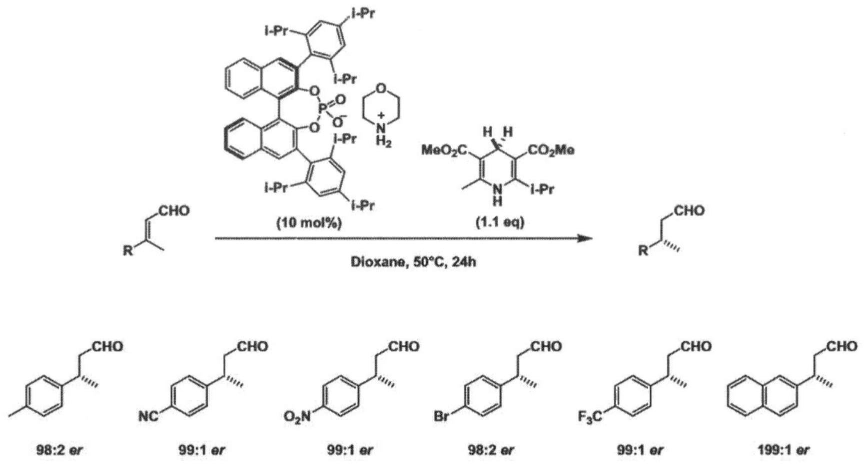

98:2 *er*

99:1 *er*

99:1 *er*

98:2 *er*

99:1 *er*

199:1 *er*

With Sonja Mayer *Angew. Chem. Int. Ed.* 2006, 4193-4195.

Asymmetric, Counteranion-Directed Catalysis: Enal Transfer Hydrogenation

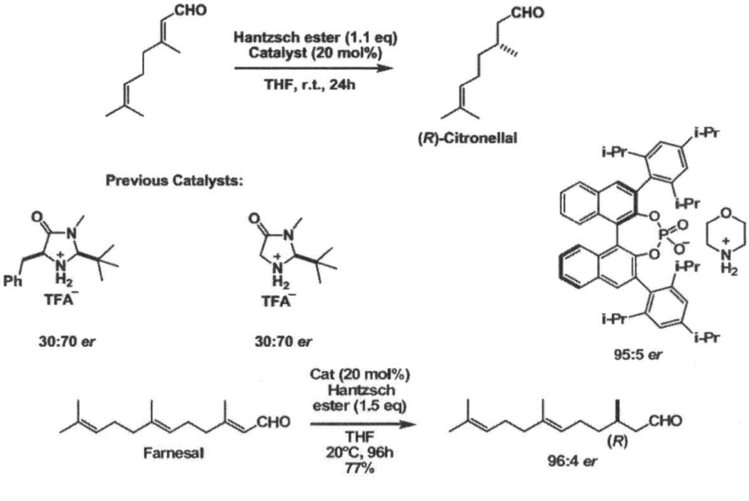

Hantzsch ester (1.1 eq)
Catalyst (20 mol%)

THF, r.t., 24h

(R)-Citronellal

Previous Catalysts:

Ph · TFA⁻

30:70 er

· TFA⁻

30:70 er

95:5 er

Cat (20 mol%)
Hantzsch
ester (1.5 eq)

THF
20°C, 96h
77%

Farnesal

(R)

96:4 er

With Sonja Mayer *Angew. Chem. Int. Ed.* 2006, 4193-4195.

Asymmetric, Counteranion-Directed Catalysis: Ketone Substrates

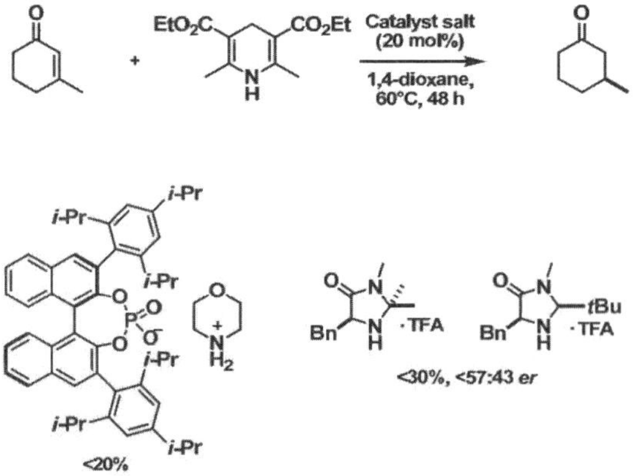

EtO_2C \ \ \ \ \ CO_2Et Catalyst salt
(20 mol%)

1,4-dioxane,
60°C, 48 h

<20%

Bn · TFA   Bn · TFA

<30%, <57:43 er

54

Asymmetric, Counteranion-Directed Catalysis: Enone Transfer Hydrogenation

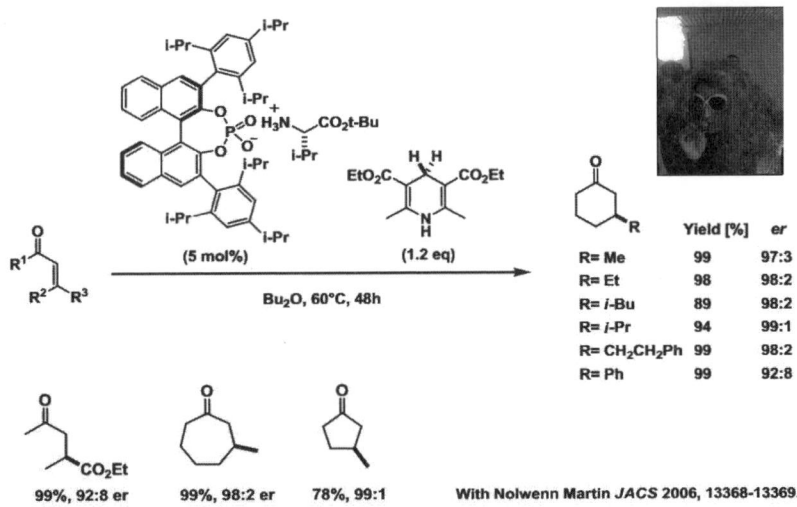

	Yield [%]	er
R= Me	99	97:3
R= Et	98	98:2
R= i-Bu	89	98:2
R= i-Pr	94	99:1
R= CH₂CH₂Ph	99	98:2
R= Ph	99	92:8

99%, 92:8 er   99%, 98:2 er   78%, 99:1

With Nolwenn Martin *JACS* 2006, 13368-13369.

Asymmetric, Counteranion-Directed Catalysis: Enal Epoxidation

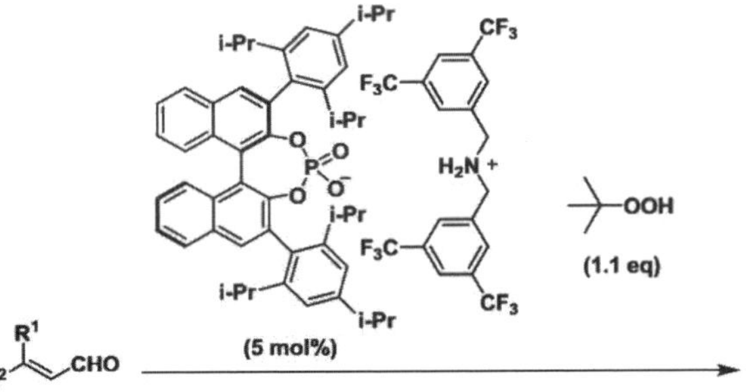

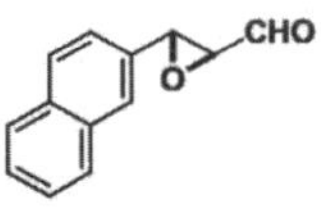

(1.1 eq)

(5 mol%)

Dioxane, 35°C, 72h

76%, >99:1 dr, 98:2 er

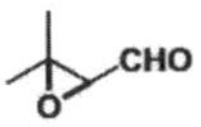

85%, 97:3 er

With Xingwang Wang *Angew. Int. Ed.* 2008, 1119.

Asymmetric, Counteranion-Directed Catalysis: Enone Epoxidation

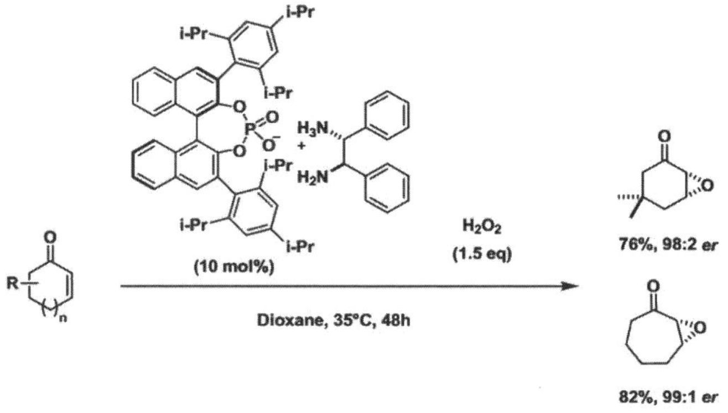

i-Pr
i-Pr
i-Pr
i-Pr
i-Pr
i-Pr
H3N
H2N
(10 mol%)
H2O2
(1.5 eq)
Dioxane, 35°C, 48h
76%, 98:2 er
82%, 99:1 er

With Wang & Reisinger *JACS* 2008, 6070-71.

ACDC Beyond Organocatalysis: Tsuji-Trost Reactions

Pd°
X⁻
Nu⁻

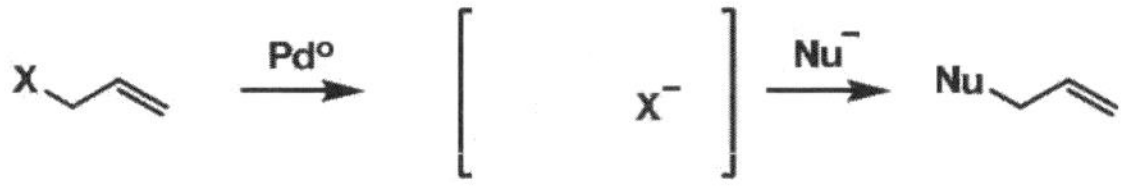

Reaction proceeds via cationic π-allyl palladium complex!

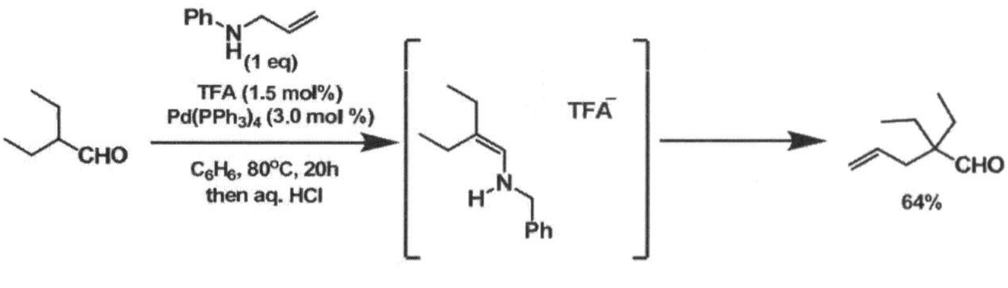

(1 eq)
TFA (1.5 mol%)
Pd(PPh3)4 (3.0 mol %)
C6H6, 80°C, 20h
then aq. HCl
TFA⁻
64%

Murahashi et al. *JOC* 1988, 4489.

Asymmetric, Counteranion-Directed Transition Metal Catalysis:

Ph⌒CHO (Me) + Ph₂CH-NH-allyl (1 eq)

(R)-TRIP (1.5 mol%), Pd(PPh₃)₄ (3.0 mol %)

MS 5Å, MTBE, 40°C

NaBH₄ 82%

2N HCl 85%

98.5:1.5 er

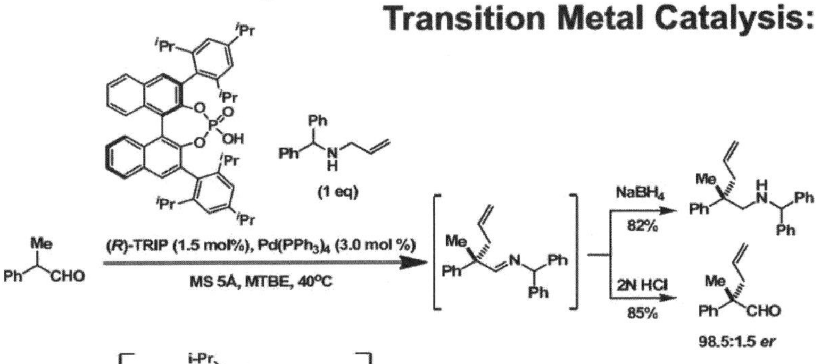

via

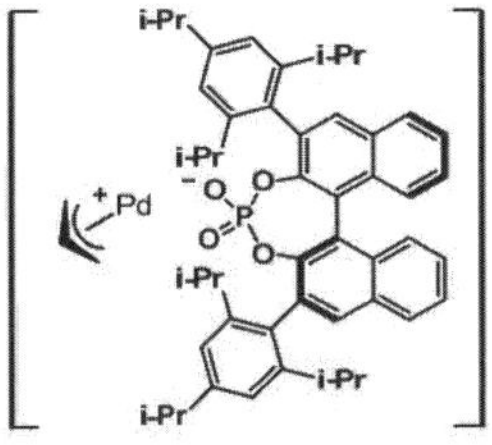

With Santanu Mukherjee *JACS* **2007, 11336-11337.**
For independent applications of Gold-TRIP salts, see:
Toste et al. *Science* **2007,** *317***, 496-499**

Asymmetric, Counteranion-Directed Transition Metal Catalysis: Jacobsen-Katsuki Epoxidation

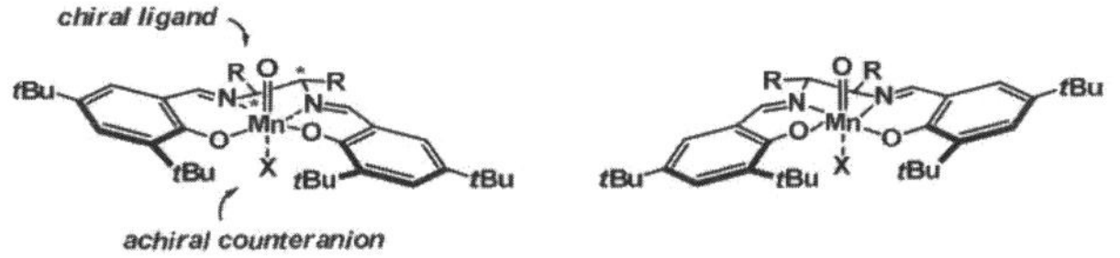

chiral ligand

achiral counteranion

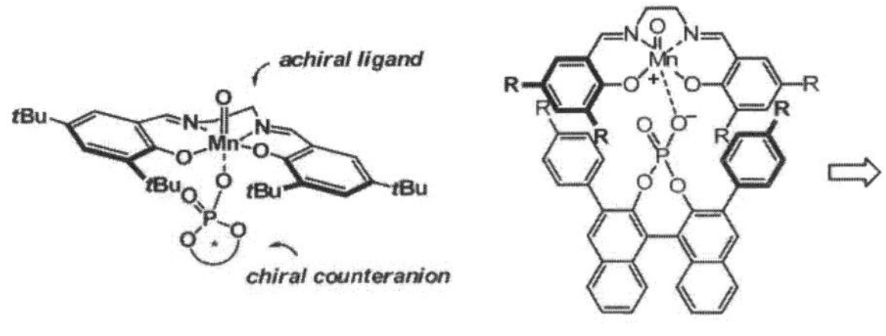

achiral ligand

chiral counteranion

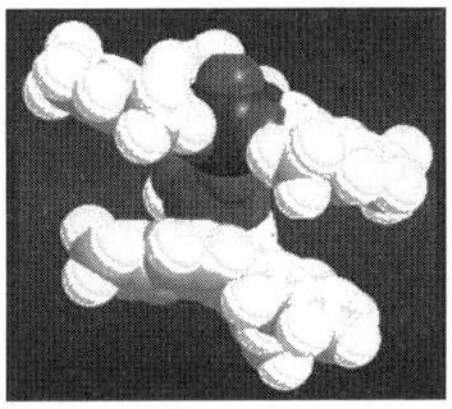

Asymmetric, Counteranion-Directed Transition Metal Catalysis: Jacobsen-Katsuki Epoxidation

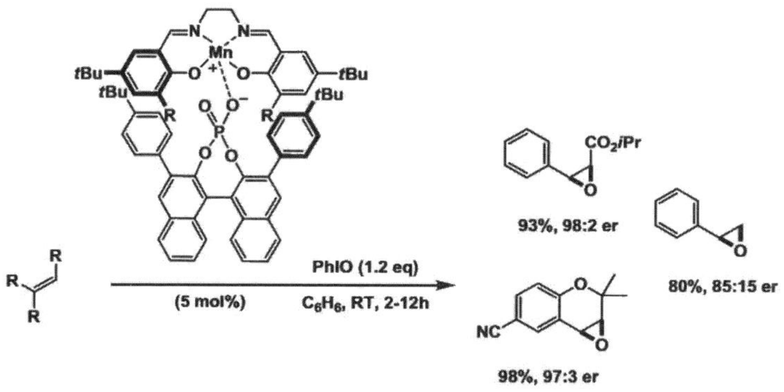

tBu
tBu
Mn
R
P
tBu
tBu
R
PhIO (1.2 eq)
(5 mol%)
C6H6, RT, 2-12h
CO2iPr
93%, 98:2 er
80%, 85:15 er
NC
98%, 97:3 er

With Saihu Liao, ACIE 2010, 49, 628-631.

Organocatalysis

- Brønsted Acid Catalysis
- Brønsted Base Catalysis
- Lewis Base Catalysis
- Lewis Acid Catalysis

Organocatalysis is the catalysis with small organic molecules that do not contain a metal as part of their active principle, and that function by donating or removing electrons or protons...

Lewis Acid Organocatalysis?

...is rare since organic Lewis acids are less common.

Carbonyl Compounds as Lewis Acids

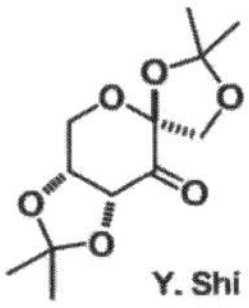

Y. Shi

Carbenium Ions as Lewis Acids

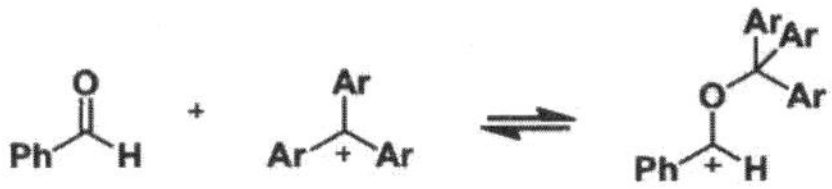

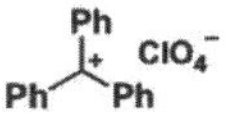

Mukaiyama

Silylium Ions as Lewis Acids

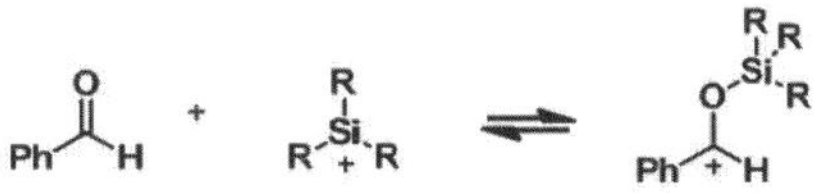

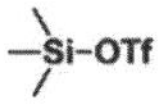

Vorbrüggen
Noyori

Asymmetric Lewis Acid Organocatalysis

(10 mol%)

CH$_3$CN, -40°C

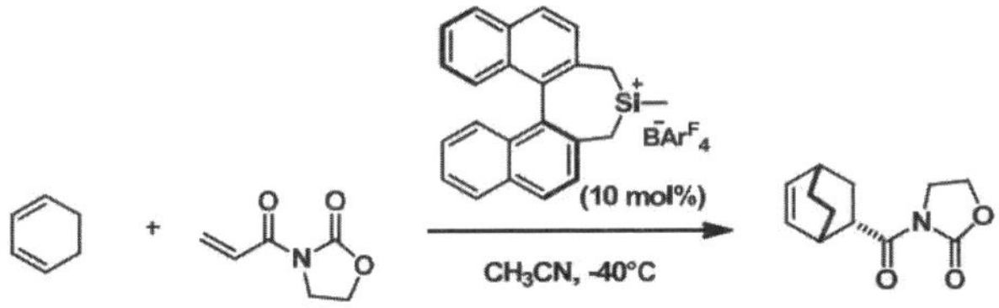

95%, 10% ee

Helmchen,
Jorgensen et al.
JACS 1998

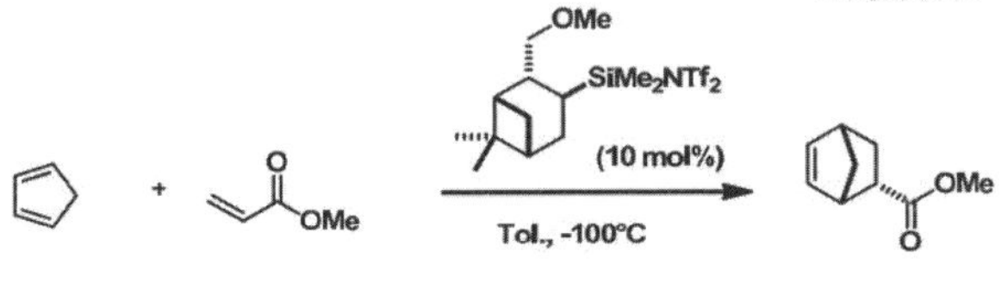

SiMe$_2$NTf$_2$

(10 mol%)

Tol., -100°C

94%, 54% ee

Ghosez et al.
TL2000
Also see Leighton et al.
Tetrahedron 2006

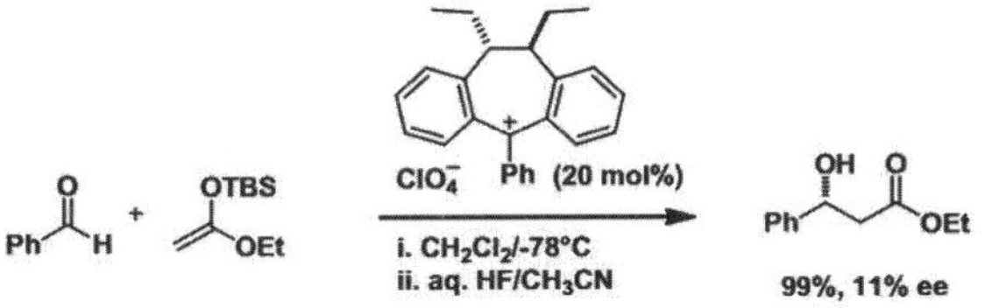

ClO$_4^-$ Ph (20 mol%)

i. CH$_2$Cl$_2$/-78°C
ii. aq. HF/CH$_3$CN

99%, 11% ee

Chen et al.
JACS 1997

Asymmetric Lewis Acid Catalysis of the Mukaiyama Aldol Reaction

...typically requires high catalyst loadings. For one exception, see Carreira et al. JACS 1994, 8837.

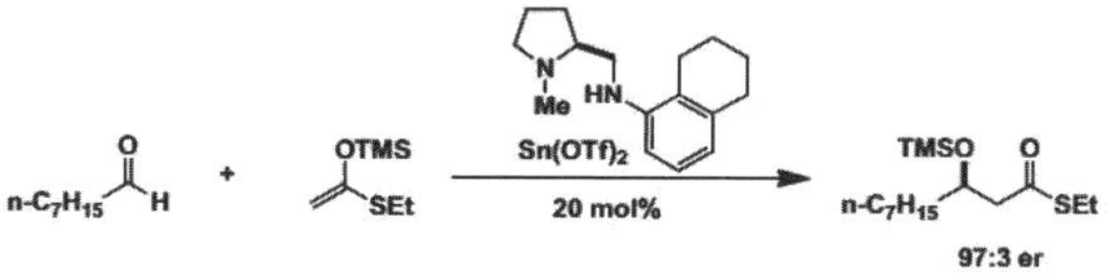

(Mukaiyama et al.)

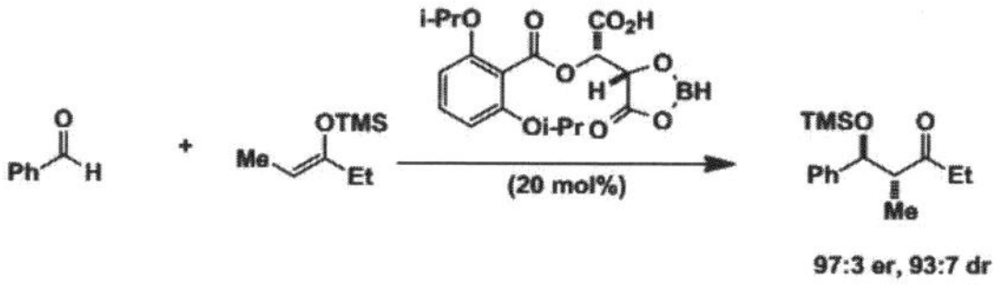

97:3 er

(Yamamoto et al.)

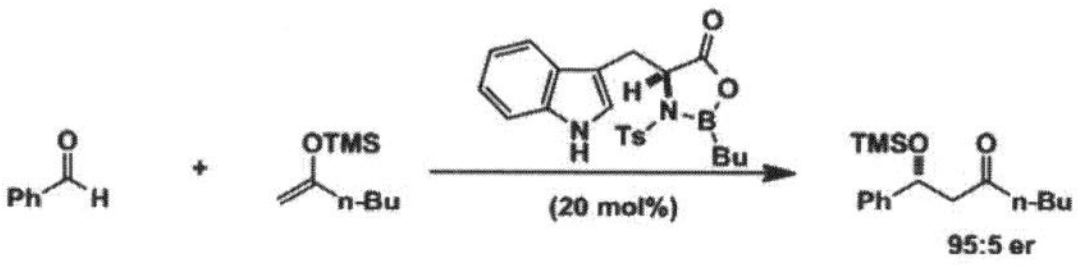

97:3 er, 93:7 dr

(Corey et al.)

95:5 er

Mechanism of the Mukaiyama Aldol Reaction: The Problem of Background Silyl-Lewis Acid Catalysis

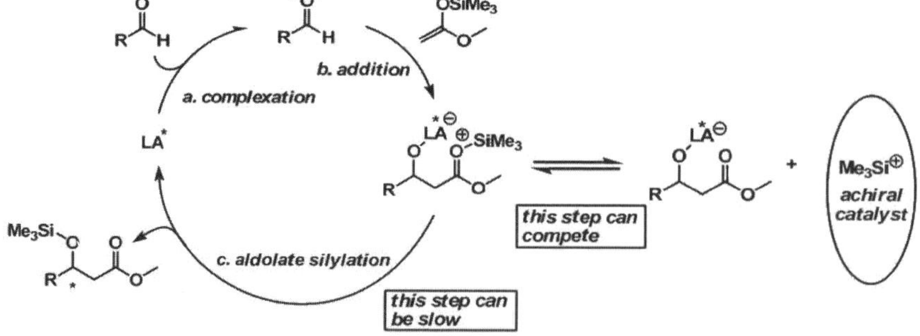

Bosnich et al. JACS 1995, 4570-4581

Disulfonimides are Powerful Silyl-Lewis Acid Catalyst Precursors

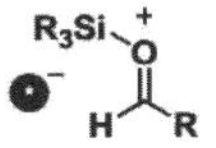

R_3Si
O
H
R

Aldehyde Activation?

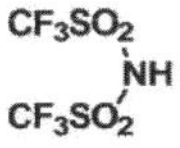

CF_3SO_2
NH
CF_3SO_2

Tf_2NH
(pKa= 1.7)

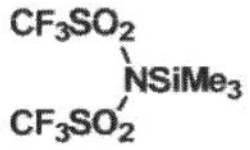

CF_3SO_2
$NSiMe_3$
CF_3SO_2

$TMSNTf_2$

(Ghosez, Mikami, Yamamoto...)

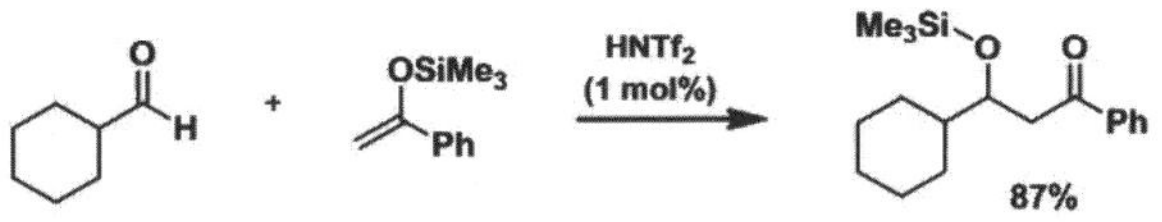

O
H
+
$OSiMe_3$
Ph
$HNTf_2$
(1 mol%)
Me_3Si
O
O
Ph
87%

Yamamoto et al. Synlett 2001, 1851.
via $TMSNTf_2$ ("self repair mechanism")

A new Motif for Organocatalysis: Chiral Disulfonimides

burried acid

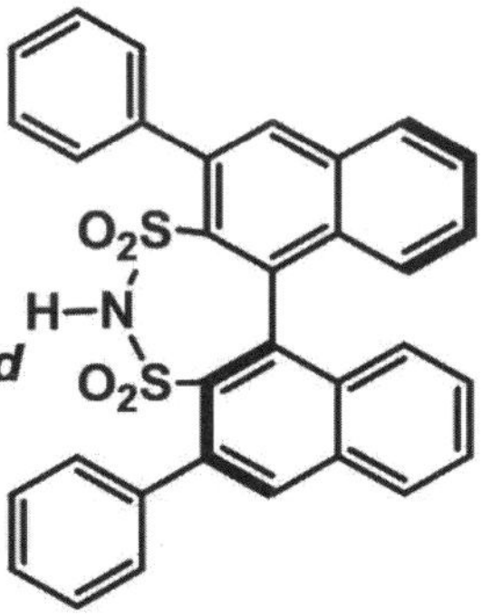

O_2S
H–N
O_2S

C_2-symmetry

Synthesis of Sulfonic Acids and Disulfonimides

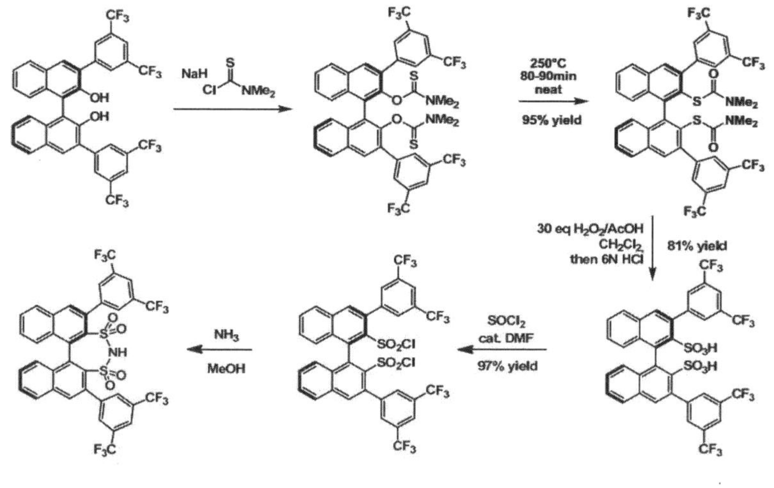
NaH
Cl(C=S)NMe₂
250°C
80-90min
neat
95% yield
30 eq H₂O₂/AcOH
CH₂Cl₂
then 6N HCl
81% yield
SOCl₂
cat. DMF
97% yield
NH₃
MeOH

With Lay & Garcia-Garcia et al., *Angew.* 2009, 4363.

Second Generation Synthesis

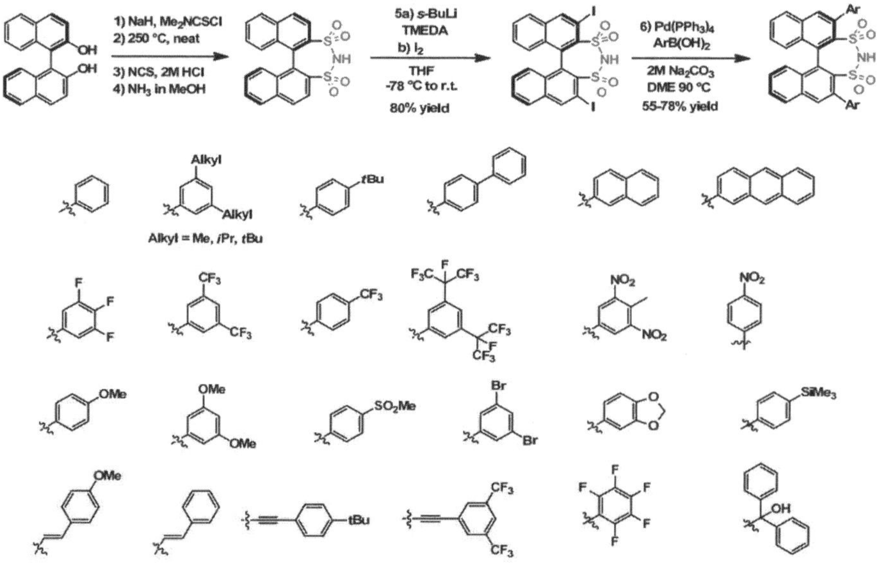
1) NaH, Me₂NCSCl
2) 250 °C, neat
3) NCS, 2M HCl
4) NH₃ in MeOH
5a) s-BuLi
TMEDA
b) I₂
THF
-78 °C to r.t.
80% yield
6) Pd(PPh₃)₄
ArB(OH)₂
2M Na₂CO₃
DME 90 °C
55-78% yield
Alkyl = Me, iPr, tBu

Lewis Acid Organocatalysis: Mukaiyama Aldol Reaction

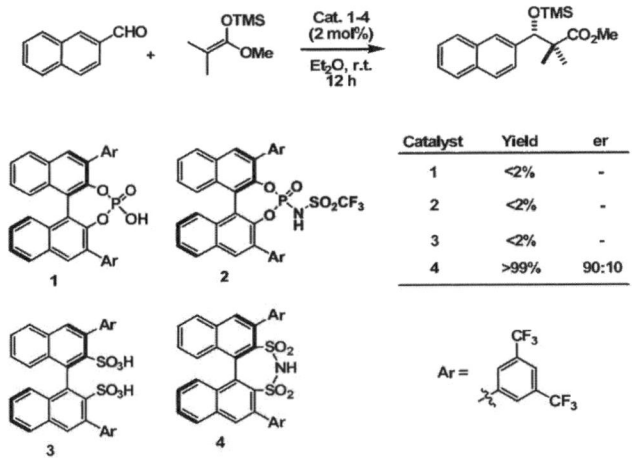

Catalyst	Yield	er
1	<2%	-
2	<2%	-
3	<2%	-
4	>99%	90:10

With Lay & Garcia-Garcia et al., *Angew.* 2009, 4363.

Lewis Acid Organocatalysis: A Powerful Disulfonimide Catalyst of the Mukaiyama Aldol Reaction

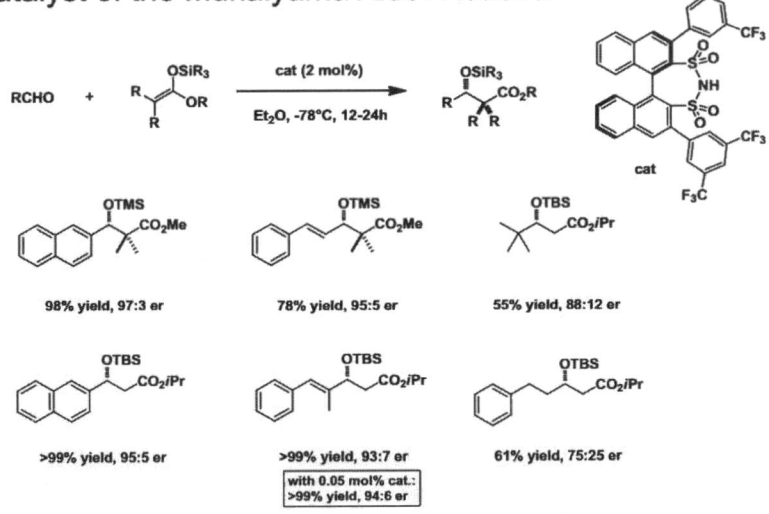

98% yield, 97:3 er	78% yield, 95:5 er	55% yield, 88:12 er
>99% yield, 95:5 er	>99% yield, 93:7 er with 0.05 mol% cat.: >99% yield, 94:6 er	61% yield, 75:25 er

Lay & Garcia-Garcia et al., *Angew.* 2009, 4363.

Lewis Acid Organocatalysis:
(„Frustrated") Lewis-Pair Catalysis Mechanism

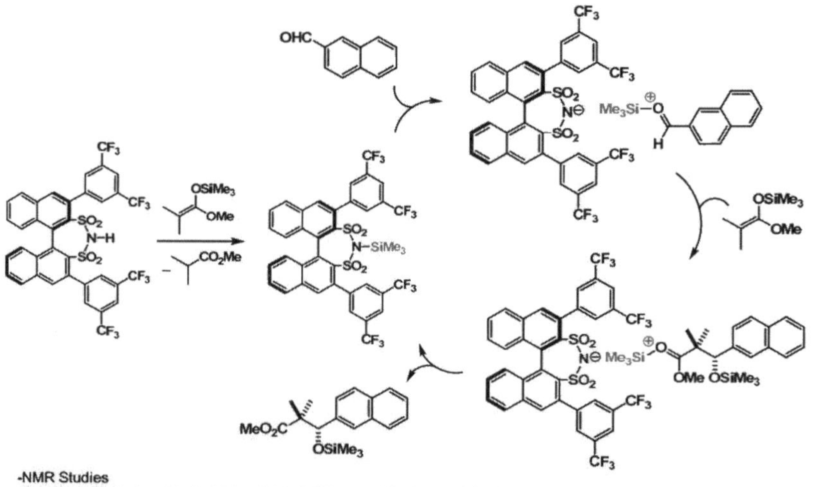

-NMR Studies
-Reaction proceeds in the presence of 2,6-di-tBu-4-methyl pyridine (DTBMP)
-Sulfonic acids & silyl esters are inactive (Ghosez TfOH vs. TfOTMS; Tf₂NH vs Tf₂NTMS)

The Potential of Chiral Disulfonimides in Asymmetric Lewis Acid Organocatalysis:
Vinylogous Mukaiyama Aldol Reaction

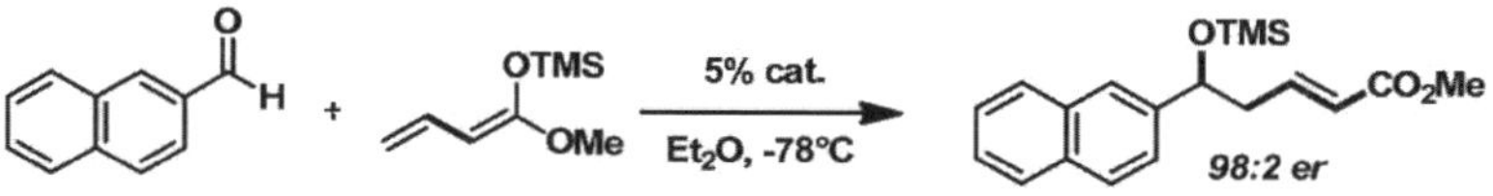

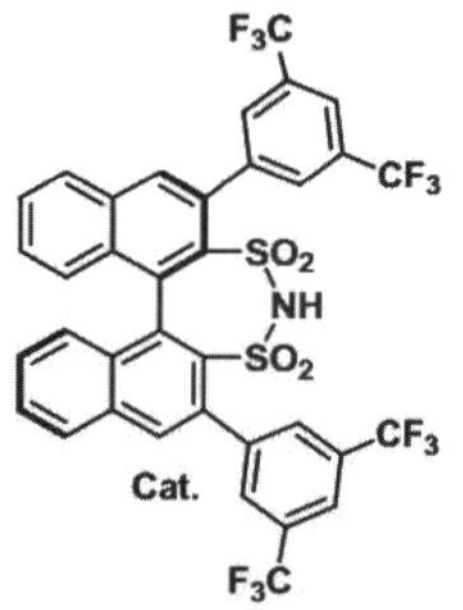

A bis-vinylogous Mukaiyama aldol reaction:

>90%
ε/γ >20:1
91:9 er

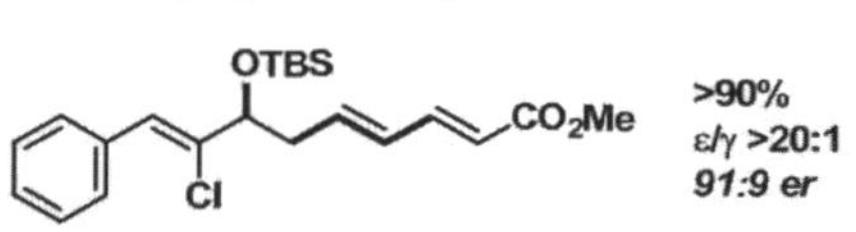

With Lars Ratjen and P. Garcia-Garcia et al.

The Potential of Chiral Disulfonimides in Asymmetric Lewis Acid Organocatalysis: Hetero Diels-Alder Reaction

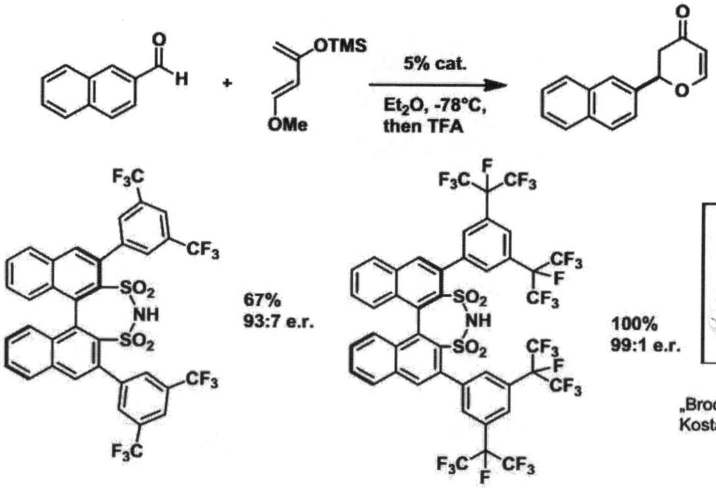

5% cat.

Et$_2$O, -78°C, then TFA

67%
93:7 e.r.

100%
99:1 e.r.

„Broccoli Catalyst"
Kostas Rampalakos

Chiral Disulfonimides: A Powerful new Motif for Asymmetric Lewis Acid Organocatalysis

97:3 er
Mukaiyama Aldol

TMSCN-Addition

Friedel-Crafts

Mukaiyama Michael

97:3 er
Hetero-Diels-Alder

Allylation

98:2 er
vinylogous
Mukaiyama Aldol

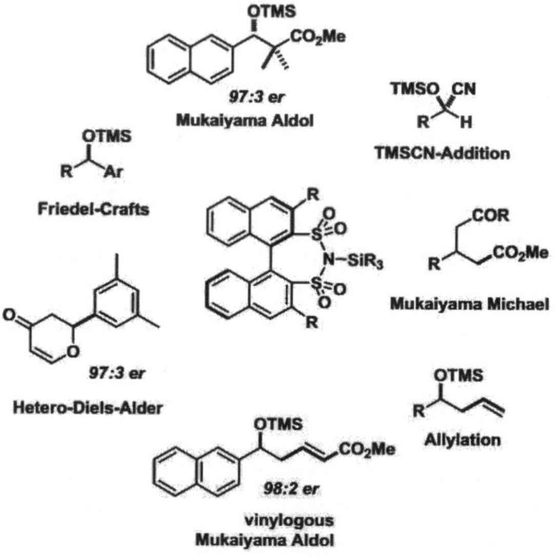

Luigi Ambrosio
Director of Institute of Composite and Biomedical Materials

Address:
Institute of Composite and Biomedical Materials, (IMCB-CNR), Piazzale Tecchio 80, 80125 Naples, Italy.
Telephone: +39-081-2425929, Fax: +39-081-2425932; Email: ambrosio@unina.it.

Education:
1982 Doctoral Degree in Chemical Engineering, University of Naples, Italy.

Professional experience:
1983-1985 Research Associate at Department of Materials and Production Engineering, University of Naples.

1985 Visiting Scientist at Kontron Medical Inc., 9 Plymouth Street, Everett, MA, USA.

1985-1986 Research Associate at Department of Chemical Engineering and Institute of Materials Science, Polymer Program, University of Connecticut, Storrs, CT, USA.

1986-1987 Research Associate at Department of Materials and Production Engineering, University of Naples, Italy.

1987-1988 Principal Research Scientist at Kontron Medical Inc., 9 Plymouth Street, Everett, MA, USA.

1989-1997, Research Scientist at Institute of Composite Materials Technology, CNR, Piazzale Tecchio 80, 80125 Naples, Italy.

1997-2000, Senior Research Scientist at Institute of Composite Materials Technology, CNR, Piazzale Tecchio 80, 80125 Naples, Italy.

1997-2003, Adjunct Professor at Institute of Materials Science, University of Connecticut, Storrs, CT, USA.

1997-2007, Appointed Professor of Biomaterials, University of Naples "Federico II".

2007-2008, Appointed Professor of Biomaterials Design, University of Naples "Federico II".

2001 – Today Research Director at Institute of Composite and Biomedical Materials, (IMCB-CNR), Piazzale Tecchio 80, 80125 Naples, Italy.

2003 – 2005. Associate Director of the Institute of Composite and Biomedical Materials, (IMCB-CNR), Piazzale Tecchio 80, 80125 Naples, Italy.

2006 – April 2008 Associate Director of the Institute of Composite and Biomedical Materials, (IMCB-CNR), Piazzale Tecchio 80, 80125 Naples, Italy.

209-2010, Appointed Professor of Artificial Organs,, University of Naples "Federico II".

April 2008 – Today. Director of the Institute of Composite and Biomedical Materials,

Bioactive polymers and composites in regenerative medicine

Vincenzo Guarino, Antonio Gloria, Marco Alvarez-Perez, Maria Grazia Raucci, Valentina Cirillo,
Roberto De Santis and Luigi Ambrosio*

Institute of Composite and Biomedical Materials, National Reserarch Council of Italy,
P.le Tecchio 80, Naples

* Corresponging author: ambrosio@unina.it

INTRODUCTION

The basic principle of tissue engineering entails the guided application and the control of cells, materials and the microenvironment into which they are delivered [1]. In a tissue engineering process, tissues or organs may be created *in vivo*, *in vitro*, or *ex vivo* and implanted into the patient [2]. In this direction, the selection of three elements firstly concurs to satisfy the basic tissue engineering principles on fabricating biological tissues: (I) viable, responsive cells; (II) a scaffold to support tissue formation; and (III) a growth-inducing stimulus.

Firstly, the sites of tissue reconstruction firstly have to contain a sufficient number of viable progenitor cells capable of producing the tissue of interest. If the number of cells is deficient, one must engineer the site to provide the necessary population of cells or provide some stimulus to recruit than to the reconstruction site [3].

Secondly, the reconstruction site should be filled with a 3D, porous scaffold that facilitates attachment, migration, and proliferation of cells. It should also stimulate the synthesis of extracellular matrix (ECM) proteins and the ingrowth of tissue throughout the site. Thirdly, the cells in the graft site must receive the appropriate stimuli (i.e., mechanical, biochemical) in the form of soluble, bioactive molecules (e.g., growth factors, cytokines, hormones) that will influence the cell phenotype and enhance the end result of tissue formation at the graft site [4,5].

Starting from these considerations, one of the most important challenges in tissue engineering is the appropriate design of open-pore biocompatible and biodegradable porous scaffolds which are able to provide a temporary substitute for the ECM. The relevance of the design of scaffolds to be used as ECM analogues should not be underestimated. Nearly 30 years ago, Bissell proposed dynamic reciprocity, which states that a tissue achieves a specific function in part through interactions of the cells with the ECM [6].

Subsequent work demonstrated that gene expression can be mediated by ECM binding to ECM receptors on the cell surface, which provides a link to the cytoskeleton and eventually the nuclear matrix [7]. The inclusion of neighbouring cell interactions and soluble signals originating systemically or from cells in the immediate or distant vicinity provides a more complete model of tissue environment [8].

For these reason, much attention has been also given to the simulation of the extracellular environment. Scaffolds for tissue regeneration occupy a fundamental role in tissue development, because they must support the proliferation and differentiation of cells as they mature into a functional tissue. In particular, the scaffold-assisted regeneration of specific tissues has been shown to be strongly dependent on morphological parameters such as surface-to-volume ratio and pore size and interconnectivity. Indeed, these micro-architectural features not only significantly influence cell morphology, cell binding and phenotypic expression, but also control the extent and nature of nutrient diffusion and tissue in-growth [9]. Furthermore, it has been suggested that the pore dimensions may directly affect some biological events; as a result, different tissues require optimal pore sizes for their regeneration [10].

Therefore, scaffolds with bimodal micron scale (l-bimodal) porosities may often be necessary for the regeneration of highly structured biological tissues, such as bone and cartilage [11]. On the other hand, transport issues, 3D cell colonization and tissue in-growth would be inhibited if the pores are not well interconnected, even if the porosity of the scaffolds is high [12].

Here, we aimed at discussing the basic functions and requirements of scaffolds in tissue engineering, so underlining the ability of specific manufacturing techniques to impart all morphological and functional features in order to satisfy the specific demands of tissue regeneration.

TAILORING 3D POROUS SCAFFOLDS TO SPECIFIC TISSUES

Porous, three-dimensional scaffolds have been used extensively as biomaterials in the field of tissue engineering for in vitro study of cell–scaffold interactions and tissue synthesis and in vivo study of induced tissue and organ regeneration. Since scaffolds structural properties play a relevant role in all processes involved in tissue genesis including cell adhesion, migration, proliferation, growth, differentiation and biosynthesis, an accurate control over the pore size and its distribution, pore shape, pore interconnectivity, and overall porosity of scaffolds is mandatory for the success of any tissue engineering approach [13]. Furthermore, the success degree of substrates for the clinical application may be obtained by optimizing key requirements such as biocompatibility, degradability, mechanical integrity and osteoconductivity [14]. For these reasons, the material design to replace lost structures, or to improve existing structures by the promotion of new tissue formation, moves toward the choice of degradable or partially degradable composite materials which show a wide survey of physical and chemical properties able to satisfy the heterogeneous demand of tissue engineering application. Besides, the design of functional materials able to direct cell activities to promote the tissue regeneration organ has to overcame the traditional idea of the scaffold as purely a structural support, also providing various biological cues, to be sagely incorporated into the scaffold, in order to guide cell and tissue growth [15]. Here, different polymer and composite scaffolds obtained by sagely mixing different material phases with specific functional cues were proposed for the regeneration of highly complex connective tissues like bone, ligaments, osteochondral defect and nerve in order to highlight the potential of tuning the polymer and composite properties through the appropriate choice of material components and process tecnologies.

a) Bone

In the natural bone tissue, the ability to sustain characteristic loads is guaranteed either by a tailored hierarchical structure at the micrometer and sub-micrometer scale, and either by a well defined chemical composition. In this context, programmed biomechanical properties need to be satisfied since the load bearing function has to be transferred from the engineered material to the growing of the mineralized extracellular matrix (mECM) [16].

Moving towards the current standing of biomaterials science and technology, the challenging way to satisfy the compelling multifunctional needs and extend material property combinations relies on the development of composite systems made of degradable and partially degradable materials. Indeed, the employment of biodegradable composites significantly hinders the stress-shielding phenomena associated with the use of traditional rigid metal-based implants. Meanwhile, the continuous degradation of the implant causes a gradual load transfer to the healing tissue, preventing the stress-shielding atrophy with the stimulation of the healing and the bone remodelling.

The use of composite materials based on calcium phosphate and apatite with intrinsic bioactivity offers the possibility to realize a strong bond with natural hard tissues through more structurally and mechanically efficient interfaces, firstly enhancing the capability of the substrate to form new bony extracellular matrix (ECM) and assuring a more rapid and efficacious integration to the implant site. In the last years, the targeted combination of resorbable ceramic and biodegradable polymer materials enable to reach the benefits required to bone regeneration substrates, by coupling the bioactive potential of highly osteogenic surfaces with a controlled degradation and mechanical response of polymer-based bulk systems. For instance, biodegradable composite scaffolds may be designed by adding hydroxyapatite (HA) particles inside the polymeric matrix composed of polyesters as polycaprolactone, polylactid acid, polyglycolic acid and their copolymers [17-19].

Indeed, HA is known to be biocompatible, bioactive, i.e., ability to form a direct chemical bond with surrounding tissues, osteoconductive, nontoxic, noninflammatory, and nonimmunogenic agent. In particular it has been demonstrated that the presence of bioactive solid signals such as HA in the polymer matrix may improve bone formation by mimicking the natural bone mineral phase [20]. However, the osteoconductive response enhancement induced by inclusion within the polymeric matrix of rigid bone-like particles is not coupled with a drastic increase of the polymer mechanical response (Fig.1) by for the use as bone tissue analogue [20,21].

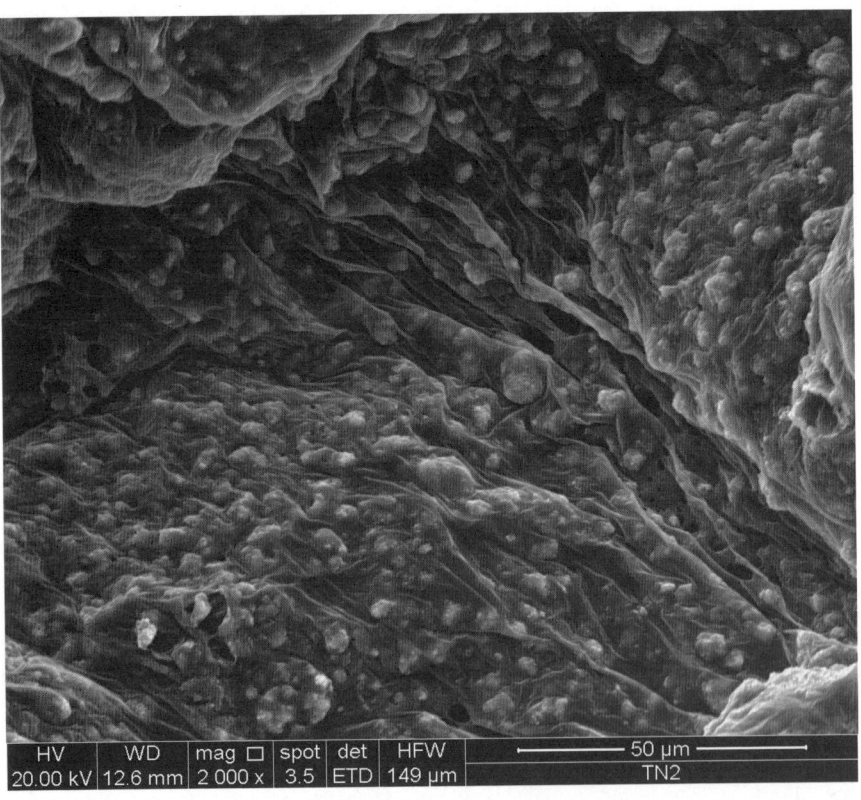

Figure 1: PCL/HA scaffolds obtained via phase inversion and salt leaching for Bon

Therefore, several composite scaffold typologies have been proposed by combining biocompatible polymeric matrices with one or more reinforcement systems in different forms – particles and long fibres – to optimize the final mechanical response of the scaffold. Three dimensional porous Poly ε-caprolactone (PCL) based composite scaffolds, tubular in shape, may be developed by the combination of the filament winding technique and a phase inversion/salt leaching process [22,23]. The employment of highly biocompatible and bioresorbable PCL and polylactid acid (PLA) assures the maintenance of sufficient physical and mechanical properties for at least six months before their degradation. The integration of a solid porogen (i.e. sodium chloride crystals) within a 3-D polymer matrix enables the creation of an interconnected pore network with well-defined pore sizes and shapes. Then, the integration of resorbable PLA fibres (19 vol.%) allows to approximately doubles the elastic modulus compared to that of an un-reinforced scaffold with the same degree of porosity [22]. Moreover, the significant increase of elastic modulus – up to an order of magnitude higher (2.21 ± 0.11 MPa) – also can be obtained by integrating both PLA fibres and a-TCP particles with the PCL matrix, suggesting a dramatic synergistic effect between the two reinforcement systems used [23].

In this case, the addition of tri-calcium phosphate (a-TCP) particles, which is able to precipitate in calcium-deficient hydroxyapatite (CDHA) form, is recognized by the host tissue as being similar to natural bone apatite. The presence of bulk signal may also significantly support also the surface bio-mineralization induced by immersion into simulated body fluid (SBF) solution. Indeed, the precipitation of apatite-like crystals is promoted by the ceramic spots which stick out of the bulk interacting with interstitial blood plasma (Fig.2). These interaction determines a core cascade of events which are directly promoted by the polar interaction of the surface functional groups with the calcium and phosphate ions in the fluid [24]. In this context the employment of supersaturated SBF solution, if compared with the standard Kokubo treatment [25] enhances the mechanism of crystal deposition, enabling to control the apatite crystal growth, and finally, surface topography and chemistry of proposed complex structures [26].

HV	WD	mag □	spot	det	HFW	1 mm
20.00 kV	9.8 mm	100 x	3.5	ETD	2.98 mm	TN2

Figure 2: PCL scaffold by rapid prototyping technique for osteochondral defect

b) Ligament

The use of tissue engineering approaches in ACL regeneration allows these scaffolds to degrade while promoting tissue growth, enabling the body fully to regenerate lost or damaged tissue without the risk of scaffold or neo-ligament rupture, or stress shielding of the new tissue. Individual studies have reported scaffolds combining good initial mechanical properties, controlled degradation, and biocompatibility when used for in vivo ACL reconstruction [27]. For instance, Ge et al. [28] proposed a "dual" composite structure characterized by two ends of structure should allow osteogenesis and be integrated with host bone after implantation, while the middle part of it should host tissue in-growth and, in time, attendant functionality. Starting from this inspired approach based on composite materials for ligament devices, it has been proposed a composite scaffold for tissue engineering of ACL [29]. The system is composed of a bioactive porous matrix, reinforced by biodegradable fibres. The matrix consists of a porous material, namely the benzyl ester of hyaluronic acid, and a fibre, poly(l-lactic acid). These were wound, using a filament winding technique, to emulate the mechanical properties and function of collagen fibres. This system permits the attainment of the right balance between bioactivation, morphological features and mechanical performance for the ACL scaffold. Hyaluronan derivatives have been shown to be bioactive, effectively reproducing the extracellular matrix domain. In particular, the benzyl ester of hyaluronic acid showed good biocompatibility and the capability to promote cell adhesion and proliferation. A porosity range from 100 to 500µm has been obtained by the solvent casting/salt leaching technique. A filament angle of 20° for PLLA fibres (24 multifilaments, 75dtex) and Hyaff 11p100 has been obtained by filament winding technique. The fibres were impregnated with the hyaluronan-derivative solution containing salt crystals (subsequently to be leached out by an aqueous solution) and wound on a mandrel. By the evaluation of mechanical properties in tensile condition, it has been demonstrated that elastic modulus depends upon the components and structural parameters of the scaffold. The in vitro degradation performed in lactate solution showed a large decay of mechanical properties after three months. Preliminary tests on adhesion and proliferation of 3T3 mouse fibroblasts on the proposed scaffold showed encouraging results by day 3, with a higher value of adhesion on the higher porosity scaffold (300-500µm) for both the PLLA and Hyaff11 fibres. In the future, it is expected that the tissue engineering approach will lead to the design and production of a new generation of scaffolds, able to mimic the mechanics of natural ligament and lead to the quick and complete regeneration of new natural tissue.

c) Osteochondral defect

Articular cartilage together with subchondral bone plays a crucial role in the natural joints [30]. In the natural joint articular cartilage and subchondral bone act as a load-bearing system providing stability large range of joint motion with excellent lubrication, uniform distribution of loads. The role of cartilage is to protect the subchondral bone from high stresses, while increasing joint congruence and reducing nominal contact pressure; moreover, articular cartilage also allows low-friction movements [30,31]. However, even though articular cartilage and subchondral bone often undergo degeneration and failure [31], no accepted method for complete repair of osteochondral defects seems to exist [32]. Furthermore, also the use of joint devices has evidenced several complications [32,33].

In this scenario, a tissue engineering approach could be very challenging for the regeneration of osteochondral defects. In particular, nanocomposite and/or hybrid scaffolds were designed and developed through rapid prototyping techniques (i.e., 3D fiber deposition) [34]. 3D PCL/hydroxyapatite nanocomposite scaffolds were built layer by layer using a 3D plotter, thus obtaining "morphologically controlled" structures with specific architectures, precise pore shape and size. Nanocomposite scaffolds were built by extruding and alternatively depositing the fibres considering specific lay-down patterns. PCL/hydroxyapatite pellets were initially placed in a stainless steel syringe and then heated at a specific temperature using a heated cartridge unit placed on the mobile arm of a bioplotter-dispensing machine (Envisiontec GmbH, Germany). Successively, a nitrogen pressure was applied to the syringe through a cap. These scaffolds were clearly characterized not only by the fibres diameter (depending on the needle diameter and/or the deposition speed), but also by the fibres spacing (strand distance, i.e. centre-to-centre distance) and layer thickness, which influence the overall pore size. Results from mechanical and biological tests suggested that the inclusion of hydroxyapatite nanoparticles within the polymer matrix improved the performances of the 3D fiber-deposited scaffolds. The possibility to create 3D heterogeneous bilayered scaffolds (PCL-PCL/hydroxyapatite), composed of two distinct but integrated layers for the cartilage and bone regions, or to suitably inject collagen and collagen/hydroxyapatite within the nanocomposite scaffolds was also highlighted for the regeneration of osteochondral defects.

d) Nerve

After trauma or nerve injury, current clinical treatments involve the use of autograft or allograft. This has many limitations, including donor scarcity, multiple surgeries, donor site morbidity, scarring, and the need for an allograft patient to take immunosuppressant indefinitely post-surgery to avoid rejection. The fabrication of artificial nerve grafts by a neural tissue engineering approach represents an alternative strategy for the restoration of nerve function after injury. In particular, the development of synthetic material for use in the engineering of nerve grafts is a promising alternative, which obviates the need for immuno-suppression or surgery involving autografts or allografts. In this context, polymeric nanostructured platforms, realized by electrospinning technology may be efficiently used as artificial grafts for nerve tissue regeneration.

The electrospinning process involves the application of a high voltage electric field to a capillary tip to produce electrically charged jets from a polymer solution. Under the influence of the electrostatic field, a pendant droplet of the polymer solution at the capillary tip is deformed into a conical shape, better known as Taylor cone [35]. Such event is determined by the interplay of the electrostatic repulsion between the surface charges of the droplet and Coulomb-like forces exerted by the strong external electric field applied [36]. If the voltage surpasses a threshold value, electrostatic forces overcome the surface tension, and a fine charged jet is ejected. The jet moves towards a ground plate acting as counter electrode, accompanied by rapid evaporation of the solvent molecules with the formation of very thin fibres [37]. The result is the deposition of a thin polymer fibre network on the grounded metallic target. The viscosity of the polymer solution and the presence of entanglements among polymer chain hinder the formation of fibre defects (beads) because of the control of jet instabilities (stretching and whipping processes).

Thus far, for selected applications it is desirable to control the fibre diameter, but also fibre pores, and the assembly of structures, their different spatial orientation (aligned and random), and finally the surface properties of electrospun fibrous networks. First, a mapping of material and process parameters as a function of the specific polymer/solvent system used, which also takes into account the jet instability phenomena, is mandatory to predict "ab initio" the morphological properties of the electrospun scaffolds [38]. It has been reported that fibre diameters can be controlled by altering the concentration of the polymer solution. More specifically, binary solutions with higher polymer concentrations produce larger diameter fibres [39]. Meanwhile, the process has to be adapted to the chemical properties of used polymers, both natural (fibrinogen [40], collagen [41] and gelatin [42]) and synthetic (poly(glycolic acid) (PGA), poly(lactic acid) (PLA) [43], polycaprolactone (PCL) [44] etc.), so that the first choice is about the polymer mixture according to solution properties respect to the attended results.

Moreover, concerning the development of temporary substrates for tissue regeneration, the electrospun system has to recapitulate the functional properties of these complex structures, including their structural anisotropy. In this context, the use of a custom-made set-up may be advantageous for inducing anisotropy in nanofibrous scaffolds to create a three-dimensional (3D) micro-pattern for tissue guidance. Rotating grounded targets may be efficiently used to control the alignment of the collected fibres. High rotational rates enable producing fibres which align parallel to the direction of rotation, while lower one promote the creation of fibrous sheets with randomly aligned fibres [45].

In particular, it has recently been demonstrated that electrospun membranes comprising biodegradable polymers, such as polylactid acid or polydioxanone, exhibit favorable interactions with neuronal cells, promoting adhesion and supporting cell differentiation. Schwann cells seeded onto polycaprolactone-based electrospun membranes showed important evidence of cell morphology and proliferation, underlining the ability of cells to form bipolar spreading onto the nanofibrous surface and acting as a positive cue to elongate neurite outgrowth [46]. However, the tissue engineering approach requires a meaningful understanding of neurite-like processes or neuritogenesis to reproduce the complex architecture of neuronal networks to design biomaterials that can be used for nerve regeneration. In this context, the investigation of mechanisms involving the neurite initiation site as well as neurite out-growth is pivotal to promote proper cell substance formation for improving the nerve cell affinity into the tissue scaffold. Recent studies demonstrated that the inclusion of bioactive polymers such as collagen or gelatin in synthetic matrices plays a crucial role in promoting nerve regeneration [47]. Here, we reports about the ability of gelatin molecules to act as biological cues for promoting nerve repair. In particular, the integration of gelatin in PCL nanofibers significantly improves the bio-interaction of PC-12 pheochromocytoma nerve cells with the substrate. Scaffold bioactivity will be the result of the synergistic contribution of scaffold material topography, i.e fiber diameter on the nanoscale, and biochemical signals offered by gelatin biopolymer. First of all, gelatin chemistry influences the morphology of the fibrous systems by affecting the interaction of the polymer solution with the electric field. Molecule polarity and density of gelatin both contribute to control the stretching of PCL chains thus

promoting fibres with bigger sizes (ca. 550 nm as diameter). Hence, chemistry, i.e more pronounced hydrophilic behavior – and fibre morphology – i.e, fibres with sub-micrometric size scale – significantly concur to characterize the biological response of PCL/Gelatin membranes PC-12 cell line. Indeed, PC-12 cells preferably differentiate in the presence of PCL and gelatin nanofibers as confirmed by the higher expression of the growth cone GAP-43 and the estimation of the average neurite length [47,48]. Therefore, these experimental evidences confirm that differentiation is related to the presence of neurite-like processes and gelatin considerably contributes to the stabilization of the neurite out-growth process.

Figure 3: PCL micro and nanofibres via electrospinning technique for nerve regeneration

CONCLUSION

Proposed strategies of scaffold design certainly proved to be efficacious to assure a proper structural reinforcement, also preserving the hierarchical organization of micro and macro-structure of natural tissue. The modulation of the material phases significantly contribute to final properties of the scaffold through the right balance between porosity requirements and mechanical response. Proposed material-based approach allows to realize tailored systems with specific functional cues which may guide cell activities involved to tissue in-growth

ACKNOWLEDGEMENTS

This work is supported by M.U.R - FIRB TissueNet Project.

REFERENCES

[1]   Vacanti C.A. and Vacanti J.P. *Orthop. Clin. North Am.* **2000**, 31, 351.

[2]   Musgrave D.S., Fu F.H., and Huard J. *J. Am. Acad. Orthop. Surg.* **2002**, 10, 6.

[3]   Fleming J.E., Cornell C.N., and Muschler G.F. *Orthop. Clin. North. Am.* **2000**, 31, 357.

[4]   Laurencin C.T., Attawia M.A., Elgendy H.E., and Herbert K.M. *Bone* **1996**, 19, 93S–99S.

[5]   Attawia M.A., Herbert K.M., and Laurencin C.T. *Biochem. Biophys. Res. Commun.* **1995**, 21, 639.

[6]   Bissell M., Hall H.G., Pary G. *J.Ther. Biol.* **1982**, 99, 31.

[7]   Nickerson J. *Journal of Cell Science* **2001**, 114 (3), 463.

[8]   Nelson C.M., Bissell M.J. *Annual Review of Cell and Developmental Biology* **2006**, 22, 287.

[9]   Yang S., Leong K., Du Z., Chua C. *Tissue Eng.* **2001**, 7(6), 679.

[10]   Zeltinger J., Sherwood J.K., Graham D.A., Mueller R., Griffith L.G. *Tissue Eng.* **2001**, 7(5), 557.

[11]   Salerno A., Guarnieri D., Iannone M., Zeppetelli S., Di Maio E., Iannace S., Netti P.A. *Acta Bio.* **2009**, 5(4), 1082.

[12]   Moore M.J., Jabbari E., Ritman E.L., Lu L., Currier B.L., Windebank A.J., et al. *J. Biomed. Mater. Res.* **2004**, 71A(2), 258.

[13]   Guarino V., Causa F., Salerno A., Ambrosio L. Netti P.A. *Mat. Sci. Tech.* **2008**, 24(9), 1111.

[14]   Guarino V, Causa F, Ambrosio. *J. Appl. Biomat. Biomech.* **2007**, 5(3), 149.

[15]   Lutolf M.P., Hubbell J.A. *Nat. Biotech.* **2005**, 23(1), 47.

[16]   Guarino V., Causa F., Ambrosio L. *Expert Rev. Med. Devices* **2007**, 4(3), 406.

[17]   Guarino V., Causa F., Netti P.A., Ciapetti G., Pagani S., Martini D., Baldini N., Ambrosio L. *J. Biomed. Mater. Res. B* **2008,** 86B, 548.

[18]   Causa F., Netti P.A., Ambrosio L., Ciapetti G., Baldini N., Pagani S., Martini D., Giunti A. *J. Biomed. Mater. Res.* **2006**, 76, 151.

[19]   Ciapetti G., Ambrosio L., Savarino L., Granchi D., Cenni E., Baldini N., Pagani S., Guizzardi S., Causa F., Giunti A. *Biomaterials* **2003**, 24, 3815.

[20]   Koh Y.H., Bae C.J., Sun J.J., Jun I.K., Kim H.E. *J. Mater. Sci: Mater. Med.* **2006**, 17, 773.

[21]   Mondrinos M.J., Dembzynski R., Lu L., Byrapogu V.K.C., Wootton D.M., Lelkes P.I., Zhou J. *Biomaterials* **2006**, 27, 4399.

[22]   Guarino V., Causa F., Taddei P., Di Foggia M., Ciapetti G., Martini D., Fagnano C., Baldini N., Ambrosio L. *Biomaterials* **2008**, 29, 3662.

[23]   Guarino V., Ambrosio L. *Acta Biomat.* **2008**, 4, 1778.

[24]   Zhang R., Ma P.X. *J. Biomed. Mater. Res.* **1999**, 4, 285.

[25]   Nagano M., Kitsugi T., Nakamura T., Kokubo T., Tanahashi M. *J. Biomed. Mater. Res.* **1996**, 31(4), 487.

[26]   Yan W.Q., Nakamura T., Kawanabe K., Nishigochi S., Oka M., Kokubo T. *Biomaterials* **1997**, 18(17), 1185.

[27]   Ambrosio L., Apicella A., Mensitieri M., Nicolais L., Huang S.J., Marcacci M., Peluso G. *Clinical Materials*, **1994**, 15, 29.

[28]   Ge Z., Yang F., Goh J.C.H., Ramakrishna S., Lee E.H.. *J. Biomed. Mater. Res.* **2006**, 77A, 639.

[29]   Causa F., Sarracino F., De Santis R., Netti P.A., Ambrosio L., Nicolais L. *J. Appl. Biomat. Biomech.* **2006**, 4, 21.

[30]   Mow V.C., Ateshian G.A. and Spilker R.L.   *J. Biomech. Eng.* **1993**, 115 (4B), 460.

[31]   Swieszkowski W., Tuan B.H.S., Kurzydlowski K. J. and Hutmacher D. W. *Biomol. Eng.* **2007**, 24, 489.

[32]   Redman S.N., Oldfield S.F. and Archer C.W. *Eur. Cell. Mater.* **2005**, 9, 23.

[33]   Jackson D.W., Scheer M.J. and Simon T.M.   *J. Am. Acad. Orthop. Surg* **2001**, 9, 37.

[34]   Wirth M.A., Rockwood Jr C.A. *Clin. Orthop. Relat. Res.* **2004**, 307, 47.

[35]   Taylor G.I., 1969, Electrically driven jets. Proc. R. Soc. London A313, 453.

[36]   Yarin A.L., Koombhongse S., Reneker, D.H. *J. Appl. Phy.* **2001**, 89, 3018.

[37]   Pham Q.P., Sharma U., Mikos A.G. *Tissue Eng.* **2006**, 12(5), 1197.

[38]   Teo W.E., Ramakrishna S. *Nanotech.* **2006**,17 R89-R106.

[39]   Boland E.D., Matthews J.A., Pawlowski K.J., Simpson D.G., Wnek G.E., Bowlin G.L. *Frontiers Biosci.* **2004**, 9, 1422.

[40]   Wnek G.E., Carr M.E., Simpson D.G., Bowlin G.L. *Nano Lett.* **2003**, 3, 213.

[41]   Matthews J.A., Wnek G.E., Simpson D.G., Bowlin G.L. *Biomacromol.* **2002**, 3, 232.

[42]   Guarino V., Alvarez-Perez M.A., Cirillo V., Ambrosio L. *J. Bioac. Comp. Pol.* **2011**, 26(2), 144.

[43]   Kim GH, Kim WD., **2007**, *J Biomed Mater Res: Appl Biomater* 81B: 104-110

[44]   Kim H.S., Kim K., Jin H.J., Chin I.J. *Macromol. Symp.* **2005**, 224, 145.

[45]   Murugan R., Ramakrishna S. *Tissue Eng.* **2006**, 12(3), 435.

[46]   Ghasemi-Mobarakeh L., Prabhakaran M.P., Morshed M., Nasr-Esfahani M.H., Ramakrishna S. *Biomaterials* **2008**, 29, 4532.

[47]   Prabhakaran M.P., Venugopal J., Chyan T.T., Hai L.B., Chan C.K., Yu-Tang A.L., Ramakrishna S. *Tissue Eng.* A **2008**, 14, 1.

[48]   Alvarez-Perez M.A., Guarino V., Cirillo V., Ambrosio L. *Biomacromolecules* **2010**, 11(9), 2238.

MAURIZIO PRATO

Professor of Organic Chemistry

Address:

Dipartimento di Scienze Chimiche e Farmaceutiche

University of Trieste

Trieste, Italy

Education:

Maurizio Prato graduated in Padova, where he was appointed Assistant Professor in 1983.
Moved to Trieste as an Associate Professor in 1992, where was promoted Full Professor in
2000. He spent a sabbatical terms at Yale University (1986-7) and at the University of
California, Santa Barbara (1991-2). He was Visiting Professor at the Ecole Normale
Supérieure de Paris (2001) and at the University of Namur, Belgium (2010).

Professional experience:

His research focuses on the functionalization chemistry of carbon nanostructures for
applications in materials science and medicinal chemistry. His scientific contributions have
been recognized by National and International awards including: the Federchimica Prize
(1995, Association of Italian Industries), the National Prize for Research (2002, Italian
Chemical Society), an Honor Mention from the University of Trieste (2004), the Ciamician
Gonzalez Prize, Spanish Royal Society of Chemistry (2008), the Mangini Gold Medal, Italian
Chemical Society (2009), the Ree-Natta Lectureship, Korean Chemical Society (2010).
He was the recipient of an *ERC Advanced Research Grant*, European Research Council, 2008
and has become a Member of the National Academy of Sciences (*Accademia Nazionale dei
Lincei*) in 2010.

Synthesis and Characterization of Functionalized Carbon Nanotubes

Chiara Fabbro, Tatiana Da Ros and Maurizio Prato*

Dipartimento di Scienze Chimiche e Farmaceutiche, Università degli Studi di Trieste, Piazzale Europa 1, 34127 Trieste

1. INTRODUCTION

The rapid development of carbon nanotube (CNT)-based technology in many fields has made this novel material very popular in the scientific community. CNTs are tubular structures made of only carbon atoms arranged in a benzenoid network similar to graphite, but rolled up in a cylindrical shape, and therefore subjected to a curvature, with important consequences in their chemical reactivity. The number of cylinders that constitute a CNT can vary from only one to several ones, giving rise to different kinds of CNTs, namely single-walled CNTs (SWCNTs), double-walled CNTs (DWCNTs) up to generic multi-walled CNTs (MWCNTs), where the distance between two walls is of ca. 0.35 nm.

The coiling up direction of the graphene sheet determines different kinds of CNT. In fact, the angle between C-C bonds and the axis of the tube can vary, and the so-called helicity of a CNT will depend on this angle. Thus, different kinds of CNTs are described according to a chiral vector and each CNT can be associated to a couple of values (n,m), identifying the corresponding vector of helicity, as in the two examples depicted in Figure 1.

The two limits of the huge variety of possibilities are zigzag and armchair CNTs, the former having $\theta = 0°$ and m=0, and the latter with $\theta = 30°$ and $n = m \neq 0$. In-between, all the CNTs with $0° < \theta < 30°$ are defined as chiral. Obviously, with increasing n and/or m, the diameter of the tube increases. The description of this feature is important not only to define structurally different CNTs using a common vocabulary, but mainly because the electronic properties vary a lot depending on the chiral vector. In fact, infinite-length armchair SWCNTs are metallic, whereas infinite-length zigzag or chiral CNTs are semi-conducting, even though these are not strict rules. In the case for example of small diameter tubes, where the curvature plays an important role, some exceptions exist.[1]

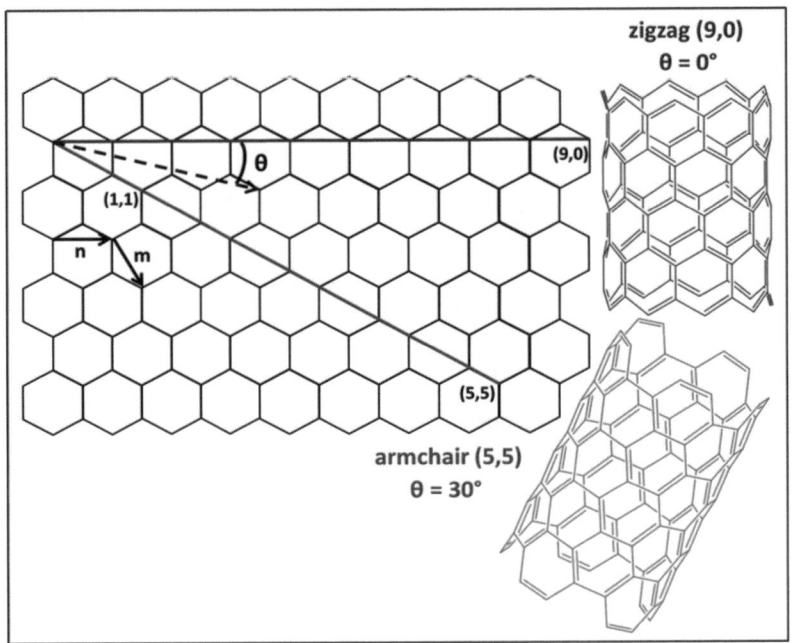

Fig. 1. Graphical explanation of the chiral vector θ of a CNT. The vector refers to the direction of rolling up of a graphene sheet.

Additionally, as a result of the 1-D nature of CNTs, electrons can be conducted without being scattered. The absence of scattering of the electrons during conduction is known as ballistic transport and allows the nanotube to conduct without dissipating energy as heat.

Besides these unique electronic properties, ideal CNTs exhibit chemical and thermal stability, together with extremely high tensile strength and elasticity.

Nevertheless they still present to date some important drawbacks that need to be addressed by the research in the field. One of the major problems is related to batch-to-batch lack of reproducibility. In fact, commercial samples of CNTs, produced by the same supplier, do not necessarily contain identical materials. This lack of reliability is probably due to the big heterogeneity of a CNT sample, that makes it subject to too many unpredictable variables during the synthetic process. This variability is associated with CNT length, diameter, helicity, impurity content (i.e. metal and carbonaceous impurities) and presence of defect. The possible defective structures are vacancies, i.e. incomplete bonding defects, topological changes, such as pentagons/heptagons interrupting the hexagonal network (Figure 2), and doping with elements other than carbon.[2]

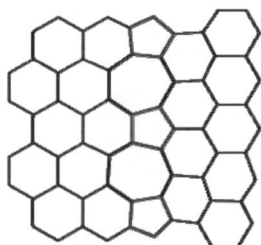

Fig. 2. Example of pentagon/heptagon defects in the hexagonal CNT network, inducing local strain. Note the change in tube helicity between left and right side of the defect.

The other big limitation associated with CNT research is their poor dispersibility. In fact pristine, as-produced CNTs are insoluble in water and in any organic solvent, and a correct functionalization is required in order to improve dispersibility for a better manipulation of the material.

Several synthetic protocols exist for CNTs. The main ones are arc discharge, originally developed for fullerene synthesis, laser ablation and catalytic carbon vapour deposition (CCVD).[3] The first two methodologies involve high-power energy to vaporize solid-state carbon (graphite) and form CNTs at high temperature (> 1000 °C), while CCVD uses carbon precursors in the gas phase and takes place at relatively low temperatures (500-1000 °C). In general the use of a proper metallic catalyst can switch from the production of MWCNTs to SWCNTs. Usual impurities deriving from the synthetic process are metallic nanoparticles and carbonaceous material other than the desired CNTs, represented mainly by amorphous carbon. Among the three different synthetic methods, CCVD provides a better control in the growth of CNTs and can be scaled up to industry level because of the lower costs. Different CCVD processes have been developed, among which one of the most popular is the high-pressure carbon monoxide disproportionation process (HiPCO), used for producing SWCNTs.[4] This technique exploits carbon monoxide (CO) as the carbon source, and $Fe(CO)_5$ as the catalyst precursor. The process takes place at 800-1000°C through the disproportionation of CO to CO_2 and C for the CNT growth on the iron nanoparticles. Optimized conditions led to the production of SWCNTs with diameters ranging from 0.7 to 1.4 nm, with an acceptable degree of purity.

2. CNT FUNCTIONALIZATION

As already mentioned, CNT usefulness is strongly hindered by the difficulty in their handling. In fact, as produced, CNTs are not soluble in water or in all common organic solvents. Therefore, a chemical modification needs to be performed to reduce strong inter-tube interactions, thus improving CNT dispersibility. Moreover, CNT functionalization offers the advantage to provide an anchor point for further modifications.

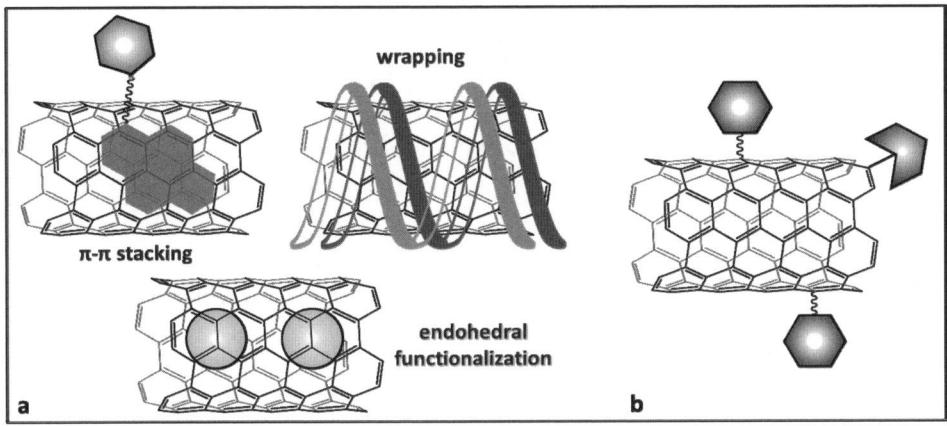

Fig. 3. Non-covalent (**a**) and covalent (**b**) approaches for CNT functionalization.

The main distinction to be underlined when speaking of CNT functionalization is between non-covalent and covalent approaches (Figure 3). They both present pros and cons, which should be carefully taken into account when deciding which is the best route to follow, according to the desired application.

2.1 NON-COVALENT FUNCTIONALIZATION

Among non-covalent CNT functionalization strategies, one possibility is based on small aromatic molecules, bound to the tubes by means of π-π stacking,[5] while another one is based on the wrapping of polymeric molecules around the tubes. The latter method can involve both biological macromolecules, such as nucleic acids,[6] lipids[7] or peptides[8] and synthetic polymers,[9] and occurs via π-π stacking and/or van der Waals interactions. Another, quite particular, approach for the non-covalent modification of CNTs is the filling of their inner cavity.[10] The main advantage of the non-covalent modification of CNTs is that the electronic and mechanical properties of the tubes are preserved. Therefore this should be the first choice when CNTs are exploited in fields such as molecular electronics. On the other hand, the drawback of a non-covalent approach is the possible reversibility of the bonds involved, which can result in the loss of the functionalization. This risk needs to be considered for example in the case of most biomedical applications, since it is difficult to exactly foresee the fate of a non-covalently bound molecule when the functionalized CNTs are administered *in vivo*. For this reason, a covalent approach should be preferred if CNTs will serve as drug delivery systems.

2.2 COVALENT FUNCTIONALIZATION

The reactivity of CNTs in terms of covalent chemistry on the C backbone is due to local strain, which is caused by two main reasons: the first one is the curvature-induced pyramidalization of the conjugated carbon atoms, and the second is the π orbital misalignment between adjacent pairs of conjugated carbon atoms.[11] Since the pyramidalization angle gives a good measure of the local weakening of π-conjugation and of the strain energy of pyramidalization, it is clear why this effect is important for fullerenes and for CNTs caps, which are half-fullerenes, while it is less relevant in the case of the flatter CNT sidewalls (Figure 4, upper part).

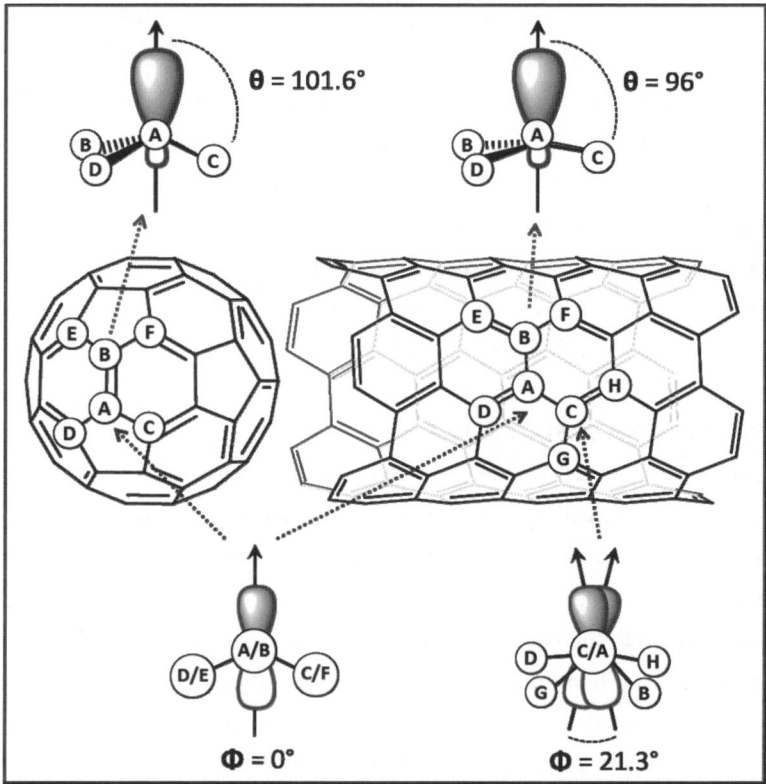

Fig. 4. Representation of pyramidalization angle ($\theta - 90°$) and of π orbital misalignment (Φ) for fullerene C_{60} and for a (5,5) SWCNT.

On the other hand, π orbital misalignment accounts for strain in CNT sidewalls much more than in fullerenes, where it is very little (in fullerene C_{60} π orbital alignment is perfect, Figure 4, lower part). Therefore C_{60} and CNT reactivity towards addition reactions are both driven by geometry-dependent strain, but for different reasons. Furthermore, since the pyramidalization angles and the π-orbital misalignment angles of CNTs scale inversely with the diameter of the tubes, a higher reactivity is expected for smaller tubes than for larger ones, if considering just strain-induced reactivity. Nevertheless, there is another parameter that could deeply influence the reactivity of CNT sidewalls: the presence of defects, such as pentagon-heptagon pairs in the hexagons network, that results in a locally enhanced chemical reactivity of the graphitic nanostructures, and can cause an unpredictable chemical behavior of CNTs with different diameters. As already anticipated, covalent sidewall functionalization may cause a partial loss of the high conductivity and of the remarkable mechanical properties of CNTs, since it generates sp^3 carbon sites on CNTs, which interrupt the conjugation of π electrons. The possibilities for a covalent modification of the unsaturated carbon network of CNTs, though being limited and requiring usually harsh conditions, have had a rapid development in the last years, starting from fluorination, one of the first covalent reactions ever performed on SWCNTs in 1998.[12] In this case the degree of functionalization could reach such high levels that in the end the tubes can become insulators. In a second step, CNTs can be further modified using Grignard reagents or with organolithium compounds.[13] Other existing possibilities include addition of both alkyl[14] and aryl[15] radicals, nucleophilic[16] or electrophilic[17] additions and cycloadditions.[18]

Herein the attention will be focused mainly on the oxidation of CNTs and their subsequent defect site chemistry, and on 1,3-dipolar cycloaddition of azomethine ylides, widely exploited in this field.

2.2.1 OXIDATION AND DEFECT SITE CHEMISTRY

The most common way to oxidize CNTs involves acid treatments, with sonication of CNTs in concentrated acid mixtures, where nitric acid or hydrogen peroxide plays the role of the oxidant.[19] Also other oxidative agents have been used, such as phosphomolybdic acid,[20] potassium permanganate,[21] or molecular oxygen, by heat treatment of the CNTs in O_2 atmospheres.[22]

CNTs are oxidized with multiple aims, mainly purification and shortening. In fact, CNTs are typically synthesized with poly-disperse micrometer lengths, bundled together into macroscopic entangled ropes, and contain metallic impurities, deriving from the catalyst used during their synthesis. Many applications, however, require individual short CNTs, devoid of metallic impurities. Oxidative treatments reduce metal content due to the removal of the amorphous carbon usually covering metallic particles. Also, oxidation causes the etching of CNT caps when the particles are still inside the tubes. The subsequent oxidation of the residual metal to soluble species, e.g., $Fe^0 \rightarrow Fe^{III}$, is therefore achieved, which are easily removed. Moreover, as a result of the chemical oxidation, the tips and the sidewalls of the CNTs contain oxygenated groups, mainly carboxylic groups, useful for further derivatization.[23] The process of the oxidation of CNTs can be described as step-wise with (i) an initial attack on the originally existing reactive sites, such the terminal fullerene-like caps, and the heptagon-pentagon defects, which carry the higher strain, (ii) a defect-generation step, when additions to hexagon π bonds take place, generating new defects, and introducing hydroxyl groups, then further oxidized to carboxyl groups, and (iii) a defect-consuming step, when the graphene structure around the defect sites is broken, and the tube is cut. Depending on the strength of the conditions, oxidation can stop at the first step, thus attacking only already existing defects, or proceed with the generation of new defects and finally with the cutting of the nanotubes. A detailed study of diameter-dependent oxidative stability confirmed a direct relationship between diameter and reactivity, as expected on the basis of what has been previously explained in this chapter. Studying the resonant Raman radial breathing mode (RBM), the authors clearly showed that smaller diameter tubes are more easily air-oxidized than larger diameter ones.[24]

As already anticipated, the introduction of carboxylic groups allows a further derivatization of CNTs. Esterification and amidation reactions are in fact widely exploited possibilities for the covalent modification of oxidized CNTs, since they usually permit to easily obtain highly functionalized material, bearing the desired molecule according to the specific envisaged application.[25]

2.2.2 1,3-DIPOLAR CYCLOADDITION OF AZOMETHINE YLIDES

The 1,3-dipolar cycloaddition of azomethine ylides, widely exploited in fullerene functionalization since 1993,[26] was translated to carbon nanotubes as well.[27] This very versatile reaction consists in the addition of a dipole to a dipolarophile, where the dipole is an azomethine ylide, formed by reaction of an aldehyde (or a ketone) with an α-amino acid, and the dipolarophile is a double bond on the fullerene cage or on the CNT (Scheme 1).

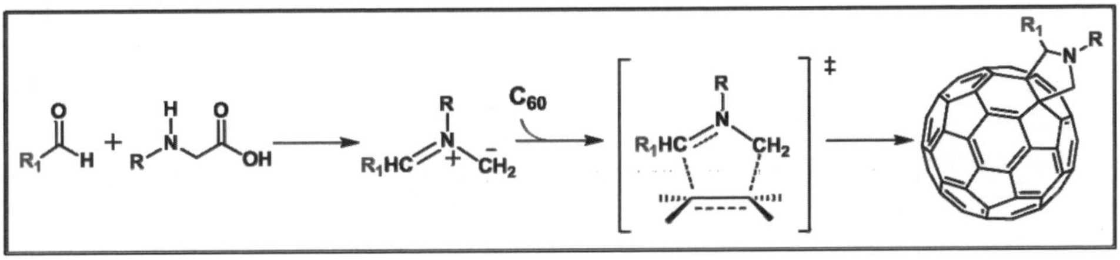

Scheme 1. General mechanism of the 1,3-dipolar cycloaddition of azomethine ylides to C_{60}.

In fullerene C_{60}, the first addition takes place in any of the double bonds, since they are all equivalent. Instead, the CNT double bonds are not all equivalent. In fact, it was found by theoretical studies that armchair tubes favor functionalization at C-C bonds tilted with respect to tube axis, (i.e. bond A-C in Figure 4), whereas zigzag CNTs prefer attachment of the ylides at segments parallel to the tube axis.[28] Since real samples of CNTs are normally made of mixtures of different tubes in terms of diameter, length, chiralities and defects, we wonder which is the behavior of the different CNTs.

Theoretical calculations showed that reactivity depends mainly on tube diameter, and only weakly on the chirality, as expected considering pyramidalization angles and the π-orbital misalignment angles.[29] Moreover, this particular kind of 1,3-dipolar cycloaddition, being the dipole a nucleophile, is in principle accelerated by the presence of electron-donating groups in the dipole, and electron-withdrawing groups in the dipolarophile, such as carboxylic groups introduced on the tubes by an oxidation step.[29,30]

3. CNT CHARACTERIZATION

A great difficulty associated with CNT functionalization, especially when a covalent approach is chosen, is their characterization. In fact all the normal techniques used in organic synthesis cannot be translated to this material. The main reason is that the amount of functional groups is usually too low, with respect to the CNT component of the sample, to be detected. Also, for techniques performed in solution, the choice of the solvent is highly limited, even for functionalized CNTs, and the required concentrations are usually far too high to be achieved. For what concerns NMR, further problems are given by the presence of ferromagnetic or paramagnetic metallic particles in CNT samples, due to the magnetic field.

Herein some of the main characterizations exploitable for different CNTs (Figure 5) will be presented.

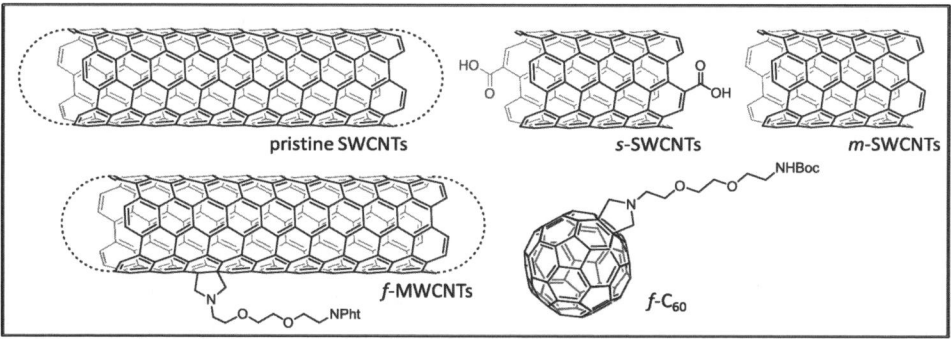

Fig. 5. Carbon nanomaterials the characterization of which are illustrated in the following paragraphs.

3.1 UV-VIS-NIR SPECTROSCOPY

The electronic density of states of SWCNTs, being one dimensional objects, is not a continuous function of energy, but possesses spike-like features known as van Hove singularities (Figure 6a). Optically allowed transitions between these features are typically observed in the UV-vis-NIR region as sharp peaks, even though the sharpness is not fully appreciable due to the overlap of similar transitions given by different SWCNTs. Moreover, interactions between nanotubes in a bundle broaden optical lines. In the specific case of commercial HiPCO (High-Pressure Carbon monOxide disproportionation process, see paragraph 1) SWCNTs, the main features represent transitions between the first pair of singularities in metallic tubes (M_{11}) and between the first and the second pairs of singularities for semiconducting tubes (S_{11} and S_{22}), as indicated in Figure 6b. Moreover, the energy of electronic transition is inversely proportional to tube diameter.[31]

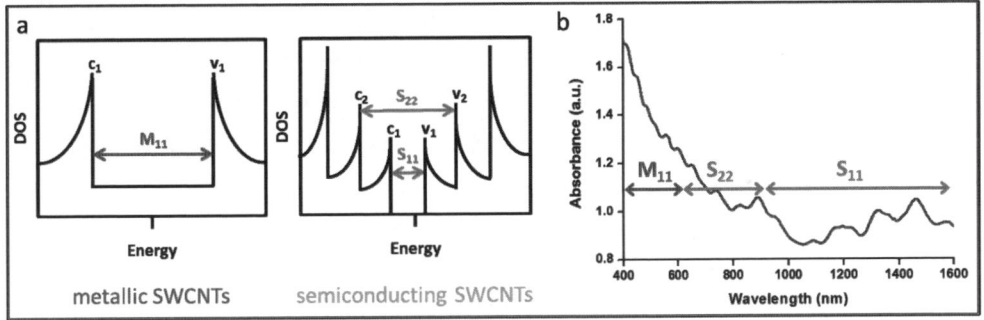

Fig. 6. a) Density of state diagrams for metallic and semiconducting SWCNTs; *b*) UV-vis-NIR spectrum of pristine HiPCO SWCNTs (DMF).

UV-vis-NIR spectroscopy can provide information on the functionalization of SWCNTs. In fact, the covalent modification can disrupt the periodicity of the SWCNTs conjugated electronic structure, and hence perturb their electronic transitions, resulting in a loss of the van Hove features.

3.2 RESONANT RAMAN SPECTROSCOPY

When a beam of light crosses a transparent sample, a small fraction of it is scattered, and a small fraction of this small fraction is scattered inelastically, i.e., at different frequencies than the incident light. In fact, the interaction between the incident photon and the sample can result in an energy exchange, namely the Raman effect. Usually the Raman scattering signal is weak. However, its intensity can be strongly enhanced if working in resonance Raman scattering, i.e., using a laser energy that matches the energy of optically allowed electronic transitions in the material studied.

SWCNTs, due to their broad range of optically allowed transitions, always give a resonance Raman response, and consequently only tubes with one of the band-gaps equal to the exciting laser energy are probed. Therefore, using different laser energies, different populations of tubes in a heterogeneous SWCNT sample are analyzed (Figure 7a). The main features of a typical SWCNT Raman spectrum are depicted in Figure 7b.

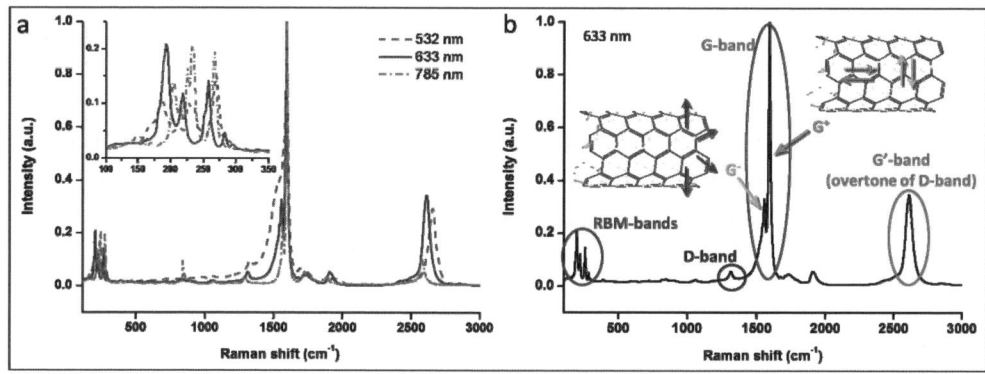

Fig. 7. a) Raman spectra of pristine HiPCO SWCNTs obtained with different lasers; *b*) main SWCNT Raman features.

Radial breathing mode (RBM) corresponds to the radial expansion-contraction of SWCNTs, exactly as if the tube was breathing. In fact, it depends on nanotube diameter: RBM frequency is inversely proportional to tube diameter. RBM range is typically 100-350 cm^{-1}, and, if its intensity is particularly high, weak overtones can be observed at double frequency.

The G-mode corresponds to the tangential planar vibrations of carbon atoms and is present also in graphite, from which the name derives. However, differently from graphite, SWCNT G-band (1580 cm^{-1}) is split into up to six peaks, due to the loss of symmetry. Just the two main ones are normally considered, i.e. the G$^+$, which is given by vibrations along tube axis, and the G$^-$, due to vibrations in the circumferential direction (Figure 7b). In this latter band, there is an evident difference in the line-shape between semiconducting SWCNTs, which give a Lorentzian-like peak (as G$^+$) and metallic ones, where the peak is broadened, due to the presence of free electrons.[32] This difference can be clearly appreciated comparing the spectra depicted in Figure 7a: when using the 532 nm laser, mainly metallic tubes are in resonance, due to their M_{11} transition (see Figure 6), and in fact the G$^-$ line-shape is much broader than in the case of the 785 nm laser, which is instead in the semiconductor S_{22} transition region.

Another important Raman feature is the D band, whose position is at around 1300 cm^{-1}, and it shifts to higher wavenumbers as the laser excitation energy increases. This mode is associated with disorder, i.e. with the sp^3-hybridized carbon atoms present in the tube or in the carbonaceous impurities. Therefore, it is often used to evaluate the structural quality of SWCNTs, expressed usually as D/G ratio. Moreover, the full-width-at-half-maximum (FWHM) intensity of the D-band of the various carbon impurities is generally much broader than that of CNT D-band, and thus D-band line-width could give an indication of the SWNCT purity level.[33]

Other less significant bands are typically present in a SWCNT Raman spectrum, due to overtone modes of single bands or combinations of different bands. Among them the strongest one is usually the G' band. The name of this band is misleading, since it derives from graphite, where this mode is usually the second strongest after the G mode, but it is actually the overtone of the defect-induced D mode, even though it can be present in defect-free nanotubes.

The main changes that could be appreciated in a SWCNT Raman spectrum upon chemical modification are the following:

- loss or decrease of some RBM bands, due to destruction or extensive modification of the corresponding CNTs;

- change in the D/G ratio, due to modification of sp^3/sp^2 carbon atoms ratio (an increased D/G ratio could give a quali-quantitative information on the degree of covalent functionalization, but it could also be due to increased amorphous carbon content);

- change in the FWHM intensity of the D-band, according to the amorphous carbon content of the sample, as already explained;

- change in the baseline height (a great content of amorphous carbon can give a strong fluorescence, perturbing the Raman spectrum).

A very strong limitation of UV-vis-NIR and Raman spectroscopies is that their usefulness is usually restricted to SWCNTs. In fact, MWCNT UV-vis-NIR absorption, deriving from the overlap of many different contributions in each tube, does not result in the typical well-defined van Hove features of a SWCNT spectrum, and thus it cannot give information on CNT covalent functionalization.

Concerning Raman, MWCNTs do not present defined RBM, being each tube made of several concentric walls. Moreover, the D band is quite intense, often more than the G-band, because of the high extent of defects usually present in MWCNTs. Consequently, it is difficult to appreciate an increase in the D/G ratio due to functionalization. For DWCNTs the situation is intermediate between single and multi-walled CNTs and it is normally possible to appreciate RBM and D/G ratio variations.

3.3 THERMOGRAVIMETRIC ANALYSIS (TGA)

TGA is an analytical technique for determination of the thermal stability of a material. The weight of the sample is recorded, while heating it in a furnace under a controlled atmosphere, which can be either inert (using gases such as N_2 or He), or oxidative (using air or pure O_2). In the latter case CNT combustion takes place usually between 400 and 600°C, depending on different factors, and resulting in the formation of CO_2. CNT combustion temperature, easily determined as the maximum in the first derivative (Figure 8), depends on different factors. For example, smaller diameter nanotubes are believed to oxidize at lower temperature due to higher curvature strain. The same effect could be given by defects in the CNT structure. Also the presence of metal particles could induce a lower combustion temperature, since they can catalyze carbon oxidation.[34] Furthermore, it is possible to use TGA under air to determine the metal content of a sample, since the residual weight after completeness of CNT combustion is ascribable to metals in their oxidized form.

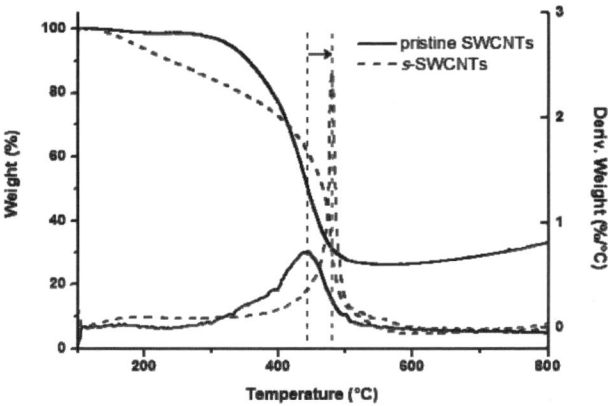

Fig. 8. TGA (air) of pristine SWCNTs and *s*-SWCNT-**1**.

On the other hand, in the case of TGA run under inert atmosphere, CNTs are stable up to ca. 800°C, and therefore it is possible to ascribe the weight loss at lower temperatures to the organic material present in the sample, thus estimating the degree of functionalization after a chemical modification of CNTs. Nevertheless, it is important to underline that also amorphous carbonaceous species possibly present in a CNT sample could burn in the same temperature range, and this fact should be taken into account when using TGA measurements for quantitative characterization.

3.4 TGA-COUPLED MASS SPECTROMETRY (TGA-MS)

The possibility offered by TGA-MS is interesting: a mass spectrometer coupled to the vent-hole of the TGA furnace, analyzes the gaseous material evolving during the experiment, thus giving precious information on what is going on. In this case, the TGA analysis should be performed under He, due to its low molecular weight, so that the mass spectrometer could screen all the masses with molecular weights bigger than 4. Otherwise the signal given by the gas would completely saturate the instrument. It is important to point out that the mass spectra obtained are not like normal spectra. In fact a massive fragmentation occurs during TGA, so that only small fragments are produced. In a typical analysis a big CO_2 peak (Mol. Wt.: 44) appears in correspondence of every combustion taking place, associated with a decrease in the O_2 (Mol. Wt.: 32) signal (inert gas cylinders always contain some), which is being consumed by the combustion itself. However, also higher molecular weight peaks appear, making it possible to determine the presence of certain molecules. If very characteristic fragments are produced, they could be ascribed to specific molecules, as in the example shown in Figure 9b, where the peaks of mass 48 and 64 correspond to SO^+ and SO_2^+ respectively, deriving from decomposition and ionization of SO_2, which is in turn generated by sulfonic acid groups introduced by a chemical modification of CNTs. Alternatively, the assignment could be done by simple comparison of the whole pattern of peaks with the one obtained by analysis of known samples, as in the example reported in Figure 10.

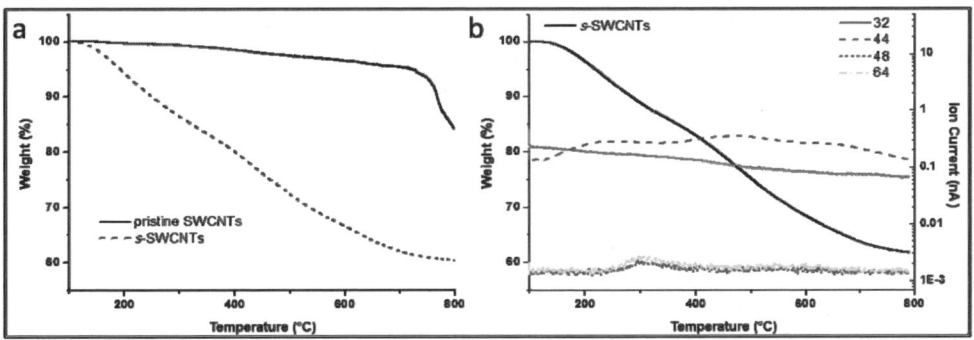

Fig. 9. a) TGA (N₂) of pristine SWCNTs and *s*-SWCNTs; *b*) TGA-MS analysis (He) of *s*-SWCNTs.

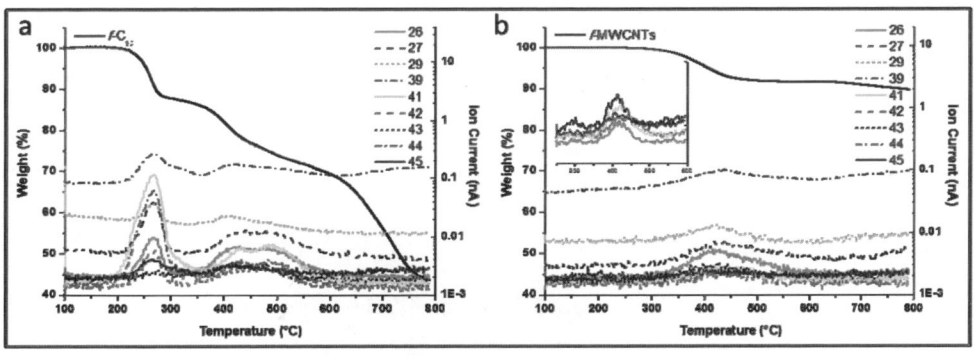

Fig. 10. TGA-MS analyses (He) of *f*-C₆₀ (*a*) and *f*-MWCNTs (*b*)

3.5 TRANSMISSION ELECTRON MICROSCOPY (TEM)

TEM is a major tool for the morphological characterization of CNTs. A TEM image derives from the interference of the sample, deposited onto a grid, with an electron beam transmitted through it. Thus, metal particles interfere more, producing spots with higher contrast. On the other hand, individual SWCNTs are usually very difficult to visualize (with a non-high resolution TEM), due to the low contrast with the grid surface, which is usually made of carbon. Bundles of SWCNTs instead, as well as DWCNTs or MWCNTs, can be easily detected providing useful information, such as CNT length and diameter, together with a rough assessment of their purity, in terms of metals and amorphous material content. Regarding the evaluation of the dispersibility of a sample, it should be underlined that, during the drying procedure that follows the deposition of the CNT dispersion on the grid, CNTs can easily re-aggregate. Therefore the TEM analysis does not provide reliable information on their original aggregation state.

3.6 ATOMIC FORCE MICROSCOPY (AFM)

AFM is a kind of Scanning Probe Microscopy, which consists in scanning a sharp tip (radius of curvature = 3-50 nm) over a surface and detect one or more probe/surface interactions. In AFM the result is a topographical image of the surface. The tip is mounted on a flexible cantilever, allowing it to follow the surface profile, and its deflection is measured using a laser spot reflected from the top surface of the cantilever into an array of photodiodes. AFM can be performed in Contact Mode, when the tip actually touches the surface, thus possibly moving objects or damaging soft samples, in Non-Contact Mode, when the tip scans the surface at a certain distance, with a consequent lower resolution, and in Tapping Mode (TM), when the tip is oscillated rapidly against the surface during scanning, and the microscope extrapolates the surface topography by the change in the oscillation amplitude, triggered by tip/surface interactions (Figure 11).

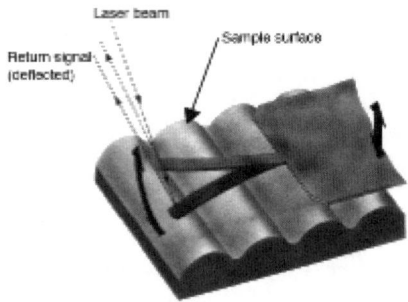

Fig. 11. Cantilever tapping on a surface.[35]

Typically, when scanning a surface in TM, two images are obtained: one is the height image, while the other one shows the amplitude of the tapping oscillation as the probe scans across the surface, being a sort of "error" image (Figure 12). In fact, this image shows where and how much the amplitude changed, giving the impression of shadows on sharp feature boundaries, which makes it look like a snapshot of the surface. However, shadows are actually areas where the amplitude had not yet been corrected back to the set value, and the real topographical image is the first one.[36]

TM-AFM analysis can be exploited to obtain structural information on a CNT sample deposited on a surface, such as length and diameter distribution. However, lateral resolution is usually not precise enough (being in the range 1-5 nm) to evaluate very short distances, such as the diameter of a SWCNT, due to the delay in the response of the cantilever to tip/surface interactions. Nevertheless, it is possible to obtain this information from the height profile of the objects on the surface, since vertical resolution does not suffer from this limitation (see Figure 12c).

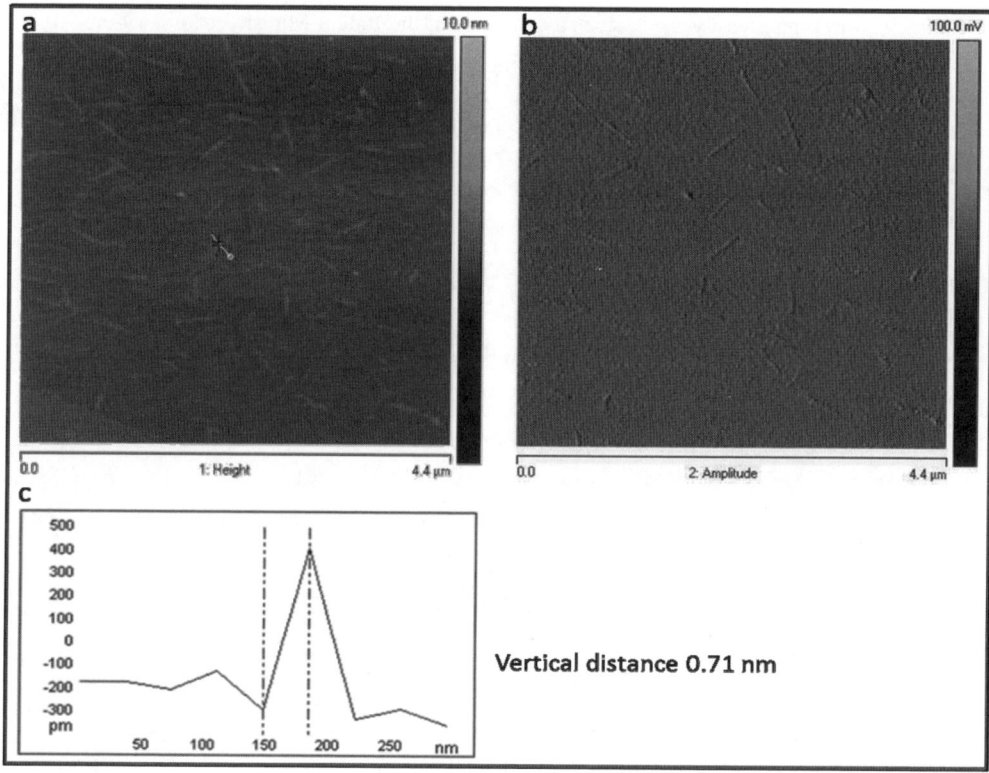

Fig. 12. Representative TM-AFM height (*a*) and amplitude (*b*) images of *s*-SWCNTs (250 rpm, 120 min, N$_2$); *c*) height profile of the line indicated in the topographic image *a*.

4. CONCLUSIONS

CNTs are a very versatile scaffold to be used for different purposes, both in the medical and materials sciences. The functionalization helps to easier handle these materials and the synthetic approaches are multiple, leading to a plethora of derivatives presenting a great potential. The applied protocols to obtain these interesting results differ from the traditional organic chemistry methodologies, because of the peculiar characteristics of solubility and reactivity of carbon nanomaterials. Therefore, it is necessary to translate the traditional synthetic procedures. Greater differences with respect to traditional organic chemistry can be found also in the characterization techniques, where the nanocarbon materials take advantage of many different techniques such as Raman spectroscopy, thermogravimetric analyses, electron microscopy.

5. ACKNOWLEDGEMENTS

This work was supported by the University of Trieste, the Italian Ministry of Education MIUR (Cofin Prot. 20085M27SS and Firb RBIN04HC3S) and the Regione Friuli Venezia-Giulia.

6. REFERENCES

1. Lu X. and Chen Z., *Chem. Rev.*, **2005**, *105*, 3643.

2. Charlier J.-C., *Acc. Chem. Res.*, **2002**, *35*, 1063.

3. a) Govindaraj A. and Rao C. N. R., In *Carbon Nanotechnology*, Eds.; L. Dai, Elsevier B.V., 2006, 15; b) Fu Q. and Liu J., In *Carbon Nanotechnology* ds.; L. Dai, Elsevier B.V., 2006, 81.

4. Nikolaev P., Bronikowski M. J., Bradley R. K., Rohmund F., Colbert D. T., Smith K. A., and Smalley R. E., *Chem. Phys. Lett.*, **1999**, *313*, 91.

5. a) Ehli C., Rahman G. M. A., Jux N., Balbinot D., Guldi D. M., Paolucci F., Marcaccio M., Paolucci D., Melle-Franco M., Zerbetto F., Campidelli S., and Prato M., *J. Am. Chem. Soc.*, **2006**, *128*, 11222; b) Chen R. J., Zhang Y., Wang D., and Dai H., *J. Am. Chem. Soc.*, **2001**, *123*, 3838.

6. Zheng M., Jagota A., Semke E. D., Diner B. A., McLean R. S., Lustig S. R., Richardson R. E., and Tassi N. G., *Nat. Mater.*, **2003**, *2*, 338.

7. a) Qiao R. and Ke P. C., *J. Am. Chem. Soc.*, **2006**, *128*, 13656; b) Richard C., Balavoine F., Schultz P., Ebbesen T. W., and Mioskowski C., *Science*, **2003**, *300*, 775.

8. a) Li X., Chen W., Zhan Q., Dai L., Sowards L., Pender M., and Naik R. R., *J. Phys. Chem. B*, **2006**, *110*, 12621; b) Zorbas V., Smith A. L., Xie H., Ortiz-Acevedo A., Dalton A. B.,

Dieckmann G. R., Draper R. K., Baughman R. H., and Musselman I. H., *J. Am. Chem. Soc.*, **2005**, *127*, 12323.

9. Star A., Stoddart J. F., Steuerman D., Diehl M., Boukai A., Wong E. W., Yang X., Chung S.-W., Choi H., and Heath J. R., *Angew. Chem. Int. Ed.*, **2001**, *40*, 1721.

10. a) Dujardin E., Ebbesen T. W., Krishnan A., and Treacy M. M. J., *Adv. Mater.*, **1998**, *10*, 1472; b) Smith B. W., Monthioux M., and Luzzi D. E., *Chem. Phys. Lett.*, **1999**, *315*, 31; c) Wilson M. and Madden P. A., *J. Am. Chem. Soc.*, **2001**, *123*, 2101.

11. Niyogi S., Hamon M. A., Hu H., Zhao B., Bhowmik P., Sen R., Itkis M. E., and Haddon R. C., *Acc. Chem. Res.*, **2002**, *35*, 1105.

12. Mickelson E. T., Huffman C. B., Rinzler A. G., Smalley R. E., Hauge R. H., and Margrave J. L., *Chem. Phys. Lett.*, **1998**, *296*, 188.

13. a) Saini R. K., Chiang I. W., Peng H., Smalley R. E., Billups W. E., Hauge R. H., and Margrave J. L., *J. Am. Chem. Soc.*, **2003**, *125*, 3617; b) Boul P., *Chem. Phys. Lett.*, **1999**, *310*, 367.

14. Ying Y., Saini R. K., Liang F., Sadana A. K., and Billups W. E., *Org. Lett.*, **2003**, *5*, 1471.

15. a) Bahr J. L., Yang J., Kosynkin D. V., M. J. Bronikowski, Smalley R. E., and Tour J. M., *J. Am. Chem. Soc.*, **2001**, *123*, 6536; b) Hudson J. L., Casavant M. J., and Tour J. M., *J. Am. Chem. Soc.*, **2004**, *126*, 11158.

16. a) Holzinger M., Vostrowsky O., Hirsch A., Hennrich F., Kappes M., Weiss R., and Jellen F., *Angew. Chem. Int. Ed.*, **2001**, *40*, 4002; b) Graupner R., Abraham J., Wunderlich D., Vencelová A., Lauffer P., Röhrl J., Hundhausen M., Ley L., and Hirsch A., *J. Am. Chem. Soc.*, **2006**, *128*, 6683; c) Syrgiannis Z., Hauke F., Röhrl J., Hundhausen M., Graupner R., Elemes Y., and Hirsch A., *Eur. J. Org. Chem.*, **2008**, *2008*, 2544.

17. Tessonnier J.-P., Villa A., Majoulet O., Su D. S., and Schlögl R., *Angew. Chem. Int. Ed.*, **2009**, *48*, 6543.

18. a) Yinghuai Z., Peng A. T., Carpenter K., Maguire J. A., Hosmane N. S., and Takagaki M., *J. Am. Chem. Soc.*, **2005**, *127*, 9875; b) Ménard-Moyon C., Dumas F., Doris E., and Mioskowski C., *J. Am. Chem. Soc.*, **2006**, *128*, 14764; c) Ogrin D., Chattopadhyay J., Sadana A. K., Billups W. E., and Barron A. R., *J. Am. Chem. Soc.*, **2006**, *128*, 11322.

19. a) Liu J., Rinzler A. G., Dai H. J., Hafner J. H., Bradley R. K., Boul P. J., Lu A., Iverson T., Shelimov K., Huffman C. B., Rodriguez-Macias F., Shon Y. S., Lee T. R., Colbert D. T., Smalley R. E., *Science* **1998**, *280*, 1253; b) Ziegler K. J., Gu Z., Peng H., Flor E. L., Hauge R. H., and Smalley, R. E. *J. Am. Chem. Soc.*, **2005**, *127*, 1541.

20. Warakulwit, C.; Majimel, J.; Delville, M.-H.; Garrigue, P.; Limtrakul, J.; and Kuhn, A. *J. Mater. Chem.*, **2008**, *18*, 4056.

21. Aitchison, T. J.; Ginic-Markovic, M.; Matisons, J. G.; Simon, G. P.; and Fredericks, P. M. *J. Phys. Chem. C*, **2007**, *111*, 2440.

22. a) Chiang, I. W.; Brinson, B. E.; Huang, A. Y.; Willis, P. A.; Bronikowski, M. J.; Margrave, J. L.; Smalley, R. E.; and Hauge, R. H. *J. Phys. Chem. B*, **2001**, *105*, 8297; b) Strong, K. L.; Anderson, D. P.; Lafdi, K.; and Kuhn, J. N. *Carbon*, **2003**, *41*, 1477.

23. Zhang, J.; Zou, H.; Qing, Q.; Yang, Y.; Li, Q.; Liu, Z.; Guo, X.; and Du, Z. *J. Phys. Chem. B*, **2003**, *107*, 3712.

24. Zhou, W.; Ooi, Y. H.; Russo, R.; Papanek, P.; Luzzi, D. E.; Fischer, J. E.; Bronikowski, M. J.; Willis, P.; and Smalley, R. E. *Chem. Phys. Lett.*, **2001**, *350*, 6.

25. a) Chen, J.; Hamon, M. A.; Hu, H.; Chen, Y.; Rao, A. M.; Eklund, P. C.; and Haddon, R. C.; *Science*, **1998**, *282*, 95; b) Hamon, M. A.; Chen, J.; Hu, H.; Chen, Y.; Itkis, M. E.; Rao, A. M.; Eklund, P. C.; and Haddon, R. C. *Adv. Mater.*, **1999**, *11*, 834; c) Del Canto, E.; Flavin, K.; Movia, D.; Navio, C.; Bittencourt, C.; and Giordani, S. *Chem. Mater.*, **2011**, *23*, 67.

26. Maggini, M.; Scorrano, G.; and Prato, M. *J. Am. Chem. Soc.*, **1993**, *115*, 9798.

27. Georgakilas, V.; Kordatos, K.; Prato, M.; Guldi, D. M.; Holzinger, M. and Hirsch, A. *J. Am. Chem. Soc.*, **2002**, *124*, 760.

28. Cho, E.; Shin, S.; and Yoon, Y.-G. *J. Phys. Chem. C*, **2008**, *112*, 11667.

29. Lu, X.; Tian, F.; Xu, X.; Wang, N.; and Zhang, Q. *J. Am. Chem. Soc.*, **2003**, *125*, 10459.

30. a) Carey, F. A.; Sundberg, R. J. In *Advanced Organic Chemistry*, 5th ed. 2007, ISBN: 978-0-387-68346-1; b) Ess D. H.; and Houk, K. N. *J. Am. Chem. Soc.*, **2008**, *130*, 10187.

31. Kataura, H.; Kumazawa, Y.; Maniwa, Y.; Umezu, I.; Suzuki, S.; Ohtsuka, Y.; and Achiba, Y. *Synth. Met.*, **1999**, *103*, 2555.

32. Jorio, A.; Souza Filho, A. G.; Dresselhaus, G.; Dresselhaus, M. S.; Swan, A. K.; Unlu, M. S.; Goldberg, B. B.; Pimenta, M. A.; Hafner, J. H.; Lieber, C. M.; and Saito, R. *Phys. Rev. B*, **2002**, *65*, 155412.

33. Dillon, A. C.; Yudasaka, M.; and Dresselhaus, M. S. *J. Nanosci. Nanotechnol.*, **2004**, *4*, 691.

34. Arepalli, S.; Nikolaev, P.; Gorelik, O.; Hadjiev, V. G.; Holmes, W.; Files, B.; and Yowell, L. *Carbon*, **2004**, *42*, 1783.

35. MultiMode SPM Instruction Manual, Copyright © 1996, 1997, 2004 Veeco Instruments Inc.

36. SPM Training Notebook Copyright © 2003 Veeco Instruments Inc.

Letizia Sambri
Assistant Professor in Organic Chemistry

Address:

   University of Bologna, Department of Organic Chemistry «A. Mangini»
v.le Risorgimento 4, 40136 Bologna, Italy; e-mail: *letizia.sambri@unibo.it*

Education:

   1993- Laurea in Industrial Chemistry (summa cum Laude), University of Bologna, Italy, under the supervision of Prof. A. Fava.
1998: Ph.D. in "Chemical Sciences", University of Bologna, under the supervision of Prof. G. Bartoli.

Professional experience:

   1998- 2000 Post-doctoral fellow Dept. Organic Chemistry "A. Mangini", University of Bologna, supervisor Prof. G. Bartoli.
Since July 2000 Assistant professor, Dept. Organic Chemistry "A. Mangini", University of Bologna.
January-April 2009: Visiting Scientist Prof. L. De Cola Group, Münster University (Germany).
From 2005 member of the Board of Directors of the Italian Chemical Society (SCI)-Emilia Romagna Section.
Current fields of research: design and synthesis of stimuli responsive materials for application in organic electronics and gels.

Versatility of Silicon-Based Compounds in Organic Synthesis

Letizia Sambri* and Andrea Baschieri

Dipartimento di Chimica Organica "A. Mangini", University of Bologna, v.le Risorgimento 4, 40136 Bologna, Italy

INTRODUCTION

Organosilicon compounds exhibit a multitude of functions in organic synthesis,[1] therefore a plenty of research has been performed in this field and the interest in these important compounds is still growing. Organosilanes own many attractive properties. They generally are much more moisture- and air-stable than various organometallic reagents and can be readily prepared from a wide range of starting materials. Silicon derivatives have low toxicity and display rich and diverse chemistry that can usually be rationalized by understanding a relatively small number of fundamental properties of silicon.[2]

The reactivity of organosilanes depends both on steric requirements and electronic contributions. Silicon is below carbon in the periodic table, but it is capable of very different bonding characteristics. It is less electronegative than carbon and hydrogen and it forms unusually strong bonds with electronegative elements of the second period bearing a lone pair of electrons; therefore most of organosilicon chemistry is driven by the formation of these bonds at the expense of weaker bonds.

In addition, the availability of relatively low energy empty 3d AOs lets Si to allow higher coordination numbers, forming the so called hypervalent silicon compounds. The ability to expand its valence state has consequences on the mechanisms of many reactions proceeding at Si.

Silicon can stabilize a positive charge in β-position, and this effect is known as the β-Si effect.[3] Moreover, silicon provides stabilization of α-negative charge,[4] in fact carbanions with an α-silicon group are more stable than their carbon analogues.

Since the versatility in reactivity, a wide range of reaction-types involving silicon compounds have been developed over the years. An exhaustive overview is almost impossible and during this lecture only some classes of reactions will be covered.

From a synthetic point of view, organosilanes possess many desirable characteristics. In fact, besides their green character, they can survive under a large variety of reactions and purification conditions, such as air, moisture, silica-gel, reducing or oxidizing conditions.[5] The silyl moiety can be introduced at almost any stage in the synthesis and can remain unaltered until the appropriate conditions make it reactive.

In 1981 Fleming reported the first total synthesis employing an organosilane to obtain a prostaglandine intermediate **7**,[6] (Scheme 1). It seems interesting to report some important steps of that synthesis to underline the stability and the selectivity in reactions of the silyl group.

Addition of dichloroketene to trimethylsilylcyclopentadiene **1** afforded dichlorosubstituted [3.2.0]bicycle **2**. This product was treated with Zn/AcOH to gave a α–chlorocyclobutanone which underwent to diazomethane-mediated ring-expansion afforded cis-fused [3.3.0]bicycle **3**, after a second dechlorination and an enolate-mediated epimerization at C(2) with a base to give the thermodynamic product. Reduction of the ketone group with NaBH₄ furnished the *endo*-alcohol which configuration was inverted by mesylation followed by a S_N2 attack of acetate ion to yield the *exo*-acetate **4**. Only at this point the silyl group took part in the synthetic sequence. The allylsilane moiety reacted with chlorosulfonylisocyanate (CSI) to furnish the amide **6**, with complete regio- and stereoselectivity. Following a series of simple known reactions the desired loganin acetate aglucone **7** was obtained.

Thus in this sequence the allylsilicon moiety was able to survive the first eight steps of the total synthesis being completely unaffected prior to its exploitation. This contrasts with the characteristics of traditional organometallic reagents, which generally have to be used as soon as they are formed.[5]

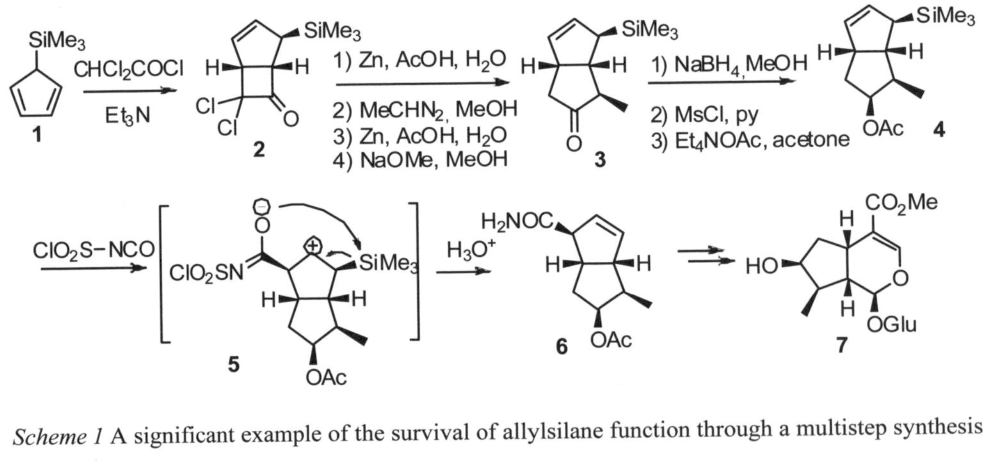

Scheme 1 A significant example of the survival of allylsilane function through a multistep synthesis

SILICON REAGENTS AS NUCLEOPHILES: ALLYL SILANES AND RELATED COMPOUNDS

Exploiting the ability of silicon atom to expand its valence, unsaturated organosilanes have been extensively employed in organic synthesis as carbon nucleophiles since the pioneering studies of Sakurai and Fleming research groups in the 1970s.[7]

Albeit over the past decade vinyl, propargyl and allenylsilanes, could find some fruitful applications, the leading unsaturated compounds widespread used in synthesis, since the introduction of silyl reagents, were allylsilanes,[8] probably since more reactive and stereoselective than the parent unsaturated derivatives together with their stability and ease of formation.

The regioselective addition of silicon allylic reagents to carbonyl compounds (Sakurai-Hosomi reaction) provides stereocontrolled access to functionalized homoallylic alcohols, a high value class of synthetic intermediates for the construction of a wide variety of complex natural products.[9]

Allyltrimethylsilane, the simplest derivative, is cheap and commercially available, but it is not a strong nucleophile,[10] thus the reaction with aldehydes generally requires an external promoter.

In order to employ allylsilanes in asymmetric synthesis, decade-long efforts have been devoted to develop enantioselective procedures. We can essentially consider three different approaches for the enantioselective allylation of aldehydes with silanes (Scheme 2): *i*) a reaction catalyzed by an external chiral Lewis acid; *ii*) a Lewis base promoted[11] addition of allylSiCl$_3$, pioneered by Kobayashi[12] and developed into an effective asymmetric method by Denmark;[13] *iii*) the addition of allylsilacyclobutanes disclosed by Oshima and Utimoto[14] and further developed by Leighton.[15]

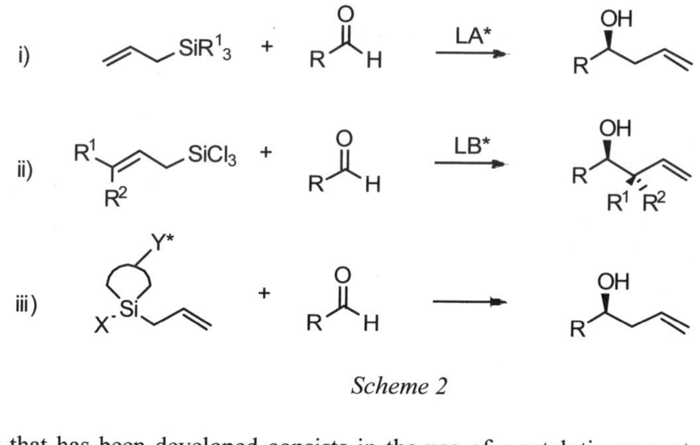

Scheme 2

The first strategy that has been developed consists in the use of a catalytic amount of a chiral Lewis acid to differentiate the enantiotopic faces of the electrophile and to activate it toward the allylsilane addition. The reaction proceeds through an open transition state and it is defined as Type II reaction and generally gives very good results in terms of yields and ee%.[16]

However, this proved not to be the best strategy, in fact the advantages of catalytic method and high enantioselectivities are significantly offset by the lack of diastereoselectivity with substituted allylating reagents because of the non-rigid nature of the transition state of the reaction (Scheme 3, eq. 1).

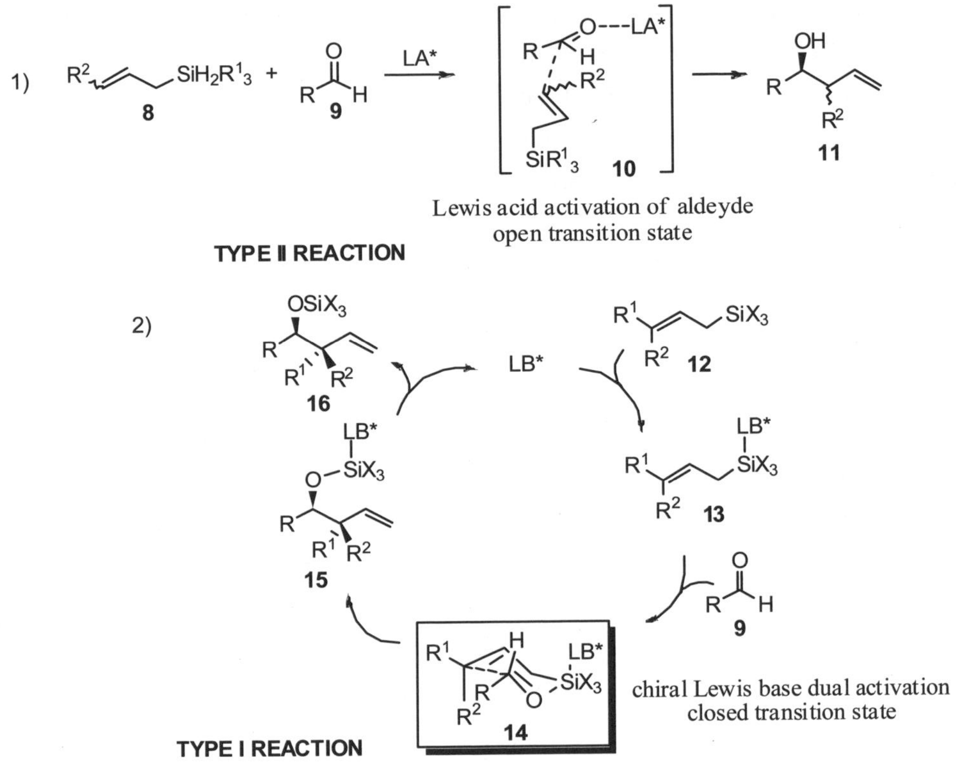

Lewis acid activation of aldeyde
open transition state

TYPE II REACTION

TYPE I REACTION

chiral Lewis base dual activation
closed transition state

Scheme 3

In the early 1990s, Kobayashi[12] reported a catalyst free reaction in which the poorly reactive (*Z*)- and (*E*)-allyltrichlorosilanes regio- and diastereoselectively added to aldehydes in *N,N*-dimethylformamide to afford the corresponding homoallylic alcohols in high yields under neutral conditions. In these reactions, DMF reacted as a base and coordinated to the silicon atom of the allyltrichlorosilanes to form hypervalent silicates, which can further coordinate the aldehyde and promote the reaction smoothly. The high stereoselectivities can be explained invoking a six-membered cyclic transition state like **14** in Scheme 3, and this reaction is usually known as a Type I reaction.

The observation that a base can activate the allylation reaction opened up the possibility of using chiral Lewis bases to effect the enantioselective allylation of aldehydes using allyltrichlorosilanes.

In contrast with the Lewis acid catalyzed reaction, the chiral Lewis base catalyzed allylation involves a dual mechanism of activation which provides for high reaction rates and excellent transfer of stereochemical information because the reaction proceeds through a closed assembly of allyltrichlorosilane, aldehyde, and the chiral Lewis base.[11] Thus, a high degree of diastereo- as well as enantioselectivity is expected. Furthermore, the non-covalent association between the chiral Lewis base and chlorosilane substrate, and its low reactivity if not activated, makes possible the use of a catalytic amount of chiral promoter.

Denmark[13] was the first that exploited chiral Lewis bases as catalysts for the enantioselective crotylation of aldehydes with trichlorosilanes. In a survey of Lewis bases as activators, phosphorus-based amides figured as the most effective ones. To achieve asymmetric induction, chiral phosphoric triamides **18**, derived from chiral diamines, have been developed and applied in the allylation reaction. One of the first examples is reported in Scheme 4. In general, good results are obtained with aromatic aldehydes, whereas aliphatic aldehydes are not good substrates for the reaction, since under the reaction conditions, a rapid formation of the α-chloro silyl ether occurs.[13e]

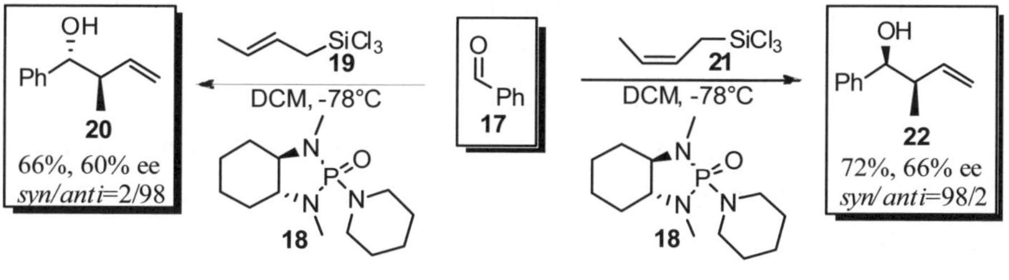

Scheme 4

Careful analysis of the mechanism of the reaction and considerations on the reactive transition state structure led to the development of improved catalysts based on a bis-phosphoramide scaffold like **23**, (Scheme 5) that improved the enantioselectivity.[17]

Over the years, various different structures have been designed, able to act as powerful and effective Lewis bases in allyltrichlorosilanes addition to aldehydes. Some examples are reported in Scheme 5.

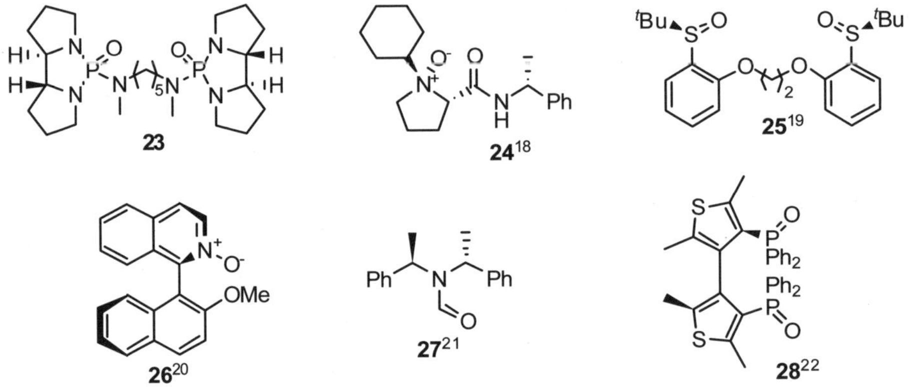

Scheme 5

The above described examples have shown how the stereoselectivity in allylation reactions can be improved and predicted by forcing the reaction to proceed *via* a closed chair-like transition state making the Si atom more Lewis acidic. Another way of increasing the Lewis acidity of the Si centre is given by its inclusion into a small ring. In fact, the Lewis acidity of silicon increases when it is incorporated into strained four-or five-membered ring system, owing to the smaller energy gap between sp^3 and dsp^3 orbitals of a strained system respect to an acyclic one.[23] As reported in Scheme 6, Oshima and Utimoto[14] found that silacyclobutane is a stronger Lewis acid than the corresponding tetraalkylsilanes, in fact the addition of **30** to benzaldehyde proceeds without any catalyst, conversely to the corresponding acyclic derivative **29**.

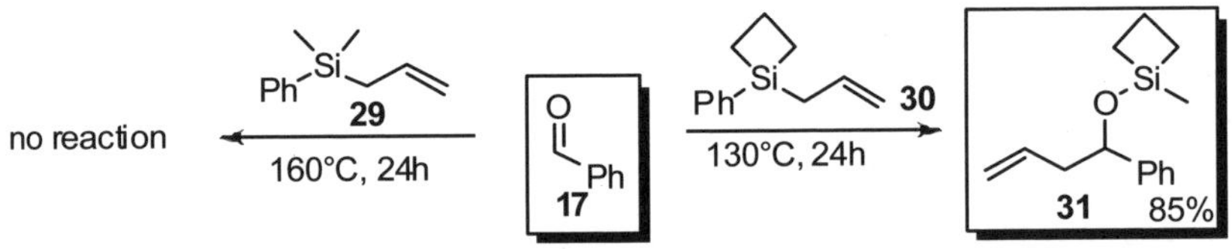

Scheme 6

Following the discovery that allyl(chloro)dioxasilacyclopentanes react with aldehydes at room temperature, Leighton and co-workers introduced a range of chiral allylsilanes in which the Si atom is contained within a five-membered ring.[15] The first successful example[24] exploited the derivative (*S,S*)-**33**, as reported in Scheme 7. The long Si−N and short C−N bonds ensure the silacycle is still strained, and the electronegative substituents further enhance the Lewis acidity of the Si centre.[25]

The reaction of (*S,S*)-**33** with pivalaldehyde gave alcohol **34** in 80 % yields and with a 96% ee; however with less hindered aliphatic aldehydes, with aromatic and α,β-unsaturated aldehydes, (*S,S*)-**33** was found to be less effective, providing enantioselectivities in the 78–88% range. Nevertheless, this results served as a proof of concept of the possibility to use a chiral allylsilane, that can be easily synthetized, as an effective allylating agent that reacts in a Type I sense.

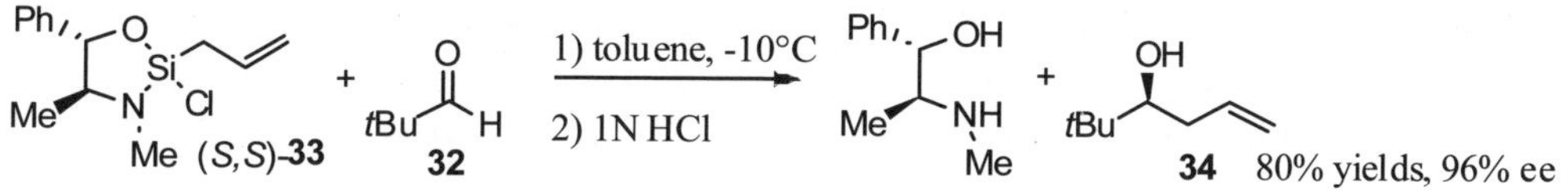

Scheme 7: Allylation of pivalaldehyde with chiralcyclic allyl silane (*S,S*)-**33**

Afterwards, a highly enantioselective and more general second-generation diamine-derived silacycles like **35** were developed for aldehyde allylation and crotylation (Scheme 8).[26]

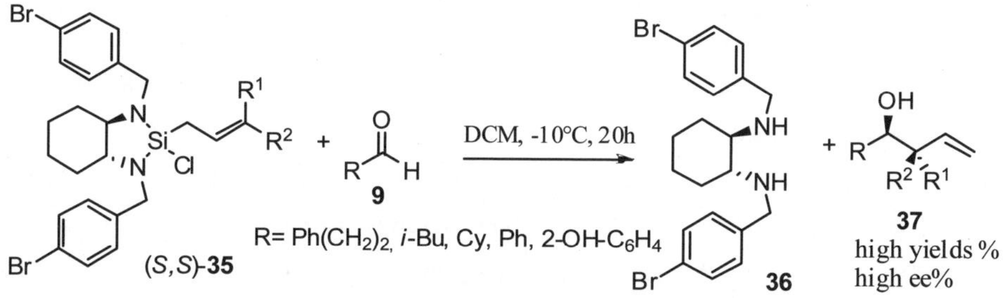

Scheme 8

Chiral cyclosilanes can be also successfully employed in the enantioselective imine allylation. It is worthy of note that the choice of nitrogen substituent in the imine determines the diastereoselectivity of the reaction,[27] as reported in the example in Scheme 9.

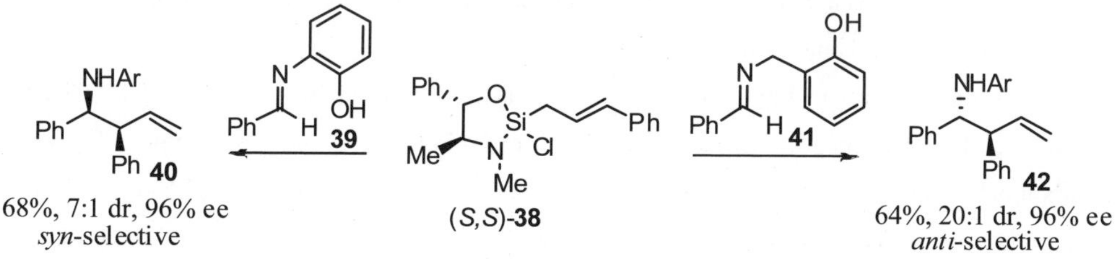

Scheme 9

The silyl reagents are crystalline, shelf-stable, and easy to prepare, even in large scale, (Scheme 10).[24,26] In fact, the reaction of allyltrichlorosilanes with the enantiopure amino-derivative in the presence of triethylamine gives the cyclo-derivative in high yields. Allylchlorosilane is commercially available, while Cross-metathesis provides an efficient route to γ-substituted allylsilanes.

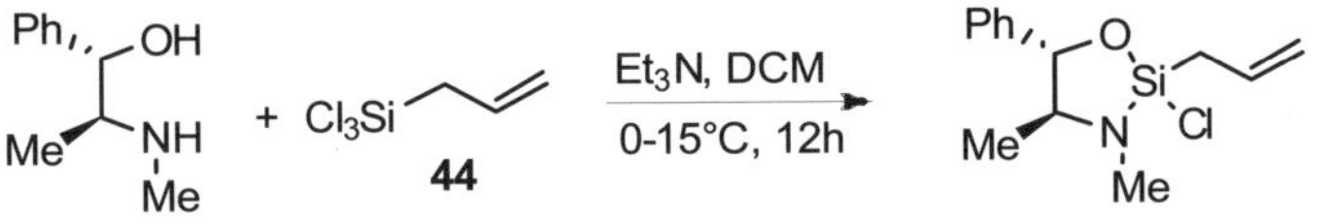

Scheme 10

As a consequence of the reaction mechanism chiral cyclic allyl silanes need to be employed in stoichiometric amounts. Nevertheless, in practical and economic terms, the use of pseudoephedrine (S,S)-43 in stoichiometric amounts is competitive with the employment of most chiral ligands employed in substoichiometric quantities since it can be easily recovered by extraction. The use of silicon has various 'green' advantages and with this strategies several otherwise difficult asymmetric transformations can be performed in several gram-scale.[15]

Even if allylsilanes are the most reactive and stereoselective between the possible unsaturated silicon-species, vinyl-, propargyl- and allenylsilicon reagents recently found interesting applications in asymmetric synthesis, as detailed in a review on these promising and green reagents.[5] Between these unsaturated silicon derivatives, allenylsilanes have been probably the most widely investigated owing to the predictable stereochemical outcomes of their transformations due to their rigid structural array.

SILICON-BASED LEWIS ACIDS (SLA)

The tendency of the silicon atom to expand its valence shell, giving rise to five- and six-coordinate intermediates, allows to consider silylating agents as Lewis acids. Therefore it is possible to use them as mediators in carbon-carbon bond forming reactions.[28]

Most examples of SLA in the literature are based on achiral species, like TMSOTf, TMSCl and TMSClO$_4$. One of the advantages of these reagents over metallic Lewis acids is the compatibility with many carbon nucleophiles. Furthermore, SLA are not prone to aggregation like metal halides, and this substantially simplifies the analysis of the reaction mechanisms, and the reactivity of SLA can be controlled by varying the steric and electronic demand of alkyl substituents. Moreover, a large scale application of eventually developed high conversion reactions should be possible, due to the green character of silicon derivatives.[28]

TMSOTf, and to a lesser extent TMSCl, are synthetically useful Lewis acids. However, recent variants that exhibit increased Lewis acidity have been introduced, such as trialkylsilyl bistrifluoromethanesulfonamides (R_3SiNTf_2), developed by Ghosez[29] and Mikami,[30] which proved to be particularly reactive and are prepared *in situ* from the corresponding Brønsted acid and an allylsilane or related species, as depicted in Scheme 11.

Scheme 11

Strong counterion effects on the strength of the catalytic activity of cationic SLA in Mukaiyama aldol and Diels-Alder reactions was reported by Sawamura.[31] Silyl borates of the form $R_3Si(solvent)BAr_4$, which contain a very weakly coordinating counteranion, were shown to be even more powerful Lewis acids than silyl bistrifluoromethanesulfonamides. A relevant example is reported in Scheme 12. In this case, the silicon Lewis acid is easily prepared by reaction of triethylsilane with trityl tetrakis(pentafluorophenyl)borate in toluene.

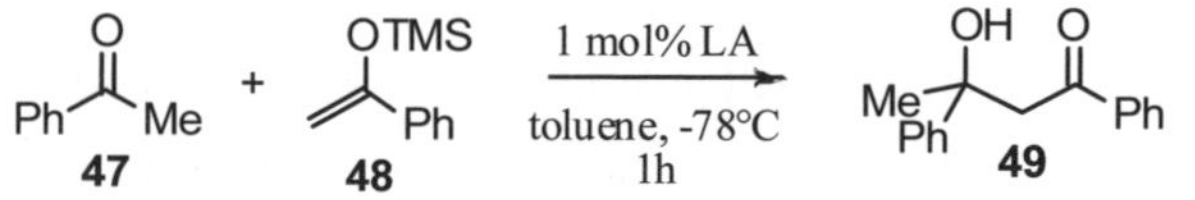

LA	yield of 49
$Et_3Si(toluene)^+ B(C_6F_5)_4^-$	97%
$TMS-NTf_2$	12%
$TMS-OTf$	0

$$Et_3SiH + Ph_3CB(C_6F_5)_4 \xrightarrow{toluene} Et_3Si(toluene)^+ B(C_6F_5)_4^- + Ph_3CH$$

Scheme 12

We have already seen in the previous section how the addition of a Lewis base can generate a more reactive silicon Lewis acid species. Recently Takenaka[32] applied this approach in an efficient desymmetrisation of *meso* epoxides **50** (Scheme 13) by using a catalytic amount of chiral pyridine *N*-oxide **51** as Lewis base to increase the Lewis acidity of $SiCl_4$, used in stoichiometric amounts.

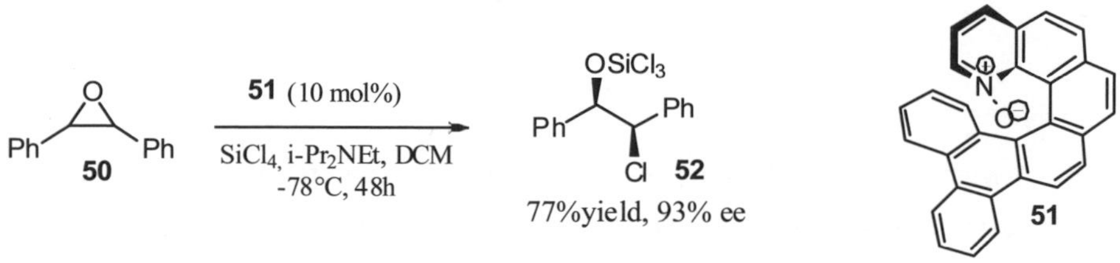

Scheme 13

The third strategy adopted for increasing the Lewis acidity of silicon derivatives is to include Si-atom in a strained cycle. Using a concept similar to that used for enantioselective allylation reactions with allylsilanes, Leighton introduced a new class of silicon Lewis acids able to promote enantioselective additions to acyl hydrazones.[15,33]

The milestone of these class of reagents is the derivative (S,S)-**53**, (Scheme 14), easily obtained by reaction of pseudoephedrine with PhSiCl₃. The reaction of (S,S)-**53** with the acyl hydrazone **54** generates an activated intermediate **55** in which the faces of the electrophile are sterically differentiated, therefore an enantioselective addition of a nucleophile can occur.

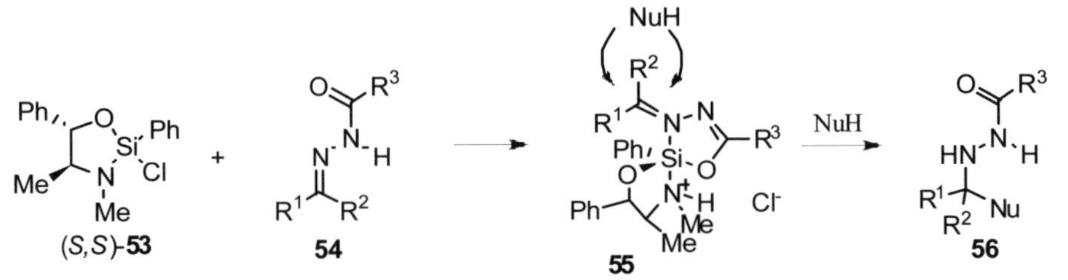

Scheme 14: Proposed Enantioselective Nucleophilic Additions to Silane–Acylhydrazone Complexes

Silane (S,S)-**53** proved to be an efficient Lewis acid for the activation of acylhydrazones and other imine derivatives in a large variety of reactions,[15] such as Friedel Crafts alkylation, cycloadditions, tandem Aza-Darzens–ring-opening reactions and Pictet–Spengler reactions.

One of the first examples of its use is reported in Scheme 15.[33] The [3 + 2] cycloaddition of acyl hydrazone **57** with *tert*-butyl vinyl ether proceeds with excellent enantioselectivity and diastereoselectivity providing a pyrazolidin derivative **58** that is primed for further reactions.

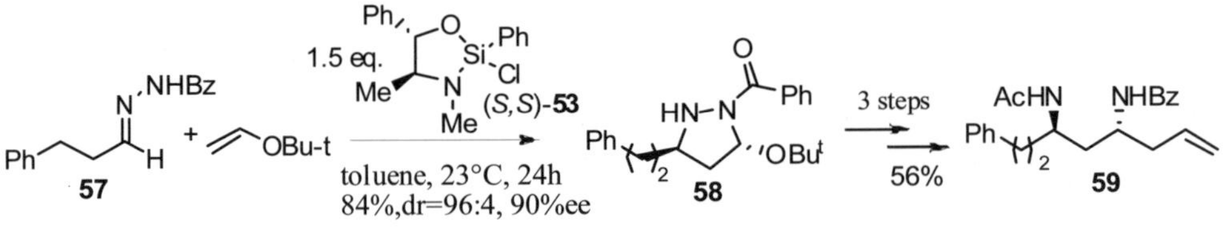

Scheme 15

More recently Leighton found[34] that the replacement of the phenyl substituent in his first generation SLA with an alkoxy group provides a more straightforward method for catalyst tuning, in fact the more electron-withdrawing alkoxy group generates a more reactive activator. A significant application is the enantioselective Mannich reaction involving aliphatic ketone-derived acyl hydrazones **60** and silyl enol ethers in the presence of 1.3 equivalents of (*S,S*)-**61** (Scheme 16).

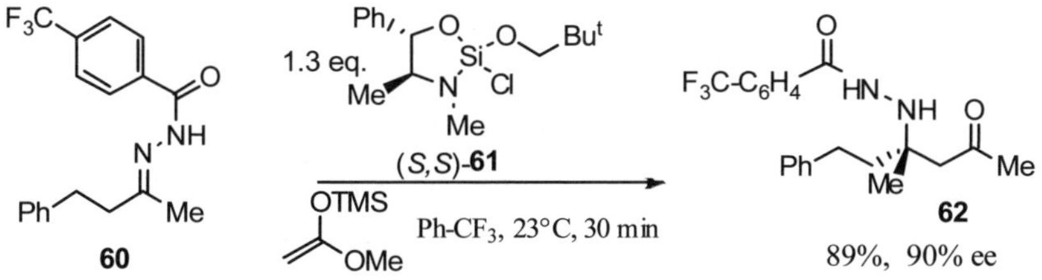

Scheme 16

Even when exploited as Lewis acids, Leighton's cyclic silyl derivatives (*S,S*)-**53** and (*S,S*)-**61** are necessarily used in stoichiometric amount. However, as discussed in the previous section, the practical and economical advantages offered by these derivatives can compete with most chiral ligands employed in substoichiometric amounts.[15]

SILYL PROTECTING GROUPS

Silyl ethers are among the most frequently used protecting groups for the alcohol moiety.[35] The reason of such popularity derives from the possibility to modulate their reactivity, both in formation and in cleavage, by the introduction of suitable substituents on silicon atom. In fact both steric and electronic effects can influence the stability of silyl ethers.

In general, the bulkier are the substituents, the greater is the stability towards acid and basic conditions, oxidation and reduction conditions, organometallic reagents and silica gel. Moreover, electronic effects play important roles and can be exploited in differentiate the stability under given conditions. For instance alkyl silyl ethers are more easily cleaved than the corresponding aryl silyl ethers under acid conditions, and the opposite situation is observed upon treatment with a base. In addition, electron withdrawing substituents on the silicon atom enhance the ether stability under acid conditions, whereas increase its susceptibility toward bases.

Silyl ethers are usually formed[35] by treating an alcohol with the corresponding silyl chloride or triflate in the presence of a base to scavenge the acid by-product. Their cleavage can be performed in a large variety of conditions, depending on the silyl ethers and on the selectivity needed, (Scheme 17).

A plenty of procedures have been developed for the deprotection step,[35] that can occur by treatment with a diluted acid (HCl, AcOH, TFA, PPTS, etc..), a base (K$_2$CO$_3$, etc.), a fluoride source (HF, TBAF, HF-py complex, NH$_4$F, etc.), or a Lewis acid (BF$_3$·Et$_2$O, FeCl$_3$, SnCl$_2$, CeCl$_3$·nH$_2$O/NaI system).

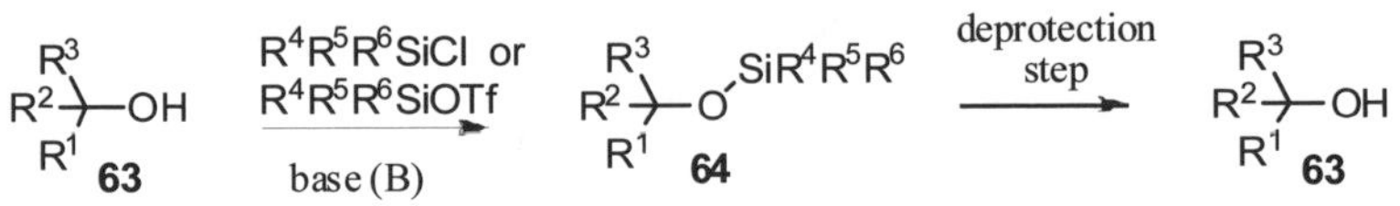

Scheme 17

Among Lewis acids, the CeCl$_3$·nH$_2$O/NaI combination[36] prove to be very efficient in the cleavage of various silyl ethers. High selectivities were recorded both between different silyl derivatives and between silyl ethers and other protecting groups.

The oxaphilicity of Ce(III) and the presence of an iodide ion can be invoked to explain the observed reactivity, (Scheme 18). The Ce(III)-coordination at the oxygen atom of **65** promotes the nucleophilic attack of the iodide to the silicon atom to form a penta-coordinate complex **66**, which irreversibly decomposes to the silicon iodide derivative **68** and the cerium alcoholate **67**. Subsequent aqueous hydrolysis led to the deprotected alcohol **69**.

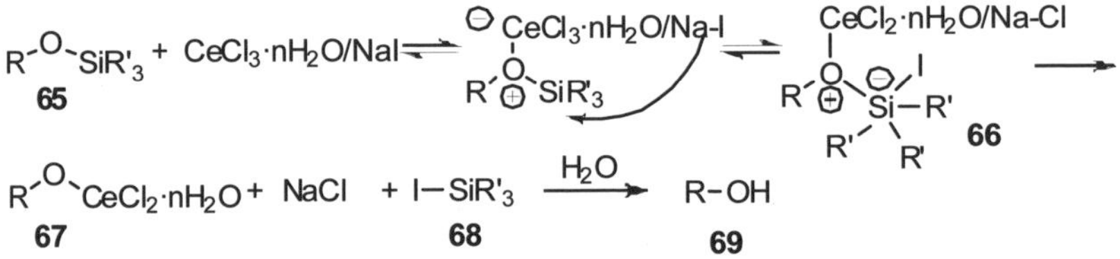

R'$_3$Si— = TBS (t-BuMe$_2$Si), TES (Et$_3$Si), TBDPS (t-BuPh$_2$Si), TIPS (i-Pr$_3$Si)

Scheme 18

A large variety of silicon protecting groups can be removed, including TBS (*t*-BuMe$_2$Si), TES (Et$_3$Si), TBDPS (*t*-BuPh$_2$Si) and TIPS (*i*-Pr$_3$Si) (Table 1). However, under controlled experimental conditions, TES and TBS ether groups can be selectively cleaved with respect to TBDPS and TIPS (for examples see Table 1, entries 5 and 7).

Moreover, the selectivity towards other alcohol protecting groups was observed: acetates, THP, and Bn ethers remain unaffected, but not MOM ethers (see selected examples in Table 1, entries 8 and 9). In addition, the stereochemistry of asymmetric carbons remains unaltered during this cleavage (Table 5, entry 2), since the mild reaction conditions required by CeCl$_3$·nH$_2$O/NaI system generally do not give rise to racemization or epimerization phenomena.

This reaction particularly suffers from steric interactions, and secondary alcohols require reaction times six-fold of that for primary alcohols. Therefore a selective deprotection of a primary silyl ether in the presence of a secondary one could be possible.

Table 1. Selected Examples of Desilylation of Silyl Ethers by $CeCl_3 \cdot nH_2O/NaI$ in CH_3CN

Entry	Starting material	Product	conditions	Yields (%)
1	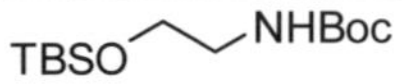 TBSO~~~NHBoc	HO~~~NHBoc	rt, 4h	99
2	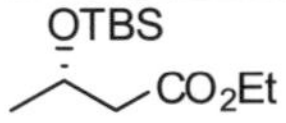 OTBS ~~~CO₂Et	OH ~~~CO₂Et	reflux, 3h	95
3	$CH_3(CH_2)_9OTBDPS$	$CH_3(CH_2)_9OH$	reflux, 6h	95
4	$CH_3(CH_2)_5CHCH_2CH_3$ OTBDPS	$CH_3(CH_2)_5CHCH_2CH_3$ OH	reflux, 36h	94
5	$TBSO(CH_2)_3OTBDPS$	$HO(CH_2)_3OTBDPS$	rt, 24h	94
6	$CH_3(CH_2)_9OTIPS$	$CH_3(CH_2)_9OH$	reflux, 3h	92
7	$TBSO(CH_2)_3OTIPS$	$HO(CH_2)_3OTIPS$	reflux, 1h	95
8	 AcO~~~OTBS	HO~~~OTBS	rt, 24h	96
9	 THPO~~~OTBS	THPO~~~OH	rt, 24h	95

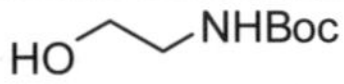

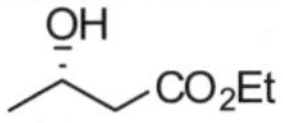

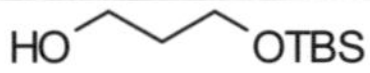

ORGANOSILANES IN CROSS- COUPLING REACTIONS

Much of the motivation behind the growing employment of organosilicon compounds in organic synthesis arises from the possibility to perform stereospecific palladium catalysed cross-coupling reactions. In fact, reactions which utilize organosilanes in this type of cross-coupling, commonly referred to as Hiyama couplings, achieved significant importance in the last three decades as a concrete alternative to the Suzuki, Stille and Negishi couplings, due to the selective reactivity, stability and non-toxicity associated with organosilicon compounds.[37]

In general, Pd-catalysed cross-coupling strategies require an 'electrophilic' coupling partner, usually an organohalide or pseudohalide (sulfonate, phosphate, *etc*) and a 'nucleophilic' coupling partner, commonly an organometallic reagent including B, Sn, Zn, Cu, Mg, Zr and Si species (Scheme 19).

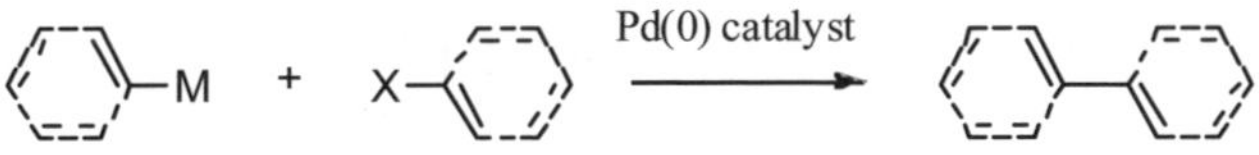

$M = SnR_3, BR_2, ZnX, CuX, MgX, SiR_3$
$X = $ halide, OTf, N_2X

Scheme 19

Owing to the low polarisation of the C−Si bond, organosilanes are relatively unreactive nucleophilic coupling partners for Pd(0)-catalysed cross-coupling reactions; therefore, reactions are usually performed in the presence of an activator, typically a fluoride source like TBAF, which give rise to the formation *in situ* of a pentacoordinate siliconate species, which undergoes a rapid transmetalation with the Pd-catalyst,[38] (see Scheme 20).

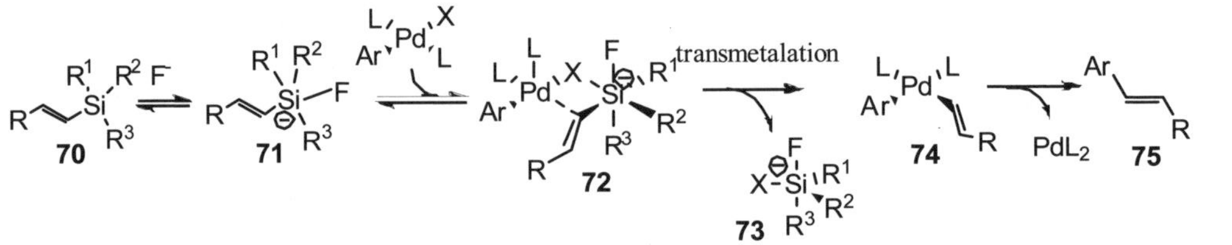

Scheme 20 Hypotheses for the generation of fluorine activated species for transmetalation

Hiyama couplings under fluoride-free conditions are also possible. Denmark provided significant contributions to this field,[39] showing that organosilanols undergo Pd-catalysed cross-coupling in the presence of a base, giving rise to the corresponding silanolate as the reactive species, (Scheme 21). Among the bases, TMSOK is a particularly efficient and mild choice.

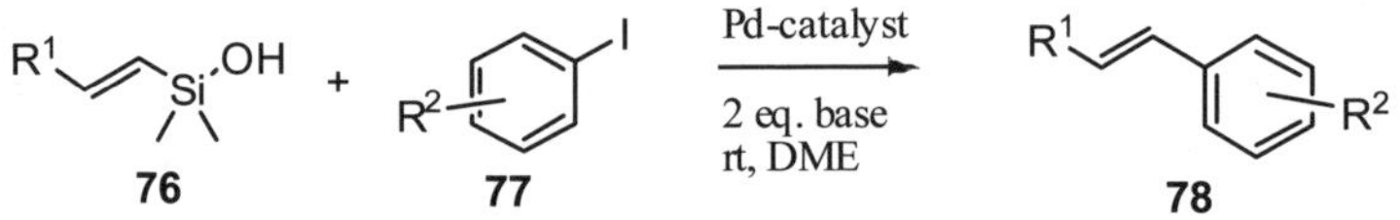

Scheme 21

Mechanistic studies[39] have revealed a different reaction pathway for this base-mediated Hiyama coupling. Specifically, reaction does not require the formation of a pentavalent siliconate species, rather transmetallation proceeds in a direct, intramolecular fashion on an intermediate tetracoordinate Pd(II) species **80**, (Scheme 22).

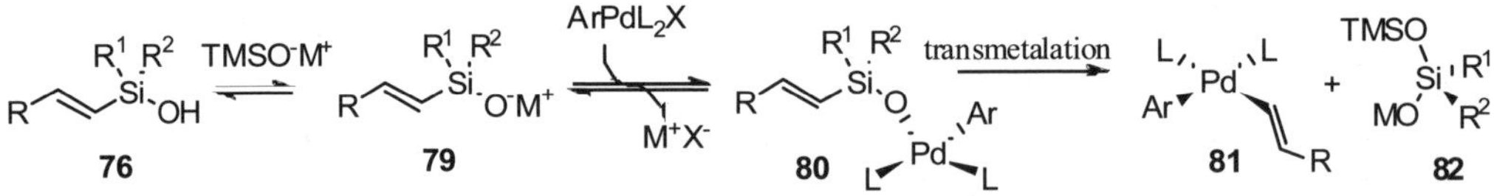

Scheme 22

Compared to the base–mediated couplings, fluoride-activated cross-couplings tend to be faster and less sensitive to structural and electronic features of the substrates.[38]

The different mode of activation of different organosilanes can be exploited in sequential Hiyama coupling reactions designing a bi-silyl compound bearing one silyl substituent which readily reacts under basic activation while the other silyl substituent is inert. Subsequent fluoride-promoted coupling of the second silyl substituent would allow for sequential coupling. In this manner, the different mechanistic pathways cross-coupling can be used to distinguish between silicon substituents. Denmark described the sequential cross-coupling of 1,4-bissilylbutadienes,[40] (Scheme 23), and Lopez[41] applied this cross-coupling strategy to a highly stereoselective synthesis of retinoids.

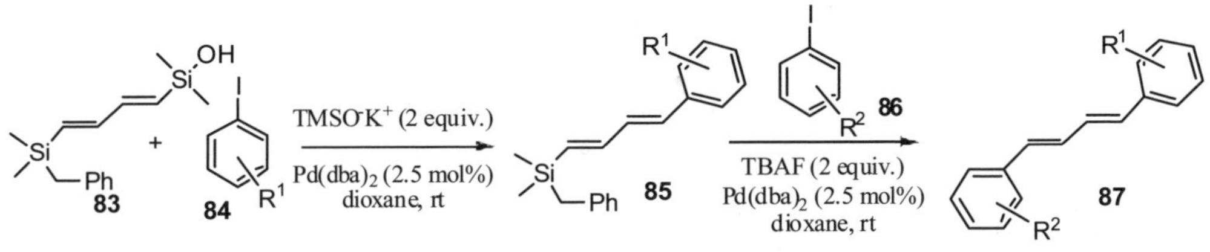

Scheme 23

Recently Denmark exploited Hiyama couplings to prepare particularly challenging 2-aryl heterocycles[42] These couplings require careful optimisation of the reaction conditions. In fact, the choice of the protecting group on the indole nitrogen, the pre-formation of the sodium silanolate, an accurate choice of Pd catalyst and in some cases the inclusion of a copper salt need to be considered, (Scheme 24).

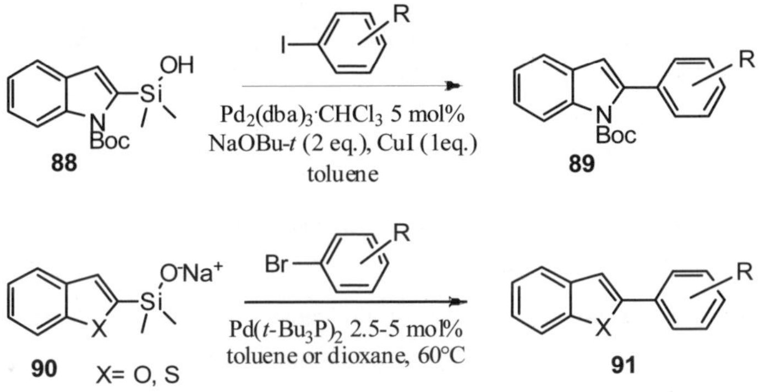

Scheme 24

Although silanols, fluorosilanes and alkoxysilanes are the most commonly employed cross-coupling agents for Hiyama reactions, a range of all-carbon substituted silane coupling partners, which generate the reactive coupling agent *in situ*, can also be used. These reagents include, 2-pyridyl-, 2-thienyl, benzyl and allylsilanes.[43]

An interesting controlled reactivity was shown by Yoshida.[43a] His group found that 2-pyridylsilanes are useful alkenyl, alkynyl and benzyl transfer agents under the classical Hiyama conditions; however, in

117

the presence of a Ag(I) salt, the allyldimethylsilyl group functions as a 2-pyridyl transfer agent (Scheme 25).

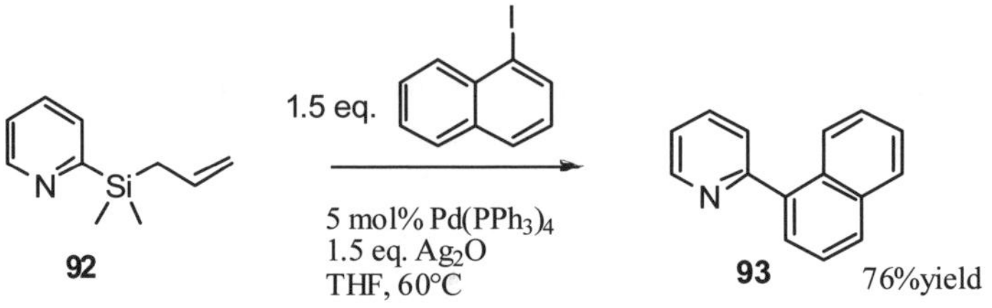

Scheme 25

The possibility to use alkyl derivatives as coupling partner was successfully obtained by Fu[44]with secondary activated and unactivated halides. A Ni(II)/amino alcohol based catalyst provides an efficient and versatile method for Hiyama reactions of secondary alkyl electrophiles and aryltrifluorosilanes. The inclusion of norephedrine as a ligand was important for obtaining good yields of product, (Scheme 26).

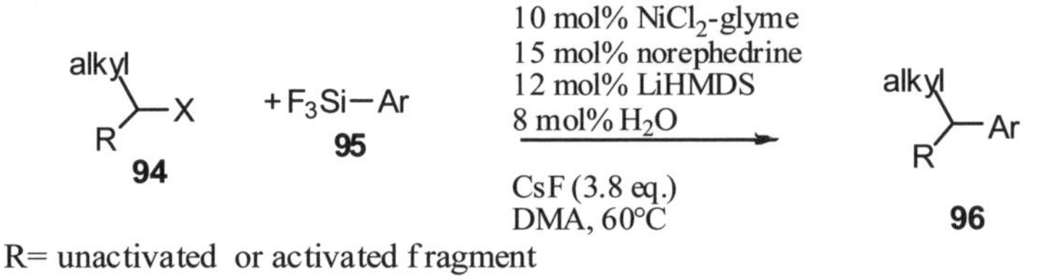

R= unactivated or activated fragment

Scheme 26

ORGANOSILICON COMPOUNDS IN ORGANIC MATERIALS

Organosilicon compounds based on silole (silacyclopentadiene) have become an interesting subject in materials science since they possess a number of favorable attributes for application in organic electronics.[45] In fact, siloles exhibit very high electron mobilities, exceeding those for the well-known tris(8-hydroxyquinoline) aluminium (Alq$_3$), which is widely used as an electron-transport (ET) material in organic light-emitting diodes (OLEDs).[46] Moreover, many siloles exhibit very high photoluminescence (PL) quantum yields in the solid state (as high as 97±3%), suggesting applications as luminophores.

Exploiting these properties, siloles have been utilized for varied applications,[45] such as in sensors,[47]organic solar cells,[48] organic field-effect transistors[49] (OFETs) and as emissive or ET layers in fabrication of OLEDs[50] since Tamao et al. reported[51] a general one-pot synthesis of 2,5-diaryl-3,4-diphenylsiloles and discovered their high efficiency in electroluminescent devices. In particular, silole derivatives have been shown to be excellent electron-transporting materials for OLEDs because they possess a low-lying LUMO due to the interaction between the σ* orbital of their two exocyclic Si-C bonds and the π* orbitals of the butadiene moiety.[52] The calculated LUMO level of a silole ring is even lower than those of other heterocyclopentadienes, such as pyrrole, furan, and thiophene.[53]

Simple structural modifications in the silole skeleton can provide significant control of the electronic properties through varying the position of those substituents on the silole ring and/or the donor/acceptor strength of the exo-cyclic substituents. Some possible structures are depicted in Scheme 27.

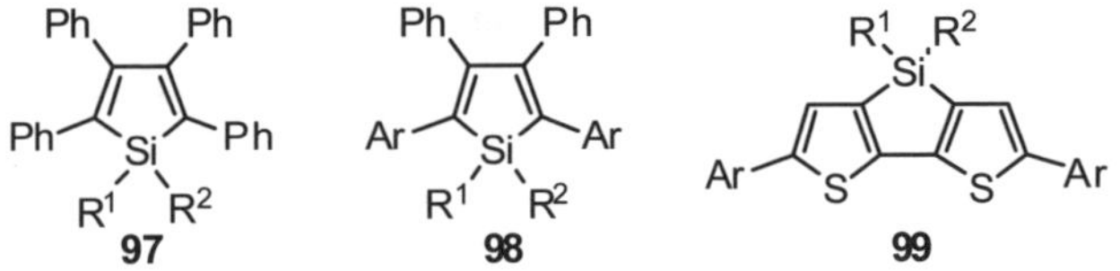

Scheme 27

The substituents on silicon atom exert primarily inductive effects upon the properties of the silole core, accordingly, only minor effects upon absorption and fluorescence spectra have been recorded.[45] On the contrary, it has been found that the photophysical properties and electronic structures of a silole significantly depend on the nature of the 2,5-substituents which can exhibit strong π-interactions with the silole core, allowing one to destabilize the HOMO through the use of strong π-donors, or stabilize the LUMO through the use of strong π-acceptors. Therefore silole derivatives that exhibit emission from blue to orange-red by the incorporation of various substituents at the 2 and 5 positions of the silole ring have been synthetized.

Although the 1,1- and 2,5-substituents of siloles can be easily changed, synthetic difficulties arose varying 3,4-substituents, thus limiting the studies on their effect on the electronic structures of siloles.[54] The few appeared papers report that the effect of 3,4-substituents is less extended than the one of 1,1- and 2,5-substituents.

The general strategy[55] for the synthesis of dialkylbis(phenylethynyl)silanes **104** is depicted in Scheme 28.

Diethynylsilanes **101**, obtained by reaction of phenylethynyllithium with the corresponding dialkyldichlorosilane, undergo intramolecular reductive cyclization in an *endo-endo* mode upon treatment with lithium naphthalenide (LiNp) in the presence of ZnCl$_2$-TMEDA to form 3,4-diphenylsilole-2,5-di(zinc chloride) **103** which can be coupled with the appropriate aromatic electrophile in the presence of a suitable Pd-catalyst to obtain **104**.

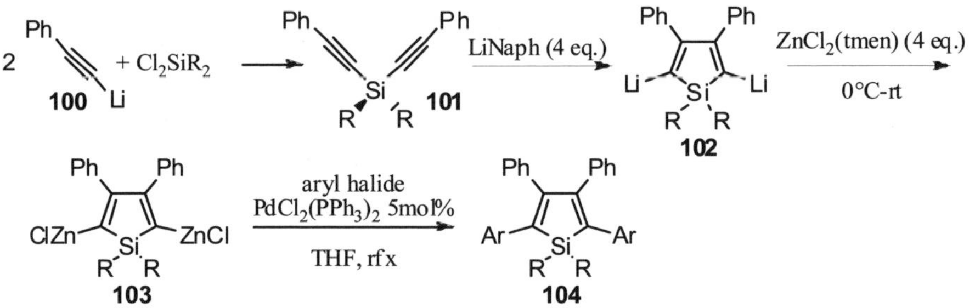

Scheme 28

Some of the synthesized siloles which proved to be good material for the construction of electroluminescence devices are reported in Figure 1.[56]

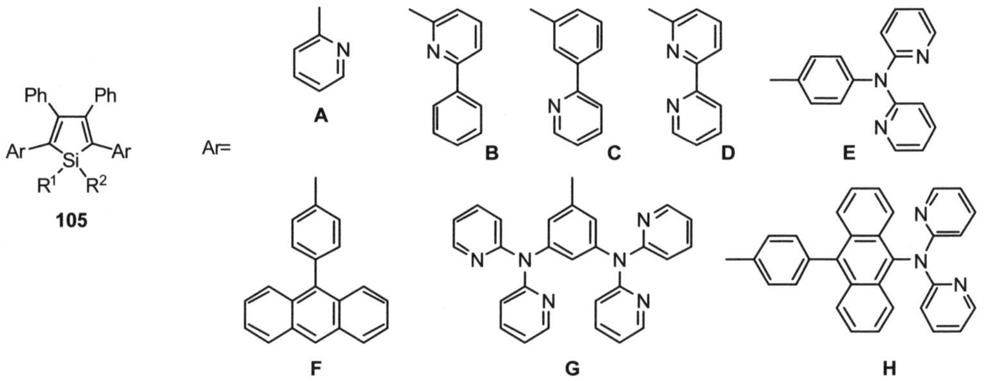

Figure 1

Some trends on the nature of 2,5-substituents can be outlined.[56b] The presence of either dipyridylamino or anthracene groups in the molecular structure of the siloles brings about a significant improvement of the balance of charge in the devices, and the conjugation between electron- and hole-transporting groups appears to be important for achieving high efficiency. As a consequence, Gerbier and co-workers synthesized a silole in which two anthracene units and two dipyridylamino groups are connected to the central silole core through a conjugated backbone (**54-H** in Figure 1).

Recently, a two thiophene fused silole, dithienosilole (DTS), like **99** in Scheme 27, has attracted much attention due to its usefulness as a building block that can be incorporated into conjugated molecules and polymers.[57] Although DTS has a number of advantages, there have been only a few reports on DTS-based OLED performance due to synthetic limitations of DTS derivatives.[58]

Kim and Kang[59]recently reported a new synthetic route to three new DTS derivatives, as reported in Scheme 29. Since classical coupling reactions gave poor to moderate yields, they set up a new cross-coupling reaction using indium-reagent as nucleophile in the presence of Pd(dppf)Cl$_2$ (dppf=1,10-bis(diphenylphosphino) ferrocene) obtaining **111a-c** in moderate to good yields. As expected, the substituent on the Si atom slightly perturbed the UV-vis absorption properties, redox-potentials, and photo- and electroluminescent efficiency. In particular, a low-lying LUMO in **111c** was observed compared with those of **111a** and **111b**, indicating that the phenyl substituents on the Si atom induced a relative diminution of the band gap. Compound **111a** used as an ET material exhibits a bright emission in a multilayered EL device.

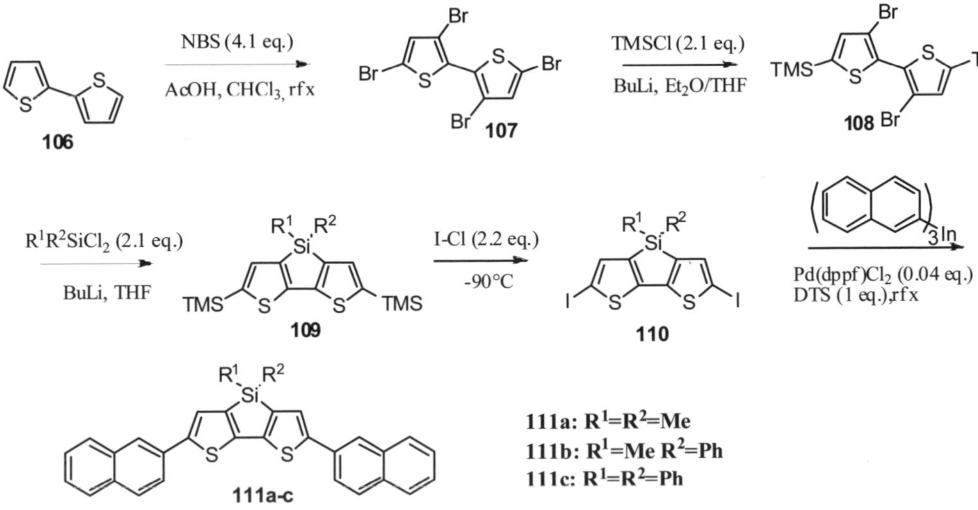

Scheme 29

The external quantum efficiency of OLEDs fabricated by using siloles as emissive or electron-transporting materials ranges from 0.0045% to 8%. Such results indicate that the efficiency of silole-based OLEDs is quite sensitive to both the device structure and the molecular structure of the siloles. Therefore, continuous efforts for the development of suitable silole derivatives for application in OLEDs and the finding of optimized device conditions are still required.[59]

Siloles found also successful employment in forming fluorescent organogels. It is well known that most chromophores exhibit aggregation induced quenching of emission, resulting in decreased emission intensity in poor solvents or in solid state devices.[60] In contrast, various 2,3,4,5-tetraphenylsilole derivatives exhibit an unusual aggregation-induced emission (AIE) behavior, and show extremely high photoluminescence (PL) quantum yields (up to 100%) in the solid state, while they are non-emissive in solution.[61]

Recently, Zhang[62] described the preparation of two new organic gelators based on silole with urea and cholesterol moieties. Compounds **114** and **115**, obtained as reported in Scheme 30, can gel some hydrocarbon solvents and a large fluorescence enhancement was observed after gelation. Moreover, their fluorescence intensities can be changed reversibly accompanying the gel–solution transition through alternating cooling and heating.

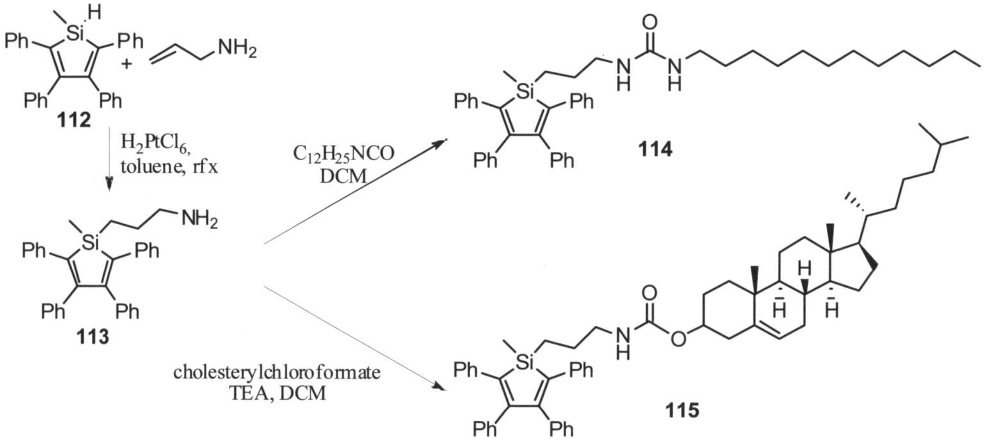

Scheme 30

An other interesting example was more recently reported by Wang[63] and coworkers who designed other novel silole derived gelators containing long-chain amide moieties (**114** and **115**, see Scheme 31). Moreover, besides gel, the 2,3,4,5-tetraphenylsilole moiety can self-assembly into liquid crystals and molecular monolayers with the aid of long alkyl chains.

The key intermediate compound, 2,5-bis(4-amidophenyl)-3,4,-diphenylsilole (**121**), obtained with a procedure similar to that reported in Scheme 31, was then reacted with acyl chlorides **118a-b** to give the derivatives **122a-b** which proved to be able to gel hydrocarbon solvents.

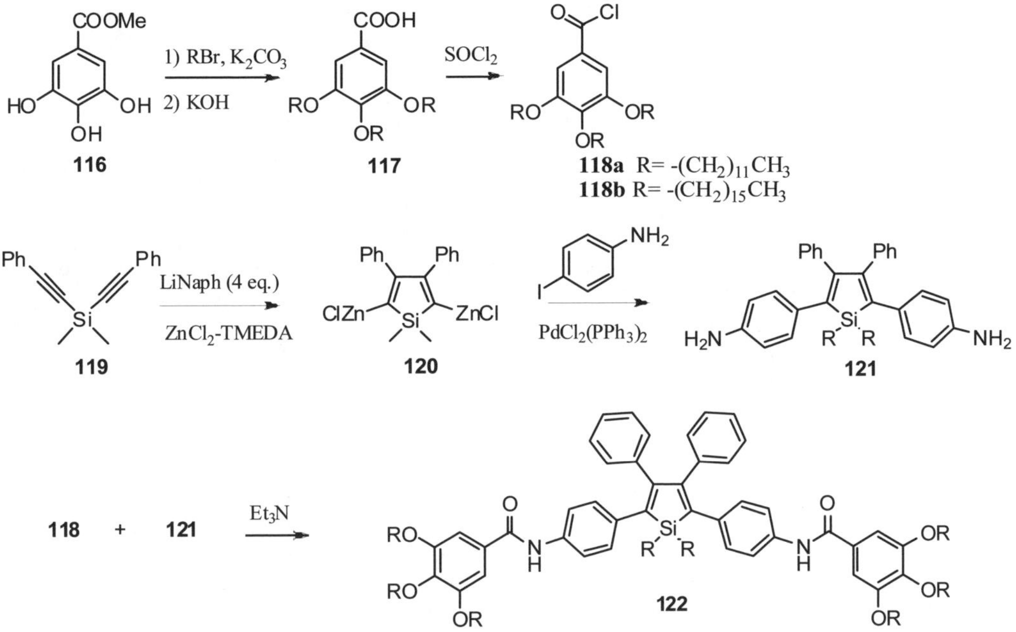

Scheme 31

REFERENCES

1. a) Panek J. S. in *Comprehensive Organic Synthesis* Eds.: B. M. Trost, I. Fleming, Pergamon Press, Oxford, 1991, Vol. 1, Chapter 2.5, pp. 579-627. b) Ojima, I. in *The Chemistry of Organic Silicon Compounds*, ed. S. Patai and Z. Rappoport, Wiley, New York, 1989, pp. 1479–1526.

2. a) Mori, A.in *Latest Frontiers of Organic Synthesis*, ed. Y. Kobayashi, Research Signpost, Kerala, India, 2002, pp. 83–102. b) Hiyama, T. in *Metal-catalyzed Cross-coupling Reactions*, ed. F. Diederich and P. J. Stang, Wiley-VCH, Weinheim, 1998, pp.421–453. c) Brook, M. A. *Silicon in Organic, Organometallic, and Polymer Chemistry*, Wiley, New York, 2000. d) Chuit, C.; Corriu, R. J. P.; Reye, C.; Young, J. C. *Chem. Rev.* **1993**, *93*, 1371. e) Holmes, R. R. *Chem. Rev.* **1996**, *96*, 927. f) Kira, M.; Zhang, L. C. In *The Chemistry of Hypervalent Compounds*; Akiba, K.-y., Ed.; Wiley-VCH: Weinheim, Germany, 1999, p 147.

3. Lambert, J. B.; Zhao, Y.; Emblidge, R. W.; Salvador, L. A.; Liu, X.; So, J.-H.; Chelius, E. C. *Acc. Chem. Res.* **1999**, *32*, 183, and references therein.

4. a) Chan, T. H.; Wang, D. *Chem. Rev.* **1995**, *95*, 1279; b) Schlosser, M.; Franzini, L. *Synthesis* **1998**, 707.

5. Curtis-Long, M. J.; Aye Y. *Chem. Eur. J.,* **2009**, *15*, 5402.

6. Fleming, I.; Au-Yeung B.-W. *Tetrahedron* **1981**, *37*, Supplement,13.

7. a) Fleming, I.; Barbero, A.; Walter, D. *Chem. Rev.* **1997**, *97*, 2063, and references therein; b) Fleming, I.; Dunoguès, J., Smithers, R. *Org. React.* **1989**, *37*, 57. c) Hosomi, A.; Sakurai, H. *Tetrahedron Lett.* **1976**, 1295.

8. a) Chabaud, L.; James, P.; Landais, Y. *Eur. J. Org. Chem.* **2004**, 3173; b) Masse, C. E.; Panek, J. S. *Chem. Rev.* **1995**, *95,* 1293. c) Langkopf, E.; Schinzer, D. *Chem. Rev.* **1995**, *95*, 1375.

9.   a) Kennedy, J. W. J.; Hall, D. G. *Angew. Chem. Int. Ed.,* **2003**, *42*, 4732. b) Chemler, S. R.; Roush, W. R. in *Modern Carbonyl Chemistry* Ed. J. Otera, Wiley-VCH, Weinheim, 2000, Chap. 11, pp. 403–490.

10.   Mayr, H.; Patz, M. *Angew. Chem. Int. Ed. Engl.,* **1994**, *33*, 938.

11.   Gawronski, J.; Wascinska, N.; Gajewy, J. *Chem. Rev.* **2008**, *108*, 5227, and cited references.

12.   a) Kobayashi, S.; Nishio, K. *Tetrahedron Lett.* **1993**, *34*, 3453. b) Kobayashi, S.; Nishio, K. *Synthesis* **1994**, 457. c) Kobayashi, S.; Nishio, K. *J. Org. Chem.* **1994**, *59*, 6620.

13.   a) Denmark, S. E.; Coe, D. M.; Pratt, N. E.; Griedel, B. D. *J. Org. Chem.* **1994**, *59*, 6161. b) Denmark, S. E.; Fu, J. *J. Am. Chem. Soc.* **2000**, *122*, 12021. c) Denmark, S. E.; Fu, J. *J. Am. Chem. Soc.* **2001**, *123*, 9488. d) Denmark, S. E.; Fu, J. *J. Am. Chem. Soc.* **2003**, *125*, 2208. e) Denmark, S. E.; Fu, J.; Coe, D. M.; Su, X.; Pratt, N. E.; Griedel B. D. *J. Org. Chem.* **2006**, *71*, 1513. f) Denmark, S. E.; Fu, J. *Chem. Commun.* **2003**, 167.

14.   Matsumoto, K.; Oshima, K.; Utimoto, K. *J. Org. Chem.* **1994**, *59*, 7152.

15.   Leighton, J. L. *Aldrichimica Acta* **2010**, *43*, 3, and references therein cited.

16.   a) For one of the first examples, see: Gauthier Jr., D. R.; Carreira E. M. *Angew. Chem. Int. Ed. Engl.* **1996**, *35*, 2363. b) Yanagisawa, A. in *Comprehensive Asymmetric Catalysis*, ed. E. N. Jacobsen, A. Pfaltz and H. Yamamoto, Springer-Verlag, Heidelberg, 1999, vol. II, Ch. 27.

17.   Denmark, S. E.; Fu, J.; Lawler, M. J. *J. Org. Chem.,* **2006**, *71*, 1523

18.   Traverse J. F. Yu Zhao, Hoveyda, A. H.; Snappe, M. L. *Org. Lett.* **2005**, *7*, 3151.

19.   Wang, P.; Chen, J.; Cun, L.; Deng, J.; Zhu, J.; Liao, J. *Org. Biomol. Chem.,* **2009**, *7*, 3741.

20.   Malkov, A. V.; Kočovský, P. *Eur. J. Org. Chem.,* **2007**, 29, and cited references.

21.   Iseki, K.; Mizuno, S.; Kuroki, Y.; Kobayashi Y. *Tetrahedron,* **1999**, *55*, 977.

22.   Simonini, V.; Benaglia, M.; Benincori T. *Adv. Synth. Catal.,* **2008**, *350,* 561.

23.   Sereda, O.; Tabassum, S.; Wilhelm, R. *Top. Curr. Chem.* **2010**, *291*, 349.

24.   Kinnaird, J. W. A.; Ng, P. Y.; Kubota, K.; Wang, X.; Leighton, J. L. *J. Am. Chem. Soc.* **2002**, *124*, 7920.

25.   Zhang, X.; Houk, K. N.; Leighton, J. L. *Angew. Chem., Int. Ed.* **2005**, *44*, 938.

26.   Kubota, K.; Leighton, J. L. *Angew. Chem., Int. Ed.* **2003**, *42*, 946.

27.   Huber, J. D.; Leighton, J. L. *J. Am. Chem. Soc.* **2007**, *129*, 14552.

28.   Dilman, A. D.; Ioffe, S. L. *Chem. Rev.* **2003**, *103,* 733

29.   Mathieu, B.; Ghosez, L. *Tetrahedron Lett.,* **1997**, *38*, 5497.

30.   Ishii, A.; Kotera, O.; Saeki, T.; Mikami, K. *Synlett,* **1997**, 1145.

31.   Hara, K.; Akiyama, R.; Sawamura, M. *Org. Lett.,* **2005**, *7*, 5621.

32.   Takenaka, N.; Sarangthem, R. S.; Captain, B. *Angew. Chem. Int. Ed.,* **2008**, *47*, 9708.

33.   Shirakawa, S.; Berger, R.; Leighton, J. L. *J. Am. Chem. Soc.* **2005**, *127*, 2858.

34.   Notte, G. T.; Leighton, J. L. *J. Am. Chem. Soc.,* **2008**, *130*, 6676.

35.   a) Green, T. W.;. Wuts, P. G. M *Protective Groups in Organic Synthesis*, Wiley-Interscience, New York, 1999. b) Kociensky, P. J. *Protecting Groups,* Georg Thieme Verlag, Stuttgart, 2000.

36.   a) Bartoli, G.; Marcantoni, E.; Sambri L. *Synlett,* **2003**, 2101. b) Bartoli, G.; Bosco, M.; Marcantoni, E.; Sambri, L.; Torregiani, E. *Synlett* **1998**, 209.

37.   For some recent reviews see: a) Denmark, S. E.; Baird, J. D. *Chem. Eur. J.,* **2006**, *12*, 4954. b) Denmark, S. E.; Regen, C.S. *Acc. Chem. Res.,* **2008**, *41*, 1486. c) Hiyama, T.. Shirakawa, E *Top. Curr. Chem.* **2002**, *219*, 61.

38.   Denmark, S. E.; Neuville, L.;. Christy, M. E. L; Tymonko, S. A. *J. Org. Chem.,* **2006**, *71*, 8500.

39.   a) Denmark, S. E.; Sweis, R. F.; Wehrli, D. *J. Am. Chem. Soc.* **2004**, *126*, 4865. b) Denmark, S. E.; Sweis, R. F. *J. Am. Chem. Soc.* **2004**, *126*, 4876.

40.   Denmark, S. E.; Tymonko, S.A. *J. Am. Chem. Soc.* **2005**, *127*, 8004.

41.   Montenegro, J.; Bergueiro, J.; Saá, C.; López, S. *Org. Lett.,* **2009**, *11*, 141.

42. a) Denmark, S. E.; Baird, J.D.; Regens, C.S. *J. Org. Chem.,* **2008**, *73*, 1440. b) Denmark, S. E. Smith, R. C.; Chang, W.-T. T.; Muhuhi, J. M *J. Am. Chem. Soc.,* **2009**, *131*, 3104.

43. a) Nokami, T.; Tomida, Y.; Kamei, T.; Itami, K.; Yoshida, J. *Org. Lett,.* **2006**, *8*, 729. b) Li, L. H.; Navasero, N. *Org. Lett.,* **2006**, *8*, 3733. c) Itami, K.; Nokami, T.; Yoshida, J. *J. Am. Chem. Soc.,* **2001**, *123*, 5600. d) Trost, B. M.; Machacek, M.R.; Ball, Z.T. *Org. Lett.,* 2003, **5**, 1895. e) Nakao, Y., Imanaka, H.; Sahoo, A.K.; Yada, A; Hiyama, T. *J. Am. Chem. Soc.,* **2005**, *127*, 6952.

44. a) Strotman, N. A.; Sommer, S.; Fu, G. C. *Angew. Chem. Int. Ed.,* **2007**, *46*, 3556. b) Powell, D. A.; Fu, G. C. *J. Am. Chem. Soc.* **2004**, *126*, 7788. c) Lee, J.-Y.; Fu, G. C. *J. Am. Chem. Soc.* **2003**, *125*, 5616.

45. Zhan, X.; Barlow, S.; Marder, S. R. *Chem. Commun.,* **2009**, 1948, and cited references.

46. Murata, H.; Malliaras, G. G.; Uchida, M.; Shen, Y.; Kafafi, Z. H. *Chem. Phys. Lett.,* **2001**, *339*, 161.

47. a) Toal, S. J.; Jones, K. A.; Magde, D.; Trogler, W. C. *J. Am. Chem. Soc.,* **2005**, *127*, 11661. b) Dong, Y. Q.; Lam, J. W. Y.; Qin, A. J.; Li, Z.;. Liu, J. Z; Sun, J. Z.; Dong, Y. P.; Tang, B. Z. *Chem. Phys. Lett.,* **2007**, *446*, 124. d) Zhao, M. C.; Wang, M.; Liu, H. J.; Liu, D. S.; Zhang, G. X.; Zhang, D. Q.; Zhu, D. B. *Langmuir,* **2009**, *25*, 676.

48. a) Kim, D.-H.; Ohshita, J.; Lee, K.-H.; Kunugi, Y.; Kunai, A. *Organometallics,* **2006**, *25*, 1511; b) Usta, H.; Lu, G.; Facchetti, A.; Marks, T. J. *J. Am. Chem. Soc.,* **2006**, *128*, 9034.

49. a) Liao, L.; Dai, L. M.; Smith, A.; Durstock, M.; Lu, J. P.; Ding, J. F.; Tao, Y. *Macromolecules,* **2007**, *40*, 9406. b) Hou, J. H.; Chen, H.-Y.; Zhang, S. Q.; Li, G.; Yang, Y. *J. Am. Chem. Soc.,* **2008**, *130*, 16144. c) DiCarminea, P. M.; Wanga, X.;. Pagenkopfa, B. L; Semenikhin, Q. A. *Electrochem. Commun.,* **2008**, *10*, 229.

50. a) Uchida, M.; Izumizawa, T.; Nakano, T.; Yamaguchi, S.; Tamao, K.; Furukawa, K. *Chem. Mater.,* **2001**, *13*, 2680. b) Lee, J. H.; Liu, Q.-D.; Motala, M.; Dane, J.; Gao, J.; Kang, Y. J.; Wang, S. N. *Chem. Mater.,* **2004**, *16*, 1869. c) Zhan, X. W.; Haldi, A.; Risko, C.; Chan, C. K.; Zhao, W.; Timofeeva, T. V.; Korlyukov, A.; Antipin, M. Y.; Montgomery, S.; Thompson, E.; An, Z. S.; Domercq, B.; Barlow, S.; Kahn, A.; Kippelen, B.; Brédas, J.-L.; Marder, S. R. *J. Mater. Chem.,* **2008**, *18*, 3157.

51. Yamaguchi, S., Endo, T.; Uchida, M.; Izumizawa, T.; Furukawa, K.; Tamao, K. *Chem.–Eur. J.,* **2000**, *6*, 1683.

52. Yamaguchi, S.; Tamao, K. *Bull. Chem. Soc. Jpn.* **1996**, *69*, 2327.

53. Tamao, K.; Uchida, M.; Izumizawa, T.; Furukawa, K.; Yamaguchi, S. *J. Am. Chem. Soc.,* **1996**, *118*, 11974.

54. a) Yamaguchi, S.; Tamao, K. *J. Chem. Soc., Dalton Trans.,* **1998**, 3693; b) Zhan, X.; Haldi, A.; Risko, C.; Chan, C. K.; Zhao, W.; Timofeeva, T. V.; Korlyukov, A.; Antipin, M. Y.; Montgomery, S.; Thompson, E.; An, Z.; Domercq, B.; Barlow, S.; Kahn, A.; Kippelen, B.; Brédas, J.-L.; Marder, S. R. *J. Mater. Chem.,* **2008**, *18*, 3157; c) Katkevics, M.; Yamaguchi, S.; Toshimitsu, A.; Tamao, K. *Organometallics,* **1998**, *17*, 5796.

55. Tamao, K.; Yamaguchi, S.; Shirot, M. *J. Am. Chem. Soc.* **1994**, *116*, 11715.

56. a) Aubouy, L.; Gerbier, P.; Huby, N.; Wantz, G.; Vignau, L.; Hirschc, L.; Janot J.-M. *New J. Chem.* **2004**, *28*, 1086; b) Aubouy, L.; Huby, N.; Hirsch, L.; van der Leec, A..; Gerbier, P. *New J. Chem.* **2009**, *33*, 1290

57. a) Kim, D.-H.; Ohshita, J.; Lee, K.-H.; Kunugi, Y.; Kunai, A. *Organometallics,* **2006**, *25*, 1511. b) Lee, T.; Jung, I.; Song, K. H.; Lee, H.; Choi, J.; Lee, K.; Lee, B. J.; Pak, J. Y.; Lee, C.; Kang, S. O.; Ko, J. Organometallics 2004, 23, 5280. (c) Ohshita, J.; Nodono, M.; Watanabe, T.; Ueno, Y.; Kunai, A.; Harima, Y.; Yamashita, K.; Ishikawa, M. J.Organomet. Chem. 1998, 553, 487. (d) Ohshita, J.; Lee, K.-H.; Hamamoto, D.; Kunugi, Y.; Ikadai, J.; Kwak, Y.-W.; Kunai, A.

Chem. Lett. 2004, 33, 892. (e) Ohshita, J.; Lee,K.-H.;Kimura, K.; Kunai, A. *Organometallics* **2004**, *23*, 5622.

58. Lee, J.; Liu, Q. D.; Bai, D. R.; Kang, Y.; Tao, Y.; Wang, S. *Organometallics* **2004**, *23*, 6205.

59. H. Jung, H. Hwang, K.-M. Park, J. Kim, D.-H. Kim, Y. Kang *Organometallics* **2010**, *29*, 2715–2723

60. (a) R. H. Friend, R. W. Gymer, A. B. Holms, J. H. Burroughes, R. N. Marks, C. Taliani, D. D. C. Bradley, D. A. Dos Santos, J. L. Bredas, M. L€ogdlund and W. R. Salaneck, Nature, 1999, 397, 121–128; (b) S. Setayesh, A. C. Grimsdale, T. Weil, V. Enkelmann, K. Mullen, F. Meghdadi, E. J. W. List and G. Leising, J. Am. Chem. Soc., 2001, 123, 946–953; (c) V. Biju, P. K. Sudeep, K. G. Thomas, M. V. George, S. Barazzouk and P. V. Kamat, *Langmuir*, **2002**, *18*, 1831–1839.

61. (a) J. Chen, B. Xu, K. Yang, Y. Cao, H. H. Y. Sung, I. D. Williamsand B. Z. Tang, J. Phys. Chem. B, 2005, 109, 17086–17093; (b) H. Murata, Z. H. Kafafi and M. Uchida, *Appl. Phys. Lett.*, **2002**, *80*, 189–191.

62. M. Wang, D. Q. Zhang, G. X. Zhang and D. B. Zhu, Chem. *Phys. Lett.*, **2009**, *475*, 64.

63. Wan, J.-H.; Mao, L.-Y.; Li, Y.-B.; Li, Z.-F.; Qiu, H.-Y.; Wang, C.; Lai, G.-Q. *Soft Matter* **2010**, *6*, 3195.

Umberto Piarulli

Prof. of Organic Chemistry

Address:

Università degli Studi dell'Insubria,
Dipartimento di Scienze Chimiche e Ambientali
Via Valleggio, 11, 222100 Como, Italy
E-mail: Umberto.Piarulli@uninsubria.it

Education:

Umberto Piarulli was born in Varese in 1965, graduated in Chemistry at Milan University in 1990 and obtained his Ph.D. in 1996 with Prof. Carlo Floriani (University of Lausanne, Switzerland).

Professional experience:

In 1996 he was appointed as researcher at Milan University in the group of professor Cesare Gennari, and in 1998 he moved to Insubria University (Como, Italy) where he became Associate Professor in 2004. In 1996 he was visiting scientist at the Cold Spring Harbor Laboratories (USA). He is author of more than 70 papers in international refereed journals. His research interests include enantioselective catalysis, organometallic chemistry, and the synthesis of combinatorial libraries of chiral ligands, the synthesis and conformational analysis of peptidomimetics and foldamers.

Recent Applications of Phosphorus Reagents: from Organic Synthesis to Stereoselective Catalysis

*Luca Pignataro,[a] Umberto Piarulli[b]**

[a]Dipartimento di Chimica Organica e Industriale, Università degli Studi di Milano, via Venezian 21, 20133 Milano, Italy
[b]Dipartimento di Scienze Chimiche e Ambientali, Università degli Studi dell'Insubria, via Valleggio 11, 22100 Como, Italy

1. INTRODUCTION

The impact of phosphorus containing compounds on modern synthetic chemistry is difficult to quantify, but one can safely assume that the study of this element has influenced all areas of chemical endeavour. Organophosphorus chemistry, as a discrete topic, is concerned with compounds containing a C-P bond and this will constitute the bulk of the applications described in this chapter. However, reagents and applications deal with oxyphosphorus and azaphosphorus species and will be also covered throughout this chapter.

Many texts on phosphorus organic chemistry have been published which would serve as a general introduction to the field.[1] A recent comprehensive work on organophosphorus compounds (including RO-P and RN-P) has appeared as part of Science of Synthesis.[2] A considerable amount of organophosphorus chemistry is present in the core literature, which can be difficult to access; the periodical Organophosphorus Chemistry, however, published annually by the Royal Society of Chemistry offers a yearly review of the highlights and key developments in the field.[3]

This chapter will focus on the use of phosphorus (III) compounds in organic synthesis, as reagents, organic catalysts, as well as ligands for transition metal catalysis. The versatility of these compounds stems from the great variety of their steric and electronic properties, which can be conveniently tuned for different applications. For this reason, after a brief presentation of the nomenclature of organic phosphorus compounds, we will discuss the methods for the determination of the steric and electronic paramenters of mainly phosphorus(III) compounds as a means to interpret their reactivity. The reactivity of P^{III} derivatives will be presented, focussing on a selection of the most recent achievements in the area of organic synthesis and stereoselective organic and transition metal catalysis.

2. NOMENCLATURE

The nomenclature of phosphorus-containing compounds is complicated to some extent by the overlap of inorganic and organic nomenclature, particularly with respect to compounds containing the P—O—H functionality. The basic nomenclature used for trisubstituted and tetrasubstituted phosphorus compounds, based on the IUPAC recommendations 1995,[4] is given in Scheme 1.

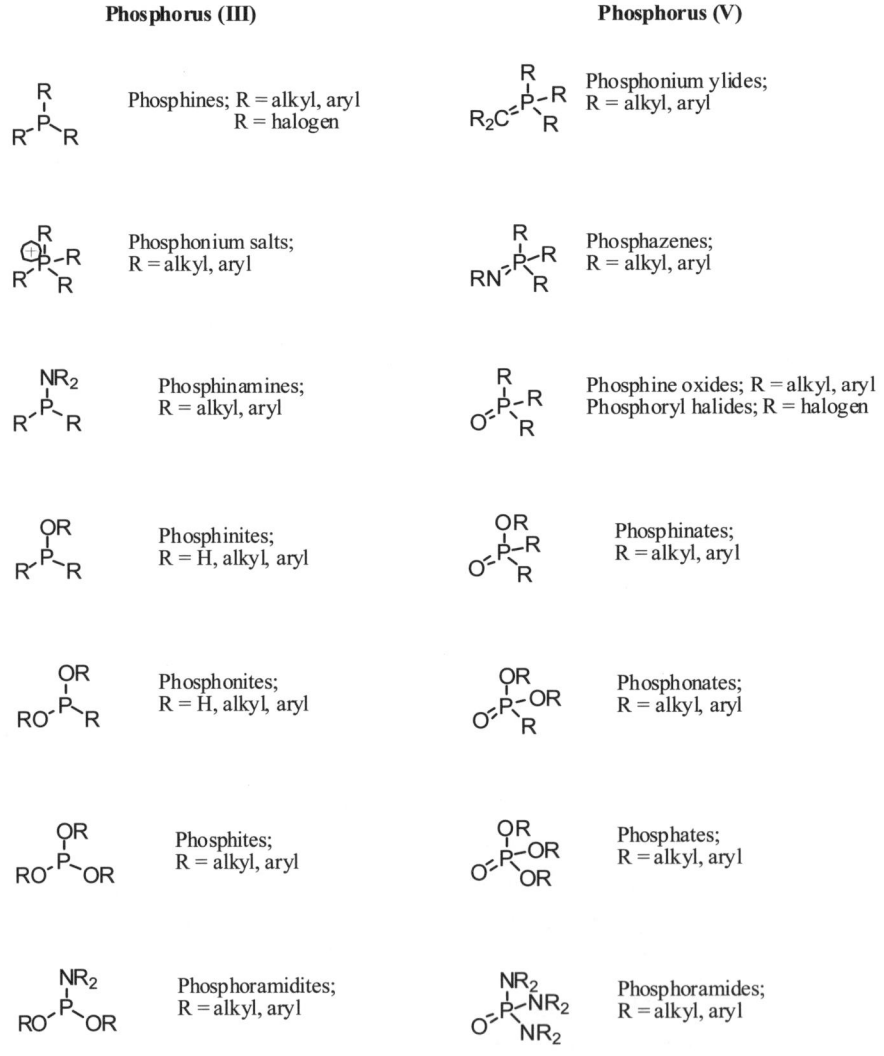

Scheme 1. Nomenclature for phosphorus compounds

Phosphines are compounds derived from PH_3 by substitution of one, two or three hydrogens with alkyl or aryl groups, and depending on the numbers of hydrogens replaced are called primary, secondary or tertiary phosphines. A specific phosphine is preferably named as a substituted phosphane [e.g. $(CH_3)_3P$ trimethylphosphane].

Phosphonium ylides, known as Wittig reagents are also named alkylidene phosphoranes (e.g. $C_6H_5CH=P(C_6H_5)_3$ benzylidene triphenyphosphorane].

3. PHOSPHINE REAGENTS IN ORGANIC SYNTHESIS

For over a century, organophosphorus chemistry was built around the interplay between tervalent tricoordinate and pentavalent tetracoordinate configurations. In this chemistry, reactions at phosphorus(III) centers are often driven by the formation of highly stable P-O single or double bonds. Indeed organophosphorus compounds are often considered reagents of choice for performing de-oxygenation, oxygen-halogen exchange reactions and Wittig-type alkene synthesis.[5] On the other hand organic phosphorus compounds are an almost endless source of easily tunable ligands for coordination chemistry and homogeneous catalysis.

In this and the following sections, rather than providing an exhaustive overview of the reactivity of phosphorus(III) reagents, which is almost impossible, we will focus on recent advances in the field of important synthetic applications of phosphines in organic synthesis, their role as organic catalysts and finally a few applications of phosphines and other phosphorus(III) derivatives as ligands for homogeneous catalytic applications.

One of the most important applications of phosphines in organic synthesis is for the Mitsunobu reaction.[6] The Mitsunobu reaction is a versatile and widely used method for the dehydrative coupling of an alcohol with an acid pronucleophile, using a combination of an oxidizing azo- reagent, most commonly an azodicarboxylate, and a reducing phosphine reagent, usually tri-phenylphosphine (TPP), under mild and virtually neutral reaction conditions.

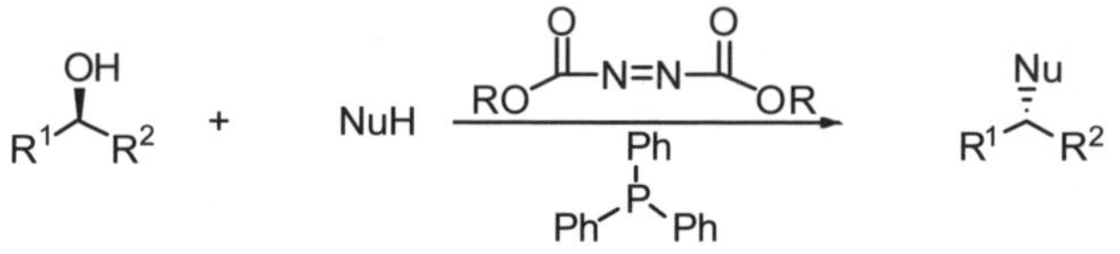

Scheme 2. General Mitsunobu reaction

The substrates are primary or secondary alcohols. Chiral secondary alcohols undergo a complete inversion of configuration unless sterically very congested. The nucleophile (or pronucleophile) is normally a relatively acidic compound containing an O-H, S-H, or an N-H group with $pK_a \leq 15$, preferably below 11. Some common nucleophiles are carboxylic acids, phenols, imides, purine/pyrimidine bases, and related heterocycles, hydrazoic acid (HN$_3$), thiocarboxylic acids, thiols, fluorinated alcohols, and hydroxamates. Phosphoric/phosphonic acids can also be used. Diphenylphosphoryl azide (DPPA), trimethylsilyl azide, or zinc azide, can be employed in place of the hazardous HN$_3$. With suitable modification of the phosphine, if necessary, one can use even C-H-based nucleophiles such as malonate esters, β-diketones, β-keto esters.

The preferred phosphine reagent is triphenylphosphine (Ph$_3$P) or tributylphosphine (nBu$_3$P), both commercially available and quite cheap. The triphenylphosphine oxide byproduct as well as unreacted Ph$_3$P are water-insoluble, and very often chromatography is required for the separation of the products. This is one of the major limitations in Mitsunobu chemistry. Trimethylphosphine (Me$_3$P) is volatile but a safety hazard. Although nBu$_3$P works well in many cases, this phosphine is still not as popular as triphenylphosphine. In alternative, diphenyl(2-pyridyl)phosphine (1), (4-dimethylaminophenyl) diphenylphosphine (2) or tris-(4-dimethylaminophenyl) phosphine have been used. In these cases, the corresponding phosphine oxide can be removed by washing the reaction mixture with dilute hydrochloric acid. If the required product is fairly soluble in the chosen organic solvent, such as toluene or THF, a diphosphine such as 1,2-diphenylphosphinoethane (DPPE, 4) may be a better choice, since the byproduct phosphine oxide is insoluble and hence can be removed by filtration.

A rather complicated mechanistic picture has been recently depicted for this reaction, which is still actively investigated.[7] The first step is the reaction of dialkyazodicarboxylate and the phosphine (e.g. triphenylphosphine) leading to the irreversible formation of the Morrison-Brunn-Huisgen (MBH) betaine 5, whose identity has been established by multinuclear NMR [^{31}P NMR: δ 44.2 (R = iPr)] and ESI-MS. In step 2, this betaine deprotonates the nucleophile to form the ionic species 6, which upon reacting with the alcohol forms the key alkoxyphosphonium salt 8 and the dialkylhydrazodicarboxylate derivative RO$_2$CNH-NHCO$_2$R (DAADH$_2$). Alternatively, betaine 5 can react first with the alcohol (depending on the order of addition) to lead to the pentacoordinate phosphorane 7; this phosphorane may also lead to the alkoxyphosphonium salt 8 upon reaction with the acid. The substitution product is then formed with inversion of configuration, when a secondary alcohol is used. In cases where the substitution product with retention is observed, the intermediacy of the acylphosphonium salt 9, which is in equilibrium with **8** and the phosphorane **10**, is invoked. Formation of the anhydride is a complicating factor in some cases.

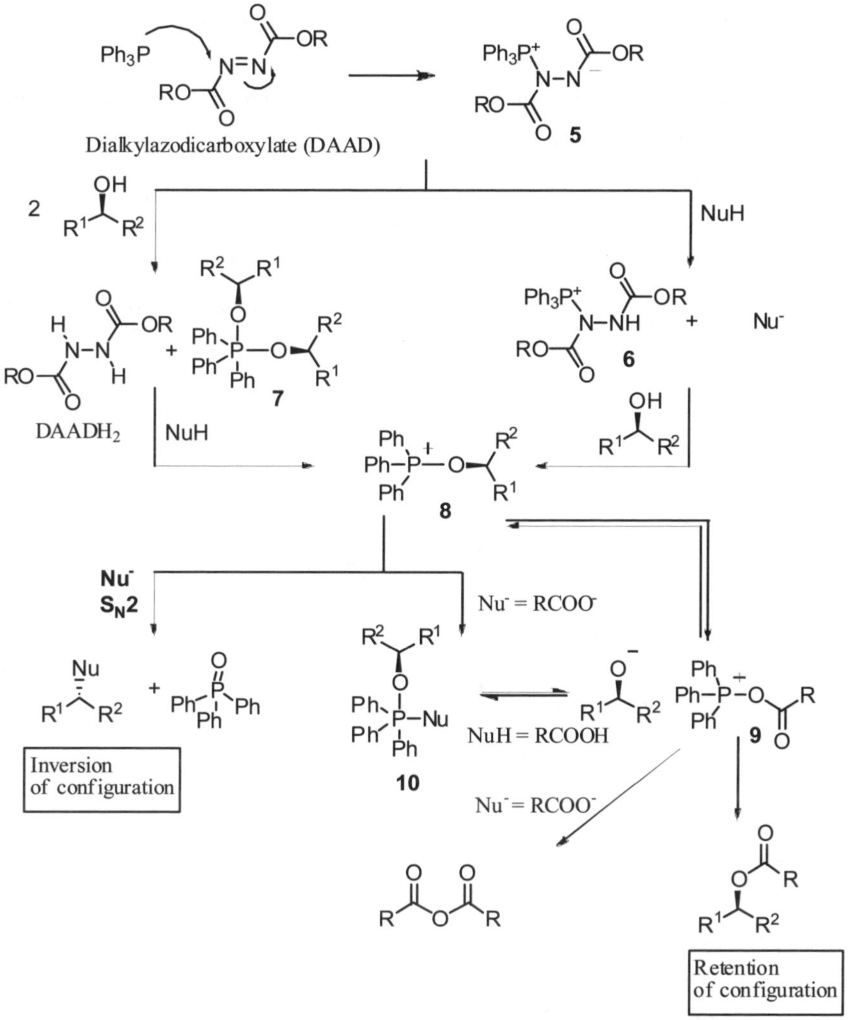

Scheme 3. Postulated mechanism for the Mitsunobu reaction

One of the major drawbacks of the Mitsunobu reaction is that stoichiometric quantities of phosphine oxide and reduced hydrazide are formed as byproducts, often making isolation of the desired product difficult, expensive and time consuming. The use of polymer supported and fluorous reagents has been explored recently, to alleviate this problems. A new strategy was also recently introduced for addressing the issue of product purification. In this approach the role of the azodicarboxylate was changed from a stoichiometric oxidant to a catalyst, using it in conjunction with a terminal oxidant [PhI(OAc)$_2$ was used in this study] whose by-products were relatively simple to separate from the desired product.[8] The major drawback of this approach was the nucleophilicity of the acetate which yielded up to 10% of the corresponding acetate ester.

Alcohols react with carboxylic acids smoothly at room temperature (or lower) to afford the esters in good yields. When a chiral secondary alcohol is used, configurational inversion of alcohol occurs under mild and essentially neutral conditions. Hydrolysis of the product subsequent to esterification affords the inverted alcohol, generally in high enantiomeric purity. In cases where the alcohol is sterically hindered, retention of configuration may be favored. The Mitsunobu reaction effectively discriminates between alcoholic and phenolic hydroxyls, and some selectivity for primary alcohols with respect to sterically encumbered secondary alcohols is also observed. Due to its mild reaction conditions, the Mitsunobu reaction has been used for lactonization and macrolactonization particularly for natural product synthesis.[9]

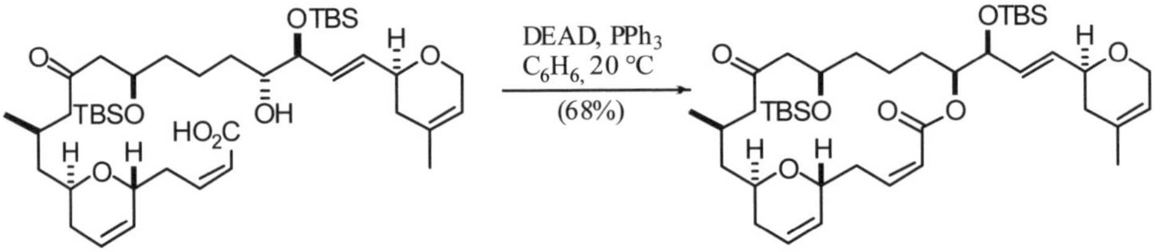

Scheme 4. Mitsunobu macrolaconization in the synthesis of (−)-laulimalide (ref.)

Organophosphates/phosphonates and sulfates can also undergo Mitsunobu esterification. In these cases, the use of the corresponding trialkyl ammonium salts can be beneficial, particularly in the presence of acid sensitive substrates.

Phenols, and alcohols with strongly electron withdrawing groups attached to the carbon, can act as nucleophiles in the Mitsunobu coupling. These reactions will lead to cyclic or acyclic ethers, depending on whether the reaction is intra- or intermolecular.

The utility of the Mitsunobu reaction in the conversion of alcohols to amines using acidic nitrogen derivatives as nucleophiles is well-known. Phthalimide and related compounds, including maleimides, can readily take part in *N*-alkylation. The resulting imides can be deprotected with hydrazine hydrate or in milder conditions by treatment with methylamine. In a reaction reported by Morita and Krause, the allenyl alcohol **11** was converted to the *R*-aminoallene **12** in good yield with complete inversion.[10]

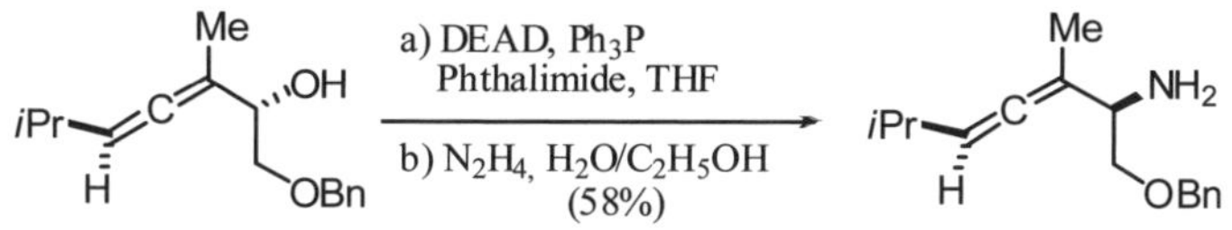

Scheme 5. *N*-alkylation by Mitsunobu reaction

In the Fukuyama-Mitsunobu reaction, the *N*-alkylation of secondary sulfonamides with alcohols is obtained using either nitrobenzenesulfonamide (2- or 4-substituted) or 2,4-dinitrobenzenesulfonamide as a pronucleophile. After the *N*-alkylation, the sulfonamide portion can be readily removed by treatment with thiols. Besides the usual combination of PPh$_3$ and diethylazodicarboxylate (DEAD) other reagent pairs have been explored: *n*Bu$_3$P plus *N,N,N',N'*-tetramethylazodicarboxylate (TMAD) in the presence of *i*Pr$_2$EtN, Me$_3$P and 1,1'-(azodicarbonyl)dipiperidine (ADDP) and finally Et$_3$P and DEAD gave better results in several cases, because of the higher reactivity of the more basic phosphines. In some cases the traditional phosphines could be replaced by a more readily hydrolytically degradable phosphite.[11] In fact, the use of phosphites would generate phosphates, which are more water soluble than their counterpart phosphine oxides.

Hydrazoic acid or a suitable azide source such as trimethylsilyl azide (Me$_3$SiN$_3$), diphenyl phosphoryl azide [(PhO)$_2$P(O)N$_3$, DPPA], zinc azide, sodium azide, or nicotinoyl azide do take part in Mitsunobu coupling with alcohols. Since the resulting azides can be readily transformed to other functional groups, this protocol has been widely utilized in various syntheses, as can be realized from the following discussion. Carbon nucleophiles with a reasonably low pK_a value also give rise to the Mitsunobu alkylation. Several pronucleophiles can be used, such as alkyl cyanoacetate, *o*-nitroarylacetonitriles and triethylmethanetricarboxylate (TEMT). In the latter case, the best results were obtained with the sterically less crowded phosphine Me$_3$P. Lithium cyanide or acetone cyanohydrin can be used in a Mitsunobu alkylation in the presence of *n*Bu$_3$P and TMAD.

The Staudinger reaction is a very mild azide reduction, involving the reaction of the azide with a phosphine.[12] Mechanistic studies of the classical reaction between a phosphine and an organic azide (R' = alkyl or aryl) suggest that the lone pair electrons of the phosphine attack the terminal nitrogen atom of the azide to yield a linear phosphazide intermediate 13. This intermediate can undergo intramolecular rearrangement via a four-membered ring transition state to yield a second intermediate, the phosphazene 14 (also known as iminophosphorane or phosphine imine), with concomitant loss of N$_2$. In the presence of water, 14 undergoes hydrolysis to yield the amine and phosphine oxide.

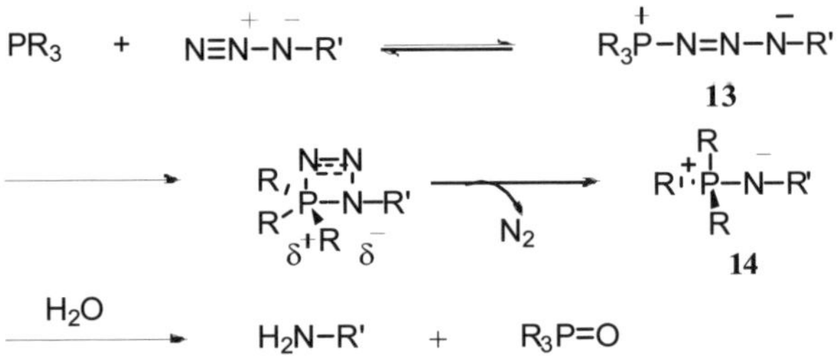

$$PR_3 \quad + \quad N{\equiv}\overset{+}{N}{-}\overset{-}{N}{-}R' \quad \rightleftharpoons \quad R_3\overset{+}{P}{-}N{=}N{-}\overset{-}{N}{-}R'$$

13

Scheme 6. Mechanism of the Staudinger azide reduction

The intermediate phosphazene **14** can undergo several other reactions depending on the electrophile: it can react with carbonyls such as aldehydes or ketones, but also with isocyanates, isothiocyanates, ketenes, CO_2 and CS_2 to substitute the C=O bond by a C=NR bond in an aza-Wittig reaction; in alternative, reactions with carboxylic derivatives (acids, esters, anhydrides, acyl halides and other reagents) afford the corresponding carboxamide derivatives and these have gained a prominent role for the chemical ligation of biomolecules and small organic molecules and the creation of bioconjugates.

In the aza-Wittig reaction,[13] the reaction between the phosphazene **14** and a carbonyl compound is likely to proceed, similarly to the Wittig reaction, through the formation of a 1,3,2-oxaazaphosphetidine which rearranges to the imine and the phosphine oxide. X-Ray structural determination of a 1,3,2-oxaazaphosphetidine showed that the phosphorus atom is in a trigonal-bipyramidal structure.[14]

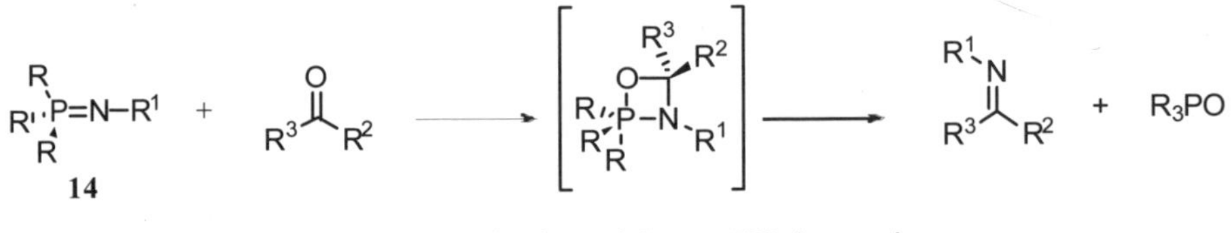

Scheme 7. Mechanism of the aza-Wittig reaction

These results were further corroborated by DFT calculations showing the formation of the 1,3,2-oxaazaphosphetidine intermediate.

From a synthetic point of view the reaction of simple or functionalized phosphazenes with aldehydes and ketones, although in this case the reactions are best performed with the more reactive and less hindered trimetylphosphine. One of the major drawbacks of this reaction is the limited stability of the imine products and their purification from the phosphine oxide by-products.

In situ reduction of the imines by hydride sources yields the corresponding secondary amines. Alcohols have been used for an indirect aza-Wittig reaction, using a dehydrogenation/hydrogenation iridium catalyst.[15]

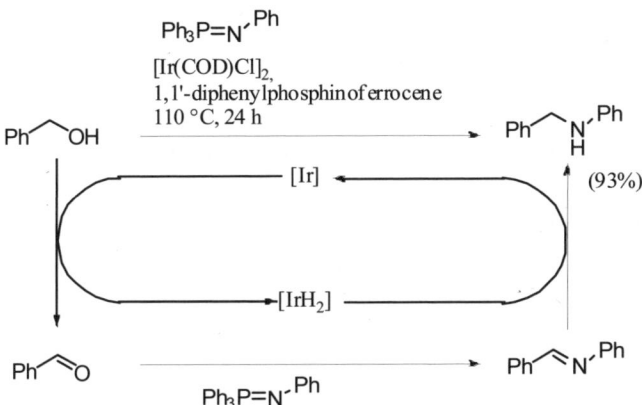

Scheme 8. Mechanism of the indirect aza-Wittig reaction

The reaction of phosphazenes with carbon dioxide or carbon disulfide has been used extensively for the formation of isocyanates or isothiocyanates, respectively; whereas the reaction with isocyanates or isothiocyanates affords carbodiimides.

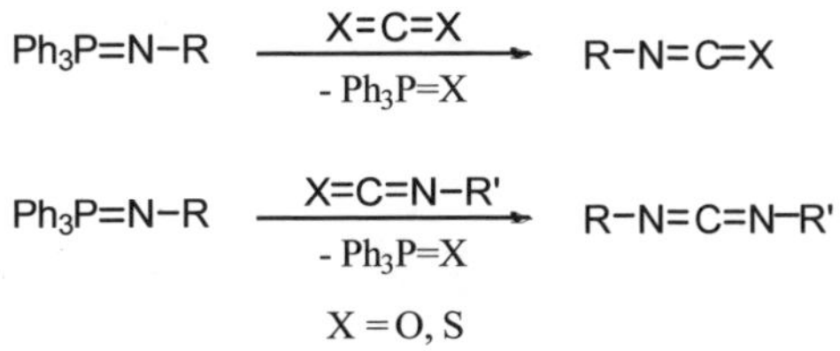

$$Ph_3P=N-R \quad \xrightarrow[-Ph_3P=X]{X=C=X} \quad R-N=C=X$$

$$Ph_3P=N-R \quad \xrightarrow[-Ph_3P=X]{X=C=N-R'} \quad R-N=C=N-R'$$

$$X = O, S$$

Scheme 9. Aza-Wittig reaction with CO_2 and CS_2

Ketenes, which are even more reactive than isocyanates, react through an aza-Wittig process to afford ketenimines. These can undergo intramolecular electrocyclic cyclization to yield heterocyclic derivatives.

As we stated above, an alternative synthetic evolution of the phosphazenes, derived from the Staudinger reduction of azides, is the reaction with carboxylic derivatives to afford the corresponding carboxamides. This methodology has more recently found important applications in the ligation techniques for the preparation of biologically active molecules. In an initial discovery, carboxylic acids, organic azides and triphenylphosphine (which could be later substituted by the more reactive nBu_3P and Et_3P) reacted slowly in benzene to give the carboxamide products together with release of N_2 and phosphine oxide (another advantage of using Et_3P is that $Et_3P=O$ is water soluble and it is therefore more easily removed).[16]

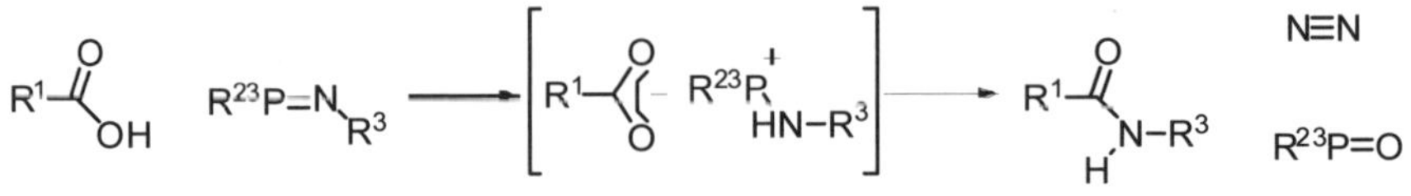

Scheme 10. Reaction of phosphazenes with carboxylic acids

A number of groups examined and optimized the reactions of Staudinger phosphazenes with carboxyl derivatives such as anhydrides,[17] thioesters,[18] acyl chlorides,[19] simple and activated esters, Boc-anhydride and BocON.[20]

This procedure was applied earlier to the synthesis of peptides and later to the direct ligation of carboxylic derivatives to azides for the construction of biologically relevant molecules.[21] The group of Bertozzi applied the Staudinger ligation to the intramolecular trapping of the phosphazene by an electrophilic ester, leading to a covalent amide bond.[22]

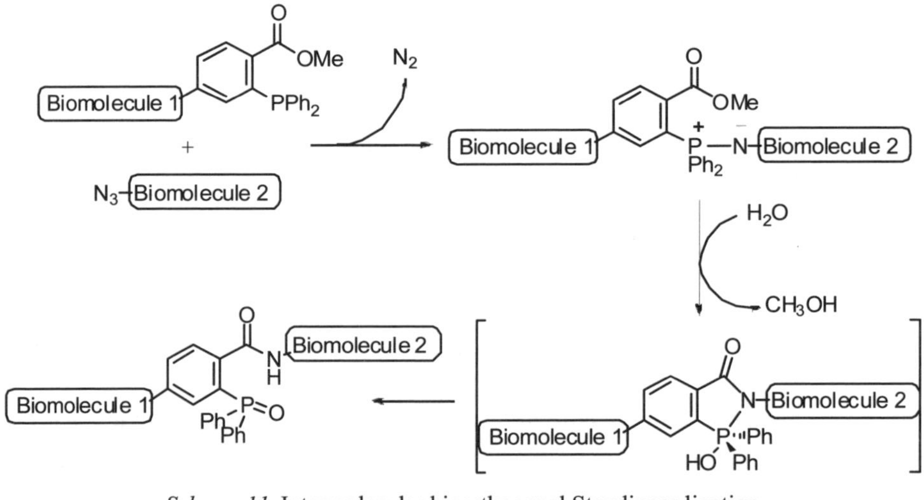

Scheme 11. Intramolecular bioorthogonal Staudinger ligation

This mutually selective reaction between two abiotic groups has also been applied to *in vivo* ligation. The intramolecularity of the process is a central feature of this reaction, which allows the reaction to occur in the aqueous environment, because otherwise the phosphazene intermediate would simply hydrolyze to afford the corresponding and phosphine oxide.

In a different approach, a very interesting case is the so-called traceless Staudinger ligation developed by Raines and co-workers,[23] which joins a peptide having a *C*-terminal phosphinothioester with a peptide having an *N*-terminal azide through an intermediate iminophosphorane. (Diphenylphosphino)methanethiol is the most often used of known reagents for effecting the traceless

Staudinger ligation, and has mediated the orthogonal assembly of a protein, site-specific immobilization of peptides and proteins, and synthesis of glycopeptides.[24]

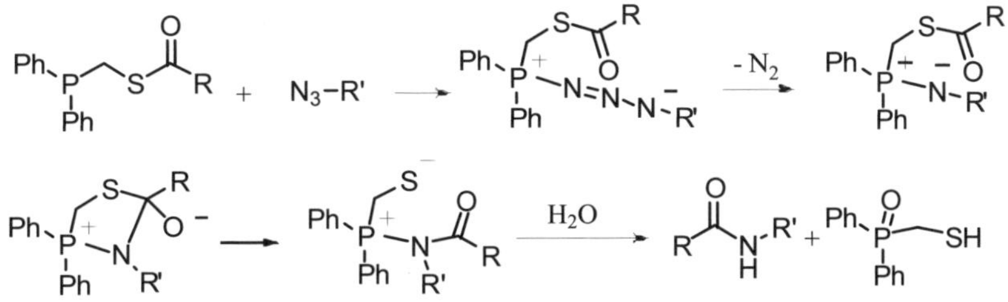

Scheme 12. Putative mechanism of the traceless Staudinger ligation

Another traceless Staudinger ligation was developed by the group of Bertozzi, using an o-diphenylphosphinophenol to promote the formation of the phosphazene and the subsequent intramolecular ligation.[25] The synthesis of the o-diphenylphosphinophenol was obtained by Pd-mediated coupling of diphenylphosphine with 2-iodophenol.

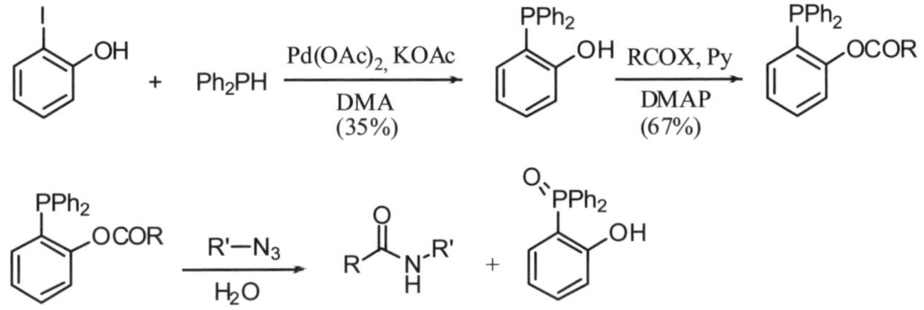

Scheme 13. Traceless Staudinger ligation developed by Bertozzi and co-workers

This reaction has been employed in a wide range of applications, including the glycoprotein synthesis[26] and the stereoselective synthesis of N-glycosyl amino acids.[27]

4. PHOSPHINES AS ORGANIC CATALYSTS

Nucleophilic phosphine organocatalysis has emerged as a versatile synthetic method.[28] Generally, tertiary phosphine-mediated reactions start via nucleophilic addition of phosphines to activated olefins, allenes and alkynes. The resultant zwitterionic intermediates react with electrophiles. Based on this mechanistic insight, nucleophilic phosphine organocatalysis has the following important features: (1) the nucleophilicity of phosphines may be easily tuned by varying the substituents, ranging from the trialkylphosphines to aryl substitution, in order to obtain a suitable catalyst for a given reaction. Furthermore, both steric and electronic properties of the phosphines may be altered, which sometimes enables fine tuning of the reaction regioselectivity; (2) in addition to intermolecular reactions, intramolecular annulation via pre-organized acyclic substrates may be accomplished; (3) chiral acyclic and cyclic phosphine ligands are readily available for screening to tackle enantioselective synthesis; (4) another important aspect of this methodology is that it is completely free from contamination by heavy metals, which is an especially attractive feature for industrial synthesis.

Relevant examples are the isomerization of alkynes into dienes (reaction a), the Rauhut–Currier reaction (reaction b), Morita–Baylis–Hillman (MBH) type reactions (reaction c), the annulation of allenic substrates with olefins (reaction d), and nucleophilic additions (reaction e).

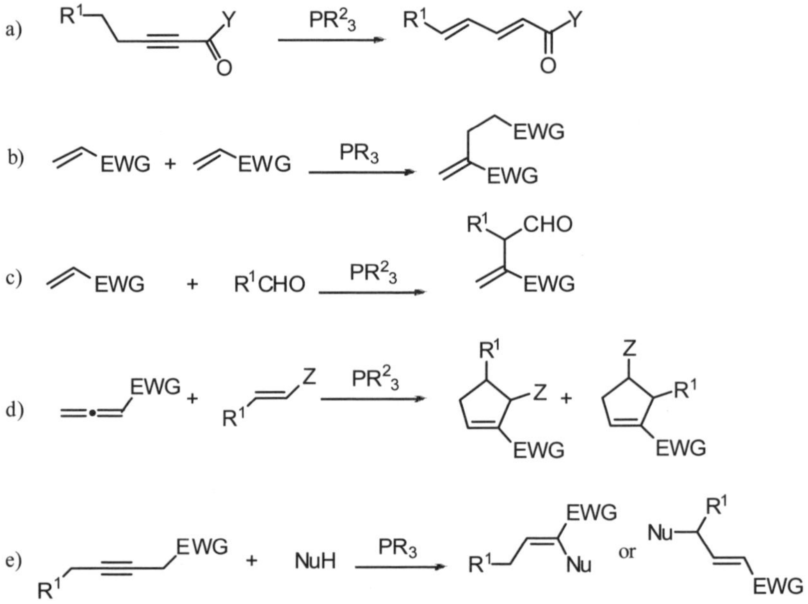

Scheme 14. Representative examples of phosphine-promoted reactions

While the isomerization of alkynones to the corresponding conjugated dienones catalyzed by transition metal complexes had been reported in the late 1980s by numerous research groups, in 1992 Trost and Kazmaier discovered that this isomerization could be catalyzed by phosphines themselves, without the need of a metal center.[29] In the presence of substoichiometric amounts (0.1 to 0.4 equivalents) of triphenylphosphine, electron-poor alkynes, including alkynones, alkynoates and alkynamides, isomerized smoothly in toluene at 80 to 110 °C to afford the corresponding conjugated dienones, dienoates and dienamides, respectively, in high yields.[30] The order of reactivity was established to be alkynones > alkynoates > alkynamides. The latter required higher temperatures and the presence of 0.5 equivalents of acetic acid, and were found to react more easily in the presence of the more basic nBu$_3$P. In addition, the isomerization reaction was found to be highly chemoselective, with no isomerization observed for electron-rich alkynes. Furthermore, it was found to be highly stereoselective, with an (E,E)-1,3-diene usually being the only product formed.

The current generally accepted mechanism for the alkyne to 1,3-diene isomerization reaction, involves an initial nucleophilic addition by the phosphine catalyst to the alkyne, followed by proton shifts, and final elimination of the catalyst to afford the final product. Allenes can be intermediates of the reaction, but they have been observed to yield the same (E,E)-1,3-diene products as do the corresponding alkynes when subjected to the same conditions.

Due to the mild conditions required, this transformation has been applied in several natural product syntheses, where the resulting diene is either an intermediate which is further transformed, or an actual structural component of the target compound.[31]

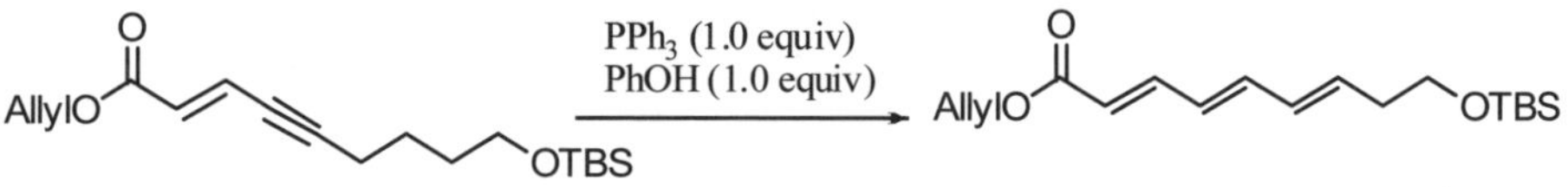

Scheme 15. Application of the alkyne-diene isomerization to the synthesis of natural products

The Rauhut-Currier reaction[32] (also known as the vinylogous MBH reaction), firstly reported in 1963, involves the coupling of one activated double bond to a second Michael acceptor, creating a new C–C bond between the α-position of one activated alkene and the β-position of a second alkene under the influence of a nucleophilic catalyst. The transformation is believed to proceed via reversible conjugate addition of a nucleophilic catalyst (e.g., either a trialkylphosphine or a triarylphosphine) to an activated alkene to generate the zwitterionic species. A Michael reaction of the enolate with a second equivalent of the activated alkene generates, after a prototropic shift and the extrusion of the phosphine catalyst, the Rauhut-Currier coupling product.

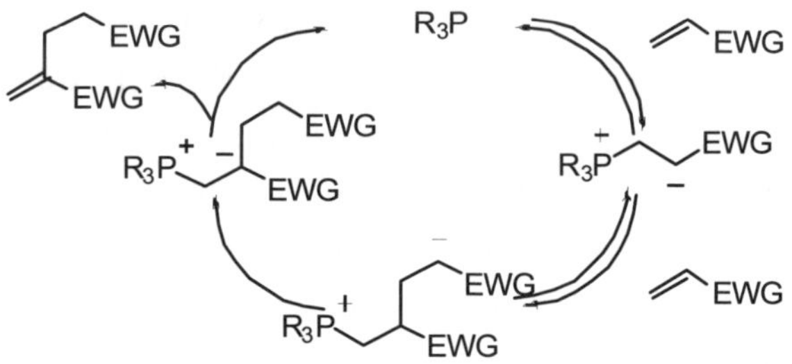

Scheme 16. Proposed mechanism of the Rauhut-Currier reaction

Several applications of this reaction have been reported involving the dimerization of activated alkenes, as well as the selective cross-coupling of different compounds and, more importantly, the intramolecular version leading to the formation of 5- and 6-membered cyclic compounds.[33] In these applications, trialkylphosphines catalyze the cycloisomerization of both symmetrical and unsymmetrical substrates with high efficiency. Amine-based nucleophiles, such as DABCO, DBU, Et₂NH and DMAP, commonly employed in the MBH reaction of activated alkenes and aldehydes were demonstrated to be much less efficient than phosphine-based nucleophiles. Additionally, while trialkylphosphines (10 mol%) were sufficiently reactive to catalyze the transformation, PPh₃ (100 mol%) demonstrated inferior reactivity.

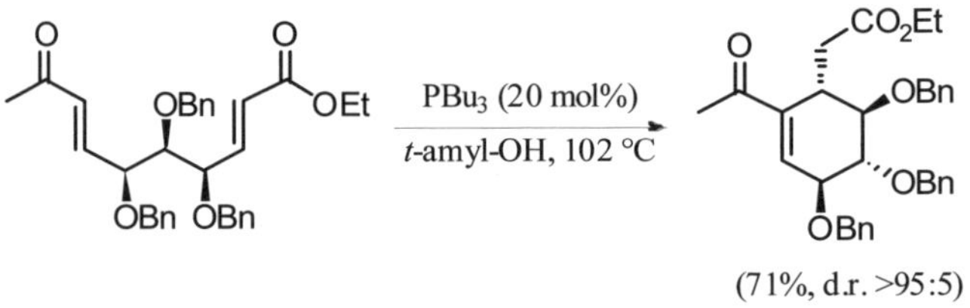

(71%, d.r. >95:5)

Scheme 17. Intramolecular Rauhut-Currier reaction

The Morita–Baylis–Hillman[34] (MBH) is described as the coupling between the α-position of an activated double bond and an sp² electrophilic carbon (typically an aldehyde, but also an imine) using an appropriate catalyst, normally Lewis bases. Great progress has been made in the execution of the MBH reaction, since the seminal report in 1972 described the reaction of acetaldehyde with ethyl acrylate and acrylonitrile in the presence of catalytic amounts of tertiary amines or tertiary phosphines and myriads of synthetic applications of this reaction are reported in several excellent reviews. Only a few phosphine organocatalysts are effective in the classical MBH reaction, whereas phosphines have found more suitable applications in the analogous couplings of electron-deficient olefins with imines, known as aza-MBH reactions.[35] From the mechanistic viewpoint, the initial addition of the nucleophilic catalyst affords a zwitterionic enolate intermediate which attacks the sp² electrophilic carbon to yield a second zwitterionic alkoxide intermediate. A proton transfer is now necessary to eliminate the catalyst, restore the double bond and protonate the alkoxide. This is the rate determining step according to recent mechanistic studies.[36] As a consequence, bifunctional catalysts possessing besides the nucleophilic site, which initiates the reaction by addition to the double bond, also a Brønsted acid functionality (e.g. a phenol, an amide, a thiourea) have in fact been found particularly effective in enhancing the reaction rates and also controlling the stereoselectivity. In the MBH reaction, chiral amines normally outperform phosphines in the stereoselectivity of the reaction; one particularly effective phosphine-thiourea catalyst was developed by Wu and co-workers,[37] and showed excellent enantiomeric excesses under mild conditions in relative short reaction time for the MBH reaction of various aromatic aldehydes with methyl vinyl ketone (MVK).

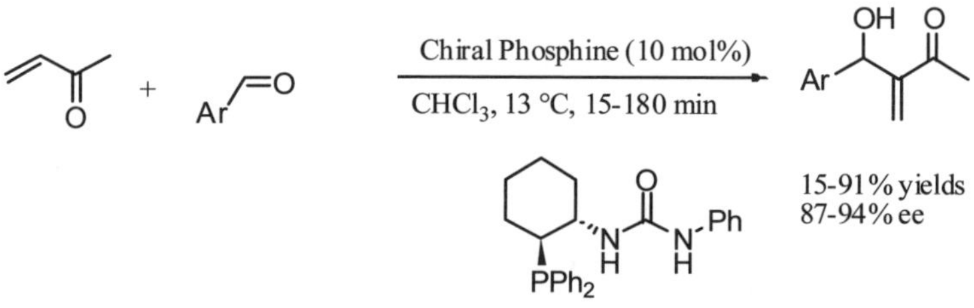

Scheme 18. Phosphine-(thio)urea bifunctional chiral phosphine-catalyzed asymmetric MBH

Chiral bifunctional phosphines have found more suitable applications in the analogous couplings of electron-deficient olefins with imines, known as aza-MBH reactions. In particular, the group of Shi[38] demonstrated that 1,1'-bi-2,2'-naphthyl-derived chiral bifunctional phosphines containing a PPh_2 Lewis base and a Ar-OH Brønsted acid moieties could be used as an effective catalyst in asymmetric aza-MBH reaction of N-tosyl imines with MVK and phenyl acrylate, providing the corresponding adducts in good yields with high ee's.

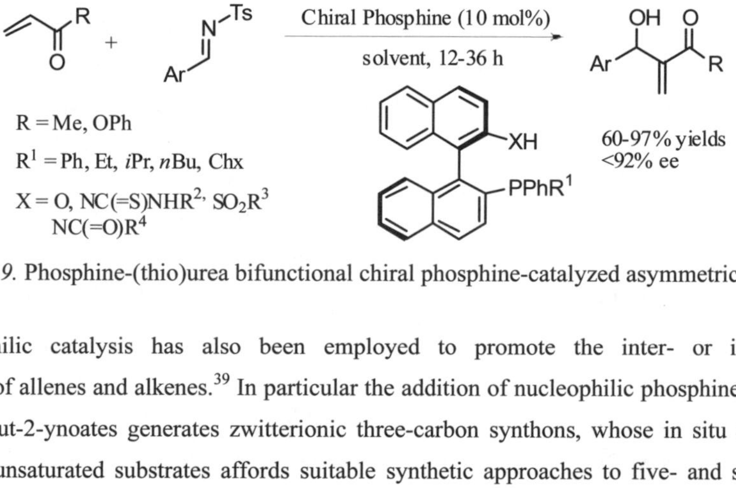

R = Me, OPh
R^1 = Ph, Et, iPr, nBu, Chx

X = O, NC(=S)NHR^2, SO$_2$R^3
NC(=O)R^4

60-97% yields
<92% ee

Scheme 19. Phosphine-(thio)urea bifunctional chiral phosphine-catalyzed asymmetric MBH

Nucleophilic catalysis has also been employed to promote the inter- or intramolecular cycloaddition of allenes and alkenes.[39] In particular the addition of nucleophilic phosphines to buta-2,3-dienoates or but-2-ynoates generates zwitterionic three-carbon synthons, whose in situ trapping with electron-poor unsaturated substrates affords suitable synthetic approaches to five- and six-membered carbo- and heterocycles. Chiral phosphines have recently been used as enantioselective catalysts with moderate to good enantiomeric excesses.[40]

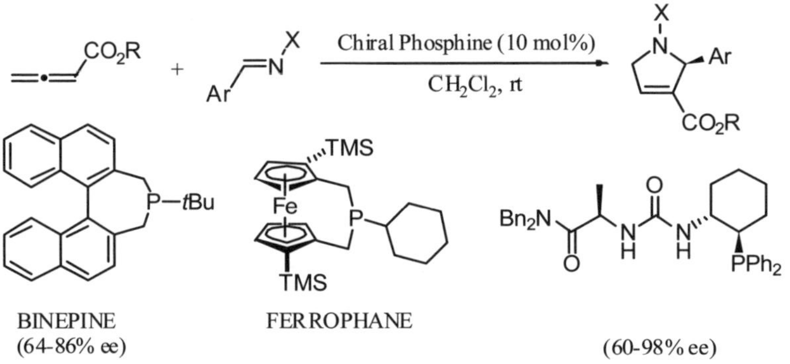

Scheme 20. Enantioselective, phosphine catalyzed cycloaddition of allenoates and double bonds

Another important though mechanistically different application of phosphine organocatalysis is the alcohol acylations with acid anhydrides. These are believed to involve the formation of *P*-acylphosphonium carboxylates, which behave as activated electrophilic intermediates. Enantioselective variants of these processes have been implemented in the form of either the kinetic resolution of chiral alcohols or the desymmetrization of *meso*-diols and are essentially related to the work of Vedejs and co-workers.[41,40b] Kinetic resolution of aryl-substituted and allylic secondary alcohols has been achieved using isobutyric anhydride (considerably more enantioselective) rather than acetic or benzoic anhydrides. Under phosphine catalysis, these reactions take place under very mild conditions (–40 °C) with enantioselectivities that may compare favorably with those obtained using enzymatic catalysts.

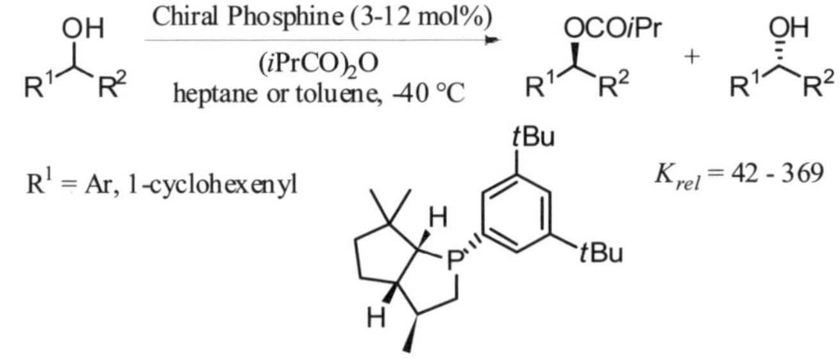

Scheme 21. Kinetic resolution of alcohols by phosphine catalysis

The desymmetrization of *meso*-hydrobenzoin via benzoylation of the alcohol functionality has also been studied. Globally, this process is less efficient than the kinetic resolution of simple alcohols, leading to an enantiomeric excess of 94% (70% yield) under conditions where, however, a significant amount of the corresponding dibenzoate is also formed.

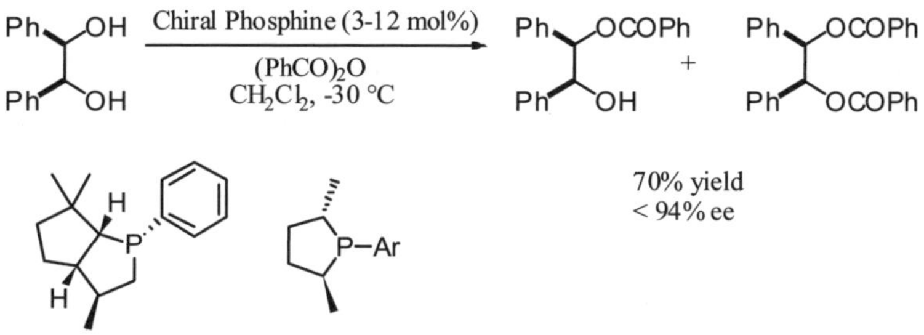

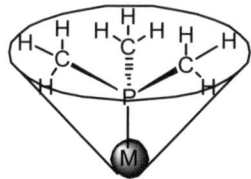

Scheme 22. Desymmetrization of hydrobenzoin by phosphine catalysis

5. PHOSPHORUS(III) LIGANDS FOR TRANSITION METAL CATALYSIS

5.1 Steric and electronic descriptors for phosphorus(III) derivatives

The reactivity of phosphorus(III) derivatives (phosphines, phosphites etc.) as reagents for organic synthesis, organocatalysts and ligands for transition metal catalysis, are highly influenced by their steric and electronic properties. This is particularly evident for the catalytically active metal complexes, whose electron density and steric saturation are actively manipulated by ligands coordinated to the metals. For this class of compounds, the analysis of ligand effects (for the metal complexes) expressed as linear free energy relationships, relying to some extent on experimentally measured descriptors, has a long tradition, building on Tolman's seminal work, and extended by the Quantitative Analysis of Ligand Effects (QALE). More recent work derived suitable descriptors from techniques of computational chemistry, allowing the theoretical evaluation of both established and novel ligands from parameters which can be calculated rapidly and efficiently.

Tolman's legacy consists in the definition of two parameters, the steric parameter (the cone angle θ) and Tolman's electronic parameter (TEP ν).[42]

The cone angle is defined as the apex angle of a cylindrical cone centred at a distance of 2.28 Å from the donor atom (P), corresponding to an average of a range of M-P bond lengths, and extended to touch the van der Waals radii of the outmost atoms (Figure 1).

Figure 1. Tolman cone angle

Cone angle of unsymmetrically substituted ligands were also defined: these used half angles $\theta_i/2$ determined for each substituent by measuring the angle between the P-M vector (M positioned at 2.28 Å from the P donor atom) and the vector from M to the outermost atom of the substituent. Tolman then assumed that contributions were additive to give:

$$\theta = \frac{2}{3}\sum_{i=1}^{3}\frac{\theta_i}{2}$$

These cone angle have been revised based on crystallographic and calculated structures.

The Tolman electronic parameter (TEP), is based on infrared spectroscopic measurements and corresponds to the vibrational frequency (ν) of the A_1 carbonyl stretching mode of [LNi(CO)$_3$] complexes in in dichloromethane solution which, according to Tolman, can be measured with an accuracy of ±0.3 cm^{-1} for a wide range of ligands. In this way, TEP is reported to capture the net electronic effect of the ligands investigated.

In addition, Tolman presented a steric and electronic map of phosphorus donor ligands by plotting ν versus θ. The cone angle and the electronic parameter have become very common characteristics when defining ligand properties, while also the notion of ligand map has become of widespread use and is certainly a guiding principle for many people when considering ligand effects.

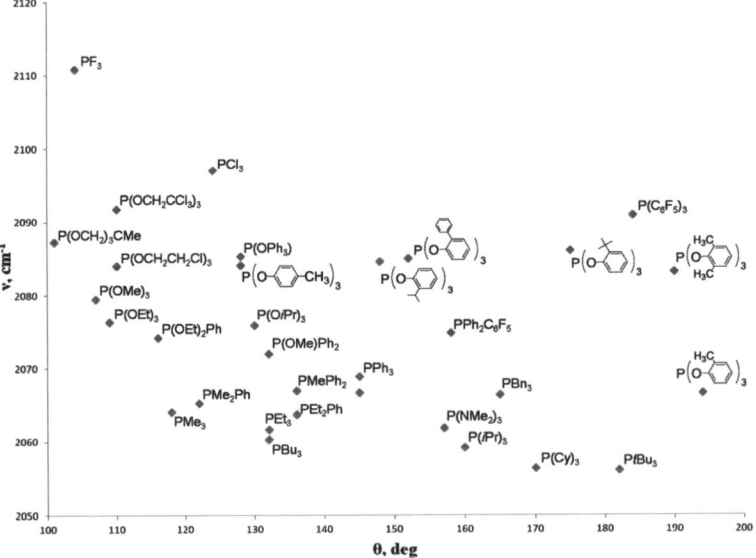

Figure 2. A scatter plot of TEP (ν) *vs* cone angle θ

Despite this widespread use, Tolman's legacy of cone angle and TEP suffer from several drawbacks. In the case of the steric parameter, Tolman's approach of minimizing cone angles by folding all substituents inward is often inappropriate when compared to calculated and crystallographically observed geometries, especially for sterically demanding and/or conformationally flexible ligands. On the electronic side, the TEP and carbonyl stretching frequencies are of limited use as electronic descriptors, due to the fact that the metal carbonyl complexes of some interesting ligands are too toxic or unstable to be readily prepared. With the advent of powerful computational methods, several approaches aimed at developing individual descriptors, derived from calculation and seeking to separate steric and electronic effect have been developed.[43] As for the ligand size, in early studies the role of computation was generally limited to providing ligand and transition metal complex geometries for cone angle measurements, along with approximate energetic weightings. More recently several steric descriptors have been reported which make use of molecular mechanics (QM) or hybrids quantum mechanics/molecular mechanics (QM/MM) calculations on the simple ligands or on their metal complexes to determine the steric contribution of the ligands. These parameters correlate rather well to Tolman's cone angle, although significant deviations are found for small ligands. While Tolman's concept of a cone angle remains perhaps the most widely used measure of ligand steric bulk, more recently proposed descriptors take better into consideration a ligand's conformational responsiveness to changes in the coordination environment. Combining calculated structural data with an exploration of conformational preferences, whether for cone angles or other descriptors, allows the consideration of novel ligand architectures before synthesis.

Computationally derived ligand electronic parameters, based on calculated vibrational frequencies in the familiar Tolman's complex [LNi(CO)$_3$], have been proposed by Clot and co-workers[44] and Cundari, White and co-workers,[45] whereas Suresh and co-workers[46] have developed a molecular electrostatic potential (MESP) approach as a measure of the electronic effects in substituted P-donor ligands. Excellent correlation factors to TEP and to the pK_a values of protonated phosphines were reported for the MESP.

Finally a combined steric and electronic approach has been recently approached by Orpen, Harvey and co-workers, who described the design of DFT-calculated descriptors for monodentate phosphorus(III) donor ligands in a range of representative complexes.[47] Using the resulting data, a ligand space was mapped, by principal component analysis (PCA) and predictive models are derived by interpretation of principal components in terms of established steric and electronic properties.

It must be underlined TEP or any other single descriptor cannot adequately represent all the electronic properties of a given ligand. For instance, two ligands may be roughly equally electron donating in a in a [Ni(CO)$_3$] complex, but one of them may be much more electron donating in a Fe(II) or Rh(III) complex. The reason for such a behaviour can be found in the nature of the ligand coordination bond to a metal which is usually accepted to consist of both σ-donation of the phosphorus lone pair into an empty orbital of the metal and π-back-donation from filled orbitals on the metal to empty orbitals of appropriate π-symmetry with respect to the metal-P bond on the ligand. The latter orbitals are now generally accepted to be a combination of P-R σ* orbitals, although earlier work speculated on the role of vacant d orbitals on the P atom. The importance of back bonding for simple alkyl and aryl phosphines has been matter of debate, also because correlation between metal-ligand bond energies or related properties, and ligand descriptors which are expected to relate only to the σ donor character, is often quite good. Computational studies provide evidence that back-bonding does indeed occur also in the case of simple trialkyl and triarylphosphines: for example in the calculated electron density of a number of metal phosphine complexes, a larger density in the region corresponding to the overlap between the metal d orbitals and the P-C σ* orbitals is expected in those cases where back-bonding occurs.[48] Another indirect proof of back bonding comes from the geometry variations of transition metal-phosphine and -phosphite complexes, whose crystal structure are known for multiple oxidation states of the complexes.[49] In these cases it is observed that metal-phosphorus bond lengths increase on oxidizing the metal, consistent with a significant back-bonding in these species. Furthermore, the reduction of the M-P π bonding, causes a decrease of the phosphorus-carbon (in the case of phosphines) or phosphorus-oxygen (for phosphites) bond length, in accord with the π acceptor orbital having a P-C (or P-O) σ* character. The electron density and structural effects show that phosphorus(III) ligands, including simple alkyl and aryl phosphines, do bind to transition metals in a complex way involving σ-bonding and π-back-bonding. A challenging problem is to calculate this latter contribution to the overall bond energy.

Several computation approaches have been used to separate bonding into different components, and these are based on indirect techniques, in which the interaction of interest is isolated from any others as part of the quantum chemical treatment. Many of these are based on stepwise processes to calculate the energy of interaction of the isolated ligand and the metal fragments, using Hartree-Fock or, more commonly, DFT methods. According to these methods, a very significant, and possibly over-estimated, contribution of back-bonding is involved in the coordination of phosphines to a metal fragment. A different approach which has also been used to attempt to quantify the contribution of back-bonding to metal-phosphorus bond, is natural bonding analysis (NBO) and in particular the use of perturbation theory to quantify secondary orbital interactions. In this representation, back-bonding then appears as a secondary orbital mixing, and the energy contribution of this mixing gives a good approximation to the energetic impact of back-bonding, providing a useful insight into the π-acceptor character of a ligand. In all these cases, the main observation is that back-bonding occurs with significant strength to all the phosphorus-based ligands, and in many cases (depending mainly on the metal fragment) it accounts for half of the bond energy for phosphines (PH_3 or PMe_3) and most of it for PX_3 species (e.g. PCl_3).

5.2 Phosphorus(III) ligands for palladium-catalyzed cross coupling

Trivalent phosphorus compounds play an outstanding role in homogeneous catalysis as ligands for transition metal. In the last 50 years, literally tens of thousands P-Ligands have been prepared and tested in various reactions. In addition, the use of chiral phosphorus ligands for enantioselective applications has seen an enormous improvement with outstanding results in terms of reactivity (measured as TON and TOF) and selectivity (ee's > 98%) and several important industrial applications.[50] This unique role among the other ligands for transition metals (e.g. amines, alcohols, Schiff bases etc.) is due to the stability of the complexes with metals catalytically active in several pivotal transformations and to the great potential for steric and electronic modifications. An exhaustive description of the myriads of ligand structures and applications to organic synthesis is far beyond the scope of this chapter. In this section, we will concentrate our attention on the recent applications of monodentate phosphines ligands to the palladium-catalyzed cross coupling reactions and of monodentate π-acidic phosphites and phosphoramidites to the enantioselective rhodium-catalyzed transformations.

Interest in palladium-catalyzed cross coupling reactions is witnessed by the enormous amount of literature published in the last few years, as well as by the Nobel prize 2010 awarded jointly to Richard F. Heck, Ei-ichi Negishi and Akira Suzuki. In palladium-catalyzed cross-coupling reactions, ligands modulate and improve the catalytic activity of the metal; key aspects, such as electron availability and steric shielding of the complexes are strongly influenced by the characteristics of the ligands.[51]

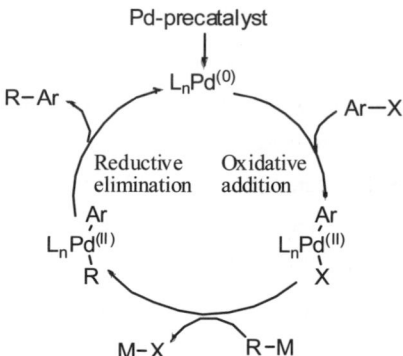

Scheme 23. General mechanism palladium-catalyzed cross couplings

A closer inspection to the catalytic cycle of these cross-coupling reactions can give some hints on how the ligand properties influence the elementary steps, and in particular the oxidative addition and the reductive elimination. It is generally accepted that steric bulk and strong electron donation of phosphines are necessary for the formation of highly efficient catalysts. In the case of very hindered phosphines, the active species in cross coupling reactions, at least with aryl chloride substrates, is generally accepted to be a PdL species, i.e. a monoligated palladium(0) complex. Hence the bulkiness of the phosphine ligands exert a stronger influence on the oxidative addition than the electronic properties of the ligands. Again, steric factors can be decisive in promoting the reductive elimination of cross coupling products from the Pd(II) complex. On the other hand, highly electron donating phosphines bind effectively to palladium, thus stabilizing the zerovalent metal complex and preventing the formation of palladium black precipitate. In the early years of Pd-catalyzed cross-couplings, PPh$_3$ was the ligand of choice, but it was soon realized that bulkier (e.g. tri-o-tolylphosphine) and later more basic phosphines (e.g. tricyclohexylphosphine, PCy$_3$) gave more satisfying results. At the turn of the century the groups of Beller,[52] Buchwald,[51b] Fu[51a] and Hartwig[53] introduced differently substituted alkylphosphines for the coupling of difficult substrates.

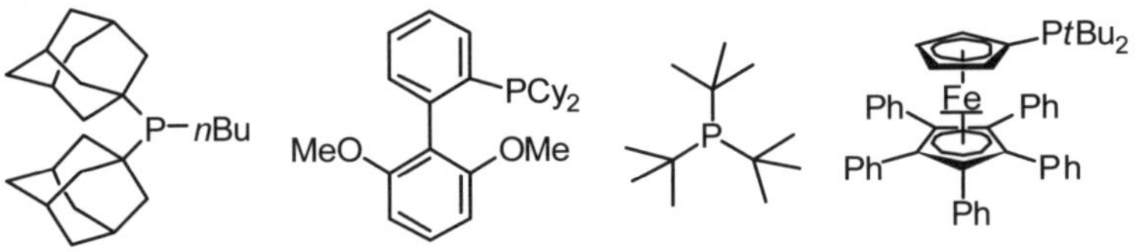

Scheme 24. Phosphines introduced by Beller, Buchwald, Fu and Hartwig

Hartwig and co-workers used a high-throughput screening (HTS) approach for the identification of efficient phosphine ligands from a library of *ca* 100 ligands for Pd-catalyzed cross-coupling reactions.[54] It was found that highly demanding and electron-rich phosphines such as tBu$_3$P, tBu$_2$FcP, tBu(1-Ad)$_2$P and tBu$_2$(2-Ad)P were among those ligands which impart the highest activity onto the palladium.

Fu and co-workers, have investigated the use of tBu$_3$P for powerful carbon-carbon bond-forming processes such as the Suzuki, Stille, Negishi, and Heck reactions under unusually mild conditions, with a wide array of challenging substrates (e.g., aryl chlorides and hindered/electronically deactivated coupling partners).[51a] These new methods have found application in a broad spectrum of settings, including natural-product synthesis, materials science, and bioorganic chemistry. The major drawback of these compounds and trialkylphosphines is that they are prone to oxidation and require gloveboxes and Schlenk techniques for handling.

In the last decade, the group of Buchwald at MIT have introduced a new class of monodentate, bulky, and electron-rich dialkylbiarylphosphines which have seen wide use as supporting ligands in a variety of transformations, especially in Pd-catalyzed carbon-carbon, carbon-nitrogen, and carbon-oxygen bond-forming processes.[51b,55]

These ligands can be prepared in a direct one-pot protocol by addition of an aryl Grignard or an aryllithium reagent to an *in situ* generated benzyne intermediate, followed by trapping the intermediate with an appropriate chlorophosphine.

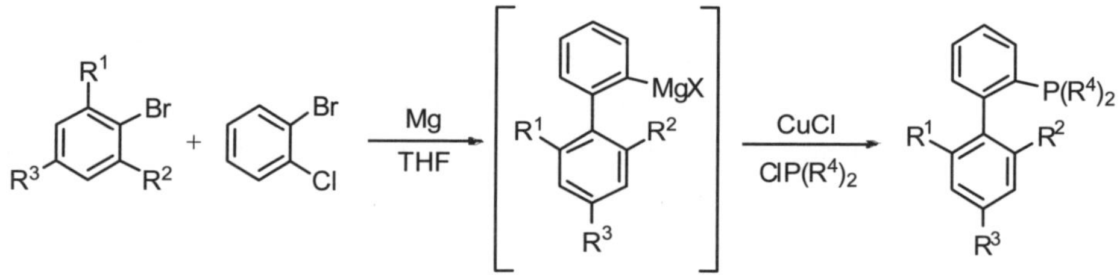

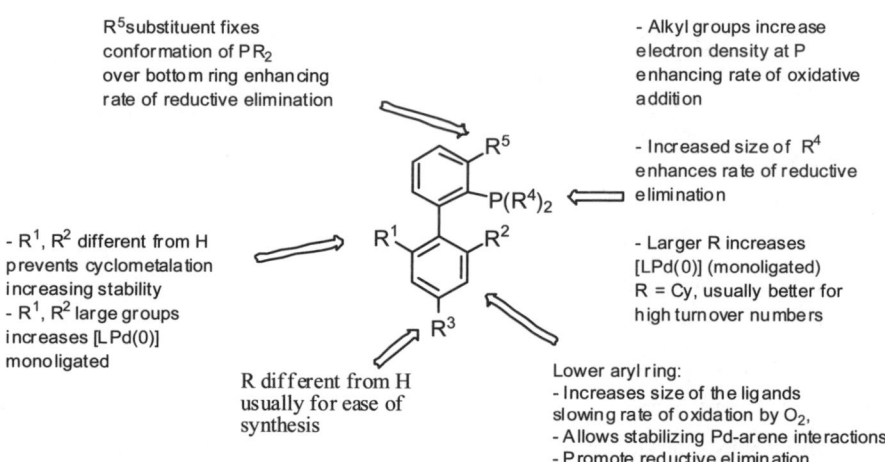

Scheme 25. Synthetic scheme for the preparation of dialkylbiarylphosphines

The main characteristics of these ligands which make them particularly attractive from a synthetic standpoint are: (a) they are crystalline materials, (b) they are air-stable, even in solution, (c) they possess a high degree of thermal stability, (d) many of these ligands are commercially available from either Strem or Aldrich, and (e) the processes that employ these ligands are operationally simple, not requiring the use of a glovebox.

The structural features that contribute to the efficiency of these ligands in their catalytic applications have been identified and are summarized in the scheme below. The modular synthesis of these ligands allow the possibility to modify their structures in order to tune their properties.

R^5 substituent fixes conformation of PR_2 over bottom ring enhancing rate of reductive elimination

- Alkyl groups increase electron density at P enhancing rate of oxidative addition

- Increased size of R^4 enhances rate of reductive elimination

- R^1, R^2 different from H prevents cyclometalation increasing stability
- R^1, R^2 large groups increases [LPd(0)] monoligated

- Larger R increases [LPd(0)] (monoligated) R = Cy, usually better for high turnover numbers

R different from H usually for ease of synthesis

Lower aryl ring:
- Increases size of the ligands slowing rate of oxidation by O_2,
- Allows stabilizing Pd-arene interactions
- Promote reductive elimination

Scheme 26. Structural features of the dialkylbiarylphosphines

The most suitable ligand for each application can vary for different substrate combinations and for this reason an optimization is normally required. Nonetheless some general guidelines have been derived for different applications. In the case of the Suzuki-Miyaura, the use of the ligand with R^1, $R^2 =$ OCH_3 afforded the best results with a number of substrates and reaction conditions. This has been ascribed to the stabilization of the precatalyst complex through favourable interaction of the aromatic π-system with the palladium center.

Dialkylbiarylphosphines have also shown to generate highly active catalysts also in the Pd-catalyzed amination originally developed by Buchwald and Hartwig. In this reaction, the nitrogen nucleophiles play an important role in the reactivity of the system, and the wide variety of these reagents in terms of electronic and steric properties requires a careful choice of the ligand and reaction conditions. A very recent perspective article by the group of Buchwald aims at providing a user's guide to select the best choice of reaction conditions and ligand for different substrate combinations.

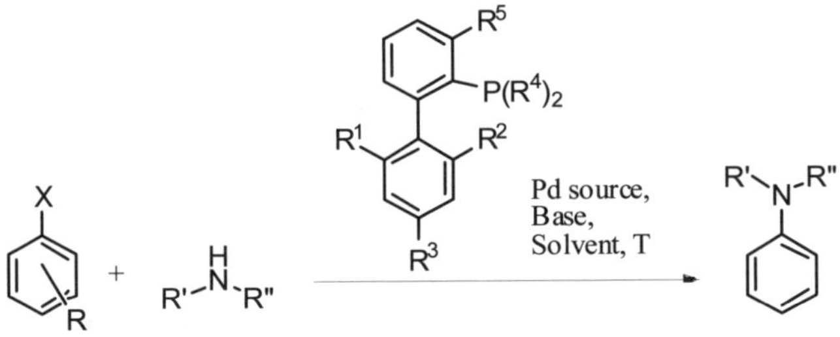

Scheme 27. Structural features of the dialkylbiarylphosphines

Another extremely relevant area of development is definitely the use of chiral for phosphorus(III) compounds as ligands for enantioselective rhodium catalyzed transformations. For some thirty years after the dawn of enantioselective transition metal catalysis (late 1960s), bidentate ligands were considered as an indispensable requirement to this end for the following reasons: (i) the *cis* coordination enforced by the covalent backbone is a critical feature in the mechanism of several important reactions; (ii) the restricted rotation of the metal-ligand bond in the chelates creates a highly defined space around the metal, which leads to increased regio- or stereochemical discrimination.[56] Only at the end of 1990s it was realised that such requirements can be met also by other kinds of ligands: chiral monodentate phosphorus-ligands such as phosphoramidites, phosphites and phosphonites were applied advantageously in the Rh-catalysed enantioselective hydrogenation of olefins and other reactions, often outperforming the bidentate ligands.[57] The easy preparation of highly diverse monodentate ligands belonging to these families makes them particularly suitable for a combinatorial search of novel enantioselective catalysts.[58]

Phosphonites   **Phospites**   **Phosphoramidites**

Scheme 28. Structural features of the monodentate chiral phosphorus ligands

The combinatorial potential of monodentate ligands was further improved by the discovery, independently reported by Reetz and co-workers and Feringa, de Vries and co-workers in 2003, that binary combinations of different monodentate ligands (L^a and L^b) in the presence of rhodium can lead to better activity and enantioselectivity than the complex of each single ligand.[59] This happens when the heterocomplex [RhL^aL^b] is more active and stereoselective than the two homocomplexes [RhL^aL^a] and [RhL^bL^b]. In a context where an increasing attention was being paid to monodentate ligands, novel approaches started to emerge for further enhancing their regio- and stereocontrol in the catalysed processes. Ligands were developed which possess, besides the atom coordinating the catalytic metal, an additional functionality capable of ligand-ligand bonding via non-covalent interactions, such as hydrogen bonding or coordinative bonding or other.

A. Non-covalent interactions for the formation of heterocomplexes

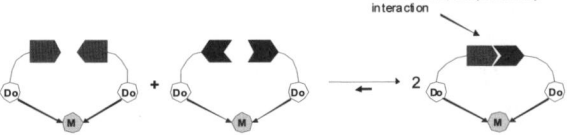

B. Non-covalent interactions for the formation of homocomplexes

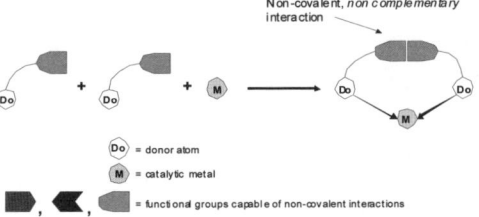

Scheme 29. Schematic representation of supramolecular bidentate ligands formed by *complementary* (A) and *non-complementary interactions* (B)

These ligands are usually referred to as "supramolecular bidentate ligands".[60] These terms also apply to those supramolecular ligands that are only capable of non-complementary interactions; indeed these systems are still capable of forming rigid and conformationally restricted complexes, although they cannot selectively form heterocomplexes when used in a mixture. Both these approaches cause reduced degrees of freedom in the respective metal coordination complexes, which is expected to result in more pre-organised systems with better capacity of controlling the metal-catalysed reaction.

Heterobidentate ligands (i.e. ligands possessing two different binding motifs) have been widely employed for forming the corresponding supramolecular bidentate ligands. The key-requirement to this end is orthogonality between the two binding moieties, which have to coordinate different metals in a mutually exclusive fashion. One metal has to be catalytically active for the reaction of choice; the other must be inert under the experimental conditions, exclusively exerting a "structural" role. In most literature examples, the former role is played by soft transition metals such as rhodium or palladium, and the latter by harder ones such as zinc or titanium. The relevant ligands are usually P,N-ligands, where phosphorus and nitrogen selectively coordinate soft and hard metals, respectively.

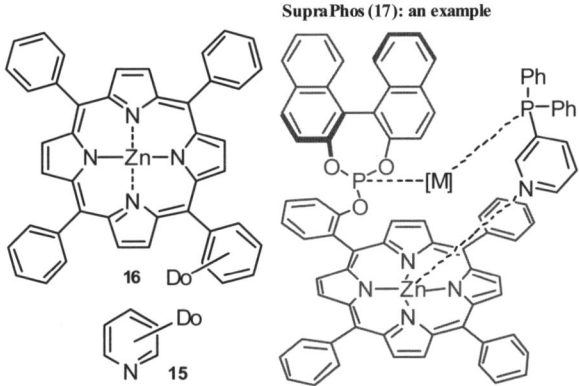

Scheme 30. 'Supraphos' supramolecular bidentate ligands and some of their applications in catalysis

An impressive application of this kind of approach, reported by Reek and co-workers in 2004, relies on the complementary interaction between a zinc-porphyrin complex bearing a phosphite group (**16**, Scheme 30) and a pyridyl phosphine or phosphite (**15**, Scheme 30).[61] A library of 48 supramolecular systems **38**, called Supraphos, was prepared from smaller libraries of **16** and **15** (14 members overall) and screened in the Pd-catalysed allylic alkylation of 1,3-diphenylallyl acetate **18a** with dimethyl malonate, in the Rh-catalysed hydroformylation of styrene, in the Rh-catalysed hydrogenation of challenging enamides and in the Pd-catalysed kinetic resolution of racemic cyclohexenyl acetate.

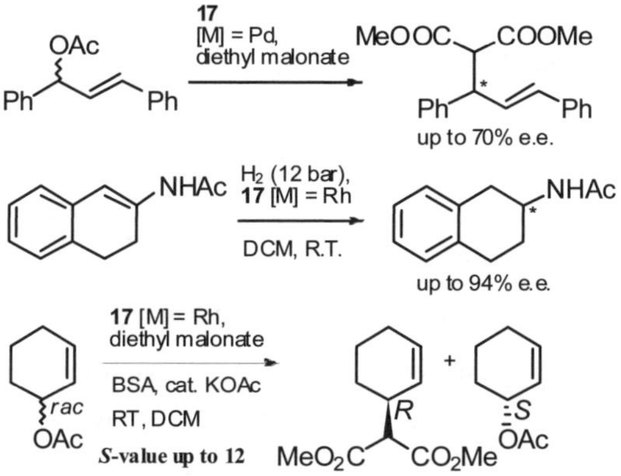

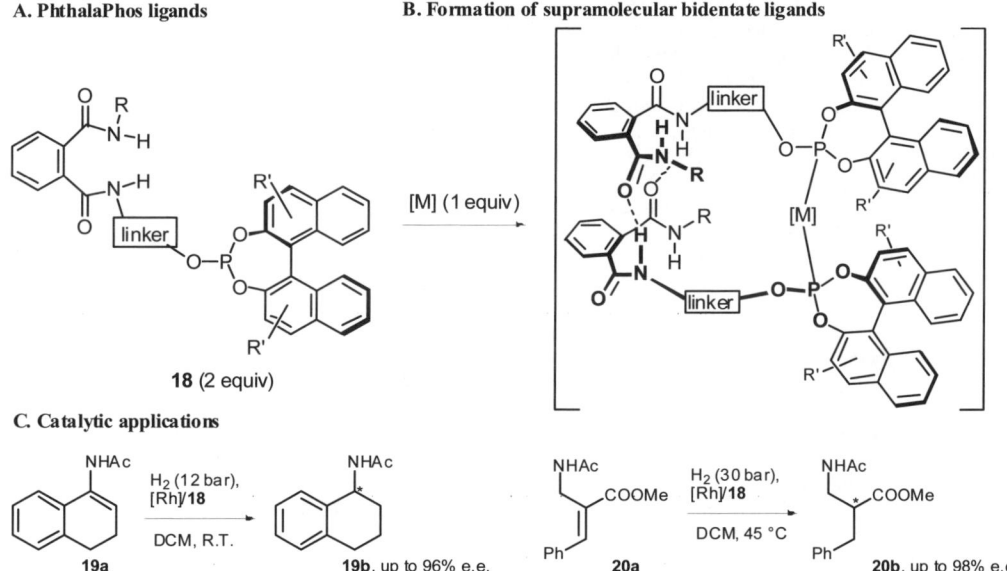

Scheme 31. Catalytic applications of 'Supraphos' supramolecular bidentate ligands.

In a second important class of supramolecular bidentate ligands hydrogen bonding interactions are exploited for achieving the self-assembly between properly functionalized monodentate ligands. Unlike several supramolecular bidentate ligands based on coordinative bonding, these systems do not require pre-formation of the supramolecular bidentate ligand, since hydrogen bonds are created dynamically and reversibly in the reaction medium where catalysis is to take place.

A. PhthalaPhos ligands

B. Formation of supramolecular bidentate ligands

18 (2 equiv)

C. Catalytic applications

Scheme 32. 'PhthalaPhos' supramolecular bidentate ligands and their application to the enantioselective hydrogenation of olefins

An example of this strategy was recently reported by our research group:[62] a library of chiral BINOL-derived phosphites (PhthalaPhos, **18** in Scheme 32) possessing a phthalic acid secondary diamide moiety was created, whose rhodium complexes showed a very good level of stereocontrol in the enantioselective hydrogenation of benchmark olefins such as methyl 2-acetamidoacrylate (up to > 99% e.e.) and *N*-(1-phenylvinyl)acetamide (up to 99% e.e.). Outstanding levels of stereocontrol could be obtained also in the hydrogenation of two challenging substrates (Scheme 32 C), such as the cyclic enamide **19a** (up to 96% e.e.) and the β-aminoacid precursor **20a** (up to 99% e.e.). Spectroscopic and computational studies confirmed the formation of the supramolecular bidentate ligands showing that, in the pre-catalytic rhodium complexes of formula $[Rh(18)_2(cod)BF_4]$, two hydrogen bonds are present between the ligands coordinated to the metal.

REFERENCES

1. *Organophosphorus Compounds. A Practical Approach in Chemistry;* Murphy, P. J. Ed; Oxford University Press: Oxford, 2004.
2. *Science of Synthesis: Houben Weyl Methods of Molecular Transformations Volume 42: Organophosphorus Compounds (incl. RO—P and RN—P)*; Francois Mathey Ed; Thieme: Stuttgart, 2008.
3. *Organophosphorus Chemistry; Specialist Periodical Reports*; The Chemical Society: London; 1970–2008, Vols. 1–37.
4. Moss, G. P.; Smith, P. A. S.; Tavernier, D. *Pure & Appl. Chem.*, **1995**, *67*, 1307.
5. (*a*) Maryanoff, B. E.; Reit, A. B. *Chem. Rev.* **1989**, *89*, 863; (*b*) Edmonds, M.; Abell, A. In *Modern Carbonyl Olefination* Takeda, T. Ed; Wiley-VCH; Weinheim, 2004.
6. (*a*) But, T. Y. S.; Toy, P. H. *Chem. Asian J.* **2007**, *2*, 1340; (*b*) Swamy, K. C. K.; Kumar, N. N. B.; Balaraman, E.; Kumar, K. *Chem. Rev.* **2009**, *109*, 2551
7. Schenk, S.; Weston, J.; Anders, E. *J. Am. Chem. Soc.* **2005**, *127*, 12566
8. But, T. Y. S.; Lu, J. N.; Toy, P. H. *Synlett* **2010**, 1115
9. Paterson, I.; De Savi, C.; Tudge, M. *Org. Lett.* **2001**, *3*, 3149
10. Morita, N.; Krause, N. *Eur. J. Org. Chem.* **2006**, 4634
11. Veliz, E. A.; Beal, P. A. *Tetrahedron Lett.* **2006**, *47*, 3153
12. Staudinger, H.; Meyer, J. *Helv. Chim. Acta* **1919**, *2*, 635
13. Palacios, F.; Alonso, C.; Aparicio, D.; Rubiales, G.; de los Santos, J. M. *Tetrahedron* **2007**, *63*, 523
14. Kano, N.; Xing, J. H.; Kawa, S.; Kawashima, T. *Polyhedron* **2002**, *21*, 657
15. Cami-Kobeci, G.; Williams, J. M. *J. Chem. Commun.* **2004**, 1072
16. (*a*) Garcia, J.; Urpí, F.; Vilarrasa, J. *Tetrahedron Lett.* **1984**, *25*, 4841; (*b*) Bures, J.; Martin, M.; Urpi, F.; Vilarrasa, J. *J. Org. Chem.* **2009**, *74*, 2203
17. Bosch, I.; Urpi, F.; Vilarrasa, J. *J. Chem. Soc. Chem. Commun.* **1995**, 91
18. Bosch, I.; Romea, P.; Urpi, F.; Vilarrasa, J. *Tetrahedron Lett.* **1993**, *34*, 4671
19. Bosch, I.; Gonzalez, A.; Urpi, F.; Vilarrasa, J. *J. Org. Chem.* **1996**, *61*, 5638
20. Ariza, X.; Urpi, F.; Viladomat, C.; Vilarrasa, J. *Tetrahedron Lett.* **1998**, *39*, 9101
21. (*a*) Sletten, E. M.; Bertozzi, C. R. *Angew. Chem. Int. Ed.* **2009**, *48*, 6974; (*b*) Debets, M. F.; van der Doelen, C. W. J.; Rutjes, F.; van Delft, F. L. *ChemBioChem* **2010**, *11*, 1168
22. Lin, F. L.; Hoyt, H. M.; van Halbeek, H.; Bergman, R. G.; Bertozzi, C. R. *J. Am. Chem. Soc.* **2005**, *127*, 2686
23. Nilsson, B. L.; Kiessling, L. L.; Raines, R. T. *Org. Lett.* **2000**, *2*, 1939
24. (*a*) Soellner, M. B.; Nilsson, B. L.; Raines, R. T. *J. Am. Chem. Soc.* **2006**, *128*, 8820; (*b*) Tam, A.; Soellner, M. B.; Raines, R. T. *Org. Biomol. Chem.* **2008**, *6*, 1173

25. Saxon, E.; Armstrong, J. I.; Bertozzi, C. R. *Org. Lett.* **2000**, *2*, 2141
26. Gamblin, D. P.; Scanlan, E. M.; Davis, B. G. *Chem. Rev.* **2009**, *109*, 131
27. (*a*) Bianchi, A.; Bernardi, A. *J. Org. Chem.* **2006**, *71*, 4565; (*b*) Nisic, F.; Andreini, M.; Bernardi, A. *Eur. J. Org. Chem.* **2009**, 5744.
28. Denmark, S. E.; Beutner, G. L. *Angew. Chem. Int. Ed.* **2008**, *47*, 1560
29. Trost, B. M.; Kazmaier, U. *J. Am. Chem. Soc.* **1992**, *114*, 7933
30. Kwong, C. K. W.; Fu, M. Y.; Lam, C. S. L.; Toy, P. H. *Synthesis* **2008**, 2307
31. Yu, S. Y.; Liu, F.; Ma, D. W. *Tetrahedron Lett.* **2006**, *47*, 9155
32. Aroyan, C. E.; Dermenci, A.; Miller, S. J. *Tetrahedron* **2009**, *65*, 4069
33. Wang, L. C.; Luis, A. L.; Agapiou, K.; Jang, H. Y.; Krische, M. J. *J. Am. Chem. Soc.* **2002**, *124*, 2402
34. Basavaiah, D.; Reddy, B. S.; Badsara, S. S. *Chem. Rev.* **2010**, *110*, 5447
35. Declerck, V.; Martinez, J.; Lamaty, F. *Chem. Rev.* **2009**, *109*, 1
36. Robiette, R.; Aggarwal, V. K.; Harvey, J. N. *J. Am. Chem. Soc.* **2007**, *129*, 15513
37. Yuan, K.; Zhang, L.; Song, H. L.; Hu, Y.; Wu, X. Y. *Tetrahedron Lett.* **2008**, *49*, 6262
38. Wei, Y.; Shi, M. *Acc. Chem. Res.* **2010**, *43*, 1005
39. Yu, S. Y.; Liu, F.; Ma, D. W. *Tetrahedron Lett.* **2006**, *47*, 9155
40. (*a*) Fang, Y. Q.; Jacobsen, E. N. *J. Am. Chem. Soc.* **2008**, *130*, 5660; (*b*) Marinetti, A.; Voituriez, A. *Synlett* **2010**, 174
41. Vedejs, E.; Daugulis, O.; Tuttle, N. *J. Org. Chem.* **2004**, *69*, 1389
42. Tolman, C. A. *Chem. Rev.* **1977**, *77*, 313
43. Fey, N.; Orpen, A. G.; Harvey, J. N. *Coord. Chem. Rev.* **2009**, *253*, 704
44. Perrin, L.; Clot, E.; Eisenstein, O.; Loch, J.; Crabtree, R. H. *Inorg. Chem.* **2001**, *40*, 5806
45. Gillespie, A. M.; Pittard, K. A.; Cundari, T. R.; White, D. P. *Internet Electr. J. Mol. Des.* **2002**, *1*, 242.
46. Suresh, C. H.; Koga, N. *Inorg. Chem.* **2002**, *41*, 1573
47. Fey, N.; Tsipis, A. C.; Harris, S. E.; Harvey, J. N.; Orpen, A. G.; Mansson, R. A. *Chem. Eur. J.* **2006**, *12*, 291
48. Marynick, D. S. *J. Am. Chem. Soc.* **1984**, *106*, 4064
49. Orpen, A. G.; Connelly, N. G. *Organometallics* **1990**, *9*, 1206
50 *Phosphorus Ligands in Asymmetric Catalysis: Synthesis and Applications;* Börner, A. Ed; Wiley-VCH; Weinheim, 2008.
51. (*a*) Fu, G. C. *Acc. Chem. Res.* **2008**, *41*, 1555; (*b*) Martin, R.; Buchwald, S. L. *Acc. Chem. Res.* **2008**, *41*, 1461; (*c*) Fleckenstein, C. A.; Plenio, H. *Chem. Soc. Rev.* **2010**, *39*, 694
52. Zapf, A.; Ehrentraut, A.; Beller, M. *Angew. Chem. Int. Ed.* **2000**, *39*, 4153
53. Hartwig, J. F.; Kawatsura, M.; Hauck, S. I.; Shaughnessy, K. H.; Alcazar-Roman, L. M. *J. Org. Chem.* **1999**, *64*, 5575
54. (*a*) Shaughnessy, K. H.; Kim, P.; Hartwig, J. F. *J. Am. Chem. Soc.* **1999**, *121*, 2123; (*b*) Stambuli, J. P.; Stauffer, S. R.; Shaughnessy, K. H.; Hartwig, J. F. *J. Am. Chem. Soc.* **2001**, *123*, 2677
55. (*a*) Surry, D. S.; Buchwald, S. L. *Angew. Chem. Int. Ed.* **2008**, *47*, 6338-6361; (*b*) Surry, D. S.; Buchwald, S. L. *Chem. Sci.* **2011**, *2*, 27
56. Jankowski, P.; McMullin, C. L.; Gridnev, I. D.; Orpen, A. G.; Pringle, P. G. *Tetrahedron: Asymmetry* **2010**, *21*, 1206.
57. (*a*) Teichert, J. F.; Feringa, B. L. *Angew. Chem. Int. Ed.* **2010**, *49*, 2486; (*b*) Minnaard, A. J.; Feringa, B. L.; Lefort L.; de Vries, J. G. *Acc. Chem. Res.* **2007**, *40*, 1267; (*c*) Jerphagnon, T.; Renaud J.-L.; Bruneau, C. *Tetrahedron: Asymmetry* **2004**, *15*, 2101, and references therein.
58. For relevant reviews, see: (*a*) Gennari, C.; Piarulli, U. *Chem. Rev.* **2003**, *103*, 3071; (*b*) Reetz, M. T. *Angew. Chem. Int. Ed.* **2008**, *47*, 2556.
59. (*a*) M. T. Reetz, T. Sell, A. Meiswinkel, G. Mehler, *Angew. Chem. Int. Ed.* **2003**, *42*, 790; (*b*) M. T. Reetz and G. Mehler, *Tetrahedron Lett.* **2003**, *44*, 4593; (*c*) D. Peña, A. J. Minnaard, J. A. F. Boogers, A. H. M. de Vries, J. G. de Vries and B. L. Feringa, *Org. Biomol. Chem.* **2003**, *1*, 1087; (*d*) A. Duursma, R. Hoen, J. Schuppan, R. Hulst, A. J. Minnaard and B. L. Feringa, *Org. Lett.* **2003**, *5*, 3111.

60.   Carboni, S.; Gennari, C.; Pignataro, L.; Piarulli U. *Dalton Trans.* **2011 DOI: 10.1039/C0DT01517B**

61.   (*a*) V. F. Slagt, M. Röder, P. C. J. Kamer, P. W. N. M. van Leeuwen, J. N. H. Reek, *J. Am. Chem. Soc.* **2004**, *126*, 4056; (*b*) J. N. H. Reek, M. Röder, P. E. Goudriaan, P. C. J. Kamer, P. W. N. M. van Leeuwen, V. F. Slagt, *J. Organomet. Chem.* **2005**, *690*, 4505; (*c*) X.-B. Jiang, L. Lefort, P. E. Goudriaan, A. H. M. de Vries, P. W. N. M. van Leeuwen, J. G. de Vries, J. N. H. Reek, *Angew. Chem. Int. Ed.* **2006**, *45*, 1223; (*d*) P. E. Goudriaan, M. Kuil, X.-B. Jiang, P. W. N. M. van Leeuwen, J. N. H. Reek, *Dalton Trans.* **2009**, 1801; (*e*) X.-B. Jiang, P. W. N. M. van Leeuwen, J. N. H. Reek, *Chem. Commun.* **2007**, 2287.

62.   L. Pignataro, S. Carboni, M. Civera, R. Colombo, U. Piarulli, C. Gennari, *Angew. Chem. Int. Ed.* **2010**, *49*, 6633.

Riccardo Giovannini
BI Research Italia S.r.l

Address:

BI Research Italia S.r.l. Via Lorenzini n. 8 – 20139 Milano, Italy

Education:
October 1997 - July 1998: training period at the University of Marburg.
Topic: reactivity of organozinc reagents in transition metal catalysed cross-coupling reactions. Supervisor: Prof. Dr. Paul Knochel.
October 1995 - October 1998 Ph.D in Chemical Sciences, University of Camerino.
Topic: Reactivity and Synthetic Applications of Polyfunctionalised Sulfones. Supervisor: Prof. Marino Petrini (University of Camerino).
November 1988 - April 1993-: Degree in Chemistry, University of Camerino.
Topic: Reactivity of 2- Nitroketones towards Organometallic Reagents. Supervisor: Prof. Giuseppe Bartoli (University of Bologna).

Professional experience:
July 2010 to date: Managing Director BI Research Italia S.r.l.
October 2007 - June 2010: Head of Chemistry Research Center- Boehringer Ingelheim Italia S.p.A.
September 2005 - September 2007: Medicinal Chemistry Group Leader at Chemistry Research Center - Boehringer Ingelheim Italia S.p.A.
September 2002 - August 2005: Investigator - GlaxoSmithKline Italia S.p.A.
February 2002 - August 2002: 6 months secondment period - GlaxoSmithKline UK
April 2001 - January 2002: Research Leader - GlaxoSmithKline Italia S.p.A.
July 2000 - March 2001: Research scientist - GlaxoWellcome Italia S.p.A.
November 1998 - June 2000: Junior Research scientist - GlaxoWellcome ItaliaS.p.A.

Research interest focussing on CNS disorders. Contribution to the identification of several clinical candidates.

The Drug Discovery Process: What a Chemist Can Contribute

Riccardo Giovannini

BI Research Italia s.p.a, via Lorenzini 8, 20139Milano. (Italy)
e-mail: riccardo.giovannini@boehringer-ingelheim.com

Laura Quaranta

Senior Team Leader, Research Chemistry, at Syngenta AG

Address:

    Syngenta Crop Protection Muenchwilen AG, WST-820.1.15

    Schaffhauserstrasse, CH-4332 Stein AG (AG), Switzerland

    e-mail: laura.quaranta@syngenta.com

Education:

    1996-2000, PhD in Organic Chemistry, University of Fribourg (CH), under the supervision of Prof Philippe Renaud, "The Chiral Relay Effect in Lewis Acid promoted Enantioselective Transformations"

    1990-1995, Laurea in Chemistry (Organic Chemistry). 110/110 *cum laude,* University of Florence, under the supervision of Prof Maurizio Taddei, "Synthesis of a Potential Inhibitor of the Tumor Necrosis Factor".

Professional experience:

    2000-2001 Postdoctoral fellow, Oregon State University, USA, working in the group of Prof. J. D. White on natural products total synthesis.

    2001-present Senior Team Leader and Project Leader in Research Chemistry at Syngenta leading a team working on the design and synthesis of active ingredients for fungal and insect control.

    She is co-author of 10 publications in refereed journals and over 20 patents in the agrochemical field. She is author of lectures and posters on international conferences.

The Discovery of Mandipropamid: a Potent and Selective Compound for the Control of Oomycetes Diseases

Laura Quaranta*

Syngenta Crop Protection Münchwilen, Schaffhauserstrasse, CH-4332 Stein, Switzerland
e-mail: laura.quaranta@syngenta.com

1. INTRODUCTION: OOMYCETES AND OOMYCETES CONTROL

1.1 The Oomycetes

The Oomycetes are a large diverse group of eukaryotic organisms which comprises circa 70 genera with more than 500 species. They have always been a threat to mankind due to their wide distribution and to the devastating epidemics which can occur in very short period of favorable weather, causing great economic losses with dramatic consequences to populations.

Although they superficially resemble fungi in mycelial growth and mode of nutrition, molecular studies and distinct morphological characteristics place them in the kingdom Stramenopila with brown and golden algae and diatoms. Their name means "egg fungus," a term coined because of the large round structures containing the female gametes, the oogonia. The other common name for some oomycetes, "water molds," refers to the fact that they were first found in freshwater habitats and their preference for conditions of high humidity and running surface water[1]. Some of them live in aquatic, but mostly in terrestrial biotopes, some are parasites other are saprophytes (feeding on dead material). Despite this ecological diversity, the oomycetes are characterized by high physiological and biochemical uniformity. Oomycetes are diploid for most of their life cycle, whereas true fungi (ascomycetes, basidiomycetes and deuteromycetes) are mainly haploid. Oomycetes and fungi have a number of enzymes which differ and different metabolic pathways (for instance lysine is synthesized via α-aminoadipic acid in fungi, via α,ε-diaminopimelic acid in oomycetes)[2]. While the cell walls of fungi contain chitin in high amounts, those of oomycetes consist mainly of cellulose, glucane and hydroxyproline. In addition they do not synthesize sterols. Members of the oomycetes group are for example *Pseuperonospora cubensis* (cucurbit downy mildew), *Plasmopara halstedii* (sunflower downy mildew), *Peronospora tabacina* (tobacco blue mold), *Pythium ultimum* (damping-off). However, the most important and economically most relevant diseases are *Phytophtora infestans* (potato and tomato late blight) and *Plasmopara viticola* (grape downy mildew). The first pathogen was responsible for the complete devastation of the potato crop in Irland in the middle of the XIX century which resulted in a famine of unbelievable dimension causing one million people to die for starvation and over two millions to emigrate to North America[3].

Even today, in particular during summers of high humidity, despite the use of modern fungicides a loss of 15% in crop yields has been calculated due to oomycetes infestation. That is why the search for novel efficient compounds for oomycetes control remains in the focus of the crop protection industry. The arsenal of resources against oomycetes diseases is rich, including broad spectrum fungicides like the strobilurins class[4]. However, due to the special position of the oomycetes between algae and fungi, chances to identify leads and develop products with high level of selectivity and environmental safety are high.

1.2 Active ingredients used for oomycetes control

Up to now three generations of fungicides for the control of oomycetes can be described[5], while a novel generation is already appearing in the future landscape.

The first generation (*figure 1*) is based on the use of heavy metals for the control of the pathogens, starting with the introduction in 1885 by Pierre-Marie Millardet of the "Bordeaux-mixture", a mixture of copper sulphate and calcium hydroxide for the control of downy mildew in grapes. For more than 50 years copper fungicides dominated the field of fungicidal plant disease control. In 1940, chemists at Rohm and Haas, Inc. discovered fungicidal activity of novel disodium ethylene-bisdithiocarbamates (EBDC) originally used as accelerators in the rubber vulcanization process. The development of new EBDCs continued with the development by Du Pont of the manganese salt (maneb) and of the zinc ion complex of maneb, known as mancozeb, which became the most important and commercially significant of all EBDCs (including present times)[6]. Copper and EBDC salts act as multisite inhibitors, having not a specific target but affecting multiple enzymes. As a consequence, the resistance risk associated is low, which currently makes them a critical partner for ant resistance management. However, due to heavy metal composition and the high rates used in the field these products are under regulatory pressure for their environmental safety profile.

CuSO$_4$+Ca(OH)$_2$
Bordeaux mixture (1885)

Cu(OH)$_2$.CuCl$_2$
copper oxychloride (1932)

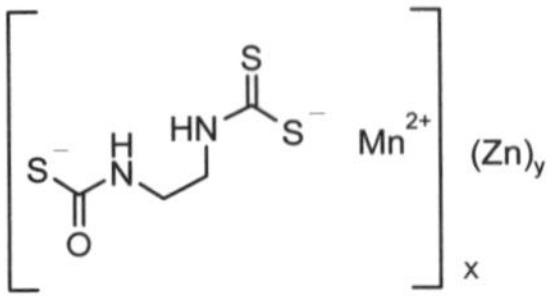

Mancozeb (1962)

Figure 1: First generation of compounds used for oomycetes control

In the years 70s, a second generation (*figure 2*) of anti-oomycetes compounds which did not contain metal ions made its appearance into the market, belonging to different chemical classes: the carbamates (ex propamocarb, prothiocarb), the isoxazoles (ex hymexazol), the cyanoacetamide-oximes (ex cymoxanil), ethyl phosphonates (ex fosetyl-Al) and the most important one, the phenylamides (ex metalaxyl). The last class of compounds interferes with the activity of RNA-polymerase I, and are currently start facing some problem of resistance development in the field as a consequence of target site mutation[1].

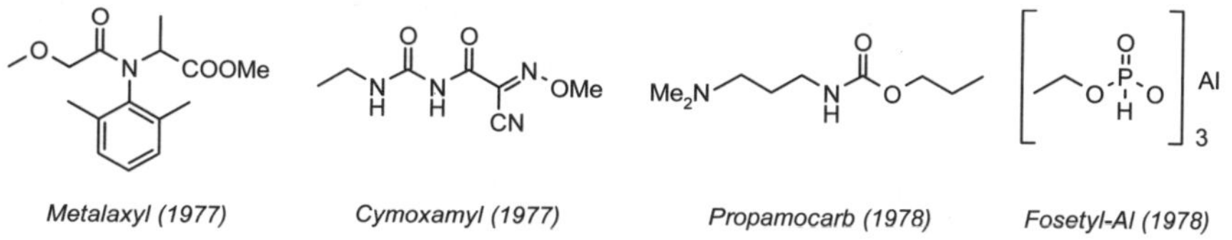

Metalaxyl (1977)   Cymoxamyl (1977)   Propamocarb (1978)   Fosetyl-Al (1978)

Figure 2: Second generation of compounds used for oomycetes control

In recent years resistance development in the cultivated fields, strong regulatory pressure and expired patent protection has stimulated a dedicated search for new agrochemicals with higher activity and a favorable environmental profile, leading in the 90s to a third generation (*figure 3*) of very active compounds capable of controlling diseases caused by oomycetes at lower rates compared to previous solutions and with good safety profiles. They mostly belong to the class of carboxylic acid amides (CAA fungicides) including cinnamic acid amides (ex dimethomorph), valinamides (ex iprovalicarb, benthiovalicarb) and mandelamides (ex mandipropamid)[7]. Due to cross-resistance observed across the three series, a common mode of biological action has been postulated, which only recently has been elucidated[8]. Fluopicolide, zoxamide and ethaboxam share with the CAA class the key amide functionality, but have different targets. Zoxamide and ethaboxam affect tubulin polymerization[9a]; fluopicolide causes redistribution of spectrin-like proteins from the membrane to the cytoplasm; these proteins are believed to maintain membrane stability[9b]. Fenamidone and famoxadone inhibit the complex III or cytochrom bc1, at the Qo site (site of ubihydroquinone oxidation) of the mitochondrial respiration (as the strobilurins fungicides)[4], while cyazofamide and probably the structurally related amisulbrom, act at the Qi site (site of ubiquinone reduction)[10]. Metalaxyl-M has been developed as the single *R*-enantiomer, the active one, of the metalaxyl active ingredient, with consequent great reduction of the application rates in the field. In addition to the lower rates, soil degradation occurs much faster than in the racemate case, leading to a better environmental profile[1].

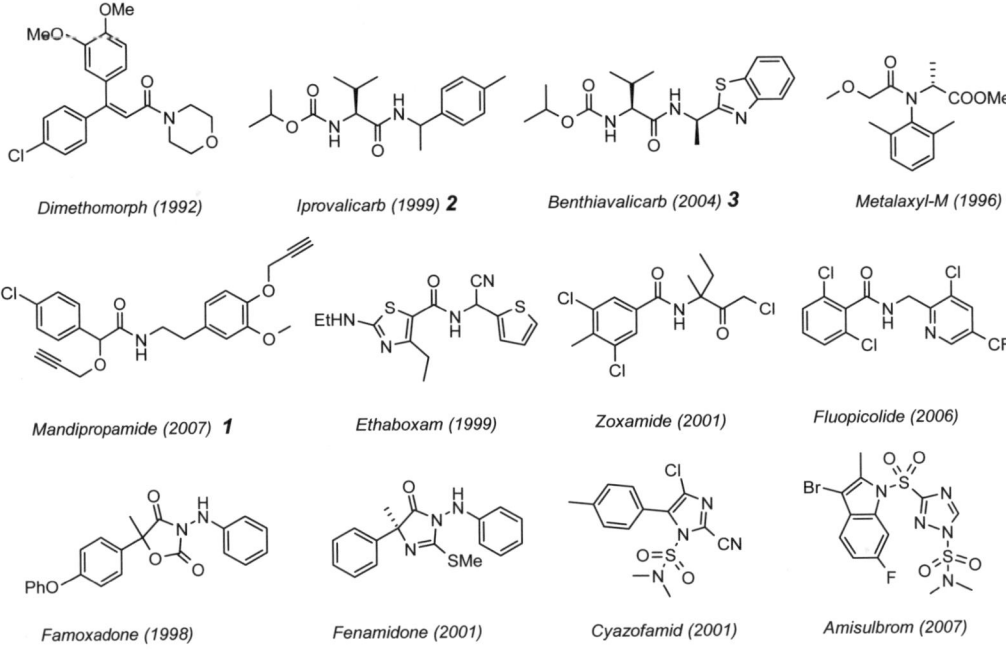

Dimethomorph (1992)   *Iprovalicarb (1999)* **2**   *Benthiavalicarb (2004)* **3**   *Metalaxyl-M (1996)*

Mandipropamide (2007) **1**   *Ethaboxam (1999)*   *Zoxamide (2001)*   *Fluopicolide (2006)*

Famoxadone (1998)   *Fenamidone (2001)*   *Cyazofamid (2001)*   *Amisulbrom (2007)*

Figure 3: Third generation of compounds used for oomycetes control

More recently two new classes of compounds have been discovered which act selectively against oomycetes (*figure 4*). Initium, discovered in 2004 by BASF, obtained its first registration in 2010 in record time due to a very favorable environmental profile. It has originated as a spin-off project during the development of a triazolopyrimidine-based class of compounds affecting tubulin polymerization which have a broader spectrum of activity, not including diseases caused by oomycetes. Initium acts by inhibition of cytochrome bc1 but does not show cross-resistance to Qo inhibitors (strobilurins)[11].

Chemists at Du Pont have identified a new piperidinyl thiazole based class of compounds with excellent preventative and curative activity against oomycetes diseases allowing low use rates and a novel mode of action[12]. The exact structure of the development compound is still unknown, however based on Du Pont patent applications published in the past years, the two compounds in figure 4 appear as likely candidates[13].

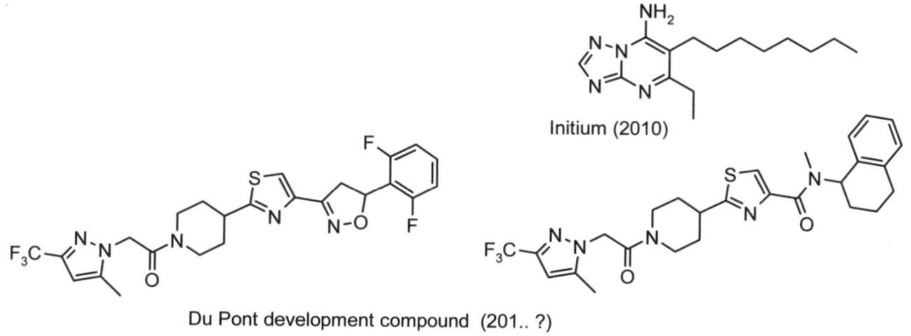

Initium (2010)

Du Pont development compound (201.. ?)

Figure 4: Fourth generation of compounds used for oomycetes control

1.3 Oomycetes market

According to the most recent data obtained by Phillips McDougal (2010) the oomycete market represents 15% of the whole fungicide market (USD 11,176 mio). The top 5 companies (Syngenta, Bayer, BASF, DuPont, Dow) occupy more than 70% of the market. A list of the major oomycete specialist is shown below (*figure 5*). Syngenta leads the market with a broad range of offers, including mefenoxam (metalaxyl-M), fluazinam (mode of action: uncoupler, broad spectrum) mandipropamid, azoxystrobin (mode of action: cytochrome bc1 inhibition, broad spectrum), chlorothalonil (mode of action: multisite, broad spectrum).

2009 Fungicide Market
$ 11,176 million USD [1]

Oomycete, 15%

2009 Oomycete Market by Crops [2]

Other Crops; 7%
Grapes; 32%
Vegetables; 28%
Potatoes; 33%

Active ingredients	2009 sales	Major companies
Metalaxyl / Metalaxyl-M	$ 275 m	Syngenta, Rallis, United Phosphorus, Sipcam
Cymoxamil	$ 145 m	DuPont, Oxon
Fosetyl	$ 140 m	Bayer
Fluazinam	$ 105 m	ISK, Syngenta
Dimethomorph	$ 85 m	BASF, SePRO
Mandipropamid	$ 57 m	Syngenta
Cyazofamid	$ 50 m	ISK
Iprovalicarb	$ 45 m	Bayer

Figure 5: Oomycete market

2. DISCOVERY OF MANDIPROPAMID: FROM VALINAMIDES TO MANDELIC ACID AMIDES

Iprovalicarb **2**, was the first fungicide out of the amino acid amide carbamate class of compounds introduced in the market in 1998 by Bayer. It was discovered during a synthesis program for new fungicidal lead structures in 1988 showing good control of oomycetes[14]. Iprovalicarb **2** was followed by Kumiai's benthiavalicarb **3** in 2003. Both compounds are characterized by the presence of the isopropyloxycarbonyl protected amino acid L-valine linked to a 1-phenylethylamine or 1-benzothiazolyl-ethylamine. Valinamides containing an N-2-(3-methoxy-4-alkoxy)phenethyl amide, like **4** were disclosed as agrochemical fungicides by American Cyanamid (now BASF) in the same years[15]. During a program aimed at improving the biological profile of valinamide derivatives chemists at Novartis (now Syngenta) identified a novel class of potent compounds by replacing the carbamate group with a sulfonyl or sulfamyl group[16]. In particular, best activities were observed for phenethylamine derivatives bearing a propargyloxy group as in **5**. Optimization of **5** led to the identification of a novel subclass of compounds where the L-valine was replaced by the non-proteogenic phenylglycine amino acid subclass; **6** with a simple propargyloxy group instead of the p-chlorophenylpropargyloxy showed the best activity in bioassays.

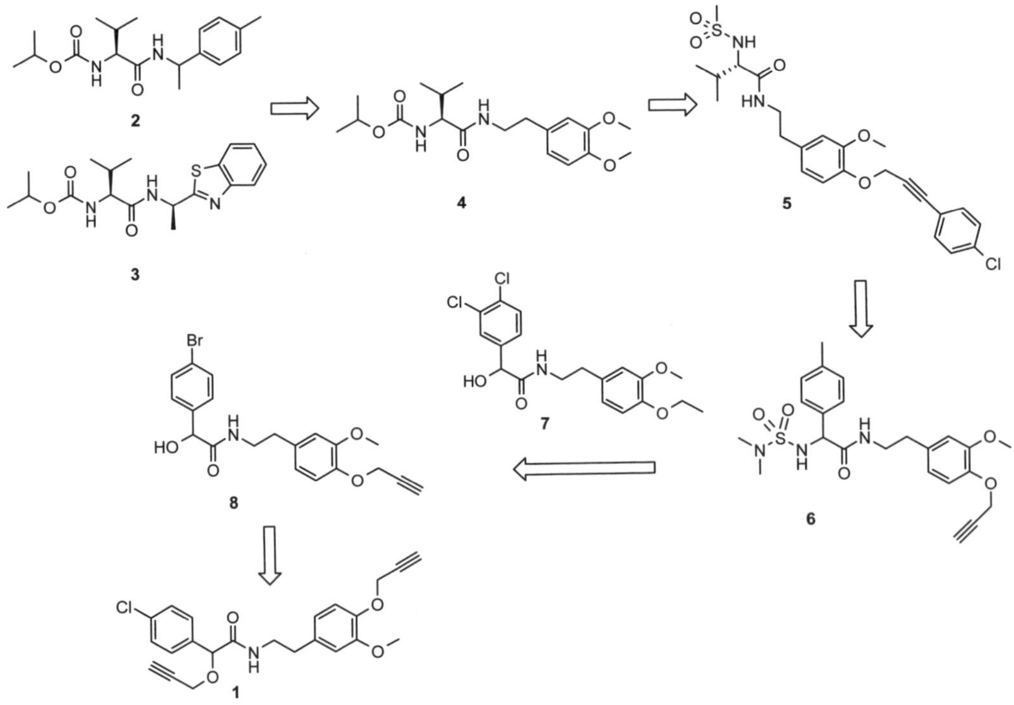

Scheme 1: From valinamides to mandipropamid

The breakthrough towards the development of the highly potent mandipropamid came with the combination of the modified dialkoxy phenethyl amine with mandelic acid derivatives. It was originally postulated that the hydroxyl function would mimic the sulfamide as H-bond donor and having pKa of similar magnitude (with the NH of the sulfamide more acidic). Supporting such a hypothesis, at the same time Agrevo's chemists (now Bayer) had identified mandelamide derivatives such as **7** having fungicidal activity against oomycetes, however only moderate[17]. Deep knowledge of the SAR (structure-activity relationship) of the sulfonylamide subclass suggested the preparation of derivatives containing an aromatic propargyloxy substituent which were found more active than the corresponding Agrevo compounds. Among them, compound **8** showed a clear improvement in the biological profile[18]. The replacement of an alkoxy group by a propargyloxy has also been reported in the pharmaceutical literature as a modification to improve the activity of antibacterial and leishmanicidal analogues[19]. The last step in the optimization of the lead structure resulted in the introduction of a second "magic" propargyl group in the mandelic acid moiety which dramatically increased the biological activity, leading to the discovery of mandipropamid **1**, the first highly potent mandelamide fungicide to be launched in the agrochemical market in 2007 by Syngenta with the trade name Revus® (against *P. infestans* in potato crops) and Pergado® (against *P. viticola* in vines). Mandipropamid is acting by preventing spore germination and by inhibition of the mycelial growth and sporulation[20]. It binds rapidly and tightly to the wax layer of the leaf surface, providing a rain-fast and long-lasting barrier against infections[21] which combined with the strong preventive and translaminar activity allows a robust control of grape downy mildew and potato late blight under severe disease pressure. Mandipropamid does not show good curative activity in the field, which could be anticipated based on the physic-chemical properties measured (solubility in H_2O: 4.2 mg/L; log P*ow*: 3.2).

2.1 FIRST GENERATION: N-SULFONYL AMINO ACID AMIDES

2.1.1 Chemistry

After the identification of the potential of compound of type **4** in providing good biological control, a program was initiated in Novartis to derivatize the lead structure. Based on an in-house expertise with sulfonyl amino acid in previous projects, the replacement of the carbamate with a sulfonamide was studied and compound **19** was synthesized showing very good activity against *Plasmopara viticola* and *Phytophtora infestans*. For further derivatization the synthesis of 2-(4-hydroxy-3-methoxy-phenethyl)amine **13**[22] was undertaken which could be amenable to large scale production (*scheme 2*). Different approaches were tested, starting from vanillin, via hydrogenation of the corresponding cyanohydrin **10** or nitrostyrene **11**. The last route, probably the most widely used suffers from scale up problems due to the hazard associated with the use of the expensive lithium aluminium hydride and to the exothermicity and unpredictable reproducibility of the catalytic hydrogenation. However, running the transformation in two steps via sodium borohydride reduction to phenylnitroethane **12** followed by catalytic hydrogenation delivered a high yield solution to the problem. A third approach is based on the reduction of the benzylcyanide **15** obtained from vanillinol **14** through a quinoid transition state [23].

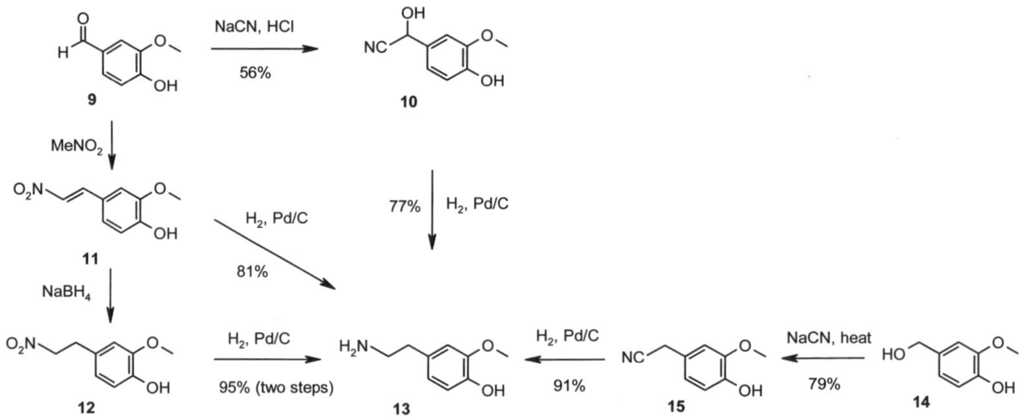

Scheme 2: Synthesis of substituted phenethylamine **13**

Two different strategies were followed for preparation of the final sulfonylated L-valinamide derivatives. The first relied on the sulfonylation of the free amine under Schotten-Baumann conditions followed by BOP mediated peptide coupling of the amino acid with amine **13** and alkylation of the free phenol to give the final compound **19** (*scheme 3*). Under conditions which involved the use of strong base (ex in the alkynylation step), the presence of the acidic NH proton of the sulfonamide group was protecting the α-carbon from racemization thus guaranteeing the sterochemical integrity of the L-valine asymmetric center in the final compound.

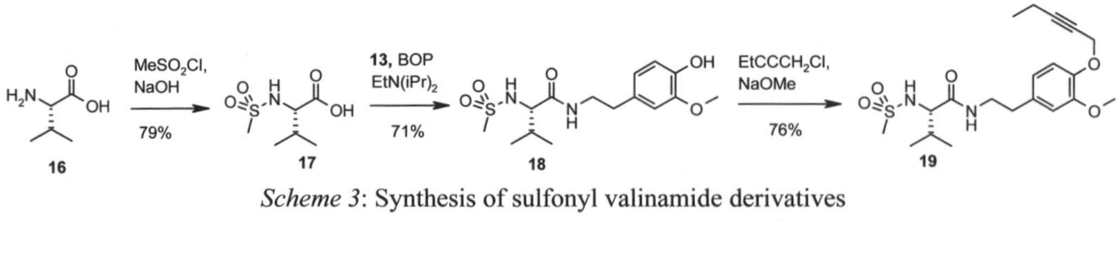

Scheme 3: Synthesis of sulfonyl valinamide derivatives

The second approach to prepare derivatives of the non-proteogenic phenylglycine amino acid was based on a Ugi four-component condensation, an efficient transformation in which a carbonyl compound, an amine, a carboxylic acid and a isocyanide are assembled in a single step to give an α-acylamino carboxamide with the concomitant formation of a chiral center α to the carbonyl (which derives from the prochiral carbonyl compound)[24]. Multicomponent reactions are nowadays well established as a powerful tool for the efficient construction of compounds with high level of complexity and diversity starting from relatively simple building blocks[25]. However, despite the increasing popularity of multicomponent reactions, this example represents one of the few applications in fungicide research known so far [26]. Phenethylamine **13** was converted into the corresponding isocyanide **22** via N-formylation, O-propargylation and dehydration (*scheme 4*). The Ugi reaction of isocyanide **22** with p-tolualdehyde and ammonium formate, which contains the amine and the carboxylic acid components, led after acid hydrolysis to the expected phenylglycinamide **23** which in turn was N-sulfonylated to give the target compound **24**.

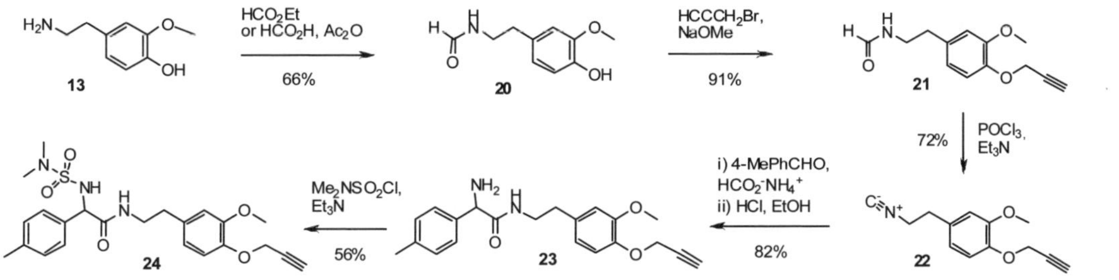

Scheme 4: Synthesis of phenylglycinamide derivatives via the Ugi reaction

Enantiopure phenylglycinamide could be obtained by carrying out the Ugi reaction with a chiral amine [27]. Indeed, the condensation of isocyanide **22** with 4-chlorobenzaldehyde, formic acid and 2,3,4,6-tetra-*O*-pivaloylgalactosamine resulted in the formation of **25** as a separable mixture of diastereoisomers. After removal of the chirality inducer sugar unit and of the formyl group, sulfonylation delivered the fungicidally active (*R*)-phenylglycinamide, as **26** (*scheme 5*)[18b].

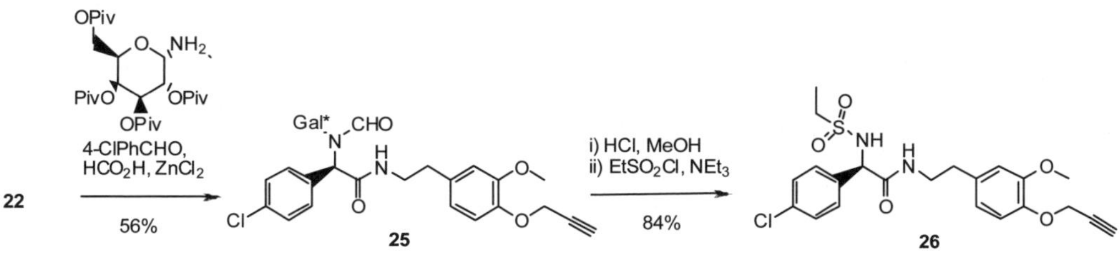

Scheme 5: Diastereoselective synthesis of phenylglycinamide derivatives *via* the Ugi reaction

2.1.2 Structure Activity Relationship (SAR)

To determine the efficacy of all derivatives against Phytophtora infestans and Plasmopara viticola, three-weeks old tomato plants and five-weeks old grape seedlings were treated with the test compounds in a spray chamber. One day (grape) or two days (tomato) after application, the plants were inoculated by spraying a sporangial suspension on the upper (tomato) or lower (grape) side of the leaves. Disease incidence was assessed after an incubation period of 4 days (tomato) or 6 days (grape) at 20°C and 95% relative humidity in a greenhouse.

Table 1 shows the influence of the amino acid portion on the activity of compounds with general structure I. The presence of polar proteogenic and non-proteogenic amino acids, like threonine (entry 10) or glutamic acid (entry11), is detrimental for activity. A lipophilic backbone is important for the activity in particular in the presence of α-branched residues, as indicated by the superior activity of compounds where **R** is *i*-propyl (entry 5) or cyclopropyl (entry 6) *vs* n-propyl (entry 4). Analogously, *s*-butyl (entry 7) derivative performs better in *in vivo* test than the *i*-butyl derivative (entry 8).

Table 1. Structure-Activity Relationship Studies of the Amino Acid Chain

Entry	R	Conf.	EC$_{80}$ value (ppm)[a] Phytophtora infestans	EC$_{80}$ value (ppm)[a] Plasmopara viticola
Iprovalicarb			42	2
1	-H	-	148	>200
2	-CH$_3$	L	>200	>200
3	-CH$_2$CH=CH$_2$	D,L	9	49
4	-CH$_2$CH$_2$CH$_3$	L	60	20
5	-CH(CH$_3$)$_2$	L	8	2
6	-CH(CH$_2$CH$_2$)	L	20	19
7	-CH(CH$_3$)CH$_2$CH$_3$	L	3	2
8	-CH$_2$CH(CH$_3$)$_2$	L	12	18
9	-CH(CH$_2$CH$_3$)$_2$	D,L	16	16
10	-CH(OH)CH$_3$	L	148	109
11	-CH$_2$CH$_2$COOH	L	>200	>200
12	-CH$_2$CH$_2$SCH$_3$	L	>200	>200
13	-CH(CH$_2$)$_5$	D,L	55	>200
14	4-ClC$_6$H$_5$	D,L	20	15

[a] EC$_{80}$: calculated conc. in ppm obtained from greenhouse trials at which the tested compound shows 80% of activity

The configuration of the chiral α–carbon proves important, with better activities seen for L-configurated natural and unnatural amino acids compared to their D-enantiomers. Interestingly,

4-chlorophenylglycine derivative (entry 14) in racemic form showed activities comparable to the best compounds. This observation, in conjunction with IP novelty, has encouraged the team in exploring further this modification, as shown later in this text.

Table 2 shows the influence of the *p*-phenol substituent within the very active subclass of dialkoxy substituted phenethylamine derivatives of type **II**. The presence of an unsaturated alkynyl group is in all cases associated with very good activities as opposite to smaller saturated chains (entry 1, 2 and 3). Surprisingly, the replacement of the initially identified 2-pentynyl group (entry 6) by 4-chlorophenylpropargyl side chain (entry 8) resulted in a high enhancement of the biological activity, which was retained by the corresponding unsaturated analogues, however with lower EC_{80} values.

II

Table 2. Structure-Activity Relationship Studies of the *p*-Phenol Substituent

Entry	R	EC_{80} value (ppm)[a] *Phytophtora infestans*	EC_{80} value (ppm)[a] *Plasmopara viticola*
Iprovalicarb		42	2
1	-H	>200	>200
2	-CH₃	>200	89
3	-CH₂CH₃	109	53
4	-CH₂CCH	55	18
5	-CH₂CCHCH₃	35	8
6	-CH₂CCHCH₂CH₃	8	2
7	-CH₂CCHCH₂CH₂CH₃	60	32
8	-CH₂CCH(4-Cl-C₆H₅)	0.02	2
9	-CH₂C=CH(4-Cl-C₆H₅)	6	2
10	-CH₂CH₂CH₂(4-Cl-C₆H₅)	10	60
11	- CH₂CH₂O(4-Cl-C₆H₅)	43	26

[a] EC_{80}: calculated conc. in ppm obtained from greenhouse trials at which the tested compound shows 80% of activity

It was soon recognized that in combination with substituted phenylglycine derivatives the very good activity of 4-chlorophenylpropargyl analogues was not retained. Instead, good activities were observed for simple propargyloxy analogues. Keeping the phenethyl amine fixed, *tables 3* and *4* highlight the effect of the sulfonyl group and the nature of the phenylglycine residue on the activity against the two key pathogens. The length of the alkyl sulfonyl chain is important with best activities seen for the linear *n*-propyl group (entry 5) which drop drastically moving to the branched *i*-Pr group (entry 6) or to shorter or longer chains. Interestingly, the best results were obtained with a dimethyl sulfamoyl cap at the amino function (entry 9). Replacement of the sulfonylamino group with sulfinylamino, alkyl amino, amide or carbamate functions let to inactive compounds.

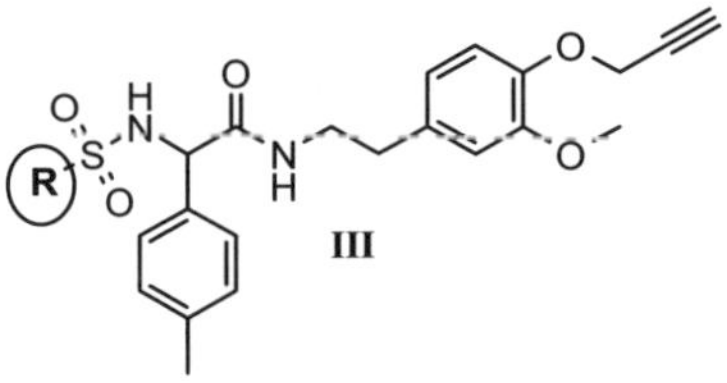

III

Table 3. Structure-Activity Relationship Studies of the Sulfonyl Group

Entry	R	EC$_{80}$ value (ppm)[a] *Phytophtora infestans*	EC$_{80}$ value (ppm)[a] *Plasmopara viticola*
Iprovalicarb		42	2
1	-CH$_3$	32	42
2	-CF$_3$	-	104
3	-CH$_2$CH$_3$	50	>200
4	-CH=CH$_2$	49	76
5	-CH$_2$CH$_2$CH$_3$	6	5
6	-CH(CH$_3$)$_2$	>200	60
7	-CH$_2$CH$_2$CH$_2$CH$_3$	45	17
8	-NHCH$_3$	10	10
9	-N(CH$_3$)$_2$	2	3
10	- (4-Me-C$_6$H$_5$)	109	16

[a] EC$_{80}$: calculated conc. in ppm obtained from greenhouse trials at which the tested compound shows 80% of activity

Substitution of the phenyl ring has a strong impact on activity, as shown in *table 4*. Small, linear alkyl groups at the *para*-position provide excellent control of *Phytophtora infestans* and *Plasmopara viticola* (entries 5,7, 9) in compounds of type **IV**, while *ortho*- and *meta*-substitution is detrimental for activity (entries 15, 16), as it is the effect of branched alkyl groups like *i*-propyl or *t*-butyl (entries 8, 10). Comparison between 3,4-disubstituted analogues (entries 17-20) and the corresponding mono 4-substituted ones confirms the weakening effect of *meta*-substituents. The role of polarity of the substituent is less clear (also because the two pathogens do not show the same trend): 4-CF$_3$ is comparable to the best alkyl substituents (entry 6), halogens show variable effects (entries 2-4), 4-OMe slightly weaker than the best (entry 11).

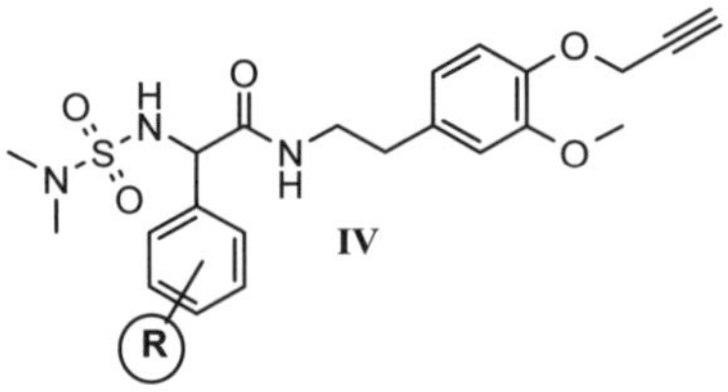

IV

Table 4. Structure-Activity Relationship Studies of the Phenylglycine Residue

Entry	R	EC$_{80}$ value (mg L^{-1})[b] Phytophtora infestans	EC$_{80}$ value (mg L^{-1})[b] Plasmopara viticola
1	H	81.2	28.5
2	4-F	45.7	78.3
3	4-Cl	10.6	13.8
4	4-Br	7.8	39.4
5	4-CH$_3$	2.5	3.2
6	4-CF$_3$	3.4	5.8
7	4-CH$_2$CH$_3$	2.6	5.1
8	4-CH(CH$_3$)$_2$	26.2	14.6
9	4-CH$_2$CH$_2$CH$_3$	3.1	16.7
10	4-C(CH$_3$)$_3$	142.8	20.2
11	4-OCH$_3$	8.0	34.9
12	4-OCF$_3$	18.8	46.4
13	4-OCH$_2$CH$_3$	24.7	35.0
14	4-SCH$_3$	34.9	16.2
15	2- CH$_3$	>200	>200
16	3- CH$_3$	149.9	116.1
17	3-Cl-4-Cl	19.2	51.5
18	3-Cl-4-F	186.0	59.8
19	3-F-4-F	109.6	46.3
20	3- CH$_3$-4- CH$_3$	12.2	28.7

[b] EC$_{80}$: calculated conc. in mg L^{-1} obtained from greenhouse trials at which the tested compound shows 80% of activity

In conclusion, two highly active N-sulfonyl amino acid amides **5** and **6** with activity superior to already marketed carboxylic acid amide (CAA) type derivative were identified as potential candidates for promotion to the development phase. However, the ease to explore diversity *via* multicomponent reactions encouraged the chemists to investigated new chemical classes. This led to the discovery of mandelamides as a very active subclass and finally to the development of the market compound mandipropamid.

2.2 SECOND GENERATION: MANDELIC ACID AMIDES

2.2.1 Chemistry

The formamide **21** precursor of the isocyanide in the Ugi synthesis of phenylglycinamides, proved to be a key intermediate for the preparation of mandelamides. Using the Seebach's variation[28] of the Passerini reaction[25c], **21** could be transformed directly into mandelamide **27**. Using titanium(IV) chloride as a Lewis acid instead of a carboxylic acid component α-hydroxycarboxamides can be directly obtained[29]. Alkylation of the hydroxyl group could be performed using activated alkyl halides. As an example the introduction on **27** of a second propargyl group with propargyl bromide led to mandipropamid **1**, which can be obtained directly from **13** and 4-chloromandelic acid *via* **28** followed by double propargylation under phase transfer catalysis. However, alkylation using inactivated alkyl halides (or pseudo-halides) proved to be difficult. An elegant alternative synthesis was designed to achieve these targets important for SAR, which is based as a key step on the reaction of an alcohol with the N-mesyloxy amide **33**[30.] Addition of a base, like triethylamine, to the mesylate results in the formation of an aziridinone (or α-lactam) intermediate, which in the presence of poor nucleophiles (ex alcohols) reacts by ring opening to an ion pair intermediate which is then trapped by the nucleophile. In the presence of *n*-propanol, compound **33** can be transformed to mandelamide **34** in high yields.

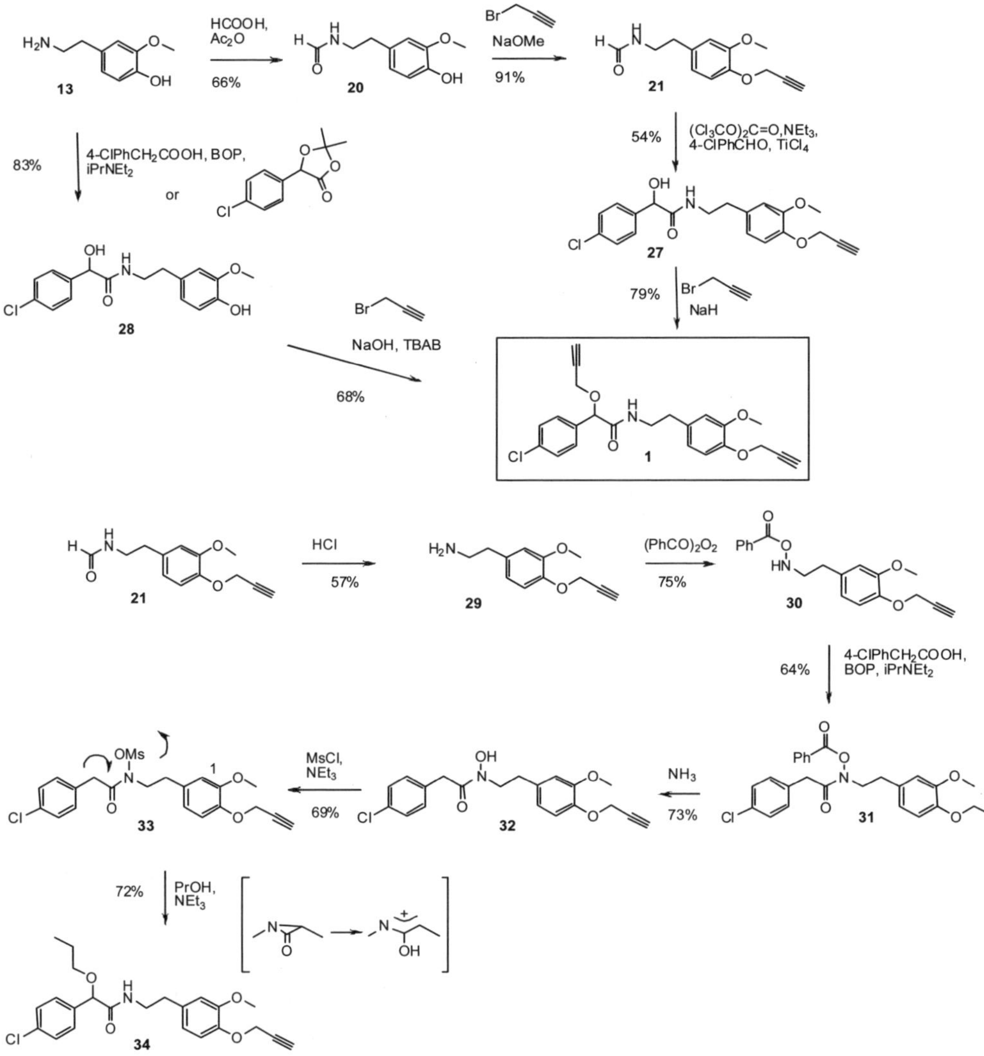

Scheme 6: Synthetic approaches to diverse mandelamide derivatives

There were indications in the literature that the two enantiomers of SX 623509 (**7**) have different biological activities and do not contribute equally to the activity of the racemate[17]. In order to check the hypothesis for compounds like **22** different approaches have been tested for the preparation of enantiopure mandelamides which resulted in the development of a new asymmetric Passerini-type three component condensation[31] and in the discovery of new catalytic hydrogenation conditions for substituted phenylglyoxylic esters[32].

Examples of Passerini reactions using chiral aldehydes[33] or chiral isocyanides[34] have been reported, however with low diastereoselectivities. Based on the postulated mechanism of the Passerini reaction, the carboxylic acid takes part in the reaction from the beginning via formation of a hydrogen-bonded adduct with the carbonyl compound followed by α-addition to the isocyanide and intramolecular transacylation to give α-acyloxycarboxamides[35]. Hence, the use of a chiral acid was considered a promising strategy for the transfer of the chirality to the final α-hydroxymandelamides after cleavage of the chiral ester. After a screening of chiral amino acid and sugar derivatives 1,2,3,4-tetra-O-acetyl-α-D-galacturonic acid[36] was chosen due to the high diastereoselectivity induced. After hydrolysis, enantiomeric ratios up to 98:2 were measured[31].

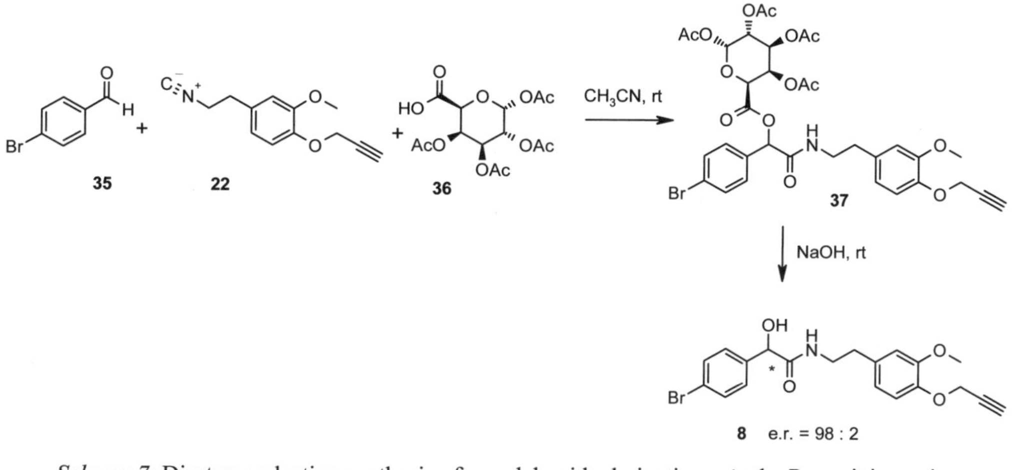

Scheme 7: Diastereoselective synthesis of mandelamide derivatives *via* the Passerini reaction

In collaboration with chemists at Solvias, the catalytic hydrogenation of phenylglyoxylic amides and esters under homogeneous and heterogeneous conditions was intensively studied. The aim was to identify a technically feasible route to mandelic amides with >90% ee. Despite some encouraging literature reports on enantioselective hydrogenation of aryl α-keto benzylamides using rhodium(I)-complexes with chiral diphosphine ligand of the AMPP-[37] or BICHEP-type[38] or using heterogeneous cinchona-modified Pt catalyst[39], in >80 different experiments with diverse substituted phenylglyoxyl phenethylamides the best result (with ees up to 87%) was obtained using Rh(I) and (*R*)-cy-oxo-pronop ligand. However, due to low catalytic activity, attention was turned to the hydrogenation of methyl phenylglyoxylate, for which several examples existed in the literature. After an extensive screening, heterogeneous and homogeneous conditions were found which fulfilled all requirements: a Pt catalyst modified with cinchona derivatives (10,11-dihydrocinchonidine, HCd) and a Ru-Me-Obi-phep catalyst gave 93% ee for the (*R*)-methyl *p*-chloromandelate. Due to high TONs (up to 4000) and TOFs (up to 210 h^{-1}) the homogeneous process was scaled up to the kg scale and the enantiomeric purity of the product enhanced to >99% by recrystallization of the *p*-chlorophenylmandelic acid[32].

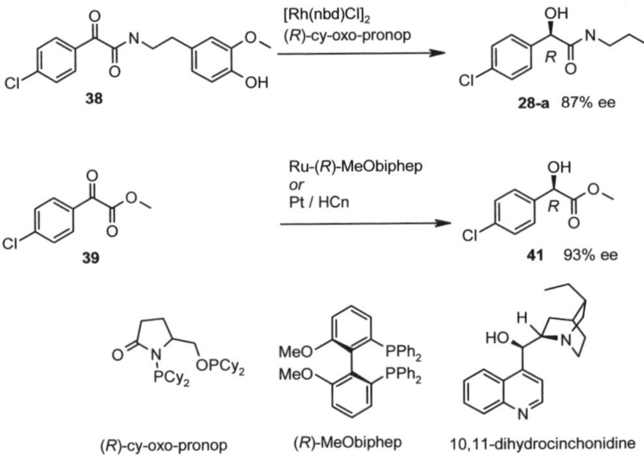

Scheme 8: Synthesis of mandelamide and mandelic ester derivatives *via* enantioselective hydrogenation

2.2.2 Structure Activity Relationship (SAR)[18b, 40]

Mandelamides possess a much higher level of oomycete activity than phenylglycinamide, as can be seen by comparing *table 5* with *table 4*[18]. Many similarities can be seen in the SAR of the two subclasses, with the highest activities seen for *para*-substituted derivatives with small lipophilic group. However, in contrast to the phenylglycinamides, the best biological results within the mandelamide subclass of type **V** were achieved with halogens as *para*-substituents in the mandelic acid ring (entries 2-4 *vs* 5). Surprisingly, the unsubstituted mandelamide (entry 1) was the second best compound behind the *para*-chloro analogue. *Meta*-substituted mandelamides retain some good activity (entries 13, 14), but are overall weaker than the corresponding *para*-derivatives, while *ortho*-substitution is not tolerated (entry 12) with fluorine as the only exception (entry 17). If the phenyl ring bears two substituents, then,

3,4-disubstituted compounds are much more potent than the corresponding 2,4- or 3,5-disubstuted ones (entries 15-19). Trisubstituted mandelamides showed weak activities (entry 20).

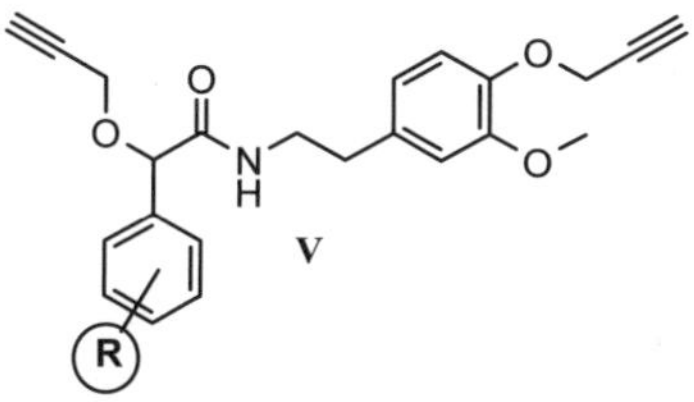

Table 5. Structure-Activity Relationship Studies of the Phenyl Residue in the Mandelamide Class

Entry	R	EC$_{80}$ value (mg L^{-1})[b] *Phytophtora infestans*	EC$_{80}$ value (mg L^{-1})[b] *Plasmopara viticola*
1	H	0.5	1.3
2	4-F	1.8	3.4
3	4-Cl	0.1	1.2
4	4-Br	3.4	2.6
5	4-CH$_3$	9.7	5.2
6	4-CF$_3$	8.0	11.6
7	4-CH$_2$CH$_2$CH$_3$	4.4	1.7
8	4-CH(CH$_3$)$_2$	6.1	2.8
9	4-CH$_2$CH$_2$CH$_2$CH$_3$	25.9	10.4
10	4-OCH$_3$	2.3	1.6
11	4-SCH$_3$	7.6	5.3
12	2-Cl	81.1	>200
13	3-Cl	3.7	2.4
14	3-CH$_3$	6.4	1.7
15	2-Cl-3-Cl	168.4	25.5
16	3-Cl-4-Cl	4.8	3.4
17	2-F-4-F	4.6	5.0
18	3-CH$_3$-4-CH$_3$	3.2	1.6
19	3-CH$_3$-5-CH$_3$	41.9	56.4
20	2-Cl-3-Cl-4-Cl	>200	44.2

[b] EC$_{80}$: calculated conc. in mg L^{-1} obtained from greenhouse trials at which the tested compound shows 80% of activity

The influence of substituents in the mandelic acid and in the phenethylamine portions of the scaffold was investigated. Alkylation of the free hydroxyl group in both cases leads to an increase in biological activity, as shown in *tables 6* and 7. In particular, C$_2$ and C$_3$ units seem to have the highest potential, with best activities observed for the derivatives bearing a propargyl group which is superior to the partially and fully saturated allyl and *n*-propyl analogous (entries 4,6 *vs* 7 in *table 6*). Branching or elongation of the propargyl unit leads to less active compounds (entries 8-9 *vs* 7 in *table 6*; entries 5-6 *vs* 4 in *table 7*). Acylation of the free hydroxy group results in surprisingly good efficacy, while introduction of a carbonate (and carbamate) functionality lead to a drop of activity (entries 10,11 in *table 6*). Benzylation was tolerated in the phenethyl hydroxyl group but with lower activity (entry 7 in *table 7*).

Table 6. SAR of O-Substituents in the Mandelic Acid Portion

Table 7. SAR of O-Substituents in the Phenethyl Portion

	R	EC_{80} value[b] P.infestans	EC_{80} value[b] P.viticola		R	EC_{80} value[b] P.infestans	EC_{80} value[b] P.viticola
1	H	34.6	10.4	1	H	45.6	100.5
2	CH_3	9.7	3.5	2	CH_3	4.2	2.0
3	CH_2CH_3	4.6	2.0	3	CH_2CH_3	3.5	1.8
4	$CH_2CH_2CH_3$	5.4	31	4	CH_2CCH	0.1	1.2
5	$CH(CH_3)_2$	3.1	10.1	5	$C(CH_3)_2CCH$	8.1	51.7
6	$CH_2CH=CH_2$	44.1	18.6	6	CH_2CCHCH_3	3.9	13.4
7	CH_2CCH	0.1	1.2	7	$CH_2C_6H_5$	25.5	19.7
8	$C(CH_3)_2CCH$	4.9	3.5				
9	CH_2CCHCH_3	18.3	14.0				
10	$C(O)CH_2CH_3$	4.2	2.8				
11	$C(O)OCH_3$	38.6	109.5				

[b] EC_{80}: calculated conc. in mg L^{-1} obtained from greenhouse trials at which the tested compound shows 80% of activity

The versatility and efficiency of the Passerini reaction had a huge importance for the SAR exploration of mandelamides derivatives, allowing the preparation of diverse compounds, as shown in *table 8*. A broad variety of aliphatic, aromatic, heterocyclic and polycyclic aldehydes could be easily converted to the corresponding *a*-propargyloxyacetamides in two steps[40]. Interestingly, the introduction of a spacer between the phenyl ring and the 2-propargyloxyacetamide group leads to active compounds. Among these "stretched" mandelamides a compound with a two-atom spacer was identified (entry 4, *table 8*) which showed an impressive efficacy against both *P. infetans* and *P. viticola*. The nature of the spacer seems to be important: one-atom carbon or oxygen spacers seem unfavorable (entries 2-3, *table 8*), a three-atom spacer including oxygen retain activity (entry 6, table 8), while four or more linker atoms seem to be unsuitable (entry 8, *table 8*). In the presence of the OCH_2 linker, the SAR of the phenyl group is very comparable to the mandelamide SAR, with *para*-substituents providing high level of activity compared to *ortho* and *meta*, with best results obtained for the *para*-chloro and *para*-methyl analogue. The 4-chlorophenyl ring of mandipropamid can be replaced with five-membered ring heterocycles (in particular thiophene) which show good biological activity (entries 2-3, *table 9*), while six-membered ring or bicyclic heterocycles lead to almost inactive compound (entries 4-5, *table 9*).

Tables 8 and 9. Effect of Different α-Propargyloxyacetamides on the Activity

	R (VIII)	EC_{80} value[b] P.infestans	EC_{80} value[b] P.viticola		R (IX)	EC_{80} value[b] P.infestans	EC_{80} value[b] P.viticola
1	bond	0.1	1.2	1	cyclohexyl	47.6	5.3
2	CH_2	26.3	18.6	2	thiophen-2-yl	2.1	1.4
3	O	>200	>200	3	furan-2-yl	17.5	12.4
4	OCH_2	0.02	0.6	4	benzo[b]thiophen-2-yl	>200	>200
5	SCH_2	33.2	20.7	5	pyrid-3-yl	186.3	>200
6	CH_2OCH_2	0.3	20.4				
7	$CH(CH_3) CH_2$	28.8	100.5				
8	$CH_2 CH_2OCH_2$	32.0	45.9				

[b] EC_{80}: calculated conc. in mg L^{-1} obtained from greenhouse trials at which the tested compound shows 80% of activity

3. MODE OF ACTION OF MANDIPROPAMID

As already mentioned in section *1.2* cinnamic acid amides, valinamides carbamates and mandelic acid amides have been classified by FRAC (Fungicide Resistance Action Commettee) as group number 40, meaning that based on the cross resistance pattern amongst members of these subclasses and the selective oomycetes spectrum they are claimed as having a common mode of action. Inhibition of cell wall deposition and phospholipid biosynthesis were described as potential targets. Only very recently Syngenta biochemists have elucidated the mode of biological action of mandipropamid[8]. Their studies indicate that the compound acts on the cell wall and does not enter the cell. Furthermore, ^{14}C glucose incorporation into cellulose was perturbed in the presence of the compound, suggesting that the primary mode of action is based on the inhibition of cellulose biosynthesis. In addition, mutagenesis experiments (a combination of forward and reverse genetics) have identified in *P. infestance* two point mutations conferring resistance to mandipropamid on the PiCesA3 gene which is involved in cellulose synthesis, thus confirming the hypothesis developed.

4. CONCLUSION

Mandipropamid is the newest active ingredients launched in the market against diseases caused by Oomycetes pathogens, in particular *Phytophtora infestans* and *Plasmopara viticola*. It belongs to the class of the carboxylic acid amides (CAA), more specifically the mandelic acid amides, and has been discovered within a chemical program aimed at exploring the diversity of the CAA scaffold through multicomponent reactions. Next to the excellent biological control which allows low application rates, it has a favorable human and environmental safety profile and based on acute and chronic dietary risk assessment all residue levels in treated crops are safe to the consumer.

The discovery of mandipropamid exemplifies the efficient approach to the discovery of new active ingredients driven by chemistry and structure-activity relationship, in the absence of knowledge of the biological target. The rapid advance in technology and different regulatory focus are rapidly changing the research approach in the crop protection industry towards a target-based one (the limitation arising by a less developed knowledge of fungal biochemistry compared to higher organisms). However, the emphasis on the low cost of active ingredients for use in agriculture is still of primary importance which confers to chemistry a key driving role in the development of new products.

References and Notes

1. (a) Gisi, U. *Advances in Downy Mildew Research*; Spencer-Phillips, P. T. N.; Gisi, U.; Lebeda, A. Ed.; Kluver:Dordrecht, 2002; 119-159; (b) Schwinn, F.; Staub, T. *Modern Selective Fungicide*, Lyr, H. Ed.; Fischer: Jena, 1995; 323-346; (c) Griffith, J. M.; Davis, A. J. ; Grant, B.R. *Target Sites of Fungicide Action*; Köller, W., Ed.; CRC Press: Boca Raton, 1992; 69-100.
2. Van der Auwera, G.; De Baere, R.; Van de Peer, Y.;, De Rijk, P.; Van den Broeck, I.; De Wachter, R. *Mol. Biol. Evol.,* **1995**, *12*, 671.
3. (a) Jordan, T.E. *J. R. Soc. Health*, **1997**, *117*, 216; (b) O'Farrel, P. *Hist. Stud.* **1982**, *20*, 1.
4. (a) Bartlett, D.W.; Clough, J.M.; Godwin, J.R.; hall, A.A.; Hamer, M.; Parr-Dobrzanski, B. *Pest. Manag. Sci.*, **2002**, *58*, 649; (b) Sauter, H.; Steglich, W.; Anke, T. *Angew. Chem.*, **1999**, *111*, 1416.
5 Lamberth, C. *Nachrichten aus der Chemie*, **2007**, *55*, 130.
6. Gullino, M.L.; Tinivella, F.; Garibaldi, A.; Kemmitt, G. M.; Bacci, L.; Sheppard, B. *Plant Disease*, **2010**, *94*, 1076.
7. Gisi, U.; Lamberth, C.; Mehl, A.; Seitz, T. *Modern Crop Protection Compounds*, Krämer, W.; Schirmer, U. Eds.; Wiley: Weinheim, 2007; 651.
8. Blum, M.; Boeheler, M.; Randall, E.; Young, V.; Csukai, M.; Kraus, S.; Moulin, F.; Scalliet, G.; Avrova, A.O.; Whisson, S.C.; Fonne-Pfister, R. *Molecular Plant Pathology*, **2010**, *11*, 227.
9. (a) Uchida, M.; Roberson, R. W.; Chun, S-J.; Kim, D-S. *Pest. Manag. Sci.* **2005**, *61*, 787.(b) Toquin, V.; Barja, F.; Sirven, C.; Gamet, S.; Latorse, M.-P.; Zundel, J.-L.; Schmitt, F.; Beffa, R.. *Pflanzenschutz-Nachrichten Bayer* **2006**, *59*, 171-1844.
10. Henningsen, M. *Chem. unserer Zeit*, **2003**, *37*, 98.
11. Gold, R.; Scherer, M.; Rether, J.; Speakman, J.; Navé, B.; Levy, T.; Storer, R.; Marris, D. BCPC Congress, November 9 – 11, 2009, Glasgow UK.
12. Pasteris, R. *12th IUPAC International Congress of Pesticide Chemistry*, June 4-8, 2010, Melbourne, Australia.
13. WO 2007/013290; WO 2008/091594; WO 2008/013925; WO 2009/055514; WO 2010/123791

14. Stenzel, K.; Pontzen, R.; Seitz, T.; Tiemann, R.; Witzenberger, A. *Brighton Crop Protection Conf.*, **1998**, 5A-7, 367-374.

15. Hunt, D.A.; Lavanish, J.M.; Asselin, M.; Los, M. (American Cyanamid) EP493683, *Chem Abstract.*, **1992**,. *117*, 145306.

16. Cederbaum, F.; De Mesmaeker, A.; Jeanguenat, A.; Kempf, H-J.; Lamberth, C.; Schnyder, A.; Zeller, M.; Zeun, R. *Chimia*, **2003**, *57*, 680.

17. (a) Ort, O.; Döller, U.; Reissel, W.; Lindell, S. D.; Hough, T.L.; Simpson, D.J.; Chung, J.P. *Pestic. Sci*, **1997**, *50*, 331; (b) WO 96/17840, *Chem. Abstr.*, **1996**, *125*, 142763; (c) WO 94/29267, *Chem Abstr.*, **1994**.

18. (a) Lamberth, C.; Cederbaum, F.; Jeanguenat, A.; Kempf, H-J.; Zeller, M.; Zeun, R. *Pest. Manag. Sci.*, **2006**, *62*, 446; (b) Lamberth, C.; Jeanguenat, A.; Cederbaum, F.; De Mesmaeker, A.; Zeller, M. ; Kempf, H-J. ; Zeun, R. *Bioorg. Med. Chem.*, **2008**, *16*, 1531-1545.

19. (a) Periers, A-M.; Laurin, P.; Ferroud, D.; Haesslein, J-L-; Klich, M.; Dupuis-Hamelin, C.; Mauvais, P.; Lassaigne, P.; Bonnefoy, A.; Musicki, B. *Bioorg. Med. Chem. Lett.*, **2000**, *10*, 161; (b) Gomes, D.; De, C.F.; Alegrio, L.V.; Freire De Lima, M.E.; Leon, L.L.; Araujo, C.A.C. *Arzneim.-Forsch./Drug. Res.*, **2002**, *52*, 120.

20. Huggenberger, F.; Lamberth, C.; Iwanzik, W.; Knauf-Beiter, G. *Proc. BCPC Internat. Congress,* BCPC: Alton, 2005, p 87.

21 Hermann, D.; Bartlett, D.W.; Fischer, W.; Kempf, H-J. *Proc. BCPC Internat. Congress,* BCPC: Alton, 2005, p 93.

22. 3-O-methyldopamine or 3—methoxytyramine **13** has been applied as versatile precursor in the synthesis of tetrahydroisoquinoline (see, *Chem. Pharm. Bull.*, **1976**, *24*, 262 and 1921) and morphine alkaloids (see, *J. Am. Chem. Soc.*, **2006**, *128*, 87), serotonin receptor antagonist (see, *J. Med. Chem.*, **1980**, *23*, 990), capsaicin-like agonists (see, *J. Med. Chem.*, **1993**, *36*, 2373) and 5-lipoxygenase inhibitors (see, *Chem. Pharm. Bull.*, **1990**, *38*, 842).

23. (a) Szantay, C.; Dörnyei, G.; Blasko, G.; Barczai-Beze M.; Pechy, P. *Arch. Pharm.*, **1981**, *314*, 983; (b) Schwartz, M.A.; Zoda, M.; Vishnuvajjala, B.; Mami, I. *J. Org. Chem.*, **1976**, *41*, 2502.

24. (a) Ugi, I. *Pure Appl. Chem.*, **2001**, *73*, 187; (b) Ugi, I. *J. Prakt. Chem.*, **1997**, *339*, 499.

25. (a) *Multicomponent Reactions*, Zhu, J.; Bienayme, H. Eds.; Wiley-VCH: Weinheim, 2005; (b) Biggs-Houck, J. E.; Younai, A.; Shaw, J. T. *Current Opinion in Chemical Biology* **2010**, *14*, 371; (c) Dömling, A. *Chem. Rev.*, **2006**, *106*, 17; (d) for asymmetric MCR: Dondoni, A.; Massi, A. *Acc. Chem. Res.,* **2006**, *39*, 351.

26. (a) Reddy, B. V. S.; Reddy, M. R.; Madan, C.; Kumar, K. P.; Rao, M. S. *Bioorg. Med. Chem. Lett.,* **2010**, *20*, 7507; (b) Haranat, P.; Anasuyamma, U.; Reddy, C. D.; Reddy, C. S. *Heterocyclic. Commun.*, **2005**, *11*, 335; (c) Dandia, A.; Singh, R.; Sarawgi, P. *J. Fluorine Chem.*, **2004**, *125*, 1835; (d) Altorfer, M.; Ermet, P.; Faessler, J.; Farooq, S.; Hillesheim, E.; Jeanguenat, A.; klumpp, K.; Maienfisch, P.; Martin, J. A.; Merrett, J. H.; Parkes, K. E. B.; Obrecht, J.-P.; Petterna, T.; Obrecht, D. *Chimia*, **2003**, *57*, 262; (e) Lu, S-M.; Chen, R-Y. *Heteroatom Chem.*, **2000**, *11*, 317; (f) ref. 17a.

27. (a) Ross, J.F.; Herdtweck, E.; Ugi, I. *Tetrahedron*, **2002**, *58*, 6127; Oertel, K.; Zech, G.; Kunz, H. *Angew. Chem., Int. Ed.*, **2000**, *39*, 1431; (c) Linderman, R.J.; Binet, S.; Petrich, S.R. *J. Org. Chem.*, **1999**, *64*, 336; (d) Lehnhoff, S.; Goebel, M.; Karl, R.M.; Klösel, R.; Ugi, I. *Angew. Chem., Int. Ed.*, **1995**, *34*, 1104; (e) Kunz, H.; Pfrengle, W.; Rück, K.; Sager, W. *Synthesis* **1991**, 1039.

28. (a) Seebach, D.; Adam, G.; Gees, T.; Schiess, M.; Weigand, W. *Chem. Ber.* **1988**, *121*, 507; (b) Schiess, M.; Seebach, D. *Helv. Chim. Acta,* **1983**, *66*, 1618.

29. Passerini-type transformations leading to α-hydroxycarboxamides have been reported with boron trifluoride, mineral acids and even water as acid component.

30. Hoffman, R.V.; Nayyar, N.K. *J. Org. Chem.*, **1995**, *60*, 7043.

31. Galbraith, S.G.; Guelfi, S.; Lamberth, C.; Zeller, M. *Synlett*, **2003**, 1536.

32. Cederbaum, F.; Lamberth, C.; Malan, C.; Naud, F.; Spindler, F.; Studer, M.; Blaser, H-U *Adv. Synth. Catal.*, **2004**, *346*, 842-848.

33. Moran, E. J.; Armstrong, R.W. *Tetrahedron Lett.,* **1991**, *32*, 3807.

34. Ziegler, T.; Kaisers, H-J.; Schlömer, R.; Koch, C. *Tetrahedron,* **1999**, *55*, 8397; Ziegler, T.; Schlömer, R.; Koch, C. *Tetrahedron Lett.,* **1998**, *39*, 5957; Bock, H.; Ugi, I. *J. Prakt. Chem.*, **1997**, *339*, 385.

35. Dömling, A.; Ugi, I. *Angew. Chem., Int. Ed.*, **2000**, *39*, 3168; Smith, M. B.; March, J. *Advanced Organic Chemistry*; Wiley: New York, 2001; p. 1251.

36. Vogel, C.; Jeschke, U.; Kramer, S.; Ott, A-J. *Liebigs Ann. Chem.,* **1997**, 737.

37. Pasquier, C.; Naili, S.; Mortreux, A.; Agbossou, F.; Pélinski, L.; Brocard, J.; Eilers, J.; Reiners, I.; Peper, V.; Marten, J. *Organometallics*, **2000**, *19*, 5723.

38. Chiba, T.; Miyashita, A.; Nohira, H.; Takaya, H. *Tetrahedron Lett.*, **1993**, *34*, 2351

39. Wang, G.Z.; Mallat, T.; Baiker, A. *Tetrahedron: Asymmetry*, **1997**, *8*, 2133.

40. Lamberth, C.; Kempf, H-J.; Kriz, Miroslav. *Pest. Manag. Sci.,* **2007**, *63*, 57.

Marcella Bonchio

First Researcher - Italian National Council of Research - CNR

Address:

Istituto CNR per la Tecnologia delle Membrane – ITM-CNR sez. di Padova-c/o Dipartimento di Scienze Chimiche, Università degli Studi di Padova, via Marzolo, 1, 35131 PADOVA.

Education:

1989:"Laurea" in Chemistry, the University of Padova 110/110 cum laude

1990:Vising scientist at Brown University, R.I, U.S.A., Prof. J. O. Edwards

1993:Ph.D Degree University of Padova, supervisor Prof. F. Di Furia

1995:Research Associate at Princeton University, New Jersey U.S.A., Prof. J.T. Groves's laboratories

Professional Experience:

In 1996 M. Bonchio joined the the National Council of Research (ITM-CNR, Padova) as Research Associate, and was then promoted to First Researcher in 2001. In 2010 she got the abilitation to Full Professor of Organic Chemistry.

M. Bonchio research interests deal with oxidation catalysis and hybrid organic-inorganic functional systems following bio-inspired guidelines. A recent breakthrough concerns the finding of stable ruthenium catalysts capable of water oxidation to dioxygen, with multi-turnover activity and fast rates. (*J. Am. Chem. Soc.*; 2008, 130, 5006; *J. Am. Chem. Soc.*; 2009, *131*, 16051; *Nature Chem.* 2010, 2, 826) The current research activity is also focused on the design of chiral derivatives, at the interface with functional materials and nanotechnology (*Angew. Chem. Int. Ed.* 2008, 38, 7275).

Bio-inspired Oxidations: Innovative Catalytic Strategies

Andrea Sartorel, Mauro Carraro and Marcella Bonchio*

ITM-CNR and Dept. of Chemical Sciences, University of Padova, via Marzolo 1, 35131 Padova, Italy

I see skies of blue and clouds of white
The bright blessed day, the dark sacred night
And I think to myself what a wonderful world
(Bob Thiele and George David Weiss, "What a Wonderful World", 1967)

INTRODUCTION

Selective and sustainable oxidation/oxygenation of organic molecules with atom-economy efficiency at low environmetal impact is one grand challenge of catalysis both in the academic and the industrial environment.[1] Oxidation catalysis plays a central role for the production of bulk and fine chemicals, the generation of key pharmaceutical intermediates and as a final step, advanced oxidation methods are essential for pollution degradation and water/environment remediation. More than 20% of all industrial organic materials are obtained by catalytic oxidation; ajor examples are the large-scale production of epoxides such as ethylene and propylene oxide, of maleic anhydride from butane, of acrolein and acrylic acid from propylene, xylene oxidation to phthalic anhydride or terephthalic acid, or cyclohexane oxidation to the so-called alchol/ketone (A/K) mixture and adipic acid. At the edge of this research frontier, one fundamental goal is the engineering of tailored catalytic systems, featuring a well-balanced interplay of selectivity, generality, performance, speed, robustness, environmental sustainability, cost-efficiency. In this quest, the archetypal paradigm of sustainable efficiency is offered by Nature.

Nature has evolved some perfect machineries to generate, store and translocate redox equivalents, powering some vital processes, such as the photosynthetic cycle of the chloroplast and its alter-ego found in the mitochondrial respiration chain. In nature, oxidation catalysis is essential for aerobic life. The chloroplast and the mitochondrion are the "natural factories" responsible for energy *transformation in-vivo* (Fig.1). Chloroplasts are not found in animal cells; they are present only in plants and some protists. By photosynthesis, chloroplasts trap light energy and convert it into the chemical bond energy of sugars. Solar energy enters the food chain of living beings through the chloroplasts, while it is converted to use through the respiratory chain of the mitochondria. The formidable complexity of processes taking place within these cellular organelles involves a cascade of redox transformations coupling electron and proton transferring with bond cleavage and forming reactions. In particular, carbohydrate/fatty acid combustion with dioxygen fuels the cellular activities. In such perspective, mitochondrial respiration releases the energy that was trapped by photosynthesis and converts it into the easily accessible, high energy phosphate bonds of adenosine triphosphate (ATP).

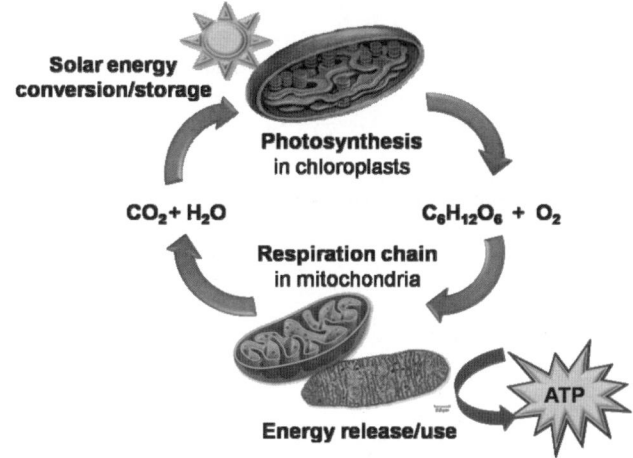

Fig. 1 The perpetual Energy Chain boosted by the Chloroplast/Mitochondrion Catalytic Systems.

Biosynthesis and metabolic reactions, or detoxification of esogenous and harmful substances also involve oxidation steps. The transport, storage, and activation of dioxygen in-vivo, is performed by key metallo-enzymes, enabling its transformation in highly reactive forms and the controlled generation of key intermediates such as the superoxide radical anion ($O_2^{-\cdot}$) or hydrogen peroxide (H_2O_2). The mastering of these reactive oxygen species (ROS) by carefully orchestrated production/depletion mechanisms is also pivotal to cell defense, by programmed death strategies (apoptosys), thus establishing an effective catalytic therapy for longevity. On the other hand, life on our planet has been powered by the perpetual cycling of oxygenic photosynthesis (Fig. 1). In this process, H_2O is the renewable multi-electron source, providing a total of four electron/mol by the oxidative half reaction ($2H_2O \rightarrow O_2 + 4H^+ + 4 e^-$; $E^0 = 1.23$ V vs. NHE).[2] This multi-electron pathway, powered by sunlight, generates O_2 as oxidation product and carbohydrates at the reduction end, whereby solar energy turns out to be stored into the chemical bonds of reactive molecules. Water oxidation occurs with unmatched efficiency at the heart of the Photosystem II (PSII) enzyme, where O_2 evolution is catalysed by a polynuclear metal-oxo cluster assembling four manganese centres and one calcium atom ($CaMn_4O_x$).[3]

Over billions of years, Nature has refined a multi-electron photocatalytic strategy to perform water oxidation at the lowest energy cost, to feed the reductive end and to collect energy in the form of carbohydrates. Our mission is now to devise its artificial mimicry, while diverting the downhill process, after water oxidation, and recombining protons into hydrogen as solar fuel. The artificial "off-leaf" transposition is a major goal of renewable energy, and of oxidation catalysis aimed at the continuous production of hydrogen through the photo-catalytic splitting of H_2O.[4]

LESSONS FROM NATURE: HYBRID CATALYSIS

Enzymes often outperform synthetic catalysts by employing multiple and complementary strategies, originating from the interplay of functional, metal-based domains and nano-structured proteic environments. This organic-inorganic hybrid approach is instrumental for boosting efficiency, while providing: low-energy pathways, substrate recognition and orientation, stabilization of reactive intermediates (Fig. 2). Moreover Nature is able to shape favorable "second sphere" interactions alleviating the fatigue of the catalytic centres, enabling their structural reorganization/self-healing, shaping distances and connections, controlling equilibria and kinetics, and regulating cascade processes.

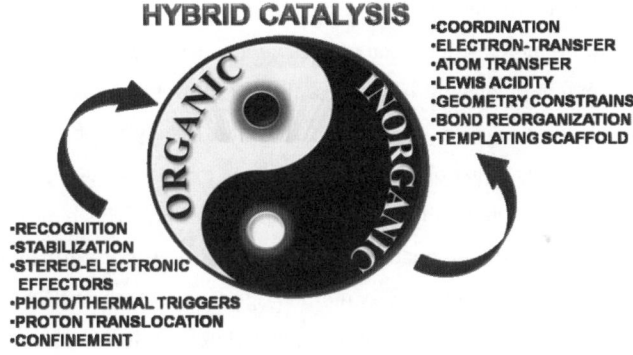

Fig. 2 Interplay of Organic and Inorganic domains in Hybrid Catalysis and Functional Interfaces.

Indeed, the sequences of oxidation/reduction catalytic events occurring at the chloroplast and mitochondrium level, are confined within different sets of membranes, also arranged in stacks, implementing a high surface area and promoting multiple-turnover catalysis (Fig.1). This is the rational behind an ever increasing effort towards the intimate comprehension of enzymatic mechanisms, thus fostering their replication in-vitro. On the other side, the direct use of natural enzymes in industrial-scale synthetic processes is possible but not available for most of the oxidation chemistry cited above often lacking substrate generality. There is an urgent call for synthetic catalysts, designed along bio-inspired guidelines, and capable to rival the natural paradigm. In this respect, the one formidable challenge is stability, and long-term performance. Natural systems can suffer from lethal damage, self-inflicted under strong-oxidizing turnover regime, as the one required for water oxidation (vide infra). However, very efficient self-repairing strategies are also adopted *in-vivo* that will be out-of-reach *in-vitro*. Therefore, the synthetic design of functional system is not intended to merely replicate the natural machinery, in some aspects too complicated and fragile, but to overcome its limitation by the engineering of more robust molecules and architectures to stand multiple turnover efficiency and be recovered and used again with no need of repair. In 1995, Ronald Breslow reported that "the design of new catalysts need not slavishly follow what we can learn from enzymes, but it is a good place to start".[5] Thus, a detailed knowledge of the composition, structure and reaction mechanism of the metal core selected by Nature and of its structural arrangement is the first mandatory requirement. The next step, as a general strategy, is the design of a tailored ligand system mimicking the proximal electronic and steric features of the natural enzyme. In this perspective, the new frontier of bio-inspired catalysis looks at the nano-scale. The introduction of a nanostructured environment is instrumental to (i) support/anchor/stabilize the inorganic cores (ii) control the morphology and surface area of the resulting nano-material (iii) funnel/promote the direction of sequential electron and proton transfer (iv) provide a spatial confinement of external catalytic triggers, such as photoactive molecules/sensitizers or redox mediators; (v) control the surface properties and the catalyts environment for substrate recognition/affinity events. This target opens up new avenues for catalyst design that benefits from the introduction of particle size and morphology as powerful parameters. This approach can benefit from the well-established progress in the production/isolation of a wide variety of nano-materials, in many cases affordable with uniform, monodisperse morphologies, also featuring improved stability and robustness as requested for functional applications. Accordingly, selected nanosized scaffolds to be considered for the shaping of the hybrid artificial enzyme, include:[6] a) carbon nanostructures;[7] b) functional polymers (star polymers or dendrimers);[8] c) biogenic protein cages like plant mosaic viruses and d) polyhedral oligomeric silsesquioxanes (POSS) often defined as molecular silica, and amenable to a large variety of media and interfaces by effect of the outer-surface functionalization with selected organic functions.[9] Following a bio-inspired outlook a combined interplay of synthetic strategies can be exploited to merge the inorganic functional cores and the organic stereo-electronic effectors within the nano-enviromnent. This implies a covalent approach but also a non-covalent, supramolecular assembly via multiple hydrogen bonding and/or complementary electrostatic interactions. In the specific field of oxidation catalysis, a further class of molecular nano-clusters is receiving considerable attention. This is the class of totally inorganic polyoxometalate complexes described in the next session

MOLECULAR METAL OXIDES AS INORGANIC SYNZYMES

Polyoxometalates (POMs) are oxo-clusters of early transition metals in their highest oxidation state, namely Mo(VI), W(VI), V(V), Nb(V), Ta(V). A simple classification is based on their chemical composition, with two types of general formula: a) $[M_mO_y]^{p-}$, *isopolyanions* b) $[X_xM_mO_y]^{q-}$, *heteropolyanions*, where M is the main transition metal constituent of the polyoxometalate, O the oxygen atoms and X can be a non-metal (as P, Si, As, Sb), another element of the p block, or a different transition metal.[10-14] POMs are generally formed only by transition metals with dimensions (cationic radius) compatible with an octahedral coordination. In addition, empty low energy d orbitals on the metal ion are required to form at least one terminal M=O double bond. The aggregation of octahedral units MO_6 occurs upon condensation between two central metal ions through μ-oxo bridged bonds, while the presence of less basic, terminal oxygen atoms prevent the linear polymerization (as in common metal oxides, silicates, germanates, tellurates) and promotes the formation of discrete molecular units. One of the most important class of polyoxometalates, is that of Keggin heteropolyanions. Their general formula is: $[XM_{12}O_{40}]^{n-}$, with M = Mo (VI) or W (VI). Keggin obtained the structure of the hexahydrated dodecatungstophosphoric acid for the first time in 1934, by powder X-ray investigation.[15] This structure is called α-Keggin and consists of a central PO_4 tetrahedron surrounded by 12 octahedrons WO_6, with a single terminal oxo group on each metal ion (Fig. 3).

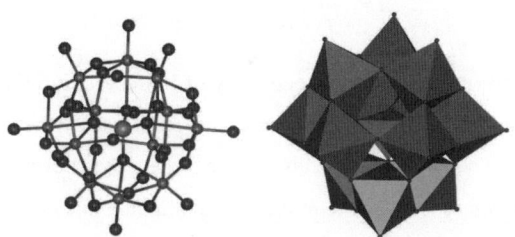

Fig. 3 Ball-and-stick (left) and polyhedral (right) model of α-Keggin $[PW_{12}O_{40}]^{3-}$ heteropolyanion.

These discrete, soluble multi-metal oxides are characterized by a formidable structural variety that can be achieved by tuning reaction parameters (such as concentrations and/or stoichiometric ratio of the reagents, temperature and pH), thus yielding nanosized complexes with different shape, charge density and surface reactivity. When considering isostructural POMs, in particular, such properties can be controlled at the molecular level, by changing their constituent elements (other transition metals or the heteroatom X). In addition, the choice of suitable counterions gives access to their solubilisation and use in a wide range of solvents, from water, with alkaline cations or protons, to apolar ones (toluene, dichloromethane), with lipophilic cations such as tetraoctylammonium. POMs are able to act as ligands for different redox active transition metals (M') such as iron, manganese, cobalt and ruthenium or d^0 metals including zirconium, hafnium and titanium. Incorporation of the transition metal within the POM structure yields Transition Metals Substituted Polyoxometalates (TMSP). POM-based ligands feature rigid polydentate binding sites with unique electron-acceptor ability, extreme robustness and interesting structural and coordination properties for oxidation catalysis (Fig.4).

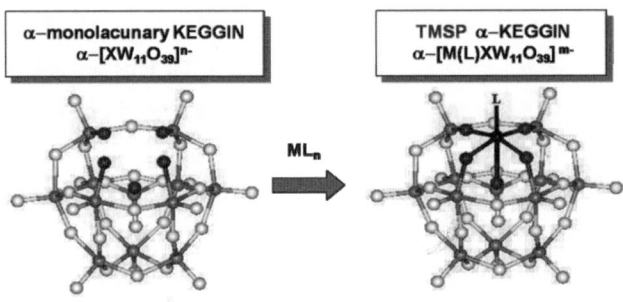

Fig. 4 Synthesis of Transition Metal Substituted POMs by metallation of lacunary POM precursors.

Worthy of notice is that the coordination geometry of representative POM structural type finds a close correspondence with the active site of natural oxygenase enzymes. Among the class of Fe-substituted polyoxotungstates some complexes are known, where the coordination geometry of the iron center exhibits a striking oxygenase synzyme motif. As highlighted in Fig. 5, Fe-substituted Keggin α-undeca- and γ-decatungstates[16,17] and the Krebs type derivatives,[18] may suggest a POM-based inorganic mimicry respectively of: *heme*-Cytochrome P450, the $Fe_2(\mu\text{-}O)_2$ diamond core of Methane Monooxygenase (MMO) and the non-*heme* dioxygenase iron site with three exchangeable coordination positions. Thus, a most desirable feature would be that such a structural diversity might pose the basis for a variety of catalytic functions paralleling the enzyme paradigm. A bio-inspired approach on the design and application of the POM-based catalysts has been proposed in the literature as highlighted in the next paragraphs.[19,20]

POM-based analogues of heme-Cytochrome P450

Cytochrome P450 are a class of metalloproteins enabling hydrocarbon oxygenation with outstanding selectivity and performance.[21,22] The active site of these enzymes is recognized to be an *heme* group, where an iron centre, coordinated by a porphyrin, is able to bind O_2 and activate a mono-oxygenase pathway. In 1973, L. C. W. Baker first recognized that monolacunary polyoxotungstates ligate hetero transition metals in a pseudo-porphyrin environment (Fig. 5).[23]

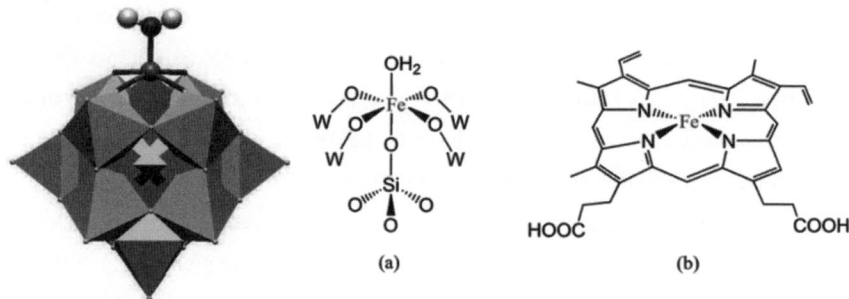

(a)   (b)

Fig. 5 Optimized geometry of $[Fe(H_2O)\alpha\text{-}SiW_{11}O_{39}]^{5-}$ and coordination geometries of the Fe centre within the totally inorganic POM environment (a) and in the heme-enzyme (b).

This key analogy opened the perspective of studying TMSP as the inorganic alternative to metalloporphyrins for oxidation catalysis. The POM environment provides the reactive centre with a highly robust framework where the metal components, in their highest oxidation states (as W^{VI} or Mo^{VI}), are not susceptible to oxidative degradation, at variance with the porphyrin moiety undergoing self-oxidation and loss of cativity under oxidation turnover. Research in this field focuses on the study of TMSPs interaction with bulk oxygen donors (OD), according to a bio-inspired pathways, leading to surface-bound, highly reactive oxene or peroxo intermediates, capable of oxygen transfer to organic substrates (Fig. 6)

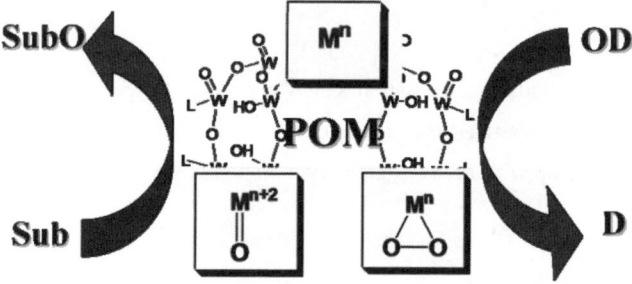

Fig. 6 Catalytic Scheme for Bio-inspired Oxidations with POMs.

Cis-cyclooctene is typically used as a model substrate for alkene epoxidation, as the process occurs with moderate to high selectivity even under auto-oxidation conditions.[24] This benchmark transformation was exploited to investigate the catalytic activity of the mono-substituted $[\alpha-(FeOH_2)SiW_{11}O_{39}]^{5-}$ (**FeSiW$_{11}$**), and compare it with the analogue porphyrin system namely iron (III) tetrakis (5,10,15,20-mesityl) porphyrin (**TMPFeCl**).[25] In this process, biphasic kinetics typical of auto-catalytic reactions are observed, showing an initial lag time (induction period) followed by a zero-order regime (full chain steps). This behavior indicates the occurrence of a radical reaction mechanism. A direct comparison of the catalytic behavior exhibited by **FeSiW$_{11}$** with respect to the organic porphyrin **FeTMPCl** has been realized under analogous reaction conditions.

The fate of the porphyrin catalyst can be conveniently monitored by UV-vis spectroscopy in the region of the Soret absorption (418 nm). At turnover regime, a progressive bleaching of the TMP ligand is observed. Most noticeably, the oxidative degradation follows an auto-catalytic behavior in close correspondence with the epoxide formation (Fig. 7). On the contrary, the stability of **FeSiW$_{11}$** is readily assessed by FT-IR analysis performed after precipitation of the catalyst at different time intervals. As shown in Fig. 7, the POM structural features are preserved after prolonged heating within the reaction mixture.

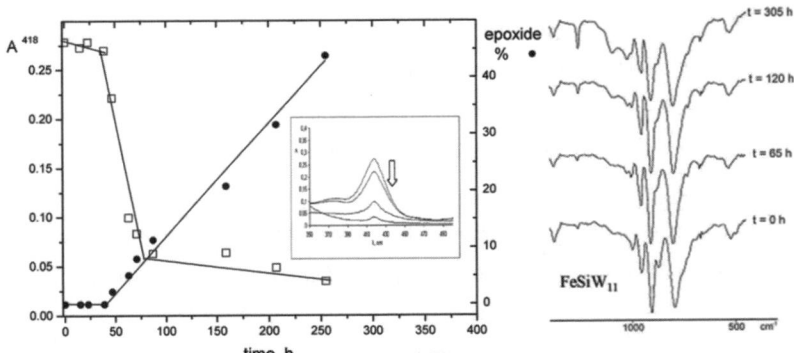

Fig. 7 TMP bleaching (empty squares) as monitored by UV-vis absorbance at 418 nm, in the Soret band region (inset), and epoxide formation (black circles) upon aerobic oxidation of *cis*-cyclooctene (left). Comparison with **FeSiW$_{11}$** stability monitored by FT-IR (right).

POM-based analogues of Methane Monooxygenase (MMO)

Aerobic oxidation of methane to methanol is accomplished at the active site of the hydroxylase component of soluble methane monooxygenase (MMOH); X-ray crystal structure of MMOH from Mc.

Capsulatus (Bath) revealed a di-iron active site, with the two metals being connected by carboxylate groups (diamond core), embedded in a four-helix bundle in each of the two identical α-subunits.[26,27] In the field of polyoxometalates, a di-iron substituted silicotungstate was prepared by reaction of the lacunary $[\gamma\text{-}SiW_{10}O_{36}]^{8-}$ with $Fe(NO_3)_3$ in acidic aqueous solution, and isolated as the tetra-n-butylammonium salt (TBA); the product was identified as $TBA_{3.5}H_{2.5}[\gamma\text{-}1,2\text{-}\{Fe(OH_2)\}_2SiW_{10}O_{38}]$ (**Fe₂SiW₁₀**).[28] Its structure was proposed to have a dinucelar iron site displaying a similar coordination geometry with respect to MMOH, with the two iron centres connected by two bent μ-oxo groups (Fig. 8).

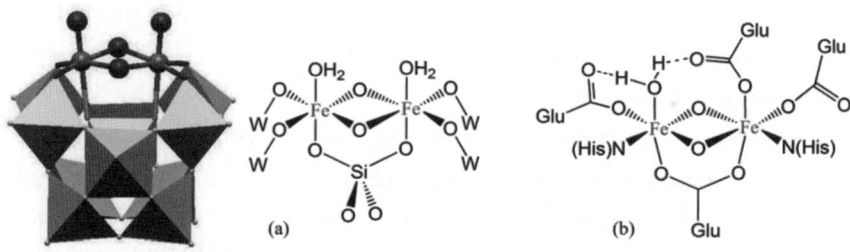

(a) (b)

Fig. 8 Optimized geometry of $[\gamma\text{-}1,2\text{-}\{Fe(OH_2)\}_2SiW_{10}O_{38}]^{6-}$ and coordination geometries of the Fe dinuclear core centre within the totally inorganic POM environment (a) and in the MMOH-enzyme (b).

Notably, **Fe₂SiW₁₀** was also reported to activate molecular oxygen for selective alkene epoxidation (in dichloroethane/acetonitrile mixture, at 356 K), performing a breathtaking record of 10000 turnovers with *cis*-cyclooctene as model olefin.[29] The kinetic law (zero-order dependence on substrate concentration, first order in both catalyst concentration and O_2 partial pressure), the stereoselectivity in *cis*-2-hexene epoxidation, and the absence of any effect upon addition of a radical scavenger to the reaction mixture, and the adamatane selectivit probe lead the authors conclude that the process was a radical-free oxidation, corresponding to a dioxygenase mechanism. This claim was then reconsidered, by a careful evaluation of the product/reagent stoichiometric balance under aerobic conditions, definitely pointing at a radical autoxidation pathway.[30] On the other hand, **Fe₂SiW₁₀** activates hydrogen peroxide for highly efficient alkane hydroxylations in organic solvents at variance with the di-manganese analogue, that is poorly reactive.[31,32]

In this chemistry, the vanadium derivative $(TBA)_4[\gamma\text{-}1,2\text{-}\{VO(\mu\text{-}OH)\}_2SiW_{10}O_{36}]$, containing a $\{VO\text{-}(\mu\text{-}OH)_2\text{-}VO\}^{4+}$ fragment, was used for the epoxidation of non-activated aliphatic terminal olefins, including propylene, in the presence of a stoichiometric amount of H_2O_2, with 99% epoxide selectivity and 87% H_2O_2 conversion in CH_3CN/t-BuOH at 20°C, in 24 h. More accessible, but less nucleophilic double bonds in non-conjugated dienes were regioselectively epoxidized in high yields.[33] Since the mono- and trivanadium substituted $[\alpha\text{-}SiVW_{11}O_{40}]^{5-}$ and $[\alpha\text{-}1,2,3\text{-}SiV_3W_9O_{40}]^{7-}$ were inactive under the adopted conditions, the key feature of the di-nuclear catalyst for achieving selective epoxidation was proposed to be the $\{VO\text{-}(\mu\text{-}OH)_2\text{-}VO\}^{4+}$ core.[34] Interestingly, the isostructural analogue of the above catalyst where the central heteroatom is posphorous instead of silicon, namely $[\gamma\text{-}1,2\text{-}\{VO(\mu\text{-}OH)\}_2PW_{10}O_{36}]^{3-}$, was found to activate hydrogen peroxide in the stereo- and regioselective oxidation of alkanes to alcohols, with > 96% selectivity and > 80% efficiency of H_2O_2 utilization. In addition, an unusual selectivity leading to the preferential oxidation of secondary rather than tertiary C-H bonds was observed, simply ascribed by the authors to steric factors.[35]

POM-based analogues of non-*heme* Dioxygenase

Catechol dioxygenase are non-heme enzymes present in soil bacteria which take part of Nature's strategy for degrading aromatic molecules in the environment. Indeed, they are responsible for the aerobic biodegradation of toxic dihydroxylated aromatic compounds into aliphatic carboxylic acids.[36,37] They are usually classified into intradiol-cleaving and extradiol-cleaving enzymes, depending on the regioselectivity of aromatic ring cleavage. Among the class of intradiol-cleaving enzymes, the protocatechuate 3,4-dioxygenase (3,4-PCD) has been the most extensivley investigated.

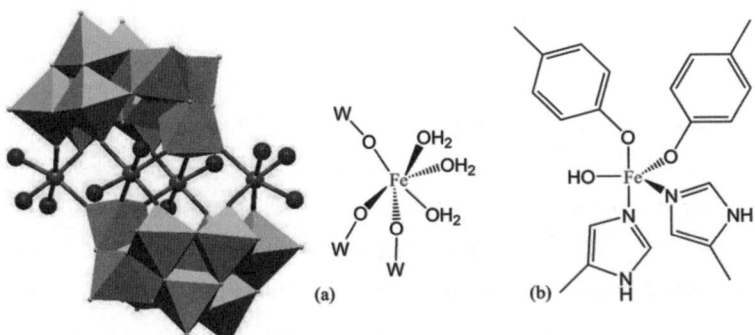

Fig. 9 Optimized geometry of $[Fe_4(H_2O)_{10}(\beta\text{-}AsW_9O_{33})_2]^{6-}$ and coordination geometries of the Fe centre displaying three exchangeable sites within the totally inorganic POM environment (a) and in the protocatechuate-3,4-dioxygenase enzyme (b).

The crystal structure of 3,4-PCD reveals a trigonal bipyramidal iron(III) centre with four amminoacidic ligands, two histidines and two tyrosines.[38,39] The fifth coordination position in the trigonal plane is occupied by a OH⁻ group. Catechol binds to the Fe(III) centre as a di-anion, by replacing the hydroxide ion and an axial tyrosine residue.[40,41] In this scenario, the one family of tetra-substituted Krebs-type POMs offers a similar coordination geometry for catechol activation (Fig. 9). Indeed, the Iron(III) derivatives with general formula $[Fe_4(H_2O)_{10}(\beta\text{-}XW_9O_{33})_2]^{n-}$ ($Fe_4X_2W_{18}$, X = As^{III}, Sb^{III}; n = 6; X = Se^{IV}, Te^{IV}; n = 4) display two POM sub-units stabilizing a belt of four Fe centres, connected through μ-oxo bridges.[42] In analogy with the natural enzyme, two of the four iron sites have three coordination sites occupied by water ligands, and are therefore available for coordination of catechols and/or dioxygen, which is a key requirement for catechol ring cleavage activity (Fig. 10).

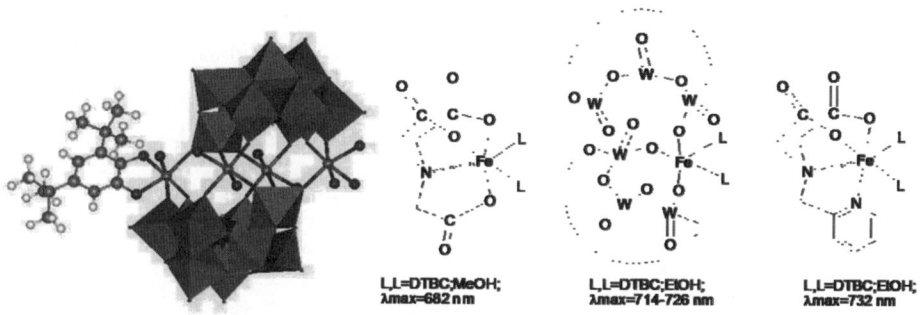

L,L=DTBC;MeOH;
λmax=682 nm

L,L=DTBC;EtOH;
λmax=714-726 nm

L,L=DTBC;EtOH;
λmax=732 nm

Fig. 10 Bio-inspired substrate activation mechanism through coordination of model 3,5-di-*tert*-butylcatechol (DTBC) toFe(III) complexes. Comparison of the inorganic POM ligand (middle) with organic tetradentate, tripod-type $N(CH_2X)_3$ ligands (left and right), influencing the spectroscopic properties of Fe-catecholate complexes.

Indeed, coordination of DTBC to the iron centres was confirmed by visible spectroscopy, by the appearance of bands in the region 570-750 nm due to ligand to metal charge transfer absorptions. Moreover, the inorganic POM ligand remarkably influences the Fe(III)–catechol interaction, resulting in diverse λ_{max} values: the energies of the Catecholate-to-Fe charge-transfer transitions can therefore be tuned by the POM composition. Red-shifted LMCT transition maxima are associated with a favourable donor-acceptor interaction, thus enhancing the semiquinone character of the bound catecholate, and its reactivity with dioxygen, in agreement with the enzyme mechanism.[43] Catechol coordination provides a sensitive probe for the Lewis acidity of the iron center, depending on the donor properties of the ligand set within its coordination sphere. A key aspect is to address the electronic character of the POM moiety vis-à-vis a well-established database available for classical polydentate organic ligands. In particular, a most informative study has been realized with tetradentate tripod ligands, of general formula $N(CH_2X)_3$, where X is a phenolate, carboxylate, pyridine or benzimidazole functionality.[44] In the POM system, the Lewis acidity of the Fe substituent is not substantially quenched by the overall negative charge of the polyanion. Indeed, within the $Fe_4X_2W_{18}$, in EtOH as solvent, the LMCT band falls at $\lambda_{max} > 700$, i.e. at lower energy compared to a tricarboxylate-based ligand, while approaching a mixed ligand set, incorporating two carboxylates and one pendant pyridine (Fig. 10). The biomimetic potential of $Fe_4X_2W_{18}$ catalysts was screened in the aerobic oxidation of 3,5-di-*tert*-butylcatechol (DTBC) as the model substrate under several reaction conditions. Fast substrate conversion was obtained in all cases, but the autoxidation process was the dominant one yielding 3,5-di-*tert*-butyl-1,2-benzoquinone (DTBQ) in 50-60% yield, along with a mixture of polymeric tars, in DCE at 1 atm O_2 pressure and 60 °C. However, in wet tetrahydrofurane as the solvent, and in the presence of 2,6-di-tert-butylcresol as a radical scavenger a cleavage selectivity up to 40% was obtained yielding both intra and extradiol products, namely 3,5-di-tert-butyl-muconic acid anhydride and 4,6-di-tert-butyl-1-oxacyclohepta-4,6.diene-2,3-dione (Fig.11).[43]

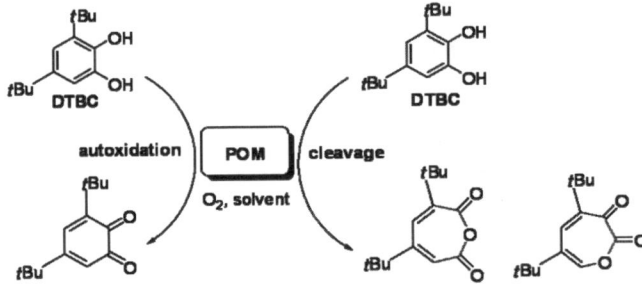

Fig. 11 Competing autoxidation and cleavage pathways in the aerobic oxidation of DTBC by $Fe_4X_2W_{18}$.

Interestingly, both the contemporary presence of Fe(III) and the Krebs like structure were necessary for such bio-inspired catalysis: indeed, isostructural complexes $[M_4(H_2O)_{10}(\beta-XW_9O_{33})_2]^{n-}$ where $M = Cr^{III}$, Mn^{II}, Co^{II}, Ni^{II}, Cu^{II}, Cd^{II} and other Iron-polyoxometalates with different coordination geometries lead to negligible DTBC conversion.[43]

SHAPING THE BEATING HEART OF ARTIFICIAL PHOTOSYNTHESIS: Efficient Water Oxidation at Polyoxometalate/Carbon Nanotube Interfaces.

Green plants convert solar energy into chemical energy by means of natural photosynthesis, that is light driven transformation of water and carbon dioxide into dioxygen and carbohydrates. Water oxidation to dioxygen occurs as one of the terminal processes of natural photosynthesis, and is accomplished at the heart of the Photosystem II enzyme(PSII), by a polynuclear metal-oxo cluster with four manganese and one calcium atom (Mn$_4$Ca). The structure of this active site has recently been revealed by crystallographic and spectroscopic techniques (Fig.12).[3]

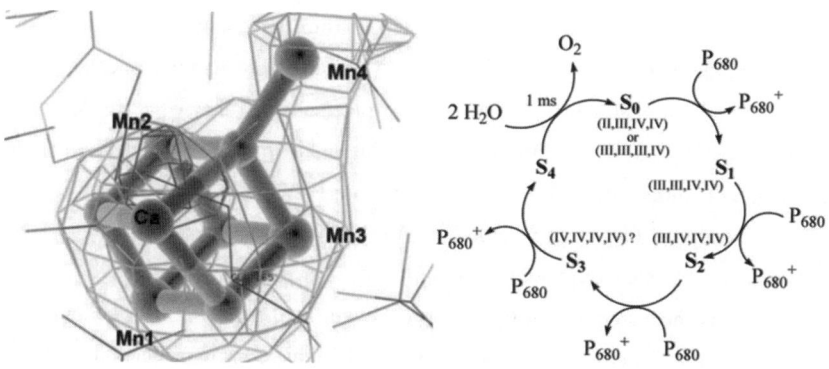

Fig. 12 Structure of the tetra-manganese Oxygen Evolving Center in PSII (courtesy of Professor James Barber, Imperial College, London, UK) based on the coordinates PDB file 1S5L for the crystal structure of PSII isolated from Cyanobacterium, Thermosynechococcus elongatus (left). The Kok cycle of the Oxygen Evolving Centre in PSII (right).

The more accredited topology involves a CaMn$_3$O$_4$ fragment, with a trigonal pyramidal arrangement, linked to a fourth "dangling" Mn center via oxo bridges. The adoption of a catalytic core, featuring adjacent multi-transition metal centers and multiple-μ-hydroxo/oxo bridging units, is a winning strategy devised by Nature to effect multiple/cascade transformations with minimal energy cost. Indeed, water oxidation occurs with unmatched efficiency catalyzed by the Oxygen Evolving Center embedded within the PSII enzyme. The PSII-OEC can master a four-electron/four-proton mechanism through sequential redox steps, thus featuring a unique flexibility of the structural metal-core by assistance of the nanostructured proteic environment.

The mechanism of water oxidation by the PSII-OEC has been extensively investigated.[45-48] Although several details still need to be clarified, a general consensus has emerged, indicating that a light induced four electron-oscillation pattern, along five oxidation states S_n (n = 0-4), is responsible for the catalytic cycling of the oxygenic $CaMn_4O_x$ core. In this scheme, S_0 is the most reduced and S_4 the four-electron oxidized state; when S_4 is generated it reacts rapidly (1 ms) releasing O_2 and returns to the initial S_0 form of the enzyme (Kok cycle, Fig. 12).[49] The photo-oxidation cycle is driven by a chlorophyll pigment radical cation, called P_{680}^+, generated upon illumination (680 nm is the optimal absorption wavelength) and which is by far the strongest oxidizing agent known in living systems ($P_{680}/P_{680}^{+\cdot} \approx 1.25$ V vs NHE). In the natural system $P_{680}^{+\cdot}$ is generated in close proximity to a redox active tyrosine (TyrZ), in turn located at ca. 7 Å distance from the manganese cluster. Such donor-acceptor sequence entails a photo-induced electron flow within nano- to microsecond time domain, whereby upon four $P_{680}/P_{680}^{+\cdot}$ enabling light flashes, electron transfer in the sequence $P_{680}^{+\cdot} \leftarrow$ TyrZ $\leftarrow$ $CaMn_4O_x$, powers the OEC S_0-S_4 evolution. Within a perfectly merged hybrid nano-environment, oxygenic photosynthesis turns out to effect a solar-activated $4e^-/4H^+$ catalytic routine, with multi-turnover (TON) efficiency yieding up to ca. 400 TON per second, with overall quantum yield close to 10% and operating at a moderate overpotential (ca. 0.3-0.4 V at physiological pH).

Herein the Achilles' heel stems from the intrinsic weakness of the functional components chosen by Nature to master such peak performance. Despite some elaborate control/protection strategies, under turnover regime the PSII enzyme undergoes to fatal damage, thus requiring a perpetual self-healing reconstruction, every ~ 30 min.[50] The artificial perspective should find its roots on more solid materials, with the aim to transcend the natural wonder, while being inspired by its key guidelines, by designing a functional system/device, with superior operation stability and long-term endurance. In the next paragraphs we will highlight a recently discovered pathway carved within the class of inorganic metal-oxides displaying a unique mimicry of PSII enzymes.

Robust, All Inorganic Polyoxometalates as Oxygen Evolving Catalysts

The first structurally characterized POM based species acting as OEC was reported in 2008.[51,52] This is a tetraruthenium polyoxometalate $\{Ru^{IV}_4(\mu\text{-}OH)_2(\mu\text{-}O)_4(H_2O)_4[\gamma\text{-}SiW_{10}O_{36}]_2\}^{10-}$ ($\mathbf{Ru_4SiW_{10}}$) easily obtained in high yield by metalation of the lacunary polyoxometalate $K_8 [\gamma\text{-}SiW_{10}O_{36}]$ with $K_4[Ru_2OCl_{10}]$ in aqueous solution.[51] The structure of $\mathbf{Ru_4SiW_{10}}$ features two staggered $[\gamma\text{-}SiW_{10}O_{36}]^{8-}$ units embedding a tetraruthenium-oxo core with adamantane like arrangement (Fig.13).

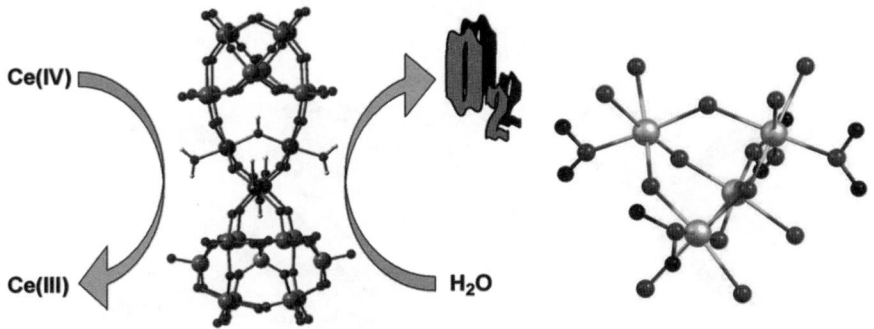

Fig. 13 Solid State Structure of **Ru₄ SiW₁₀** and oxygenic activity in the presence of bulk oxidants as Ce(IV)(left); details of the tetraruthenium-oxo core (right).

The presence of a tetranuclear active bears a stringent analogy with the tetramanganese active site of PSII. Moreover, the Ru-O connectivities in the tetraruthenate core resemble the surface defects of the rutile structural motif of RuO_2, which was one of the first OEC reported to date. Thus, the tetraruthenium core of **Ru₄SiW₁₀** can be considered as an active fragment of an extended inorganic oxide material, but featuring a molecular nature. Oxygenic Activity of **Ru₄SiW₁₀** was initially assessed in water and in the presence of sacrificial oxidants as Ce(IV), under non illuminated, dark conditions (Fig.13). Under the condition explored, excess of Ce(IV), yielded up to 500 catalytic cycles for oxygen evolution, with an initial turnover frequency (TOF) of 0.125 s^{-1}, while a second recharge in Ce(IV) induces an equivalent oxygen production. The full comprehension of the reaction mechanism of water oxidation catalyzed by **Ru₄SiW₁₀**is still under investigation. A unifying picture foresees at least four consecutive single electron oxidation steps, finally yielding a high valent intermediate responsible for oxygen production, and collapsing back in a single four electron step. The proton coupled electron transfer seems to be supported by cyclic voltammetry and also by computational raman spectroscopy addressing high valent derivatives generated *in situ* by stoichiometric addition of Ce(IV).[53]

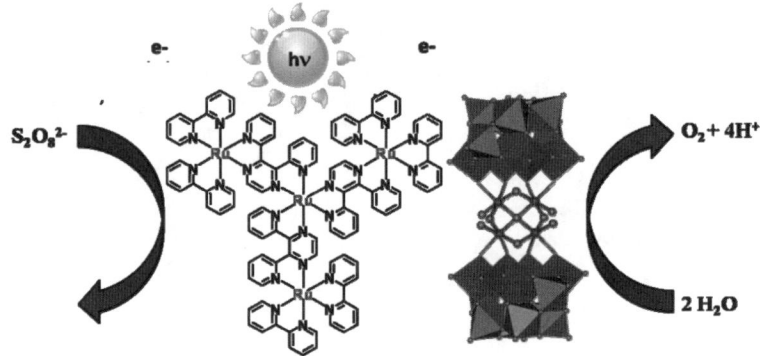

Fig. 14 Photocatalytic water oxidation powered by the interplay of tetra-ruthenium sensitizer/catalyst. Photoinduced electron flow and scavenging by persulfate as sacrificial acceptor is shown by arrows.

Besides its totally inorganic nature and consequent high stability under oxidation regime, a further key property of **Ru$_4$SiW$_{10}$** is that it works efficiently with photogenerated oxidants bearing cationic charges in the presence of a sacrificial electron sink (e.g. persulfate $S_2O_8^{2-}$), paving the way towards light induced water oxidation. The best results have been achieved with the tetranuclear Ru(II) antenna sensitizer, $[Ru\{(\mu\text{-dpp})Ru(bpy)_2\}_3]^{8+}$ (dpp = 2,3-bis(2'-pyridyl)pyrazine) (Fig.14). This molecule has enhanced absorption properties with respect to the parent mononuclear $[Ru(bpy)_3]^{2+}$, with a maximum of absorption at $\lambda = 550$ nm, then with the possibility of accessing to a larger fraction of visible light, where solar irradiation is more intense. Combining this photosensitizer and **Ru$_4$SiW$_{10}$**, water oxidation to dioxygen is observed at 550 nm, with a maximum TOF = $8\cdot10^{-3}$ s^{-1}. The quantum yield for this process, reaches the breakthrough value of 0.30.[54] One of the reasons for such high performance is the fast reaction kinetics of the photogenerated oxidant with the catalyst; indeed, a bimolecular rate constant k = $2.1\cdot10^9$ M^{-1}s^{-1}, within one order of magnitude of diffusion-controlled rate, was measured by laser flash photolysis when $[Ru(bpy)_3]^{2+}$ was used as the photosensitizer. Such very high value may be probably explained by the complementary electrostatic charge of the cationic photogenerated oxidant and the polyanionic catalyst.[55]

Nano-structured oxygen evolving anodes

The unique properties of **Ru$_4$SiW$_{10}$** as a bio-inspired water oxidation catalyst makes it extremely attractive for the support on heterogeneous materials and the development of oxygen evolving anodes, finalized to the assembly of a photoelectrocatalytic cell and the continuous production of hydrogen. A proof-of-principle was recently achieved by integrating the POM domains on modified multi-walled carbon nanotubes (MWCNT). These latter are also widely used to fabricate nanostructured electrodes, due to the possibility (i) to increase the (active) surface area, and (ii) to act as electrical wires by virtue of their electron conducing properties. To this purpose MWCNT were functionalized by covalent grafting of cationic pendant PAMAM dendrimers.[7] In details, the authors proceeded with the divergent growth of a polyamidoamine dendrimer anchored on the tube surface, achieving a second generation bearing free amino groups, further reacted with positively charged ammonium moieties.

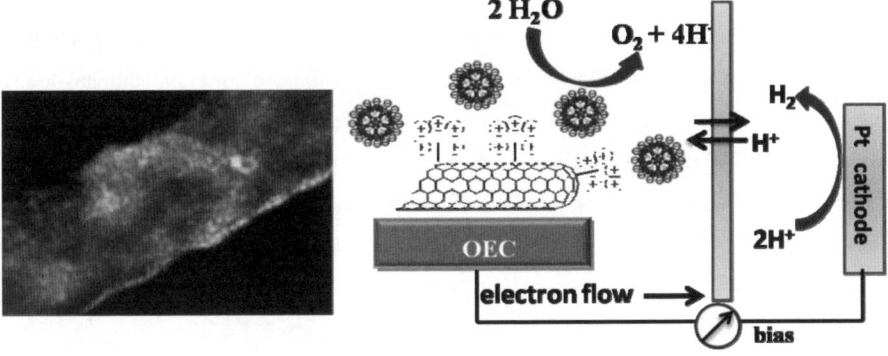

Fig. 15 **Ru$_4$SiW$_{10}$@MWCNT** composite material for water splitting anodes. TEM image (left) and scheme of a complete electrochemical cell for water splitting, integrating the nanostructured anode.

Assembling of the POM catalyst with the tubes is held by electrostatic interactions between the anionic polyoxometalate and the cationic pendants (Fig. 15). The maintenance of the **Ru$_4$SiW$_{10}$** structure integrated in the nano-material was confirmed by resonant raman spectroscopy and small angle X-ray scattering spectroscopy.[56] Highly efficient nano-structured oxygen evolving anodes are obtained by drop casting the hybrid material on Indium Tin Oxide (ITO) electrodes. Indeed, a catalytic current due to water oxidation is observed when an increasing potential is applied to the electrode, in aqueous phosphate buffer at pH = 7.0. Indeed, turnover frequency values in the range 36-306 h^{-1} were achieved depending on the applied overpotential: an appreciable catalytic current with a remarkable TOF (36 h^{-1}) was observed beginning at η = 0.35 V, and reached a peak performance of 306 h^{-1} at η = 0.60 V. These results address the importance of hybrid interfaces/contacts to control and promote electron transfer events at heterogeneous surfaces. Furthermore, the shaping of their functions by specifically tailored carbon nano-structures and/or polymeric scaffolds opens a vast scenario for tuning electron/proton transfer mechanisms in term of rates, distance, geometries and communication between donor/acceptor centers.

TAKE-HOME LESSON

Due to their remarkable stability, oxidation catalysis by POMs continues to be a timely field of investigation. The possibility to tune their composition at the molecular level, offers the advantage of controlling fundamental properties of interest for selective oxidations. Steric hindrance, redox potential, single-site activation via coordination/Lewis acid catalysis are instrumental to direct the chemo- and regioselectivity required for the processing of polyfunctional substrates. In addition, the easy access to a large structural/compositional variety, combined with their unique molecular nature, and nano-sized dimensions offers straightforward tools for mechanistic investigation under turnover regime, with the possibility to trace fundamental structure-activity descriptors, validated by computational tools, and representing a functional model of extended metal-oxides heterogeneous surfaces.

In conclusion, the use of polyoxometalate catalysts, along with benign oxidants represent a true possibilty for the development of novel and selective synthetic procedures. Finally, Artificial Photosynthesis for energy applications represents the next challenging frontier in TMSP-based catalysis.

Acknowledgments. Financial support from University of Padova (Progetto Strategico 2008, HELIOS, prot. STPD08RCX, PRAT CPDA084893/08), MIUR (PRIN contract No. 20085M27SS) and Fondazione Cariparo, (Nano-Mode, progetti di eccellenza 2010) is gratefully acknowledged.

REFERENCES

[1] Deuss, P. J.; den Heeten, R.; Laan, W.; Kamer, P. C. J. *Chem. Eur. J.*, **2011**, *17*, 4680.

[2] a) Gray, H. B. *Nature Chem.*, **2009**, *1*, 7; b) Lewis, N. S.; Nocera, D. G. *Proc. Natl. Acad. Sci. U.S.A.* **2006**, *103*, 15729.

[3] a) Loll, B.; Kern, J.; Saenger, W.; Zouni, A.; Biesiadka, J. *Nature* **2005**, *483*, 1040; b) Ferreira, K. N.; Iverson, T. M.; Maghlaoui, K.; Barber, J.; Iwata, S. *Science* **2004,** *303,* 1831; c) Yano, J.; Kern, J.; Sauer, K.; Latimer, M. J.; Pushkar, Y.; Biesiadka, J.; Loll, B.; Saenger, W.; Messinger, J.; Zouni, A.; Yachandra, V. K. *Science* **2006**, *324*, 821.

[4] Balzani, V.; Credi, A.; Venturi, M. *ChemSusChem* **2008**, *1*, 26.

[5] Breslow, R. *Acc. Chem. Res.* **1995**, *28*, 146.

[6] Ukida, M.; Klem, M. T.; Allen, M.; Suci, P.; Flenniken, M.; Gillitzer, E.; Varpness, Z.; Liepold, L. O.; Young, M.; Douglas, T. *Adv. Mater.* **2007**, *19*, 1025.

[7] Herrero, M. A.; Toma, F. M.; Khuloud, T. A. J.; Kostarelos, K.; Bianco, A.; Da Ros, T.; Bano, F.; Casalis, L.; Scoles, G.; Prato, M. *J. Am. Chem. Soc.* **2009**, *131*, 9843.

[8] Iha, R. K.; Wooley, K. L.; Nyström, A. M.; Burke, D. J.; Kade, M. J.; Hawker, C. J. *Chem. Rev.* **2009**, *109*, 5620.

[9] Li, G.; Wang, L.; Ni, H.; Pittmann, C. U. *J. Inorg. Organomet. Polym.* **2002**, *11*, 123.

[10] Pope, M. T.; Müller A. *Heteropoly and Isopoly Oxometalates,* Springer Verlag, New York, **1983**.

[11] Pope, M. T.; Müller, A. *Angew. Chem., Int. Ed.* **1991**, *30*, 34.

[12] Hill, C. L. *Polyoxometalates, Chem. Rev.* (Special Issue), **1998**, *98*, 1.

[13] Kepert D. L. *The Early Transition Metals,* Academic Press Inc., London, **1972**.

[14] Jeannin, Y. P. *Chem. Rev.* **1998**, 98, 51.

[15] Keggin, J. F. *Proc. R. Soc. London Scr. A.* **1934**, *144*, 75.

[16] Zonnevijlle, F.; Tourné, C. M.; Tourné, G. F.; *Inorg. Chem.* **1982**, *21*, 2751.

[17] Nozaki, C.; Kiyoto, I.; Minai, Y.; Misono, M.; Mizuno, N. *Inorg. Chem.* **1999**, *38*, 5724.

[18] Kortz, U.; Savelieff, M. G.; Bassil, B. S.; Keita, B.; Nadjo, L. *Inorg. Chem.* **2002**, *41*, 783.

[19] N. Mizuno, M. Hashimoto, Y. Sumida, Y. Nakagawa, K. Kamata. *Polyoxometalate Chemistry for Nano-Composite Design,* Kluwer Academic/Plenum, New York, 2002, p. 197.

[20] (a) Neumann, R.; Dahan, M. *Nature* **1997**, *388,* 353. (b) Neumann, R.; Dahan, M. *J. Am. Chem. Soc.* **1998**, *120*, 11969.

[21] de Montellano P. R. O. *Cytochrome P450: Structure, Mechanism, and Biochemistry, 3rd ed.* Kluwer Academic/Plenum Publishers: New York 2005.

[22] de Montellano, P. R. O *Chem. Rev.* **2010**, *110*, 932.

[23] Baker, L. C. W. Plenary Lecture, XV Int. Conf. on Coord. Chem., Proceedings, Moscow, 1973.

[24] Van Sickle, D. E.; Mayo, F. R.; Arluck, R. M. *J. Am. Chem. Soc.* **1965**, *87*, 4824.

[25] Bonchio, M.; Carraro, M.; Farinazzo, A.; Sartorel, A.; Scorrano, G.; Kortz, U. *J. Mol. Catal. A: Chem.,* **2007**, *262*, 36.

[26] Rosenzweig, A. C.; Frederick, C. A.; Lippard, S. J.; Nordlund, P. *Nature* **1993**, *366*, 537.

[27] Rosenzweig, A. C.; Nordlund, P.; Takahara, P. M.; Frederick, C. A.; Lippard, S. J. *Chem. Biol.* **1995**, *2*, 409.

[28] Nozaki, C.; Kiyoto, I.; Minai, Y.; Misono, M.; Mizuno, N. *Inorg. Chem.* **1999**, *38*, 5724.

[29] Nishiyama, Y.; Nakagawa, Y.; Mizuno, N. *Angew. Chem., Int. Ed.* **2001**, *40*, 3639.

[30] Nishiyama, Y.; Nakagawa, Y.; Mizuno, N. *Angew. Chem., Int. Ed.* **2007**, *46*, 4006.

[31] Mizuno, N.; Nozaki, C.; Kiyoto, I.; Misono, M. *J. Am. Chem. Soc.* **1998**, *120*, 9267.

[32] Mizuno, N.; Kiyoto, I.; Nozaki, C.; Misono, M. *J. Catal.* **1999**, *181*, 171.

[33] (a) Nakagawa, Y.; Kamata; K., Kotani, M.; Yamaguchi, K.; Mizuno, N. *Angew. Chem. Int. Ed.,***2005**, *44*, 5136. (b) Nakagawa, Y.; Mizuno, N. *Inorg. Chem.,* **2007**, *46*, 1727.

[34] Nakagawa, Y.; Uehara, K.; Mizuno, N. *Inorg. Chem.* **2005**, *44*, 9068.

[35] Kamata, K.; Yonehara, K.; Nakagawa, Y.; Uehara, K.; Mizuno, N. *Nature Chem.* **2010**, *2*, 478.

[36] Funabiki, T. *Oxygenases and Model Systems.* Kluwer Academic Publishers: Dordrecht (1997).

[37] Que Jr., L.; Ho, R. Y. N. *Chem. Rev.* **1996**, *96*, 2607.

[38] Ohlendorf, D. H.; Lipscomb, J. D.; Weber, P. C. *Nature* **1988**, *336*, 403.

[39] Ohlendorf, D. H.; Orville, A. M.; Lipscomb, J. D. *J. Mol. Biol.* **1994**, *244*, 586.

[40] Orville, A. M.; Lipscomb, J. D.; Ohlendorf, D. H. *Biochemistry* **1997**, *36*, 10052.

[41] Elgren, T. E.; Orville, A. M.; Kelly, K. A.; Lipscomb, J. D.; Ohlendorf, D. H.; Que Jr., L. *Biochemistry* **1997**, *36*, 11504.

[42] Kortz, U.; Savelieff, M. G.; Bassil, B. S.; Keita, B.; Nadjo, L. *Inorg. Chem.* **2002**, *41*, 783.

[43] Sartorel, A.; Carraro, M.; Scorrano, G.; Bassil B. S.; Dickman, M. H.; Keita, B.; Nadjo, L.; Kortz, U.; Bonchio, M. *Chem. Eur. J.* **2009**, *15*, 7854.

[44] a) Jang, H. G.; Cox, D. D.; Que Jr., L. *J. Am. Chem. Soc.* **1991**, *113*, 9200; b) Cox, D. D.; Que Jr., L. *J. Am. Chem.. Soc.* **1988**, *110*, 8085.

[45] Haumann, M.; Liebisch, P.; Müller, C.; Barra, M.; Grabolle, M.; Dau, H. *Science* **2005**, *310*, 1019.

[46] McEvoy, J. P.; Brudvig, G. W. *Chem. Rev.* **2006**, *106*, 4455.

[47] Brudvig, G. *Coord. Chem. Rev.* **2008**, *252*, 231.

[48] Hammarstrom, L.; Hammes-Schiffer, S. *Acc. Chem. Res.* **2009**, *42*, 1859.

[49] Kok, B.; Forbush, B.; McGloin, M. *Photochem. Photobiol.* **1970**, *11*, 457.

[50] Aro, E.-M.; Suorsa, M.; Rokka, A.; Allahverdiyeva, Y.; Paakkarinen, V.; Saleem, A. et al. *J. Exp. Bot.* **2005**, *56*, 347.

[51] Sartorel, A.; Carraro, M.; Scorrano, G.; De Zorzi, R.; Geremia, S.; McDaniel, N. D.; Bernhard, S.; Bonchio, M. *J. Am. Chem. Soc.* **2008**, *130*, 5006.

[52] Geletii, Y. V.; Botar, B.; Kögerler, P.; Hillesheim, D. A.; Musaev, D. G.; Hill, C. L. *Angew. Chem. Int. Ed.* **2008**, *47*, 3896.

[53] Sartorel, A.; Mirò, P.; Salvadori, E.; Romain, S.; Carraro, M.; Scorrano, G.; Di Valentin, M.; Llobet, A.; Bo, C.; Bonchio, M. *J. Am. Chem. Soc.* **2009**, *131*, 16051.

[54] Puntoriero, F.; La Ganga, G.; Sartorel, A.; Carraro, M.; Scorrano, G.; Bonchio, M.; Campagna, S. *Chem. Commun.* **2010**, *46*, 4725.

[55] Orlandi, M.; Argazzi, R.; Sartorel, A.; Carraro, M.; Scorrano, G.; Bonchio, M.; Scandola, F. *Chem. Commun.* **2010**, *46*, 3152.

[56] Toma, F. M.; Sartorel, A.; Iurlo, M.; Carraro, M.; Parisse, P.; Maccato, C.; Rapino, S.; Gonzalez, B. R.; Amenitsch, H.; Da Ros, T.; Casalis, L.; Goldoni, A.; Marcaccio, M.; Scorrano, G.; Scoles, G.; Paolucci, F.; Prato M.; Bonchio, M. *Nature Chem.* **2010**, *2*, 826.

Francesca Cardona

Assistant Professor in Organic Chemistry

Address:

Dipartimento di Chimica "Ugo Schiff", Polo Scientifico e Tecnologico, Firenze, Italy; e-mail: francesca.cardona@unifi.it

Education:

1995-1999, PhD in Chemical Sciences, Firenze. Tutor: Prof. Brandi.

1990-1995, Laurea in Chemistry (Organic Synthesis). 110/110 et laude.

Professional experience:

1999-2000 (Post-Doc), Institut of Organic Chemistry, University of Lausanne, CH, Prof. Vogel group. 2000-2001, CNR (Consiglio Nazionale delle Ricerche) fellowship, Department of Organic Chemistry, Florence. From 2002 researcher at the University of Florence. Supervisor of several Thesis Research Works from 2005, in 2006 F. Cardona receives the "G. Ciamician" silver medal of the Organic Chemistry Division of the Italian Chemical Society.

She is co-author of 51 publications in refereed journals and 4 chapters in scientific books (h-index 21). Courses given at the University of Florence, Laurea in Chemistry: Laboratory of Organic Chemistry (1st course), Green Chemistry.

Research Interests: stereoselective syntheses of iminosugars and their inhibition of glycosidases, study of new environmentally friendly oxidation methods of nitrogen containing compounds, parallel kinetic resolutions (PKR) by cycloaddition reactions.

Aerobic Oxidations of Alcohols

Francesca Cardona[a]* and Camilla Parmeggiani,[a,b]

[a]Dipartimento di Chimica "Ugo Schiff", Via della Lastruccia 13, Polo Scientifico e Tecnologico, Sesto Fiorentino, 50019, Firenze, Italy e-mail: francesca.cardona@unifi.it
[b]Current address: CNR-INO and European Laboratory for Non-linear Spectroscopy (LENS), via N. Carrara 1, Polo Scientifico Sesto Fiorentino, 50019, Firenze, Italy

INTRODUCTION

Oxidation reactions are among the most useful and used reactions in the industrial processes. However, at the same time, they are among the most polluting and hazardous processes, often occurring with high E-factor (mass of waste per mass unit of product)[1] and delivering considerable amount of toxic waste, for instance metal salts in oxidations employing stoichiometric Cr(VI) or Mn(VII) derivatives or nitrogen oxides in oxidations carried out with HNO_3. In particular, the oxidation of primary and secondary alcohols to the corresponding carbonyl compounds is of fundamental importance in organic synthesis, due to the wide ranging utility of these products as important precursors and intermediates for many drugs, vitamins and fragrances. A recent publication by Pfizer's medicinal chemists[2] showed that the three most popular oxidants used in Pfizer for the oxidation of primary alcohols to the corresponding aldehydes are Dess-Martin periodinane[3] or its precursor IBX, the Swern reagent[4] and TPAP/NMO[5] system. All of these methods still have poor atom efficiencies[6] and significant scale-up issues. As a result, the oxidation of an alcohol to a carbonyl compound, in spite of being a fundamentally important reaction, yet is actually avoided by the pharmaceutical industry.[7] From an environmentally point of view, it is of particular importance the development of methods which use cleaner oxidants and minimize the amount and toxicity of the released waste. Moreover, the use of catalysis, that allows processes to occur under mild conditions in order to save the overall implied energy, is strongly encouraged.[8] In this respect, the recovery and reuse of the catalyst is a further important goal. In this review, we will present an overview of the recent advances made by the international scientific community in this field. Oxygen (or even better air) is among the cheaper and less polluting stoichiometric oxidants, since it produces no waste or water as the sole by-product.[9] The implementation of a catalyst in combination with molecular oxygen represents an emerging alternative process to the traditional procedures.

In the development of transition metal-catalyzed aerobic alcohol oxidations, several challenges exist, as the need of low pressures of O_2 especially in flammable organic solvents, mild reaction conditions, low catalyst loadings, and avoidance of costly or toxic additives.

Another main issue is the functional group tolerance and the chemoselectivity of the alcohol transformation when other groups susceptible to oxidation are present. A further goal is the development of methods able to oxidize one class of alcohols in the presence of another. Finally, an ultimate goal is the development of diastereo- and/or enantioselective alcohol oxidations. Many homogeneous and heterogeneous catalytic systems have been developed;[10] this review, that does not pretend to be exhaustive, aims to give an overview on the most significant procedures developed in this extremely investigated field of the research. The most versatile and studied metal catalysts (copper-, ruthenium-, palladium- and gold-based) will be analyzed, highlighting their synthetic potential and always taking in account the previously mentioned synthetic challenges.

COPPER-BASED HOMOGENOUS CATALYSTS

Copper seems an appropriate metal for the catalytic oxidation of alcohols with O_2 since it is present in Nature as the catalytic centre in a variety of enzymes (e.g. galactose oxidase) that catalyze this conversion. Some catalytically active biomimetic models for these enzymes have been designed and constitute seminal examples in this area.[11] In 1984, Semmelhack reported the first practical Cu-catalyzed aerobic oxidation of alcohols, using Cu in combination with the stable nitroxyl radical TEMPO (2,2,6,6-tetramethyl-1-piperidine-N-oxyl) in DMF as solvent; however, this system was efficient only for activated primary alcohols.[12] Markó and co-workers pioneered much of the catalyst development. In their initial report, a combination of CuCl (5 mol %), phenantroline (5 mol%) and di-tert-butylazodicarboxylate, DBAD (5 mol%) allowed oxidation of alcohols with great tolerance of other functional groups.[13] However, this system required the presence of 2 equivalents of a base (K_2CO_3) and was not consistent for the oxidation of primary aliphatic alcohols. In these basic conditions, alcohols bearing α-stereogenic centres could be oxidized with no racemisation. In these first reports, the active catalyst was postulated to be heterogeneous, and absorbed on the insoluble K_2CO_3, since filtration of the mixture gave a solution devoid of any oxidizing ability. Indeed, it appeared that K_2CO_3 may also served as a solid support on which the copper catalyst could be anchored. A change of the solvent from toluene to fluorobenzene allowed to use catalytic base,[14] and further investigations led to the discovery that addition of catalytic N-methylimidazole dramatically enhanced the activity of the system allowing efficient conversion of primary aliphatic alcohols (Scheme 1).[15]

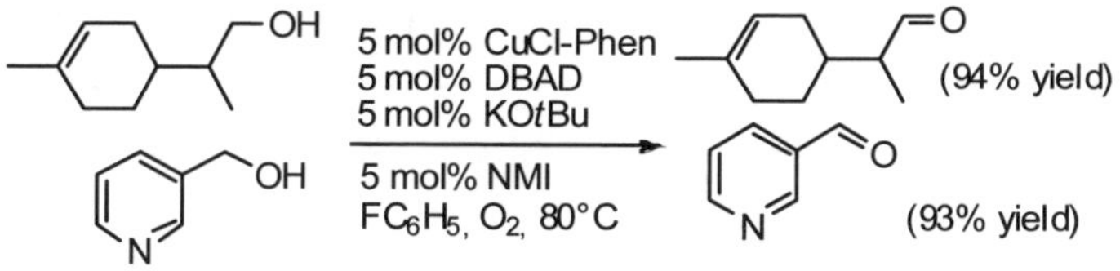

Scheme 1. The Cu-catalytic system developed by Markó and co-workers

The use of such system in combination with a diazo reagent and triphenyl phosphine recently allowed a domino one-pot oxidation-olefination process that could be applied to a wide variety of alcohol substrates including aliphatic secondary alcohols. α-Chiral alcohols could be converted into olefins without any detectable racemization (Scheme 2).[16]

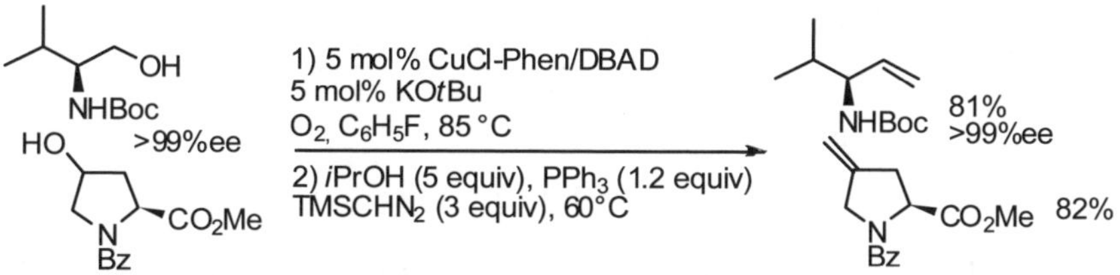

Scheme 2. Copper-catalyzed tandem oxidation-olefination process

In addition to Markó's work, other groups reported chemoselective oxidations of primary alcohols with Cu in combination with TEMPO. Sheldon and co-workers showed that $CuBr_2$ and TEMPO in the presence of 2,2'-bipyridine as a ligand for Cu led to the oxidation of several primary alcohols with no overoxidation to carboxylic acids. When mixtures of primary and secondary alcohols were reacted in these conditions, only primary alcohols were converted. The advantage of this very mild procedure is that excellent conversions were obtained with air at room temperature (Scheme 3).[17]

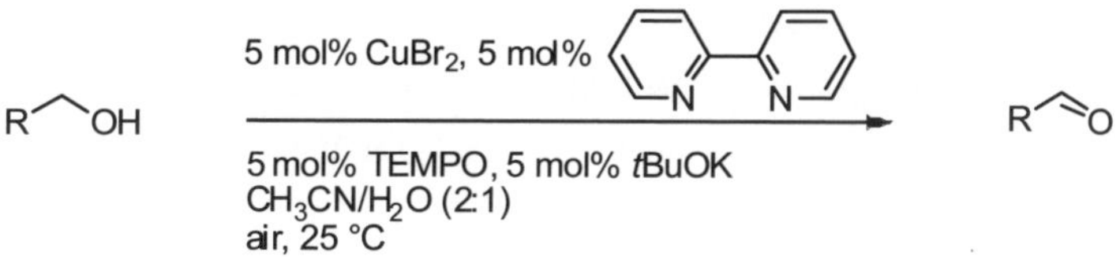

Scheme 3. The Cu(II)-TEMPO catalyzed aerobic oxidation of primary alcohols by Sheldon et al.

In the presence of an enantiopure bidentate ligand, Seckar and co-workers achieved an efficient non enzymatic kinetic resolution of several secondary benzylic amino alcohols (Scheme 4).[18]

Several other groups reported the use of alternative solvents with the aim of allowing catalyst recycling and simple product purification. In 2002 Ansari and Gree reported a CuCl-TEMPO catalyzed aerobic oxidation of several primary and secondary benzylic and allylic alcohols in the 1-butyl-3-methylimidazolium hexafluorophosphate ([bmim]PF_6) ionic liquid.

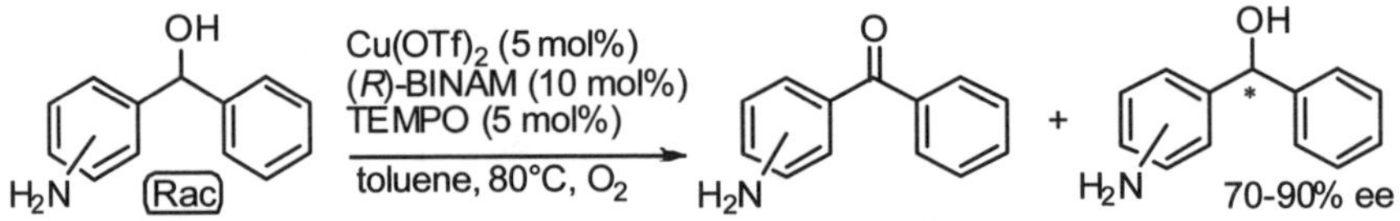

Scheme 4. Oxidative kinetic resolution of secondary benzylic amino alcohols

The authors could recycle the ionic liquid but not the catalyst.[19] Jiang and Ragauskas reported the use of a pyridil based ionic liquid, 1-butyl-4-methylpyridinium hexafluorophosphate ([bmpy]PF$_6$) in a room-temperature aerobic oxidation of primary alcohols catalyzed by a three-component system acetamido-TEMPO/Cu-(ClO$_4$)$_2$/DMAP, that allowed the recover and reuse of catalyst up to five runs without loss of activity (Scheme 5).[20]

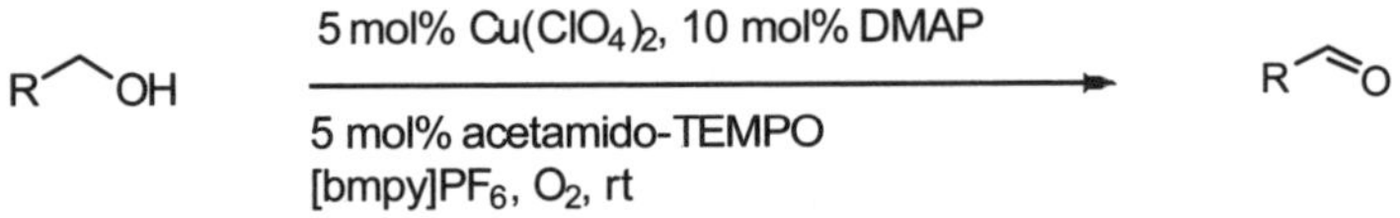

Scheme 5. Cu(II)-TEMPO-catalyzed oxidation in ionic liquid

More recently, they reported a similar three component system for the oxidation of primary alcohols to aldehydes under solvent-free conditions, and simply recovered the three catalyst components by addition of a non polar solvent (hexanes) that selectively dissolved the product aldehydes.[21] In case of solid alcohols, PEG-200 (not oxidized in these reaction conditions) was used as solvent.

To enhance catalyst recyclability, Knochel and co-workers also used a biphasic solvent system, (chlorobenzene/perfluoroctyl bromide), and a pyridine ligand containing fluorinated ponytails for the CuBr-Me$_2$S-TEMPO catalyzed oxidation of several alcohols, and they could recover and reuse the fluorous layer containing the catalyst up to eight times with little loss of activity.[22] Furthermore, the selective aerobic oxidation of benzyl alcohols to the corresponding benzaldehydes could be achieved using the sole water as solvent without the need of any organic or alternative solvent, employing a multinuclear copper (II) compound in combination with TEMPO at 25-80 °C.[23]

COPPER-BASED HETEROGENOUS CATALYSTS

In contrast to the great development of homogenous copper-based catalysts, heterogeneous systems are still largely unexplored. One of the few examples is a recyclable Cu-Mn mixed oxide supported on active carbon that was employed in combination with TEMPO as co-catalyst for the aerobic oxidation of several benzylic primary alcohols.[24]

RUTHENIUM-BASED HOMOGENOUS CATALYSTS

Ruthenium compounds have been extensively studied as catalysts for the aerobic oxidation of alcohols.[25] This metal gives the widest range of oxidation states from +2 to +8, therefore a large variety of oxidative transformations has been developed. The activity of common low valent ruthenium precursors such as $RuCl_2(PPh_3)_3$ can be increased by the use of ionic liquids as solvents.[26] Ruthenium-based compounds have been extensively investigated as catalysts for hydrogen transfer reactions. These systems, in combination with a hydrogen acceptor as co-catalyst and dioxygen as oxidant, can be readily adapted to the aerobic oxidation of alcohols in a multicatalytic process. For example Bäckvall and co-workers, employing a benzoquinone and a cobalt-Schiff's base complex, developed one of the fastest catalytic systems reported for the oxidation of secondary alcohols (Scheme 6).[27]

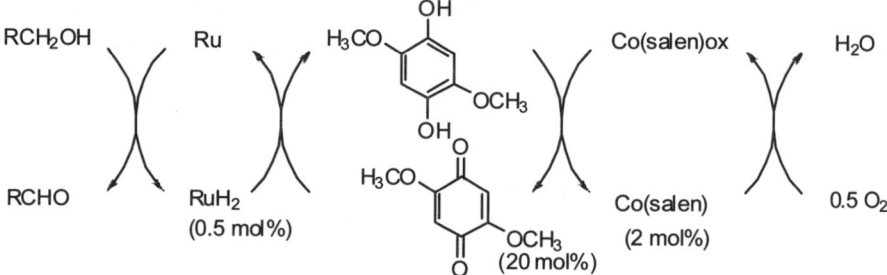

Scheme 6. Bäckvall's multicatalytic system for the aerobic oxidation of alcohols

The sole weakness of this procedure was the requirement of high loading of 2,6-dimethoxy-1,4-benzoquinone (20 mol%) which served as electron transfer mediator (ETM). Recently Bäckvall and co-workers reported, in combination with Ru Shvo's catalyst, a second generation Co hybrid catalyst that comprises cobalt salens and pendant hydroquinone groups, thus avoiding the use of benzoquinone and affording excellent conversions of secondary alcohols.[28] On the other hand, Ishii and co-workers demonstrated that the regeneration of benzoquinone can also be achieved in the absence of the cobalt co-catalyst in $PhCF_3$ as solvent.[29] In these conditions, primary alcohols could be chemoselectively oxidized in the presence of secondary alcohols.

Sheldon and co-workers developed one of the most efficient systems for the aerobic oxidation of non activated primary and secondary alcohols using $RuCl_2(PPh_3)_3$ in combination with TEMPO in PhCl at 100 °C (Scheme 7).[30]

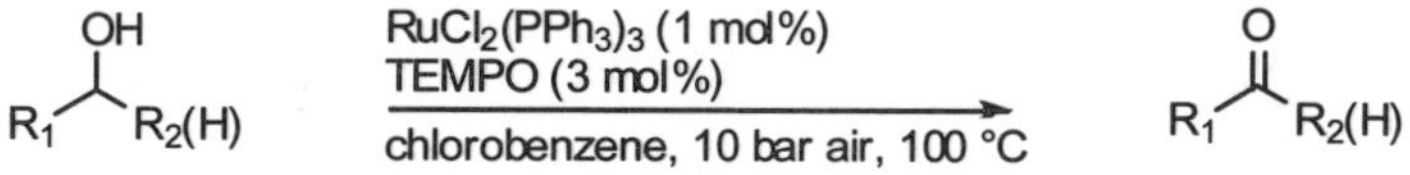

$$\text{R}_1\underset{\text{R}_2(\text{H})}{\overset{\text{OH}}{\diagup}}\xrightarrow[\text{chlorobenzene, 10 bar air, 100 °C}]{\substack{RuCl_2(PPh_3)_3\ (1\ mol\%)\\ TEMPO\ (3\ mol\%)}}\text{R}_1\underset{\text{R}_2(\text{H})}{\overset{\text{O}}{\diagup}}$$

Scheme 7. Ruthenium-TEMPO catalyzed oxidation of alcohols

Overoxidation of primary alcohols to carboxylic acids was completely suppressed by catalytic TEMPO, which avoided the autooxidation of aldeydes by efficiently scavenging free radical intermediates. Unfortunately, this system required 10 bar pressure and a number of alcohols containing heteroatoms (O, N, S) still remained unreactive, probably due to their coordination to the ruthenium metal centre and subsequent catalyst inactivation. An oxidative hydrogenation mechanism, analogous to that proposed by Bäckvall and co-workers for the Ru-quinone system, can be envisaged for the Ru/TEMPO system.

High valent perruthenate catalysts, i. e. tetra-n-propylammonium perruthenate (TPAP), are excellent air-stable Ru-catalysts, non volatile and soluble in a wide range of organic solvents. In 1997 Markó[31] and Ley[32] simultaneously showed that TPAP is able to perform the aerobic oxidation of alcohols; however, both systems had some drawbacks such as the need for high catalyst loading in a chlorinated solvent[32] or the need for high temperature (70-80 °C) (Scheme 8)[31] and were not effective for the oxidation of primary aliphatic alcohols in contrast to using NMO as the stoichiometric oxidant.[5]

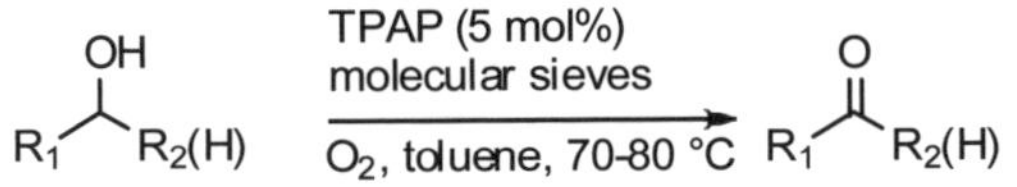

Scheme 8. TPAP-catalyzed aerobic oxidation of alcohols

More recently, Katsuki and co-workers have published several papers on Ru-salen catalysts for aerobic alcohol oxidations. They designed an efficient catalyst for the photo-induced chemoselective oxidation of primary alcohols in the presence of secondary alcohols,[33] and upon further derivatization of chiral ligands they could accomplish efficient kinetic resolutions of secondary alcohols[34] and desymmetrization of *meso*-diols (Scheme 9).[35] Moreover, recently, they developed a chemoselective oxidation of primary alcohols using a Ru-salen catalyst that did not need further irradiation conditions.[36]

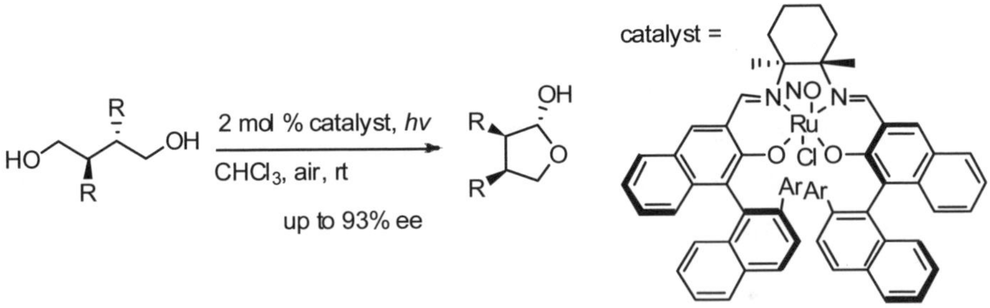

Scheme 9. Ru-salen catalyzed oxidative desymmetrization of *meso*-diols

In conclusion, ruthenium has proved to be effective for the homogenous aerobic oxidation of alcohols, but some further work remains to develop general alcohol oxidation catalysts that employ low catalyst loading and perform under mild conditions.

RUTHENIUM-BASED HETEROGENOUS CATALYSTS

Especially in industrial chemistry, heterogeneous catalytic systems are preferred over homogenous ones due to easy recyclability and separability. However, they usually suffer from low catalytic activity relative to their homogenous counterparts. Much effort has been made to overcome the difficulties involved because reduction of environmental loading due to easy separation and reuse of the catalyst could result.

A pioneering work by Ley and co-workers dates back to 1997 and reports the use of polymer-supported perruthenate (PSP) in the aerobic oxidation of alcohols; however, this catalyst suffered from oxidative degradation of the polymer support.[37] Soon later the same authors found a mesoporous silicate (MCM-41) as an efficient alternative support for TPAP and showed the recyclability of this catalyst up to 12 times (Scheme 10);[38] this material was used in a ten-step linear synthesis of the powerful analgesic natural product epibatidine, which employed only solid supported reagents.[39]

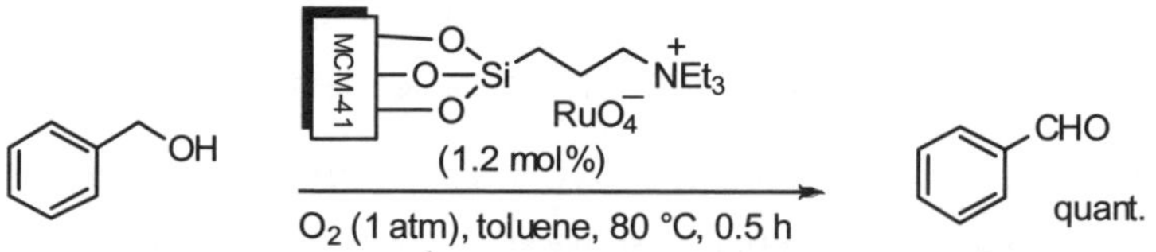

Scheme 10. Ley's modified mesoporous silicate materials MCM-41

The grafting of an organic moiety onto solid surfaces allows the building of organic-inorganic hybrid materials, which are promising supports for catalyst design.[40] For example, organically modified silicates (ORMOSIL) were studied by Pagliaro and Ciriminna for the encapsulation of the TPAP via a sol-gel process (SG-TPAP).[41] However, the first reports did not show a wide substrate scope; in order to broaden the application of the SG-TPAP catalyst, alternative conditions were investigated such as supercritical carbon dioxide (scCO$_2$)[42] and the introduction of ionic moieties[43] or of fluoroalkyl chains[44] into the silica matrix of SG-TPAP. They reported the use of hybrid fluorinated silica glass doped with TPAP (denoted FluoRuGel) as a versatile catalyst for the aerobic oxidation of different alcohols in dense CO$_2$ (Scheme 11).[44,45,46]

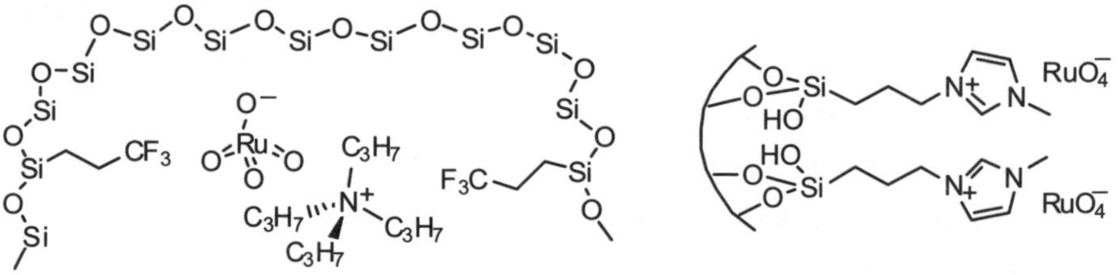

Scheme 11. TPAP-heterogeneous catalysts developed by Pagliaro and co-workers

Low valent ruthenium species have been also supported on solid matrices. Zeolites impregnated with RuO_2 nanoclusters (RuO_2-FAU) were found to be effective and selective catalysts for a wide variety of both activated and unactivated alcohols. These materials display a strong shape selectivity due to uniform pore size, and in a competitive experiment benzyl alcohol was reacted in the presence of unreacted 9-hydroxyfluorene.[47] Kaneda and co-workers developed a monomeric ruthenium cation on the surface of hydroxyapatite (Ru/HAP), which gave efficient aerobic oxidations of primary, secondary and functionalized alcohols (Scheme 12). The main disadvantage of this process was the need for a high catalyst loading (17 mol%).[48]

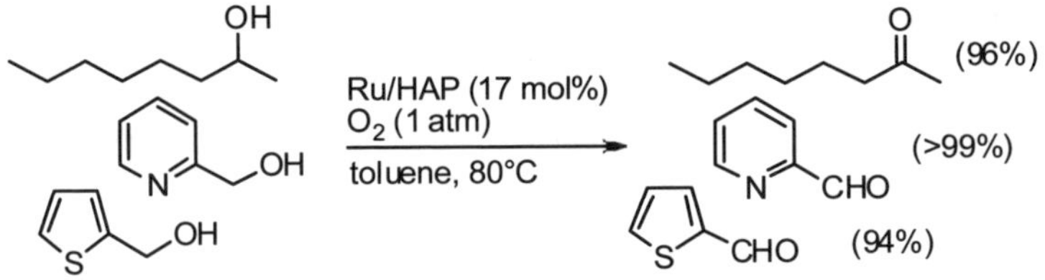

Scheme 12. Heterogeneous Ru/HAP catalyst developed by Kaneda and co-workers

Ruthenium supported on alumina ($Ru(OH)_x/Al_2O_3$) was developed by Yamaguchi and Mizuno and demonstrated the ability to oxidize both primary, secondary and activated alcohols in $PhCF_3$ as solvent or even in solvent-free conditions.[49] Recently, the use of superparamagnetic nanoparticles as a supporting material for immobilized metal catalysts was reported. For example, Mizuno and co-workers showed that a ruthenium hydroxide species on magnetite ($Ru(OH)_x/Fe_3O_4$) performed very well in the aerobic oxidation of alcohols, and catalyst/product(s) separation was extremely simple. Indeed, after completion of the oxidation reaction, a permanent magnet was attached to the outside wall of the glass reactor to magnetically "hold" the catalyst, and the reaction solution including the product(s) was separated by simple decantation.[50]

In contrast to the many reports of inorganic supports or organic-inorganic hybrid materials, after the pioneering PSP by Ley and co-workers[37] only a few polymer-supported catalysts for the aerobic oxidation of alcohols were reported. Kobayashi and co-workers developed a polymer incarcerated ruthenium (PI Ru) capable of oxidizing alcohols in aerobic conditions, based on the technique of microencapsulation and cross-linking from a polystyrene-based copolymer and ruthenium chloride hydrate as the metal source. However, this catalyst needed the presence of 15 mol% of TEMPO to show wide applicability to various alcohols (Scheme 13)[51] and leaching of Ru metal was observed in some cases (never exceeding 0.72%), which is typical of polymer supported catalysts, that suffer from low chemical and/or mechanical resistance.

The authors later showed that introduction of inorganic species to organic moieties, thus going back to the creation of organic-inorganic hybrid catalysts generated by the sol-gel approached, allowed the synthesis of an effective heterogenous catalyst which worked well for the aerobic oxidation of alcohols without the need for any additive, avoiding the leaching of Ru.[52]

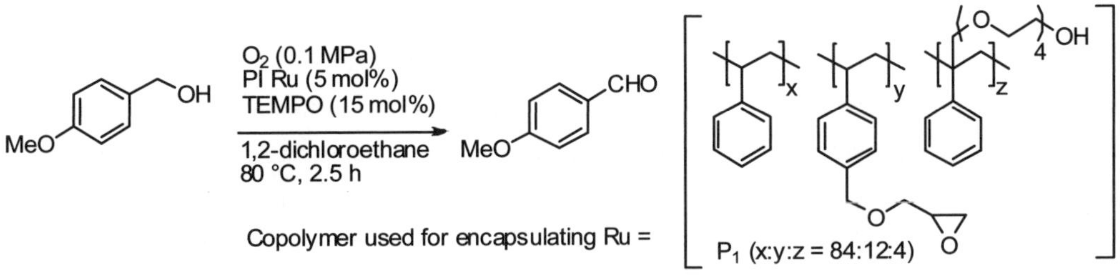

Scheme 13. Polymer Incarcerated (PI) Ruthenium catalyst

The main problem with all these heterogeneous catalysts is that they can be accessed with some difficulties since they are handmade and/or expensive. Much effort has been devoted to develop efficient oxidation methods for alcohols using the readily available carbon-supported metal catalysts.[10] Ruthenium is less expensive than Au, Pd or Pt; however, one of the few reports of Ru/C-catalyzed oxidation methods for alcohols was presented recently by Sajiki and co-workers, who showed that 10% Ru/C as a catalyst in toluene (at 50 °C) under an oxygen atmosphere was able to convert various secondary and primary benzylic alcohols to the corresponding carbonyl compounds and primary aliphatic alcohols to carboxylic acids when water was added as a co-solvent (Scheme 14).[53]

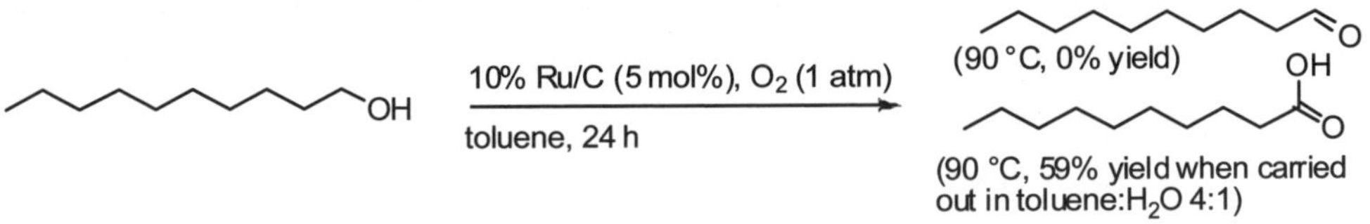

Scheme 14. Ru/C-catalyzed aerobic oxidation of alcohols

PALLADIUM-BASED HOMOGENOUS CATALYSTS

Overall, Pd(II) catalysis represents one of the most mature fields in the aerobic oxidation of alcohols. Much effort has been devoted to finding synthetically useful methods for the palladium-catalyzed aerobic oxidation of alcohols, and some excellent reviews on this topic have appeared.[54] Many mechanistic studies have been undertaken and a generally accepted mechanism for the Pd-catalyzed aerobic oxidations involves the formation of Pd-hydride species.[54b,55,56] The palladium is then reduced to the zerovalent state by the alcohol substrate and is reoxidized to palladium(II) by dioxygen. The transient Pd(0) species is metastable and prone to aggregation to bulk palladium metal (Pd black) with concomitant loss of catalytic activity. One approach to avoid this is to add coordinating ligands, which stabilize the transient Pd(0) species.

The first synthetically useful system was reported in 1998 by Peterson and Larock, who showed that simple Pd(OAc)$_2$ in combination with NaHCO$_3$ as a base in DMSO as solvent catalyzed the aerobic oxidation of primary and secondary allylic and benzylic alcohols to the corresponding aldehydes and ketones, respectively.[57] The replacement of the non-green DMSO by an imidazole-type ionic liquid resulted recently in a higher activity of the Pd-catalyst.[58] However, this method suffered from narrow substrate scope. Uemura and co-workers reported an improved procedure using Pd(OAc)$_2$ (5 mol%) in combination with pyridine and 3Å molecular sieves in toluene at 80°C,[59] that allowed oxidation of primary and secondary aliphatic alcohols in addition to benzylic and allylic ones. When applied to *tert*-cyclobutanols, this reaction proceeded with cleavage of the C-C bond (Scheme 15).[60] This approach could also be employed under fluorous biphasic conditions.[61]

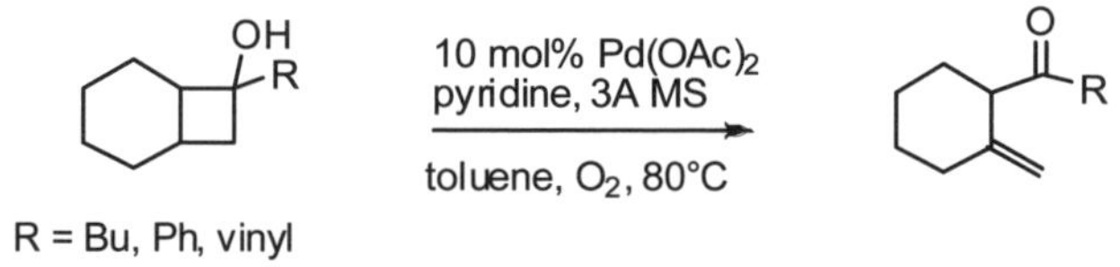

R = Bu, Ph, vinyl

Scheme 15. Pd(II)-catalyzed oxidative ring cleavage of *tert*-cyclobutanols under O$_2$ atmosphere

A much more active catalyst is represented by a water-soluble palladium(II) complex of sulfonated bathophenantroline introduced by Sheldon and co-workers.[62] This stable, recyclable catalyst allowed oxidation in a two-phase acqueous-organic medium in 5h at 100 °C/30 bar air with 0.25 mol% catalyst. No organic solvent was required (except for solid alcohols) and the carbonyl product was recovered easily by phase separation. Primary alcohols afforded the corresponding carboxylic acids via further oxidation of the aldehyde intermediate; otherwise, in the presence of 1 mol% of TEMPO, the aldehyde was obtained in high yield (Scheme 16).[62] Pd-neocuproine (in the presence of ethylene carbonate as cosolvent) was found to be even more active and exceptionally tolerant to many functional groups such as C=C bonds, triple bonds, halides, ethers, amines etc, thus showing a broad synthetic utility.[63] However, a more detailed recent investigation of this latter ligand proved that in this case formation of Pd nanoparticles, which are presumably the active catalytic species, occurs (see later for a more detailed discussion).[64]

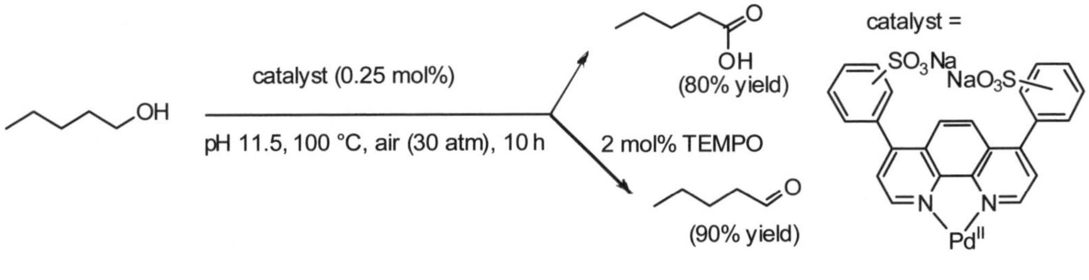

Scheme 16. Sheldon's Pd-catalyzed aerobic oxidation of alcohols

One of the main problems associated with homogeneous Pd(II)-catalyzed aerobic oxidations is often represented by Pd black formation. Tsuji and co-workers used substituted pyridines as ligands to prevent formation of Pd black, allowing oxidations to be performed under air and using low catalyst loading.[65] Sigman and co-workers also developed three novel Pd(II)-catalysts for the aerobic oxidations of alcohols,[66] and in a recent publication[67] they reported a comparison study in which they evaluated the substrate scope and the reaction conditions of each of them, concluding that the Pd(OAc)$_2$/TEA system represents the most convenient of the three developed catalytic systems. For example, this catalyst was employed for the direct conversion of α-hydroxy ketones into quinoxalines (Scheme 17).[68]

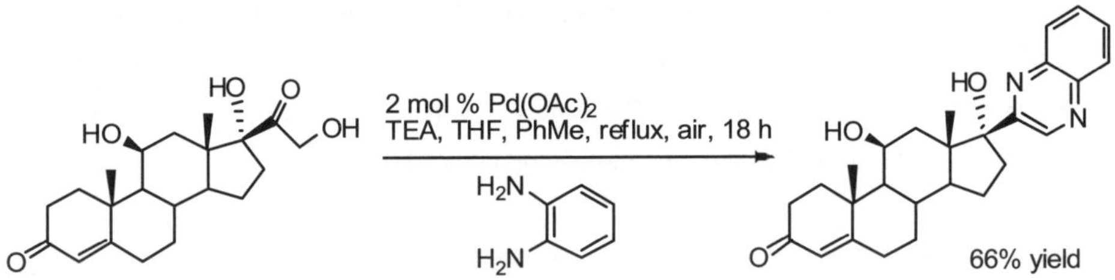

Scheme 17. Quinoxaline synthesis via a tandem oxidation process

Another nice example of Pd oxidation catalysis in tandem reactions was shown by Lebel and Paquet, who applied the catalyst developed by Sigman[66b] to the one-pot synthesis of alkenes through a tandem oxidation/olefination process (Scheme 18).[69]

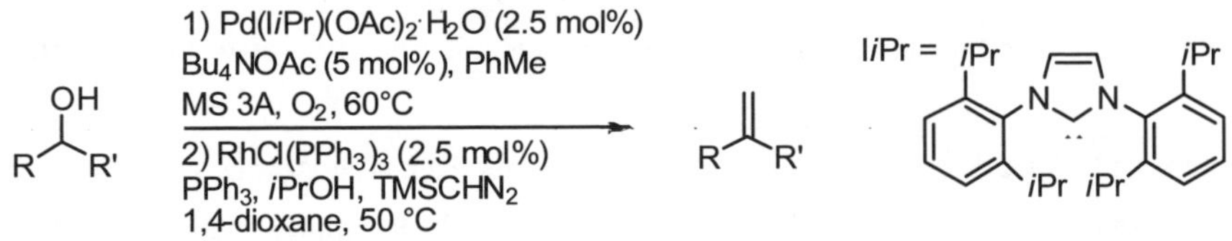

Scheme 18. One-pot Pd-catalyzed oxidation and Rh-catalyzed methylenation reaction

In the presence of a chiral diamine, the scope of these oxidations can be expanded to asymmetric catalysis, as for example the oxidative kinetic resolution of racemic secondary alcohols or the oxidative desymmetrization of *meso*-diols. Sigman and Stoltz independently discovered[70] that in the presence of the chiral diamine (−)-sparteine, which plays a dual role of chiral ligand for Pd and exogenous chiral base,[71] the Pd(II)-catalyzed aerobic oxidation of alcohols afforded efficient oxidative kinetic resolution of secondary alcohols, with enantiomeric excesses up to 99.8% (Scheme 19). This methodology was recently applied to the enantioselective total synthesis of various alkaloids,[72] and to the kinetic resolution of key pharmaceutical building blocks, relevant to the enantioselective preparation of Prozac®, Singulair® and the promising hNK-1 receptor antagonist from Merck.[73]

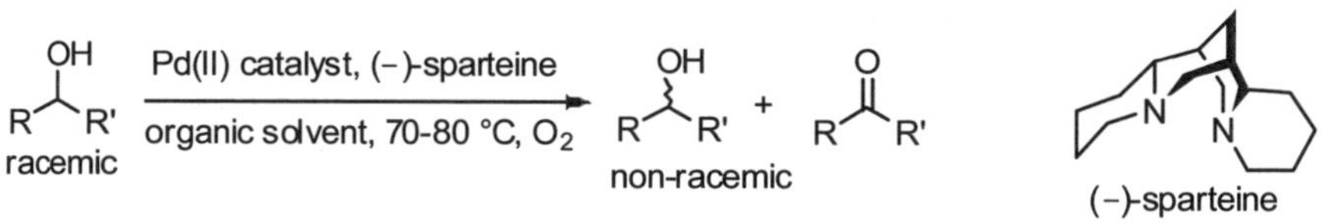

Scheme 19. Pd(II)-catalyzed oxidation kinetic resolutions of alcohols

This is an excellent method for the aerobic kinetic resolution of alcohols, leading to remarkably high ee's under optimized conditions; the main limitation of sparteine as a chiral ligand is that only the (−)-enantiomer is available in large quantities, and this will remain a problem until an effective method is found for the preparation of quantities of its enantiomer or a surrogate thereof.[74] However, all the Pd(II)-catalysts reported to date are not widely used on a larger industrial scale. Catalysts with improved stability and activity need to be developed and the research is still very active in this field. A recent study investigated the use of N,O-ligated Pd(II) complexes, which compared well with the previously reported N,N-ligands in the aerobic oxidation of 2-octanol on the gram scale.[75]

PALLADIUM-BASED HETEROGENEOUS CATALYSTS

Besides the aerobic oxidation of alcohols, palladium catalyzes many oxidative transformations including epoxidation of alkenes, oxidation of terminal alkenes to ketones and other Wacker-type reactions, oxidation of alkanes, hydroxylation of benzenes, and oxidative coupling reactions.[76] Among the transition metals, palladium shows very promising catalytic properties in the form of heterogeneous metal catalysts or nanoparticles. The major problem related to the use of palladium-based catalysts is that palladium agglomeration and formation of palladium black can cause catalyst deactivation in many cases.

Uemura and co-workers attempted heterogenization of their homogenous catalytic system based on Pd(OAc)$_2$, pyridine and 3Å molecular sieves[59] employing hydrotalcite, a naturally produced basic clay mineral, and they found that the Pd/HT system had a higher activity.[77] Especially in the oxidation of geometrically isomerizable allylic alcohols such as geraniol and nerol this catalyst proved to be efficient without any isomerization of the alkenic part (Scheme 20). However, an excess of pyridine instead of the usual 20 mol% was required with such alcohols. Pd/HT was reused at least three times, although a gradual decrease in catalytic activity was observed.

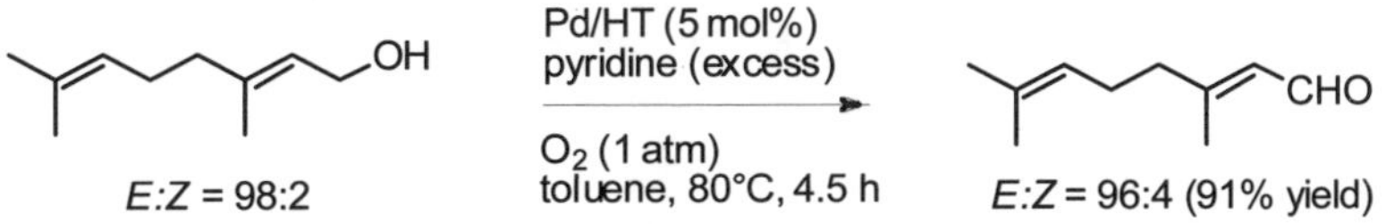

Pd/HT (5 mol%)
pyridine (excess)

O₂ (1 atm)
toluene, 80°C, 4.5 h

E:Z = 98:2   E:Z = 96:4 (91% yield)

Scheme 20. Pd(II)-supported hydrotalcite-catalyzed aerobic oxidation of geraniol

The general routes to nanoclusters/nanoparticles synthesis are based on the chemical reduction of transition metal salts with the appropriate reducing agent in the presence of a stabilizer for the metal. The resulting stabilized metal nanoclusters dispersed in solution can be used as catalysts as such or subsequently heterogeneized on solid supports by different means (e.g surface adsorption, covalent anchoring, embedding by sol-gel techniques). For example, Pd nanoclusters stabilized by *N,N*-dimethylacrylamid-based soluble cross-linked polymers (microgel) were tested as catalysts in the selective oxidation of secondary alcohols to the corresponding ketones with molecular oxygen in water.[78]

Kaneda and co-workers reported hydroxyapatite–supported palladium nanoclusters (Pd/HAP-0) prepared from stoichiometric HAP with [PdCl₂(PhCN)₂] as a metal source.[79] Fresh Pd/HAP-0 had an induction period of about 10 minutes, in which Pd(II) species were converted into Pd(0) nanoparticles. A wide variety of alcohols, also bearing heteroatoms, were oxidized with this heterogeneous catalyst in trifluorotoluene at 90°C, in water at 110°C or in solvent-free conditions. 1-Phenyl ethanol was oxidized on a quite large scale (30 grams) without any solvent at 160°C.

An amphiphilic resin dispersion of palladium nanoparticles (ARP-Pd) was reported by Uozumi and Nakao, readily prepared by reduction of a PS-PEG resin-supported Pd(II) complex with benzyl alcohol (Scheme 21). This catalyst was applied to the aerobic oxidation of benzylic, allylic and secondary aliphatic alcohols in refluxing water.[80] In the case of primary aliphatic alcohols, the corresponding carboxylic acids were obtained in excellent yields in the presence of K₂CO₃.

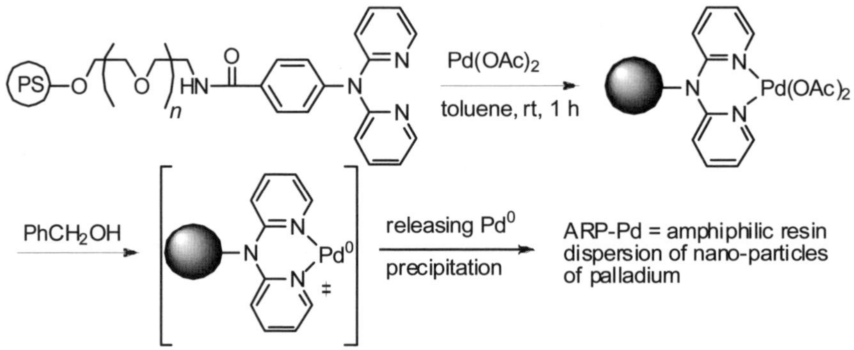

Scheme 21. Preparation of amphiphilic resin-dispersion of nanoparticles of palladium (ARP-Pd)

However, organic polymers used as support for Pd nanoclusters are potentially susceptible to oxidative degradation under aerobic oxidative conditions. Besides the already mentioned hydrotalcite and hydroxyapatite minerals, an inorganic alternative for forming a scaffold in which three-dimensional dispersions of nanoparticles can be supported is represented by ordered mesoporous structures (such as MCM-41 and SBA-15) with regular channel and pore diameters in the range of 2 to 30 nm. Karimi and co-workers developed a new type of palladium catalyst immobilized on functionalized SBA-15 and applied it to the oxidation of various alcohols in toluene at 80°C in the presence of K_2CO_3 (1 equivalent), which was found to be essential to avoid formation of Pd black.[81] Primary alcohols were converted to the corresponding esters, presumably by previous selective oxidation to carboxylic acids. This example showed that the combination of an organic ligand and ordered mesoporous channels (Scheme 22) resulted in an interesting synergistic effect that led to enhanced reactivity, prevention of the agglomeration of the Pd nanoparticles and generation of a durable catalyst.

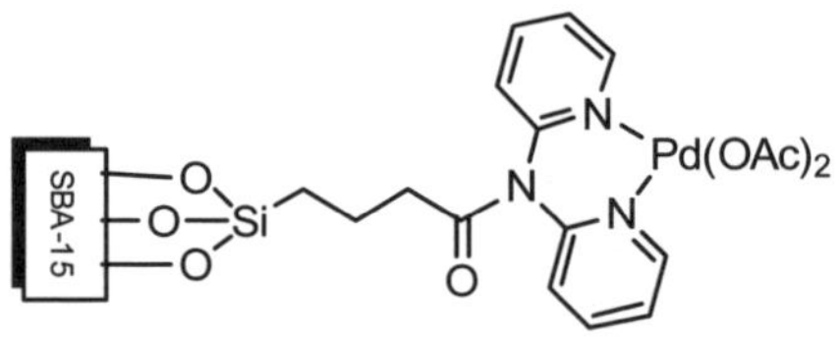

Scheme 22. Palladium nanoparticles stabilized on mesoporous channels of SBA-15

Park and co-workers reported aluminium hydroxide supported-palladium nanoparticles (Pd/AlO(OH)) prepared from [Pd(Ph_3)_4], tetra(ethylene glycol), 1-butanol, and aluminum tri-*sec*-butoxide.[82] This catalyst displayed dual catalytic activity for both alkene hydrogenation and aerobic oxidation of alcohols. Successful hydrogenation of cholesterol followed by aerobic oxidation to give cholestan-3-one was demonstrated in a one-pot manner (Scheme 23).

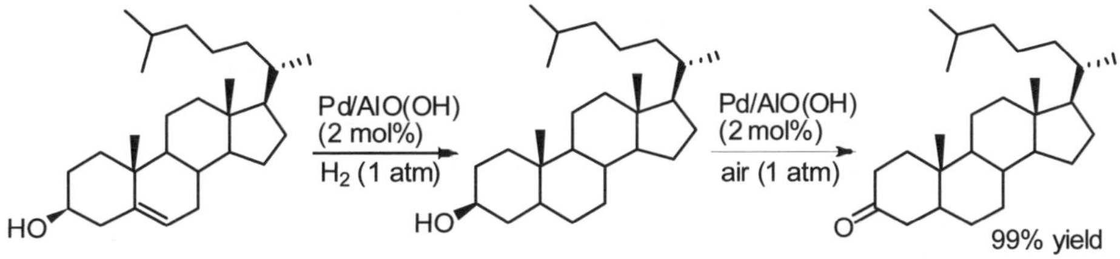

Scheme 23. Cholesterol hydrogenation followed by aerobic oxidation

However, Pd/AlO(OH) did not catalyze the oxidation or primary aliphatic alcohols such as 1-octanol. Some other examples of palladium-based heterogeneous catalysts obtained by dispersion of the metal on an inorganic support have been recently reported, such as Pd/MgO[83] or Pd/Al$_2$O$_3$[84] (this latter was found to be selective for the aerobic oxidation of allylic alcohols). Moreover, Wang and co-workers demonstrated that the preparation method of palladium catalyst on aluminum oxide was important for high catalytic performance. Indeed, the activity of Pd/Al$_2$O$_3$ catalyst prepared by an adsorption method (Pd/Al$_2$O$_3$-ads) was higher than that prepared by an impregnation method.[85] The authors postulate the formation of Pd nanoparticles, which are probably the true active species, during the course of alcohol oxidation.

A quite different approach was developed by Leitner and co-workers. They found that the giant palladium cluster, [Pd$_{561}$phen$_{60}$(OAc)$_{180}$], dispersed in poly(ethylene glycol) (PEG), efficiently catalyzes the aerobic oxidation of alcohols in scCO$_2$ (Scheme 24).

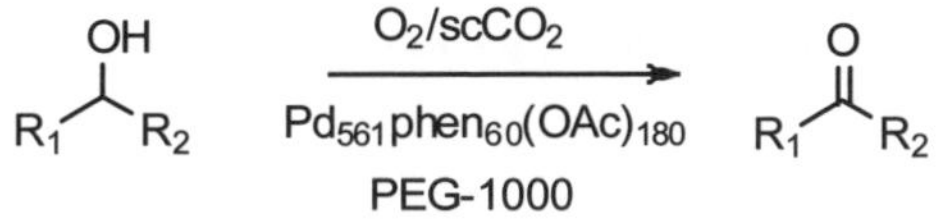

Scheme 24. Aerobic oxidation of alcohols catalyzed by PEG-stabilized Pd-nanoparticles in scCO$_2$

In this biphasic system, the PEG matrix contains the catalyst (helping in preventing aggregation and deactivation of the catalytically active nanoparticles) while the supercritical carbon dioxide phase dissolves the substrate and the product (thus providing a safe environment for the use of molecular oxygen under essential solvent-free conditions and allowing continuous operation, even with substrates of low volatility).[86] The authors postulate that the high activity and long term stability of the new catalytic system is due to formation of Pd-nanoparticles during the reaction. A variety of alcohols were oxidized in these conditions. Both the catalyst matrix and the mobile phase used in this approach are toxicologically innocuous and environmentally benign materials, thus making this approach particularly appealing for "green" nanoparticle catalysis.

Interestingly, as previously mentioned, Sheldon and co-workers recently demonstrated that, contrarily to the catalytic system based on the bathophenanthroline disulfonate ligand (Scheme 16),[62]

their previously described homogeneous catalytic system based on Pd(II) acetate in combination with the more hindered neocuproine ligand[63] actually involves palladium nanoparticles. The susbstrate alcohol acts indeed as the reducing agent and in situ forms Pd-nanoparticles which are the effective catalysts. The catalytic system based on neocuproine-stabilized palladium nanoparticles was applied to the oxidation of nandrolone (Scheme 25).[64]

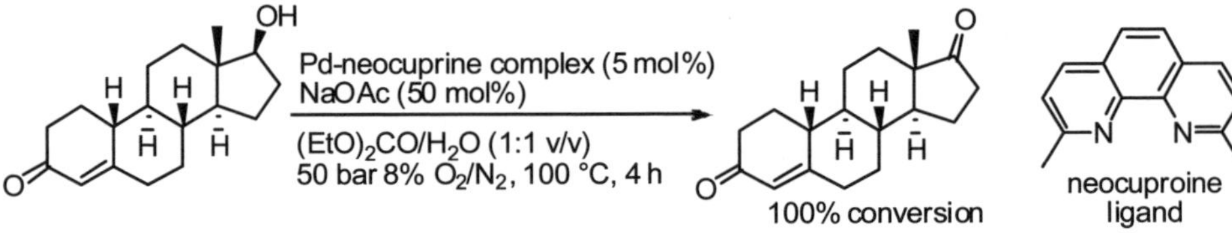

Scheme 25. Aerobic oxidation of nandrolone with Pd nanoparticles in acqueous media

Much work has to be done yet in order to investigate in detail the mechanisms involved when nanoparticles are formed in the reaction. Indeed, it is difficult to attribute the actual catalytic activity solely to the ligand bound Pd or to the Pd nanoparticles.

GOLD-BASED HOMOGENOUS CATALYSTS

The homogeneous oxidation of alcohol catalyzed by gold has rarely been reported. Shi and co-workers,[87] in early evaluations, optimized the aerobic oxidation of primary and secondary benzylic and allylic alcohols using AuCl (5 mol%) and ligand (6.3 mol%) in toluene at 90°C under oxygen atmosphere (Scheme 26).

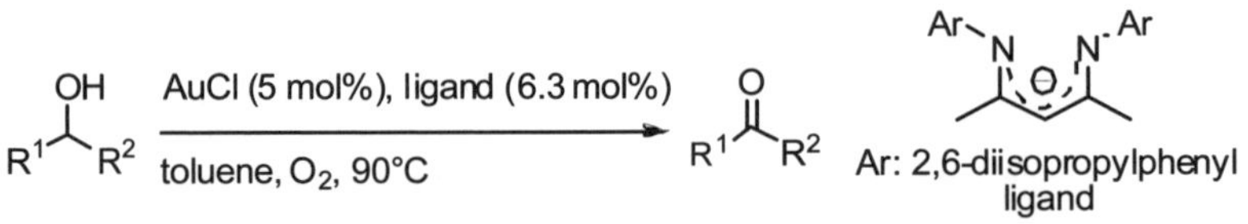

Scheme 26. Oxidation of alcohols with gold(I) complexes

Addition of 4Å molecular sieves proved to be beneficial for the reaction. With activated benzyl and allylic alcohols as substrates, both conversions and yields were very high, only aldehydes were produced with excellent selectivity with no overoxidation to carboxylic acids; however, primary aliphatic alcohols were slowly oxidized and formed aldol byproducts. This system was then improved, in terms of sustainability of the process, using water as solvent.[88] Different oxidants, bases and ligands were studied and a final optimized system using 1:1 ratio of gold(I)-neocuproine as catalyst in aqueous basic solution under O_2 atmosphere was found. Surprisingly, the oxidation run efficiently only with $NaHCO_3$ as base, NaOH or organic bases, such as Et_3N, being inefficient. The limitation of this oxidation procedure is the same as the previous: the narrow substrate scope, being limited to secondary benzyl or allyl alcohols. Moreover, given Sheldon's results with Pd-neocuproine, we cannot exclude that the effective catalysts here are gold nanoparticles formed by gold reduction by the alcohol substrate.

GOLD-BASED HETEROGENOUS CATALYSTS

Although bulk gold has for a long time being regarded as a poorly active metal, the surprisingly high activity of gold nanoparticles has initiated intensive research into their use for aerobic oxidation reactions. Moreover, the recent findings related to the synergic activity of bimetallic nanocluster catalysis has further expanded the possibilities for the design of new efficient gold-based heterogeneous catalysts.

The general procedure for the synthesis of gold nanoparticles is based on reduction of Au salts by a reducing agent or by the support itself in some cases. The first to clearly demonstrate that supported gold nanoparticles can be very effective catalysts for the oxidation of alcohol were Rossi, Prati and co-workers. They employed Au/carbon catalysts, which were effective for a wide range of substrates like diols, glucose and aminoalcohols, and found that the presence of a base was essential for catalysis.[89] Similar Au/SiO$_2$ catalysts were found to be effective with gas-phase reactants and, in this case, no base addition was required.[90] These pioneering studies using Au/carbon catalysts were extended by Hutchings and co-workers[91] who showed that Au supported on graphite could oxidize glycerol to glycerate with 100% selectivity using dioxygen as the oxidant in water with yields approaching to 60%. It was observed that the selectivity to glyceric acid and the glycerol conversion were strongly dependent upon the glycerol/NaOH ratio (Scheme 27).

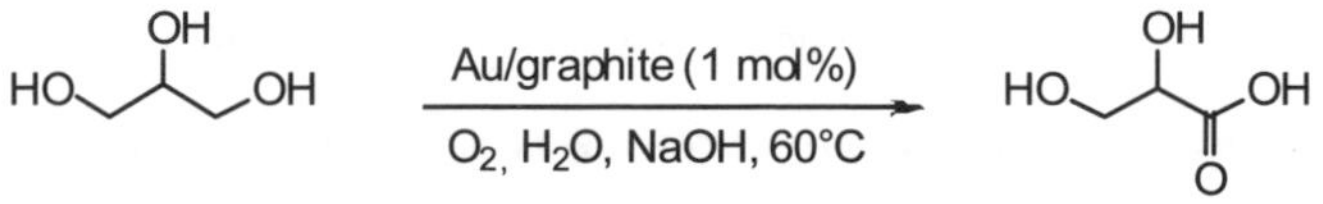

Scheme 27. Oxidation of glycerol using Au/graphite catalysts

Afterwards, Qiu et al. demonstrated that water has dual promotional functions in the catalytic activity of Au/TiO$_2$ for the selective oxidation of benzyl alcohol: it helps to form microdroplets in a multiphase reaction system and it assists the oxygen adsorption and activation.[92]

One of the most significant advances in the field of alcohol oxidations has been the observations by Corma and co-workers who showed that Au/CeO$_2$ catalyst is active for the selective oxidation of alcohols to aldehydes and ketones and the oxidation of aldehydes to acids.[93] In these studies, the catalysts are active in solvent-free conditions, using O$_2$ as oxidant without the requirement for the addition of NaOH to achieve high activity. Subsequently they showed that for the relevant oxidation of allylic alcohols, gold presented unique selectivity when compared with Pd (Scheme 28).[94]

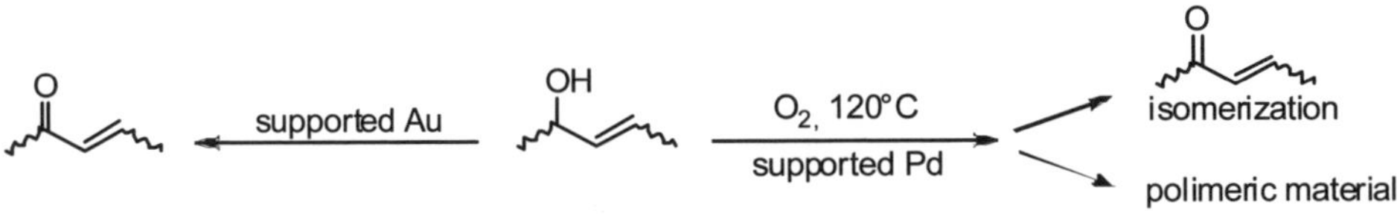

Scheme 28. Aerobic oxidation of allylic alcohols under solvent-free conditions

In many catalytic studies, the support-catalyst interaction is a crucial factor for controlling reactivity.[93,95] Interestingly, Rossi and co-workers have shown that water-dispersed "naked" gold colloidal particles can be very effective catalysts for the oxidation of glucose to gluconic acid.[96] These particles were produced as a colloidal sol by reducing $HAuCl_4$ in the presence of a large excess of glucose acting either as reagent or protector. Christensen, Riisager et al. have made a number of significant advances in the direct oxidation of primary alcohols using supported gold nanocrystals and they have concentrated their efforts on decreasing the amount of base required in these oxidations. They have shown that $Au-MgAl_2O_4$ can catalyze the oxidation of aqueous solutions of ethanol to give acetic acid in high yields.[97] This provides a potential new route to a key commodity chemical that is based on a bio-renewable feedstock using a substantially green technology approach. Recently, they presented the one-pot conversion of alcohols to imines (Scheme 29) by aerobic oxidation with Au/TiO_2 followed by condensation with primary amines in methanol.[98]

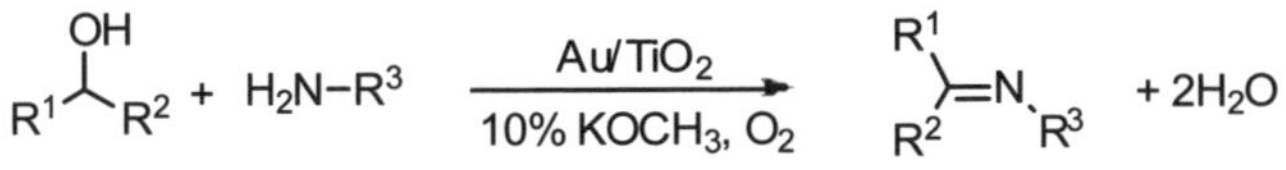

Scheme 29. Au-catalyzed one-pot formation of imines

Polymer compounds are attractive because they work as quasi-homogeneous catalysts and provide a large contact area to organic substrates. Examples of these catalysts are represented by the gold-nanocluster stabilized by water-soluble polymer poly(*N*-vinyl-2-pyrrolidone) (Au:PVP),[99] and the microgel-stabilized Au nanoclusters prepared by gold nanoclusters of small size and an appropriate microgel like vinylpyridine with a narrow size.[100] Both catalysts performed very well for the aerobic oxidation of alcohols in water. Kobayashi and co-workers developed, besides PI Ru (Scheme 13), also a polymer-incarcerated gold catalyst (PI Au) for the selective oxidation of alcohols at room temperature and atmospheric pressure of molecular oxygen on air.[101] Aromatic and aliphatic secondary alcohols. also containing S and N heteroatoms, were oxidized to the corresponding carbonyl compounds in good yields and the catalyst could be reused at least 10 times without loss of activity. The same PI Au was used for oxidative esterification of alcohols under ambient conditions[102] and could be applied to triphasic (gas-liquid-solid) reactions by using a microchannel reactor and a capillary column also with molecular oxygen.[103] Recently, Kobayashi et al. included carbon black (CB) to the composition of the PI Au to enhance the stability of gold nanoclusters probably *via* synergistic π-π interactions between the three components which enables they to increase the metal loading amount up to 0.60 mmol g^{-1}.[104]

The oxidation of benzyl alcohols was also reported in $scCO_2$ using Au-catalysts supported on TiO_2, Fe_2O_3 and C.[105] At 100°C and a total gas pressure of 15.0 mPa, benzyl alcohol, O_2 and CO_2 formed a single homogeneous phase and benzaldheyde was obtained with 99.0% selectivity but only 16.0% conversion using 1% Au/TiO_2. Higher conversions were achieved by Kawanami et al. using 2% Au/TiO_2, prepared by a deposition-precipitation (DP) method.[106]

Recently, the synergic activity of bimetallic nanoclusters has been presented. Hatchings et al. showed that alloying Pd with Au in supported Au/TiO_2 catalysts, that was found the best support in previously studies,[95] the activity for alcohol oxidation was enhanced under solvent-free conditions by a factor of over 25.[107] Then, Zheng and Stucky demonstrated that with the use of low-cost promoting agents (i.e. K_2CO_3) instead of Pd and under similar conditions, oxide-supported pure gold nanoparticles could catalyze the oxidation of alcohols even more efficiently; in particular, in contrast to $Au/Pd-TiO_2$ catalysts which were inactive for the oxidation of 2-octanol, pure supported gold nanoparticles were highly active with the help of a small amount of K_2CO_3.[108] Recently, Prati and co-workers reported that Au-Pt nanoparticles supported on the zeolite H-mordenite were able to selectively oxidize glycerol directly to glyceric acid without the use of basic conditions.[109] Also Kobayashi et al. studied the aerobic oxidation of alcohols under ambient condition with gold-platinum bimetallic clusters[110] and in a recent publication they compared the different selectivities using different metal combination catalysts.[111]

Recently, using pyrrolidone-modified SBA-15 supported Au nanoparticles, Xiao, Meng et al. presented a catalytic system which synergistically combined the advantages of both homogeneous catalysts (high activity) and heterogeneous catalysts (good recyclability). This system could oxidized 1-butanol with 50.7% conversion.[112]

Hirao and coworkers reported the first example of catalyst design for the alcohol oxidation reaction using a redox-active polymer as the catalyst support (Scheme 30).[113] They demonstrated that the redox-active PMAS (poly(2-methoxyaniline-5-sulfonic acid)) can work in a multi-catalytic process as both a stabilizer of Au NPs and a redox mediator for aerobic alcohol oxidation in water (Scheme 30). This design concept provides a new type of redox catalyst system for transferring protons and electrons.

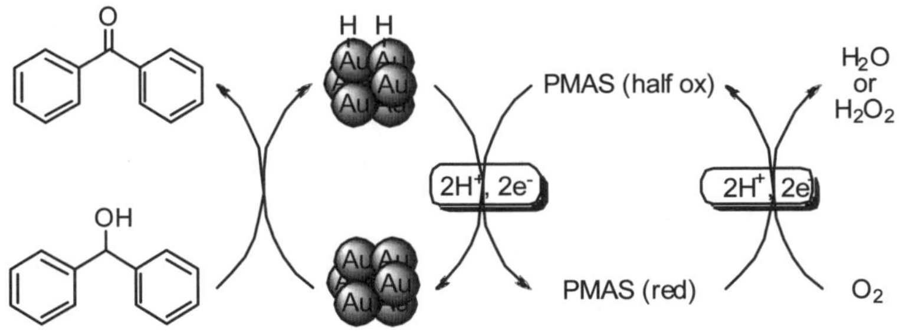

Scheme 30. Proposed multi-catalytic cycles for the oxidation with Au NPs and PMAS

CONCLUSIONS

During the last 15 years there has been a considerable increase of interest in the area of metal-catalyzed aerobic alcohol oxidations. In the field of homogeneous alcohol oxidations, the Marko's Cu-(phen), the Sheldon's Pd-(sulfonated bathophenantroline) and the Sigman's Pd(OAc$_2$)/TEA systems are the most mature. Ruthenium-based catalysts often suffer from the need for high catalyst loading. A considerable effort has been also made to replace common organic solvents with alternative solvents such as ionic liquids, fluorinated solvents or supercritical CO$_2$ or to perform the oxidation reactions in water. Regarding the chemoselectivity of the reaction, it is interesting to note the complementarity between Cu-based catalysts, who better work with primary alcohols, and gold-based catalysts, who better perform the oxidation of secondary alcohols. Selective methods to obtain aldeydes or carboxylic acids from primary alcohols were also developed. Moreover, elegant examples of efficient kinetic resolutions of racemic secondary alcohols and desymmetrization of meso diols were achieved with Pd-based catalysts in the presence of (−)-sparteine as the chiral ligand. The discovery that Pd and Au nanoparticles are effective catalysts for the oxidation of alcohol moieties has further expanded this research field in the search for new heterogeneous systems, that can allow recovery and reuse of the metal catalyst and the obtainment of pure products. Mechanistically, not much work has been done to elucidate the fine details for many of the metal-catalyzed aerobic alcohol oxidations, except for Pd-catalyzed aerobic alcohol oxidations. Especially for the new heterogeneous procedures involving nanoparticles, the exact nature of the active catalyst has still to be understood. While there has been a tremendous amount of effort applied to the development and improvement of metal-catalyzed aerobic alcohol oxidations, many improvements can be still envisioned. For instance, in order to use these methods in target synthesis, the scope of the individual catalytic systems must be broadened to include more complex alcohols that are synthetically relevant. Moreover, each method should be tested on a larger scale to explore its potential utility in the industrial processes.

REFERENCES

1. Sheldon, R. A. *Chem. & Ind.* **1992**, 903.
2. Alfonsi, K.; Colberg, J.; Dunn, P. J.; Fevig, T.; Jennings, S.; Johnson, T. A.; Kleine, H. P.; Knight, C.; Nagy, M. A.; Perry, D. A.; Stefaniak, M. *Green Chem.* **2008**, *10*, 31.
3. Dess, D. B.; Martin, J. C. *J. Org. Chem.* **1983**, *48*, 4155.
4. Mancuso, A. J.; Huang, S. L.; Swern, D. *J. Org. Chem.* **1978**, *12*, 2480.
5. Ley, S. V.; Norman, J.; Griffith, W. P.; Marsden, S. P. *Synthesis* **1984**, 639.
6. Trost, B. M. *Science* **1991**, *254*, 1471.
7. Carey, J. S.; Laffan, D.; Thomson, C.; Williams, M. T. *Org. & Biomol. Chem.* **2006**, *4*, 2337.
8. Sheldon, R. A.; Arends, I.; Hanefeld, U. *Green Chemistry and Catalysis*; Wiley-VCH 2007.
9. Bäckvall, J.-E. Modern *Oxidation Methods*; Wiley-VCH 2004.
10. For some excellent reviews see: a) Mallat, T.; Baiker, A. *Chem. Rev.* **2004**, *104*, 3037; b) Schultz, M. J.; Sigman, M. S. *Tetrahedron* **2006**, *62*, 8227; c) Matsumoto, T.; Ueno, M.; Wang, N.; Kobayashi, S. *Chem. Asian J.* **2008**, *3*, 196.
11. Wang, Y.; DuBois, J. L.; Hedman, B.; Hodgson, K. O.; Stack, T. D. P. *Science* **1998**, *279*, 537.
12. Semmelhack, M. F.; Schmid, C. R.; Cortés, D. A.; Chou, C. S. *J. Am. Chem. Soc.* **1984**, *106*, 3374.
13. Markó, I. E.; Giles, P. R.; Tsukazaki, M.; Brown, S. M.; Urch, C. J. *Science* **1996**, *274*, 2044.
14. Markó, I. E.; Gautier, A.; Chellé-Regnaut, I.; Giles, P. R.; Tsukazaki, M.; Urch, C. J.; Brown, S. M. *J. Org. Chem.* **1998**, *63*, 7576.
15. Markó, I. E.; Gautier, A.; Dumeunier, R., Doda, K.; Philippart, F.; Brown, S. M.; Urch, C. J. *Angew. Chem. Int. Ed.* **2004**, *43*, 1588.
16. Davi, M.; Lebel, H. *Org. Lett.* **2009**, *11*, 41.
17. Gamez, P.; Arends, I. W. C. E.; Reedijk, J.; Sheldon, R. A. *Chem. Commun.* **2003**, 2414.
18. Mannam, S.; Sekar, G. *Tetrahedron:Asymmetry* **2009**, *20*, 497.
19. Ansari, I. A.; Gree, R. *Org. Lett.* **2002**, *4*, 1507.
20. Jiang, N.; Ragauskas, A. J. *Org. Lett.* **2005**, *7*, 3689.
21. Jiang, N.; Ragauskas, A. J. *ChemSusChem* **2008**, *1*, 823.
22. Ragagnin, G.; Betzemeier, B.; Quici, S.; Knochel, P. *Tetrahedron* **2002**, *58*, 3985.
23. Figiel, P. J.; Kirillov, A. M.; Karabach, Y. Y.; Kopylovich, M. N.; Pombeiro, A. J. L. *J. Mol. Catal. A: Chem.* **2009**, 178.
24. Yang, G.; Zhu, W.; Zhang, P.; Xue, H.; Wang, W.; Tian, J.; Song, M. *Adv. Synth. & Catal.* **2008**, *350*, 542.
25. Sheldon, R. A.; Arends, I. W. C. E.; Dijksman, A. *Catal. Today* **2000**, *57*, 157.
26. Wolfson, A.; Wuyts, S.; De Vos, D. E.; Vankelecom, I. F. J.; Jacobs, P. A. *Tetrahedron Lett.* **2002**, *43*, 8107.
27. Csjernyik, G.; Éll, A. E.; Fadini, L.; Pugin, B.; Bäckvall, J.-E. *J. Org. Chem.* **2002**, *67*, 1657 and references cited therein.
28. Johnston, E. V.; Karlsson, E. A.; Tran, L.-H.; Åkermark, B.; Bäckvall, J.-E. *Eur. J. Org. Chem.* **2010**, 1971 and references cited therein.
29. Hanyu, A.; Takezawa, E.; Sakaguchi, S.; Ishii, Y. *Tetrahedron Lett.* **1998**, *39*, 5557.

30. a) Dijksman, A.; Arends, I. W. C. E.; Sheldon, R. A. *Chem. Commun.* **1999**, 1591; b) Dijksman, A.; Marino-González, A.; Mairata I Payeras, A.; Arends, I. W. C. E.; Sheldon, R. A. *J. Am. Chem. Soc.* **2001**, *123*, 6826.

31. Markó, I. E.; Giles, P. R.; Tsukazaki, M.; Chellé-Regnaut, I.; Urch, C. J.; Brown, S. M. *J. Am. Chem. Soc.* **1997**, *119*, 12661.

32. Lenz, R.; Ley, S. V. *J. Chem. Soc., Perkin Trans 1* **1997**, 3291.

33. Egami, H.; Shimizu, H.; Katsuki, T. *Tetrahedron Lett.* **2005**, *46*, 783.

34. Masutani, K.; Uchida, T.; Irie, R.; Katsuki, T. *Tetrahedron Lett.* **2000**, *41*, 5119.

35. Shimizu, H.; Onitsuka, S.; Egami, H.; Katsuki, T. *J. Am. Chem. Soc.* **2005**, *127*, 5396.

36. Mizoguchi, H.; Uchida, T.; Ishida, K.; Katsuki, T. *Tetrahedron Lett.* **2009**, *50*, 3432.

37. Hinzen, B.; Ley, S. V. *J. Chem. Soc. Perkin Trans 1*, **1997**, 1907.

38. Bleloch, A.; Johnson, B. F. G.; Ley, S. V.; Price, A. J.; Shepard, D. S.; Thomas, A. W. *Chem. Commun.* **1999**, 1907.

39 Haberman, J.; Ley, S. V.; Scott, J. S. *J. Chem. Soc., Perkin Trans 1*, **1999**, 1253.

40. Wight, A. P.; Davis, M. E. *Chem. Rev.* **2002**, *102*, 3589.

41. a) Pagliaro, M.; Ciriminna, R. *Tetrahedron Lett.* **2001**, *42*, 4511; b) Ciriminna, R.; Pagliaro, M. *Chem. Eur. J.* **2003**, *9*, 5067.

42. Ciriminna, R.; Campestrini, S.; Pagliaro, M. *Adv. Synth. Catal.* **2003**, *345*, 1261.

43. Ciriminna, R.; Hesemann, P.; Moreau, J. J. E.; Carraro, M.; Campestrini, S.; Pagliaro, M. *Chem. Eur. J.* **2006**, *12*, 5220.

44. Ciriminna, R.; Campestrini, S.; Pagliaro, M. *Adv. Synth. Catal.* **2004**, *346*, 231.

45. Ciriminna, R.; Campestrini, S.; Pagliaro, M. *Org. Biomol. Chem.* **2006**, *4*, 2637.

46. For a recent review on catalytic oxidations in dense carbon dioxide see: Seki, T.; Baiker, A. *Chem. Rev.* **2009**, 2409.

47. Zhan, B.-Z.; White, M. A.; Sham, T.-K.; Pincock, J. A.; Doucet, R. J.; Ramana Rao, K. V.; Robertson, K. N.; Cameron, T. S. *J. Am. Chem. Soc.* **2003**, *125*, 2195.

48. Yamaguchi, K.; Mori, K.; Mizugaki, T., Ebitani, K.; Kaneda, K. *J. Am. Chem. Soc.* **2000**, *122*, 7144.

49. Yamaguchi, K.; Mizuno, N. *Angew. Chem. Int. Ed.* **2002**, *41*, 4538.

50. Kotani, M.; Koike, T.; Yamaguchi, K.; Mizuno, N. *Green Chem.* **2006**, *8*, 735.

51. Matsumoto, T.; Ueno, M.; Kobayashi, J.; Miyamura, H.; Mori, Y.; Kobayashi, S. *Adv. Synth. Catal.* **2007**, *349*, 531.

52. Matsumoto, T.; Ueno, M.; Wang, N.; Kobayashi, S. *Chem. Asian J.* **2008**, *3*, 239.

53 Mori, S.; Takubo, M.; Makida, K.; Yanase, T.; Aoyagi, S.; Maegawa, T.; Monguchi, Y.; Sajiki, H. *Chem. Commun.* **2009**, 5159.

54. a) Muzart, J. *Tetrahedron* **2003**, *59*, 5789.; b) Stahl, S. S. *Angew. Chem. Int. Ed.* **2004**, *43*, 3400; c) Stahl, S. S. *Science* **2005**, *309*, 1824.

55. Sigman, M. S.; Jensen, D. R. *Acc. Chem. Res.* **2006**, *39*, 221 and references cited therein.

56. For a recent review on Pd[II]-catalyzed aerobic oxidations see: Gligorich, K. M.; Sigman, M. S. *Chem. Commun.* **2009**, 3854.

57. Peterson, K. P.; Larock, R. C. *J. Org. Chem.* **1998**, *63*, 3185.

58. Seddon, K. R.; Stark, A. *Green Chem.* **2002**, *4*, 119.

59   a) Nishimura, T.; Onoue, T.; Ohe K.; Uemura, S. *Tetrahedron Lett.* **1998**, *39*, 6011; b) Nishimura, T.; Onoue, T.; Ohe, K.; Uemura, S. *J. Org. Chem.* **1999**, *64*, 6750.

60.   Nishimura, T.; Ohe, K.; Uemura, S. *J. Am. Chem. Soc.* **1999**, *121*, 2645.

61.   Nishimura, T., Maeda, Y.; Kakiuchi, N.; Uemura, S. *J. Chem. Soc., Perkin Trans 1* **2000**, 4301.

62.   a) ten Brink, G.-J.; Arends, I. W. C. E.; Sheldon, R. A. *Science* **2000**, *287*, 1636; b) ten Brink, G.-J.; Arends, I. W. C. E.; Sheldon, R. A. *Adv. Synth. & Catal.* **2002**, *344*, 355.

63.   ten Brink, G.-J.; Arends, I. W. C. E.; Hoogenraad, M.; Verspui, G., Sheldon, R. A. *Adv. Synth. & Catal.* **2003**, *345*, 1341.

64.   Mifsud, M.; Parkhomenko, K. V.; Arends, I. W. C. W.; Sheldon, R. A. *Tetrahedron* **2010**, *66*, 1040.

65.   Iwasawa, T.; Tokunaga, M.; Obora, Y.; Tsuji, Y. *J. Am. Chem. Soc.* **2004**, *126*, 6554.

66.   a) Schultz, M. J.; Park,C. C.; Sigman, M. S. *Chem. Commun.* **2002**, 3034; b) Jensen, D. R.; Schultz, M. J.; Mueller, J. A.; Sigman, M. S. *Angew. Chem. Int. Ed.* **2003**, *42*, 3810.

67.   Schultz, M. J.; Hamilton, S. S.; Jensen, D. R.; Sigman, M. S. *J. Org. Chem.* **2005**, *70*, 3343.

68.   Robinson, R. S.; Taylor, R. J. K. *Synlett* **2005**, 1003.

69.   Lebel, H.; Paquet, V. *J. Am. Chem. Soc.* **2004**, *126*, 11152.

70.   a) Jensen, D. R.; Pugsley, J. S.; Sigman; M. S. *J. Am. Chem. Soc.* **2001**, *123*, 7475; b) Ferreira, E. M.; Stoltz, B. M. *J. Am. Chem. Soc.* **2001**, 123, 7725.

71.   a) Mueller, J. A.; Jensen, D. R.; Sigman, M. S. *J. Am. Chem. Soc.* **2002**, *124*, 8202; b) Mueller, J. A.; Sigman, M. S. *J. Am. Chem. Soc.* **2003**, *125*, 7005.

72.   Krishnan, S.; Bagdanoff, J. T.; Ebner, D. C.; Ramtohul, Y. K.; Tambar, U. K.; Stoltz, B. M. *J. Am. Chem. Soc.* **2008**, *130*, 13745.

73.   Caspi, D. D.; Ebner, D. C.; Bagdanoff, J. T.; Stoltz, B. M. *Adv. Synth. & Catal.* **2004**, *346*, 185.

74.   O'Brien, P. *Chem. Commun.* **2008**, 655.

75.   Bailie, D. S.; Clendenning, G. M. A.; McNamee, L.; Muldoon, M. J. *Chem. Commun.* **2010**, *46*, 7238.

76.   Punniyamurthy, T.; Velusamy, S.; Iqbal, J. *Chem. Rev.* **2005**, *105*, 2329.

77.   a) Nishimura, T.; Kakiuchi, N.; Inoue, M.; Uemura, S. *Chem. Commun.* **2000**, *1245*; b) Kakiuchi, N.; Maeda, Y.; Nishimura, T.; Uemura, S. *J. Org. Chem.* **2001**, *66*, 6620.

78.   Biffis, A.; Minati, L. *J. Catal.* **2005**, *236*, 405.

79.   a) Mori, K.; Yamaguchi, K.; Hara, T.; Mizugaki, T.; Ebitani, K.; Kaneda, K. *J. Am. Chem. Soc.* **2002**, *124*, 11572; b) Mori, K.; Hara, T.; Mizugaki, T.; Ebitani, K.; Kaneda, K. *J. Am. Chem. Soc.* **2004**, *126*, 10657.

80.   Uozumi, Y.; Nakao, R. *Angew. Chem. Int. Ed.* **2003**, *42*, 194.

81.   Karimi, B.; Abedi, S.; Clark, J. H.; Budarin, V. *Angew. Chem. Int. Ed.* **2006**, *45*, 4776.

82.   Kwon, M. S.; Kim, N.; Park, C. M.; Lee, J. S.; Kang, K. Y.; Park, J. *Org. Lett.* **2005**, *7*, 1077.

83.   Pillai, U. R.; Sahle-Demessie, E. *Green Chemistry* **2004**, *6*, 161.

84.   Hackett, S. F. J.; Brydson, R. M.; Gass, M. H.; Harvey, I.; Newman, A. D.; Wilson, K.; Lee, A. F. *Angew. Chem. Int. Ed.* **2007**, *46*, 8593.

85.   Wu, H.; Zhang, Q.; Wang, Y. *Adv. Synth. Catal.* **2005**, *347*, 1356.

86.   a) Hou, Z.; Theyssen, N.; Brinkmann, A.; Leitner, W. *Angew. Chem. Int. Ed.* **2005**, *44*, 1346; b) Hou, Z.; Theyssen, N.; Leitner, W. *Green Chem.* **2007**, *9*, 127.

87   Guan, B.; Xing, D.; Cai, G.; Wan, X.; Yu, N.; Fang, Z.; Yang, L.; Shi, Z. *J. Am. Chem. Soc.* **2005**, *127*, 18004.

88   Li, H.; Guan, B.; Wang, W.; Xing, D.; Fang, Z.; Wan, X.; Yang, L.; Shi, Z. *Tetrahedron* **2007**, *63*, 8430.

89   a) Prati, L.; Rossi, M. *J. Catal.* **1998**, *176*, 552. b) Biella, S.; Prati, L.; Rossi, M. *J. Catal.* **2002**, *206*, 242. c) Biella, S.; Castiglioni, G.L.; Fumagalli, C.; Prati, L.; Rossi, M. *Catal. Today* **2002**, *72*, 43.

90.   Biella, S.; Rossi, M. *Chem. Commun.* **2003**, 378.

91   Carrettin, S.; McMorn, P.; Johnston, P.; Griffin, K.; Hutchings, G. J. *Chem. Commun.* **2002**, 696.

92   Yang, X.; Wang, X.; Liang, C.; Su, W.; Wang, C.; Feng, Z.; Li, C.; Qiu, J. *Catal. Commun.* **2008**, *9*, 2278.

93   Abad, A.; Concepción, P.; Corma, A.; García, H. *Angew. Chem. Int. Ed.* **2005**, *44*, 4066.

94   Abad, A.; Almela, c.; Corma, A.; García, H. *Chem. Commun.* **2006**, 3178.

95   Enache, D. I.; Knight, D. W.; Hutchings, G. J. *Catal. Lett.* **2005**, *103*, 43.

96   Comotti, M.; Della Pina, C.; Matarrese, R.; Rossi, M. *Angew. Chem. Int. Ed.* **2004**, *43*, 5812.

97   Christensen, C. H.; Jørgensen, B.; Rass-Hansen, J.; Egebland, K.; Madsen, R.; Klitgaard, S. K.; Hansesn, S. M.; Hansen, M. R.; Andersen, H. C.; Riisager, A. *Angew. Chem. Int. Ed.* **2006**, *45*, 4648.

98   Kegn□s, S.; Mielby, J.; Mentzel, U. V.; Christensen, C. H.; Riisanger, A. *Green Chem.* **2010**, *12*, 1437.

99   a) Tsunoyama, H.; Sakurai, H.; Negishi, Y.; Tsukuda, T. *J. Am. Chem. Soc.* **2005**, *127*, 9374. b) Tsunoyama, H.; Tsukuda, T.; Sakurai, H. *Chem. Lett.* **2007**, *36*, 212. c) Kanaoka, S.; Yagi, N.; Kukuyama, Y.; Aoshima, S.; Tsunoyama, H.; Tsukuda, T. Sakurai, H. *J. Am. Chem. Soc.* **2007**, *129*, 12060. d) Tsunoyama, H.; Ichikuni, N.; Sakurai, H.; Tsukuda, T. *J. Am. Chem. Soc.* **2009**, *131*, 7086.

100   Biffis, A.; Cunial, S.; Spontoni, P.; Prati, L. *J. Catal.* **2007**, *251*, 1.

101   Miyamura, H.; Matsubara, R.; Miyazaki, Y.; Kobayashi, S. *Angew. Chem. Int. Ed.* **2007**, *46*, 4151.

102   Miyamura, H.; Yasukawa, T.; Kobayashi, S. *Green Chem.* **2010**, *12*, 776.

103   a) Wang, N.; Matsumoto, T.; Uoeno, M.; Miyamura, H.; Kobayashi, S. *Angew, Chem. Int. Ed.* **2009**, *48*, 4744.

104   Lucchesi, C.; Inassaki, T.; Miyamura, H.; Matsubara, R.; Kobayashi, S. *Adv. Synth. Catal.* **2008**, *350*, 1996.

105   Kimmerle, B.; Grunwaldt, J.-D.; Baiker, A. *Topics Catal* **2007**, *44*, 285.

106   Wang, X.; Kawanami, H.; Dapurkar, S. E.; Venkataramanan, N. S.; Chatterjee, M.; Yokoyama, T.; Ikushima, Y. *Appl. Catal. A* **2008**, *349*, 86.

107   Enache, D. I.; Edwards, J. K.; Landon, P.; Solsona-Espriu, B.; Carley, A. F.; Herzing, A. A.; Watanabe, M.; Kiely, C. J.; Knight, D. W.; Hutchings, G. J. *Science* **2006**, *311*, 362.

108   Zheng, N.; Stucky, G. D. *Chem. Commun.* **2007**, 3862.

109   Villa, A.; Veith, G. M.; Prati, L. *Angew. Chem. Int. Ed.* **2010**, *49*, 4499.

110   Miyamura, H.; Matsubara, R.; Kobayashi, S. *Chem. Commun.* **2008**, 2031.

111   Kaizuka, K.; Miyamura, H.; Kobayashi, S. *J. Am. Chem. Soc.* **2010**, *132*, 15096.

112   Wang, L.; Meng, X.; Wang, B.; Chi, W.; Xiao, F.-S. *Chem. Commun.* **2010**, *46*, 5003.

113   Saio, D.; Amaya, T.; Hirao, T. *Adv. Synth. Catal.* **2010**, *352*, 2177.

Ilaria D'Acquarica
Professor of Organic Chemistry

Address:

Dipartimento di Chimica e Tecnologie del Farmaco
Sapienza Università di Roma
P.le Aldo Moro 5
00185 Roma

Education:

She was born in 1969 and graduated in Medicinal Chemistry with full marks ("magna cum laude") at the Sapienza University of Rome in 1993. In 1998, she received her Ph.D. degree in Pharmaceutical Sciences. From 2000 to 2004 she was temporary research fellow of Organic Chemistry at the same University.

Professional experience:

Assistant Professor of Organic Chemistry since 2004, she is author of 50 papers published in International Journals, 2 patents, and has presented her results during several national and international meetings. She is referee of Journal of Chromatography A and B since 2001.

Research interests: separation chemistry; HPLC analysis of complex biological samples; synthesis, characterization and evaluation of novel chiral stationary phases for HPLC; enantioselective molecular recognition in the gas-phase of resorc[4]arenes receptors.

The Combination of High Performance Liquid Chromatography with Mass Spectrometry (HPLC/MS) as a Powerful Tool for Modern Organic Chemistry

Ilaria D'Acquarica

Dipartimento di Chimica e Tecnologie del Farmaco, Sapienza Università di Roma
P.le Aldo Moro 5, 00185 Roma (Italy)

1. HYPHENATED ANALYTICAL TECHNIQUES

1.1 Introduction

Chemical profiling of complex mixtures and identification of trace level impurities, when required, is a challenging task for modern organic chemistry. Over the past thirty years, specific analytical technologies have evolved which greatly facilitate the characterization of complex organic mixtures and the unambiguous identification of trace level chemical entities in complex materials. The most useful of these analytical technologies are those based on the combination of chromatography with Mass Spectrometry (MS)[1] and, more recently, chromatography with Nuclear Magnetic Resonance (NMR) Spectroscopy,[2] which have created a family of "hyphenated" analytical techniques with significant capability for molecular structure elucidation and determination of elemental species at trace levels.[3]

The direct coupling of Gas Chromatography (GC) with MS (GC/MS) was achieved for the first time in the late 1950s[4,5] and, since then, the GC/MS combination is routinely applied to trace organic analysis, including manufacturing quality control, environmental testing, forensics, drug monitoring, and pharmacological studies. Current instrumentation, which combines the high efficiency separation capability of GC with the high sensitivity of Electron Impact (EI) ionization mass spectrometry makes GC/MS the technique of choice for compositional profiling and impurity identification in relatively volatile compounds such as propellants, alcohols, and fatty acids.[3] Since EI is a "kinetically controlled" ionization process, spectra of the same chemical entity are nearly identical from one instrument to another and, over time, on the same instrument. This allows for the creation of extensive libraries of EI spectra which can be rapidly searched by modern computers to potentially identify unknown impurities. Identification by GC/EI-MS is usually accomplished with a combination of automated spectral library searching, manual interpretation of mass spectral fragmentation patterns, and analysis of authentic reference samples.

The first coupling of High Performance Liquid Chromatography (HPLC) with MS did not appear until the 1970s (see Section 1.2), because of the problem of introducing large volumes of mobile phase along with the compounds to be analyzed into the high vacuum environment of the MS source.

Other chromatographic and/or separation techniques which have been coupled with NMR are Supercritical Fluid Chromatography (SFC), Gel-Permeation Chromatography (GPC), and Capillary Electrophoresis (CE).[6]

1.2 Coupling of High Performance Liquid Chromatography (HPLC) with Mass Spectrometry (MS)

Direct coupling of HPLC to MS was considered for many years incompatible, because the mass spectrometer requires a high vacuum environment and the analytes of interest to be in the gas phase, whereas HPLC provides separation of non-volatile compounds in a liquid mobile phase. LC/MS coupling was preceded by several off-line approaches, mainly based on the MS analysis of chromatographic fractions transferred into the mass spectrometer by a probe sampling technique. Such procedures were time consuming, particularly when a large number of fractions were collected, since the probe sampling technique was a 10–20-min operation for each fraction. For this reason, a more efficient approach seemed to be to link the liquid chromatograph directly to the mass spectrometer.

A number of different LC/MS interfaces have been constructed over the years. Similarities in strategy and technology used in these interfaces make a simple classification difficult.[7] Four basic approaches have been studied to link the mass spectrometer to the liquid chromatograph:

(1) Direct introduction or a simple split of the LC effluent in which a maximum of 0.01 ml of liquid is vaporized into the ion-chamber.[8–12]

(2) Enrichment of the LC effluent using a molecular separator.[13–18]

(3) Enrichment of the LC effluent by evaporation of solvent in vacuum lock interface chambers during mechanical transport of the solutions.[19–21]

(4) Use of an atmospheric pressure ionization (API) source through which the entire LC effluent is vaporized (see Section 1.3).

The use of a *direct liquid introduction* interface (1) involves certain disadvantages. If the mass spectrometer is operated in an electron impact (EI) mode, less than 0.1% of the sample can be utilized and such waste cannot generally be tolerated. Even when used in a chemical ionization (CI) mode, only 1–2 % of the sample can be injected and the solvent carrier liquid must function as the CI reagent gas. This poses restrictions on both the MS ionization process and the LC solvent selection, but it has been demonstrated that these restrictions are not serious for most sample/solvent systems. In both cases (EI or CI), the sample must be vaporized, but McLafferty has pointed-out that vaporization from a solution does not require as high a temperature as vaporization of pure sample.[8,9] The main disadvantages of this splitter technique are the low utilization of sample and the frequent clogging of the diaphragms.

A natural extension of the direct inlet system was the development of the *thermospray* (TSP) interface, which involves the introduction and vaporization of high liquid flow-rates (1–2 ml min^{-1}) into the ion source and ionization of the vaporized material by an electron beam. The accidental observation that, in the presence of a volatile buffer in the solvent, ions were obtained without any further ionizing source led to the development of an LC/MS interface based on this so-called thermospray ionization.[10,11] Under typical operating conditions, about 95% of the liquid is vaporized, and the rest is in the form of a highly collimated particle beam which is accelerated to approximately sonic velocity by the rapid expansion of the vapor.

A major problem in the introduction of liquids through a capillary is that the high vacuum in the ion source causes rapid evaporation of the liquid inside the capillary, eventually leading to flow stoppage by solvent freezing. The use of restricted capillaries and the heating of the capillary partially solved this problem. Restricted capillaries, however, were prone to plug easily and had only limited success. An original approach to continuously flowing aqueous solutions into the mass spectrometer involved the development of an interface based on fast atom bombardment (FAB) ionization, namely the *continuous-flow* FAB interface.[12] In such system, the LC effluent, mixed with a FAB matrix (generally 5% aqueous glycerol) is continuously deposited on the tip of a FAB probe permanently situated inside the ion source. The liquid mixture is usually transported to the tip through a fused-silica capillary coaxial to a modified FAB probe. The FAB matrix prevents the liquid evaporation inside the capillary and the mixture can be submitted to atom bombardment in the FAB probe target. Despite the appearance of techniques such as electrospray ionization (ESI, see Section 1.5), which competes advantageously with FAB for the analysis of large biomolecules, *continuous-flow* FAB is still actively used in several laboratories.

Molecular separators (2) are based on the concept of the enrichment devices developed for GC/MS interfaces. Two different systems will be described here, the membrane separator[13] and the momentum separator.[14–18] The first attempt to develop a membrane interface was the use of a flash evaporator chamber coupled to a polymeric dimethylsiloxane membrane that selectively transmits non-polar molecules (*e.g.*, anthracene, etc.) to the mass spectrometer, while rejecting the more polar solutes. The original interface[13] was built of a three-stage membrane separator and was able to deal with flow-rates as high as 2.0 ml min^{-1}. Application of this membrane separator interface was restricted, however, to non-polar molecules, volatile at the working temperature of the interface (about 25 °C).

An example of a momentum separator is the *particle beam* (PB) interface, which uses a jet separator to eliminate volatile solvents and to transport analyte in the form of solid micro-aggregates to the ion source of the mass spectrometer. Jet separators were applied very early for LC/MS coupling. For example, Takeuchi et al.[14] used this device to directly couple a microbore chromatographic column (7 cm × 0.5 mm, length × internal diameter) to the CI source of a quadrupole mass spectrometer. The eluate from the column, delivered at flow-rates in the 2–16 μl min^{-1} range, was evaporated by heating the separator at 150 °C. The momentum separator provided a six-fold analyte enrichment in the vapor, and ionization was carried out by CI, where the solvent acted as the reagent gas. Although improvements have been made to the PB interface,[15–17] the concept of using stages of momentum separation to enrich the beam with particles and to remove solvent has remained unchanged. In modern PB systems,[17] the enrichment region consists of three-stages momentum separators. A liquid solution is nebulized into microdroplets of very homogeneous size which are partially desolvated and accelerated through a nozzle. Once through the nozzle, high-velocity particles are momentum separated from solvent by skimmers and are then transferred to the ion source, where they are expected to collide with the hot hyperbolic surface of the end cap. The end cap serves both as a heated target used to vaporize the particles and as an ion-trapping electrode. Despite the non-linearity, the limited sensitivity, and the limited applicability range, PB interface is a powerful tool, especially in environmental analysis of medium-polarity low-molecular-mass compounds, or in other areas where on-line LC-EI mass spectra are of importance.[18] In fact, PB interface has the unique ability to yield library-matchable EI spectra from compounds separated by conventional LC instrumentation.

The third approach to LC-MS interfacing (3) involves transport of the sample into the mass spectrometer on a continuously *moving wire* train system. In the first moving wire interface developed,[19] a 0.12 mm diameter stainless steel wire was passed through a small trough containing the LC effluent and then into two vacuum lock interface chambers where the solvent was evaporated. The residual solute was then carried into the ion source region where it was thermally vaporized and ionized. Such a system could provide EI or CI spectra as required, but would have been completely independent of the solvent used for chromatography and thus maintaining the versatility and capabilities of the separation technique. The low surface area of the wire, however, allowed the deposition of just 1% (10 μl min^{-1}) of the total eluate from a conventional chromatographic column. An extension of Scott's moving wire system was developed two years later,[20] in which increased sample utilization was attained by transporting the LC effluent on a ribbon rather than on a wire. In the so-called *moving belt* interface, the substitution by a ribbon made of stainless steel (3.2 × 0.05 mm) or polyimide (Kapton) yielded a larger deposition surface area, and 30–50-fold more sample could be managed. The first commercial moving belt interface, marketed by Finnigan, was able to use both EI and CI, and showed detection limits in the nanogram range. Mechanical transport interface development and applications have been reviewed by Alcock et al.[21]

1.3 Atmospheric Pressure Ionization (API) Interfaces

Despite elaborate research efforts in LC/MS interfacing in the past 40 years, it is only in the last 20 years that LC/MS has become a technique which can be used routinely in a wide variety of analytical laboratories and application areas. Some LC/MS instrumentation is now actually being sold as an integrated detector for HPLC, *i.e.*, to enter the chromatography rather than the MS laboratory. This can be attributed to the major progress that has been achieved in the development of reliable and user-friendly instrumentation for LC/MS and related liquid-introduction techniques for MS, which in turn is attributable to the vast advent of atmospheric pressure ionization (API) devices. Despite the fact that API interfaces has a different application area from the PB interface (see Section 1.2), these systems have much in common, as they both introduce liquid effluents in a desolvation chamber, which is held at (nearly) atmospheric pressure.

The current API technology comprises two different interfaces for LC/MS, based on atmospheric pressure chemical ionization (APCI)[22–24] and electrospray ionization (ESI).[25–27] Several dedicated reviews on the application of API technology for LC/MS have been published, some of which primarily paying attention to instrumental aspects of such technology. [28–31]

Generally, in an API source, ionization of the LC effluent is carried out at atmospheric pressure by any of several procedures including a radioactive source, electrical discharges and high-voltage electric fields. The ions produced are continuously sampled through a small aperture and pass into the spectrometer where they are mass analyzed.

An API interface/source consists of five parts: (1) the liquid introduction device, (2) the actual atmospheric pressure ion source region, where the ions are generated, (3) an ion sampling aperture, (4) an atmospheric pressure to vacuum interface, and (5) an ion optical system, where the ions are subsequently transported into the mass analyzer.[32]

The operational principle of the most widely applied API interface and ion source design is as follows. The LC effluent is nebulized into an atmospheric pressure ion source region. Nebulization is either performed pneumatically, as it happens in APCI, or by means of a strong electrical field, as it happens in ESI, or by a combination of both. Ions are produced from the evaporating droplets, either by gas phase ion-molecule reactions initiated by electrons from a corona discharge (in the case of APCI), or by the formation of microdroplets by solvent evaporation and repetitive electrohydrodynamic explosions and the desorption, evaporation or soft desolvation of ions from these droplets into the gas phase (in the case of ESI, see also Section 1.5). The ions generated, together with solvent vapour and nitrogen bath gas, are sampled by an ion sampling device into first pumping stage. The mixture of gas, solvent vapour and ions is supersonically expanding into this low-pressure region. The core of the expansion is sampled by a skimmer into a second pumping stage, containing an ion focusing and transfer device to optimally transport the ions in a suitable manner to the mass analyzer. From the vacuum point-of-view, it is not important whether a high or a low flow-rate of liquid is nebulized, because the sampling orifice actually acts as the restrictor between the atmospheric-pressure region and the first pumping stage. From the ionization point-of-view, it is not important as well whether the ions are generated by ESI or APCI, although slightly different tuning of voltages in the ion optics might be needed, due to some differences in the ion kinetic energies.

The ionization mechanisms in APCI are identical to those in conventional medium-pressure chemical ionization (CI). Positive-ion formation can be achieved by proton donation, adduct formation, or charge exchange reactions, while in negative-ion mode ions are formed due to proton abstraction, anion attachment, or electron-capture reactions. Differences in mass spectra made by APCI and conventional CI can be explained by the fact that in APCI ion formation reflects equilibrium conditions rather than reaction rates, while the latter dominates the ion products in medium-pressure CI.

APCI is currently being applied to the analysis of a great variety of compounds, especially in the environmental and pharmaceutical fields. As indicated below, it is not as effective as ESI for the analysis of biopolymers. However, APCI is very effective in the analysis of medium- and low-polarity compounds or when relatively non-polar solvents have to be used.

1.4 Reversed Phase HPLC (RP-HPLC)

Adsorption chromatography depends on the chemical interactions between solute molecules and specifically designed ligands chemically grafted onto a chromatographic matrix, *i.e.*, the stationary phase. Over the years, many different types of ligands have been immobilized to chromatographic supports for biomolecule purification, exploiting a variety of biochemical properties ranging from electronic charge to biological affinity.[33] HPLC separations that use polar stationary phases and non polar mobile phases are referred to as normal phase HPLC (NP-HPLC). Solutes have a typical elution order in NP-HPLC, *i.e.*, from less polar to more polar analytes. The opposite of NP-HPLC, or reversed phase HPLC (RP-HPLC), results from the adsorption of hydrophobic molecules onto a hydrophobic solid support in a water-based polar mobile phase. The retention order of solutes separated by this approach is opposite to that of NP-HPLC, *i.e.*, from more polar analytes to less polar ones. The base matrix for commercially available reversed phase media is generally composed of microparticulate silica or a synthetic organic polymer such as polystyrene, and chemically grafted hydrophobic ligands consist of linear hydrocarbon chains (*n*-alkyl groups) such as, for example, octadecylsilane (C18) and octylsilane (C8). A large number of stationary phases for RP-HPLC have been developed over the years,[34] and a detailed description of such phases is beyond the scope of this paper.

During an RP-chromatographic run, the solute molecules partition (*i.e.*, an equilibrium is established) between the mobile phase and the stationary phase. The distribution of the solute between the two phases depends on the binding properties of the medium, the hydrophobicity of the solute and the composition of the mobile phase. Initially, experimental conditions are designed to favor adsorption of the solute from the mobile phase to the stationary phase. Subsequently, the mobile phase composition is modified to favor desorption of the solute from the stationary phase back into the mobile phase. In this case, adsorption is considered the extreme equilibrium state where the distribution of solute molecules is essentially 100% in the stationary phase. Conversely, desorption is an extreme equilibrium state where the solute is essentially 100% distributed in the mobile phase.[35]

When biomolecules are analyzed by RP-HPLC, gradient elution is generally used instead of isocratic elution. In fact, while biomolecules strongly adsorb to the surface of a reversed phase matrix under aqueous conditions, they desorb from the matrix within a very narrow window of organic modifier (typically methanol, acetonitrile) concentration. Along with these high molecular weight biomolecules with their unique adsorption properties, the typical biological samples usually contain a broad mixture of biomolecules with a correspondingly diverse range of adsorption affinities. The only practical method for RP separation of complex biological samples, therefore, is gradient elution.

RP-HPLC is much more widely used today than NP-HPLC because of its broad selectivity, reproducibility, compatibility with biological samples, and its suitability for hyphenation with MS. In particular, RP-HPLC is compatible with ESI-MS since such ionization source favors the formation of ions in solution; conversely, APCI is compatible with non-polar mobile phases, *i.e.*, NP-HPLC since analytes are neutral in solution. In other words, the compatibility of the various HPLC modes with ESI and APCI is governed by the fact that each mode relies on different ion production mechanisms.

1.5 Electrospray Ionization (ESI)

The history of electrospray as a sample introduction method for MS goes back to the pioneering efforts of Malcom Dole and his colleagues in the late 1960s, which actually were pretty much ignored for a long time.[36–38] In fact, the technology of electrospray interfacing and ionization was received a more convincing impetus only since 1984 by Fenn and co-workers,[25–27] who demonstrated the possibility to achieve multiple charging of proteins by electrospray ionization. This observation widely opened a highly important and almost completely new application area for MS, *i.e.*, the characterization of biomacromolecules. While in former years the analysis of proteins by MS was a specialized area for FAB ionization on expensive sector instruments, electrospray ionization brought the analysis of proteins within the reach of a simple quadrupole instrument. Furthermore, the sensitivity and ease-of-operation in electrospray is considerably better than in FAB. Obviously, a further impetus to the protein field was also given by the discovery of matrix-assisted laser desorption ionization (MALDI).[39]

The scenario of electrospray ionization, much as Dole first described it, includes both the nebulization of a liquid into an aerosol of highly charged droplets and the ionization of solvated analyte species after desolvation of the charged droplets. The electrospray process is initiated by applying an electrical potential of several kV to a liquid in a stainless steel narrow-bore capillary or electrospray needle (Fig. 1). By increasing the potential difference between the needle and the counter electrode (or sampling plate), a series of processes takes place. The transitions between these processes can be monitored by measuring the current between needle and counter electrode. The four stages of electrospray nebulization[18] are:

(i) At low electric fields, the liquid drops almost vertically out of the needle, because the field effect is insufficient. Afterward, at higher fields, a nearly horizontal liquid column is formed, extending beyond the end of the needle.

(ii) With an increase in potential, the liquid column elongates, resulting in the formation of an electrically induced axial spray of charged microdroplets (the so called Taylor cone); this is due to the reaching of the Rayleigh limit, at which the forces owing to electrostatic repulsion approach the surface tension of the eluent liquid. This is the onset of the electrospray process.

(iii) Further increasing the potential causes a sudden transition to take place: the liquid cone vanishes and a fine mist of droplets is produced from a number of points on the edge of the capillary tip. This is known as rim emission mode. The potential at which this transition takes place is related to the eluent surface tension and the wettability of the capillary needle walls.

(iv) At still higher potentials, a stable corona discharge is established between the needle tip and the counter electrode. Discharge at the capillary tip disrupts the electrospray process. In practice, ESI-MS is performed between the axial and rim emission modes of spraying.

The mechanism by which the ions contained into the above microdroplets enter the gas phase was a matter of considerable discussion in the past. Two popular theories have been proposed in the years: the "charge residue" mechanism, developed by Dole,[36] and the "ion evaporation" model, proposed later by Iribarne and Thomson.[40] Dole's idea was that a sequence of droplet fissions took place until a very small droplet containing only the charged residue, *i.e.*, the ion of interest, remained. According to the ion evaporation mechanism, which actually gives a more likely explanation for the formation of single ions, small individual ions were somehow able to evaporate from the liquid. The main supporting evidence to this theory came from ion mobility measurements, which showed the production of significant amounts of gas phase ions at times where most of the charged droplets were expected to have relatively large radii and multiple charges. On the basis of a very simple model, Iribarne and Thomson estimated values for solvation energies of the ions and calculated the fields necessary for their desorption. They found that for droplets in the right size range (*i.e.*, below 10 nm), the required fields could be achieved without exceeding the Rayleigh limit. Subsequent reports provided mass spectrometric data that seemed to provide evidence of ion desorption for a wide variety of solute species.

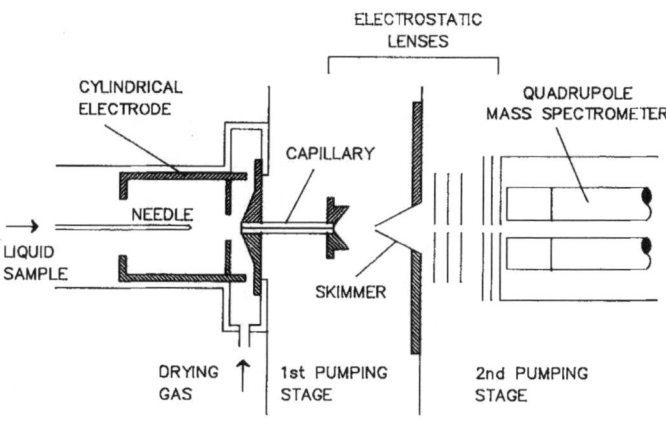

Fig. 1. Schematic Diagram of an Electrospray Mass Spectrometer Apparatus. Reproduced from Figure 1 in Ref. [27] with Permission from Wiley. Copyright 1990.

Three main features of ESI are noteworthy: first, it is a very soft ionization technique that produces protonated (or deprotonated) molecule ions of polar, non-volatile, high molecular mass and thermolabile compounds with amazing ease and efficiency. Second, multiply charged ions of the type $[M + nH]^{n+}$ are formed by protonation of molecules with multiple basic sites (the equivalent $[M - nH]^{n-}$ are observed in the negative ion mode). Third, the ions observed, generally formed by protonation or deprotonation of the molecule or by adduct formation with solvent, directly reflect acid-base equilibria in solution. In this sense, ESI is especially suited for the analysis of compounds ionized in the liquid phase.

The production of multiply charged ions has important implications in the analysis of peptides and proteins. Mass spectrometers measure mass-to-charge ratios (*m/z*) so that multiply charged ions appear in the spectrum at *m/z* values that are fractions of the actual mass of the ion. This allows one to observe signals from compounds well beyond the working mass range of the mass spectrometer, at typically *m/z* 200–2000. The molecular mass of one of these molecules can be calculated from the $[M + nH]^{n+}$ ions envelop in the spectra. This is carried out by automatic computational (deconvolution) methods.[41] A number of computer algorithms have been developed that are designed to take the multiple ion charge state pattern in a spectrum plotted as m/z ratios and convert it into a single value corresponding to the molecular mass of that component.[42]

1.6 Applications of HPLC/ESI-MS techniques

Beside the apparent flow-rate incompatibility as expressed in the need to introduce a liquid effluent from a conventional LC column into the high vacuum of the mass spectrometer, a major problem in LC/MS interfacing is the solvent composition incompatibility, as result of the frequent use of non-volatile mobile phase additives in LC separation development. In fact, long-term use of non-volatile buffers such as phosphate and borate is prohibited by all current LC/MS interface techniques, although some interfaces, such as APCI, show a higher tolerance than others. The only general solution to the above incompatibility is a change of the phase system in the LC separation, *i.e.*, the removal of all non-volatile additives from the mobile phase. However, simple changes in the mobile phase composition often results in diminished or lost chromatographic efficiency and resolution. Some general rules of thumb when converting HPLC methods developed using UV detection to MS amenable methodologies are gathered in a recent study for a series of chiral compounds of pharmaceutical interest.[43] In general, non-volatile phosphate, sulphate, borate and citrate buffers must be replaced with ammonium acetate, formate or hydroxide; formic and acetic acids can also be used. Volatile buffers are essential to give extended periods between source cleaning and to ensure maximum sensitivity of the source. Moreover, buffer concentrations of less than 5–10 mM are preferred.

If analyte retention is problematical when changing mobile phase composition, extensive post-column addition of MS-compatible solvents or reagents can be used to change the ionization state of the analyte(s), while keeping the previously developed chromatographic conditions. However, there are some disadvantages to this technique: post-column addition, in fact, requires an extra pump and can broaden chromatographic peaks due to the dead volumes introduced. Analyte concentration is also lowered, which may reduce the sensitivity of the analysis. For these reasons, post-column dilution is only acceptable when one is not sample-limited and has very good separations.[43]

The dramatic increase in the use of ESI interfaces and their current dominant position over other LC/MS techniques is mainly due to the excellent results provided by this method in the analysis of biopolymers like proteins and nucleotides, where TSP and APCI are not very effective. In fact, APCI, actually an old technique, has enjoyed a rebirth as the result of the success of ESI sources, and is often delivered as a supplemental and to some extent complementary tool. The impact of ESI in biochemical applications is difficult to review, as over 300 papers per annum appear in the literature. Further progress in this field is achieved by the vast amount of papers where MALDI is applied. The developments in this area can be considered as a continuous exploration of the potential of ESI. For this reason, it is impossible to properly give credit to all individual researchers who have made contributions.

A special mention in the application field of ESI must be made of the successful efforts to miniaturize the ES interface.[44] The rationale for this type of research lies in significant improvements of sensitivity, compatibility with low flow-rate separation techniques like capillary or nano-HPLC, and applications in sample-limited situations. The advent of nano-electrospray ionization (nano-ESI) has considerably extended the usability of ESI in the mass spectrometric laboratory,[45] significantly enhancing the area of ESI-MS applications.

2. HYDROPHILIC INTERACTION CHROMATOGRAPHY (HILIC)

2.1 Considerations on HILIC and Polar Organic Solvent-Based Separations

Polar compounds are poorly retained on a reversed phase HPLC column even with high aqueous mobile phases. For bioanalytical LC/MS applications, especially under ESI mode, poor analyte on-column retention may result in detrimental matrix effects due to the ion suppression to the analyte by co-eluting matrix components. High aqueous content mobile phases are also not beneficial to achieve the optimal spray conditions that are critical for sensitivity. Moreover, ionization of polar analytes decreases the analyte on-column retention on a RP column.[46] In the 1970s, a special mode of LC was developed to assay carbohydrates, using polar stationary phases and mobile phases containing some water and a higher percentage of acetonitrile.[47] Despite the use of commonly used RP mobile phases, this specific mode was referred to as the aqueous NP mode, because retention increased with increasing polarity of the analytes, which was analogous to NP retention behavior (see Section 1.4). It was not until the early 1990s, however, that such LC separation approach started emerging and Alpert[48] gave the practice a name by calling it Hydrophilic Interaction Chromatography (HILIC) to emphasize the presence of water in the mobile phase as the stronger eluting member, and the partition retention mechanism involved. Following twenty years of continuous development, HILIC is nowadays accepted as a common separation mode,[49] essentially dedicated to the separation of very polar compounds, such as glycopeptides, amino acids, oligonucleotides, and highly polar natural products (see also Section 2.3). Another major advantage of HILIC seems to be its facile coupling with MS, which extends its applicability to impurity detection. In fact, the use of a low aqueous/high acetonitrile mobile phase significantly improves detection sensitivity for compounds analyzed by LC/ESI-MS, thus overcoming the mismatch between NP and ESI-MS.

The exact retention mechanism for HILIC is still open to considerable debate. The partitioning mechanism arises from the preferential adsorption of water on the polar stationary phase, which results in a relative higher water content in the stagnant liquid phase of the stationary phase support than in the mobile phase eluent. Others reported that the separation in the HILIC mode is mainly governed by polar-polar interactions (*i.e.*, hydrogen bonding, dipole-dipole, charge-dipole interactions) because of the strong dependence of the elution order on the number of polar functional groups presented on the stationary phase. In some cases, a combination of both partitioning and hydrogen bonding can take place. A comprehensive summary of the ongoing discussion on the separation mechanism in HILIC is given in a recent review,[49] together with an overview of the most commonly used stationary phases (see also Section 2.2).

Acetonitrile-water mixtures are the typical mobile phases for HILIC separations. The water content for the optimized separation depends on the polarity of both stationary phases and analytes to be separated. In general, the more polar the stationary phase and the analyte, the higher water content is needed for the separation. In other words, water content must be low enough to achieve a separation, but be high enough for the mobile phase to effectively dissolve the analytes and elute them in a reasonable time. Since acetonitrile/water is commonly used also for RP-HPLC, some confusion may arise about exactly what ratio constitutes one *versus* the other mode. The nature of both the solutes and the stationary phases, and the properties of the organic solvents will affect the turning point between the RP and HILIC mode. Notably, beside NP and RP, a third elution mode exists which uses neat organic solvents, and is called polar-organic mode (POM). The POM may be considered as a bridge between the HILIC and the NP elution mode: in fact, both NP and the POM use neat organic solvents; both HILIC and the POM provide better dissolving power for polar solutes and are MS compatible; the polarity of analytes that are separated increases from NP to the POM, and to HILIC, largely due to the solubility of solutes in the corresponding mobile phases. On the other hand, these three modes share the same retention behavior (elution from less to more polar analytes).

2.2 HILIC Stationary Phases Developed in the Last Years

Although the number of commercially available columns designed specially for HILIC is growing, there is still not a substantial variety in stationary phases composition.[49] Since HILIC was considered as the aqueous NP mode, it turns out that the first separations were performed using conventional unmodified silica gels, and actually a large fraction of the recently published works are still using such materials. A review article was dedicated to the use of underivatized silica in bioanalytical HPLC/MS/MS separations.[46] Because of the high organic content in the mobile phase, significant sensitivity improvements can be achieved when using silica columns with respect to conventional RP columns. Moreover, when this technique is used in conjunction with commonly used sample extraction methods, such as solid-phase extraction (SPE), direct injection of the SPE eluate onto the separation column is often possible; in fact, the strongly eluting solvent used to desorb the analytes from the SPE device is a weakly eluting solvent for HILIC or *vice versa*. One of the properties that make underivatized silica inherently attractive in HPLC/MS is the lack of ligands that may detach and show up as spurious peaks in the mass spectra. However, severe irreversible adsorption has been observed on bare silica in HILIC mode. Recent bonded phase materials for HILIC seem to have addressed this issue, and various functionalized silica gels are today starting to become as common as unmodified silica in HILIC/MS applications.

Aminopropyl-functionalized silica gel (APS) was the first bonded stationary phase to be routinely used for carbohydrate separations under HILIC mode. A serious concern with APS is the limited stability in aqueous eluents that leads to fast release of ligands with accompanying peak shape deterioration under HILIC conditions and the concomitant exposure of free silanols.[50] The main difficulties in the chromatographic separation of sugars, beside the low UV detectability, include the problem of anomer separation and the existence of a huge number of stereoisomers. In fact, sugars exist in two anomeric forms that can interconvert slowly on the chromatographic time-scale at acidic and neutral pH values, thus giving two peaks for each solute that can lead to overcrowding in the chromatogram. Moreover, an interference regime (plateau) between the two anomeric peaks was detected in some cases, which is diagnostic of an active interconversion process during their chromatographic separation (anomerization). Such phenomena are observed when underivatized silica columns are used, whereas with amino-silica columns they are normally not observed, due to the basicity of such phases. At high pH values, in fact, the interconversion of sugar anomers is rapid and just a single peak is detected. In some cases, however, the complete separation of sugar anomers may be highly attractive, and suitable conditions should be found to prevent anomerization (see Section 2.4).

Diol-modified silica gels were among the first bonded silicas to be developed, and one of the purposes of their synthesis was to address the troublesome adsorption properties of underivatized silica. Diol silica is prepared by reaction with an alkoxy-activated glycidoxypropylsilane, followed by acid-catalyzed ring-opening hydrolysis of the oxirane group to form a siloxane-linked 2,3-dihydroxypropyl ligand. Generally, the diol-modified silica is nearly the ideal phase for HILIC applications owing to its high polarity and hydrogen-bonding properties. However, few HILIC applications of diol silica have been published.

Cyclodextrins (CDs), a family of cyclooligosaccharides connected by α-(1,4) linkages as in amylose, have been introduced as ligands for HILIC separation as early as 1989, and used to separate natural products, phosphorylated carbohydrates, and derivatized amino acids. It is the high density of exposed hydroxyl groups that makes the CD phases attractive for HILIC. Moreover, they have also chiral recognition properties, and hence represent an intriguing phase when the desire is to accomplish chiral separations by HILIC mode. CD phases opened the way to a series of carbohydrate-based materials for HILIC applications, although the immobilization of saccharides on solid supports is a difficult task to achieve with the traditional conjugation strategies (see Section 2.4).

A recent addition to the portfolio of HILIC phases is based on a grafted polymeric layer with sulfoalkylbetaine zwitterionic moieties of 3-sulfopropyldimethylalkylammonio inner salt as functional groups on wide-pore silica. The concurrent presence of a quaternary ammonium and a sulfonic acid group in a 1:1 ratio on the same pendant moiety yields a zero net charge, but sulfoalkylbetaine zwitterions are still potent osmolytes with a strong ability of binding water to surfaces. This material selectivity follows the general pattern in HILIC with increased retention as a function of solute polarity, but is largely free from adsorption phenomena of underivatized silica and from electrostatic interactions of amino-functionalized silica gels.

More recently, monolithic silica capillary columns were prepared by on-column polymerization of acrylamide.[51] Such columns showed HILIC mode retention features with three-times greater permeability and slightly higher column efficiency compared to commercially available amide-type HILIC columns packed with 5-μm silica particles.

2.3 HILIC/ESI-MS Analysis of Polar Compounds

For the drug discovery efforts currently taking place within the pharmaceutical industry, natural product extracts have been found to provide a valuable source of molecular diversity which is complementary to that provided by traditional synthetic organic methods or combinatorial chemistry.[52] However, there exists a need for analytical tools that can facilitate the separation and characterization of components from these sources in a rapid manner. The evaluation of highly polar compounds has particularly been challenging, and many efforts have been made to address this item. Surprisingly, HILIC separation mode remained confined for a long induction period to the field of carbohydrate analysis, and it was not until the early 2000s that such approach started taking off outside the above field.

HILIC excels with small and very polar analytes and in particular basic analytes where ion pairing has been necessary to obtain retention in RP mode. Only a brief summary of the analytes can be given here and Fig. 2 shows the structures of some of the small and basic molecules separated by HILIC. An overview of the main compound classes which have been analyzed by HILIC/ESI-MS is given in a recent review,[53] where applications are presented with a short comment on major drawbacks and advantages of the different chromatographic conditions employed. Beside the expected applications to amino acids and peptides, unusual employment of HILIC was shown in the modern drug discovery (anticancer drugs and antibiotics, drugs of abuse, cytotoxic saponins), in naturally occurring toxins identification, and in agricultural and food chemistry. HILIC has also been recognized as an important tool in metabolomic analysis, where the primary goal is the unbiased relative quantification of all metabolites in a biological system.

Notably, HILIC is successfully emerging as first dimension in a 2D-HPLC combination with other systems, when extreme sample complexity faced in the search for low abundance components in biological samples, most especially in proteomics, metabolomics, and glycomics. The key to success in 2D-HPLC is *orthogonality*, *i.e.*, the selectivities in the two separation dimensions should be the results of pronounced difference of retention mechanisms. Furthermore, 2D-HPLC is nowadays used almost exclusively in combination with MS detection, and the eluate from the second dimension should therefore be compatible with ESI.

What it should be clear about HILIC technique is the dramatic potential it shows in the field of the life sciences related to genomics, proteomics, metabolomics, pharmacology, agrochemistry and so on.

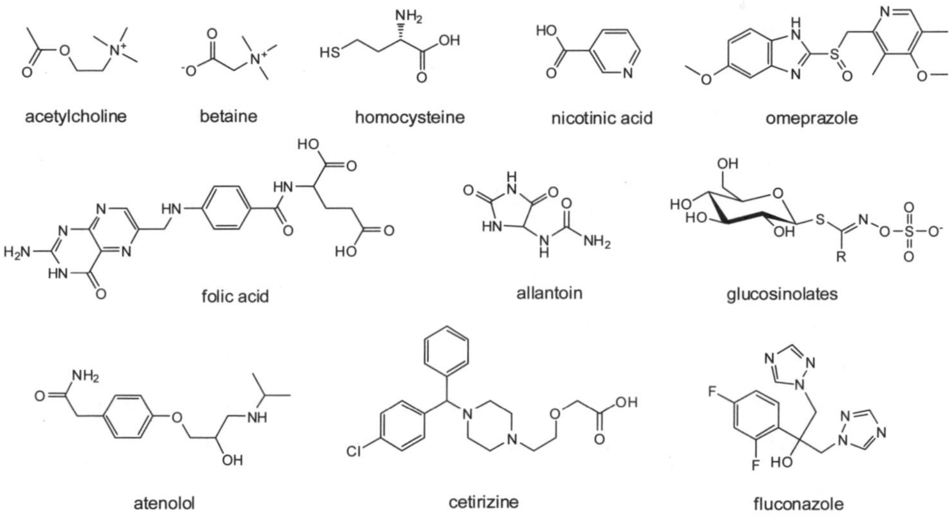

Fig. 2. Some Representative Polar and Basic Compounds Separated by HILIC.

2.4 Synthesis of Sugar-Based Silica Gels and their Use in HILIC

Carbohydrates are critical components of living systems and mediate a vast number of fundamental biological events – from complex cellular processes required to initiate and sustain life, to recognition phenomena that are responsible for autoimmune diseases, organ rejection, and inflammation. Such functional diversity is reflected in the diversity of carbohydrate structures. Hence, there is an increasing demand for easy access to glycoconjugates which can serve as probes in biochemical studies and also as potential leads for new drugs against carbohydrate-based metabolic disorders. Unlike other biomolecules such as proteins and nucleic acids, the synthesis of carbohydrates is not template-driven. Thus, whilst biotechnology can often be employed to synthesize proteins or nucleic acids, carbohydrate-containing molecules are difficult to synthesize because orthogonal hydroxyl group protection strategies are required and stereoselective glycosidic bond formation is often challenging. However, recent advances in chemoselective ligation have provided efficient techniques for the covalent immobilization of carbohydrates on a solid surface, aimed at simplifying the chemical synthesis of carbohydrate-containing molecules.[54]

Carbohydrates-based materials should be nearly ideal for HILIC applications because of the high density of exposed hydroxyl groups and hydrogen bonding properties. In 2007, immobilization of saccharides on solid supports has been developed for the first time via click chemistry,[55] which is in agreement with the demonstrated potential of this reaction in surface modification. The term click chemistry refers to a class of selective, covalent bond-formating reactions that have a large thermodynamic driving force and must *i*) provide high yields and clean products, *ii*) employ benign, easily removable solvents (or no solvent at all), and *iii*) tolerate the presence of air and ideally water. The novel separation materials were given the name of "click saccharides", to emphasize the employment of the click copper-catalyzed azide–alkyne cycloaddition (CuAAC) in their synthesis.[56]

Two types of CuAAC click reaction have been described in the literature for the immobilization of carbohydrates. In the first type, alkyne units installed on the solid scaffold underwent reaction with the azido group of the sugar (*i.e.*, azido sugars were employed), whereas, it was the other way around in the second type (*i.e.*, sugar alkynes were used). In both cases, multiple azide–alkyne coupling allowed the sugar fragments to be linked to the scaffold through 1,4-disubstituted-1,2,3-triazole tethers, thus leading to the metaphoric view of the triazole as a formidable keystone in glycosylated molecular architectures.

For the synthesis of "click saccharides", azido-modified silica gels were prepared by a two-step procedure, involving a nucleophilic halide/azide exchange on (3-chloropropyl)triethoxysilane, followed by the covalent linking of the azido-functionalized silane to the silica surface (Scheme 1). Afterwards, the CuAAC coupling with sugar alkynes took place.

Last year, an original single-step procedure was developed for the synthesis of azido-activated silica gels,[57] which were later exploited in the CuAAC of different sugar alkynes, to yield galactose- and lactose-based supports suitable for HILIC separations.

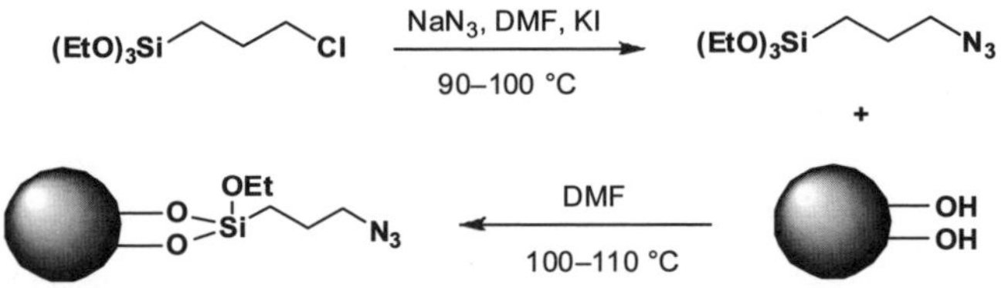

Scheme 1. Two-step Procedure for the Synthesis of Azido-functionalized Silica Gels.[56]

A distinctive feature of this activation process was the single-step introduction of azido-terminated groups onto the silica surface, spaced-out by a dipropyl ether bridge, and the simultaneous formation of a polar carbinol function one bond further away. For this purpose, the silica surface was oxirane-activated by silanization with (3-glycidoxypropyl)trimethoxysilane, concurrent with *in situ* epoxide ring-opening performed by excess (7:1) added sodium azide (Scheme 2).

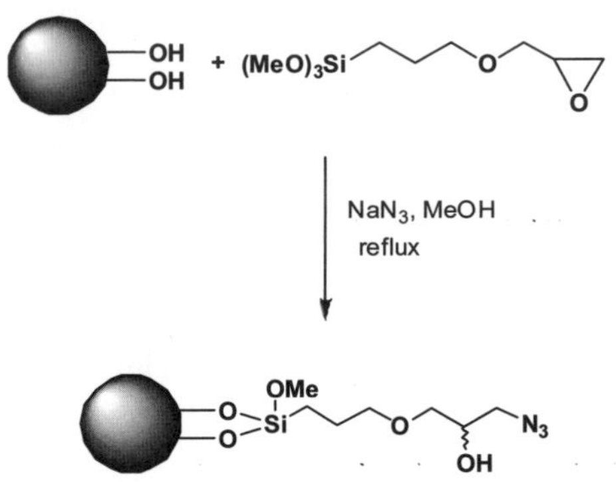

Scheme 2. One-step Procedure for the Synthesis of Azido-functionalized Silica Gels.[57]

The best support contained a lactoside unit (Scheme 3) and was used in the low-temperature separation of monosaccharides and other highly polar compounds, including amino acids and flavonoids. Notably, peak shape deformations were observed at temperatures above 5 °C, due to the interconversion of sugar anomers (anomerization) on the chromatographic time-scale. At column temperatures close to 0 °C, on-column anomerization was completely suppressed, since the lactose-based silica, being neutral, did not show any catalytic effect towards anomerization, unlike APS phases typically employed in the analysis of carbohydrates (see Section 2.2).

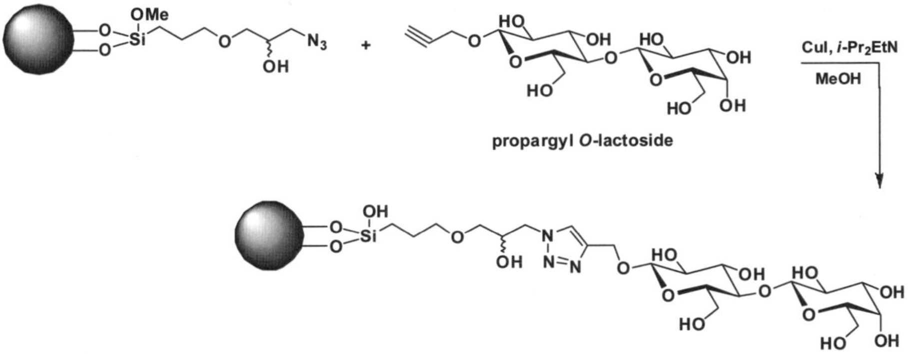

Scheme 3. Synthetic Pathway to a Lactose-based Silica Gel by CuAAC.[57]

3. STATIONARY PHASES ENDOWED WITH UNIQUE SELECTIVITY

3.1 Chirality and its Implications in Modern Organic Chemistry

The chiral nature of living systems has evident implications on biologically active compounds interacting with them.[58] On a molecular level, chirality represents an intrinsic property of the "building blocks of life", such as amino acids and sugars, and therefore, of peptides, proteins and polysaccharides. As a consequence, metabolic and regulatory processes mediated by biological systems are sensitive to stereochemistry and different responses can be often observed when comparing the activities of a couple of enantiomers. Stereoselectivity is often a typical feature of enzymatic reactions, messenger–receptor interactions and metabolic processes; it can vary even from one individual to the other. Therefore, stereochemistry has to be considered when studying xenobiotics, such as drugs, agrochemicals, food additives, flavours or fragrances.

The interest in chirality and its effects is not indeed an innovative event. However, during the last twenty years it has raised increasing expectations due to scientific and economic reasons, being the pharmaceutical industry the main contributor and driving force.

There is a broad range of examples where the stereoisomers of drugs show differences in terms of their bioavailability, distribution, metabolic and excretion behavior and where stereochemical parameters have a fundamental significance in their action and availability in biological systems. In fact, very often one stereoisomer represents the more active isomer for a given action (*eutomer*), while the other one (*distomer*) might be even active in a different way, contributing to side-effects, displaying toxicity, or acting as antagonist. Therefore, the high degree of stereoselectivity of many biological processes implies that when the racemate of a given chiral drug is administered, both enantiomers should not have to be equally potent. As a result, health and regulatory authorities, such as the Food and Drug Administration (FDA), have defined more strict requirements to patent new racemic drugs, demanding a full documentation of the separate pharmacological and pharmacokinetic profiles of the individual enantiomers, as well as their combination.[59] Even though the FDA did not mandate development of single isomers (racemates may be appropriate in certain cases), pharmaceutical companies took this announcement as an indication of things to come, and began careful studying both isomers in potential drugs. The logic behind this development is clear: a tremendous increase in the production of chiral drugs was registered after the FDA statement, and the necessity for pure enantiomers required a critical assessment of the most cost-effective way to accomplish their analysis and preparation.

Although a number of stereoselective syntheses have been described and applied to the production of single-enantiomeric substances, relatively few are suitable for large-scale preparations, particularly at the early stages of development of new drugs. Time constraints to have some amounts of pure enantiomers for the first pharmacological tests are usually crucial before a manufacture route (protocol) has to be selected. At these early stages, the development of an asymmetric synthesis would be both expensive and time consuming and thus, preparative techniques for the separation of enantiomers have an interesting potential.

The resolution of racemates requires the presence of a chiral environment. For this purpose, chiral auxiliaries, catalysts or selectors are necessary. The formation of the corresponding diastereomeric species implies an energetic difference between them which allows in many cases their enantiodiscrimination. These diastereomers can be covalently formed and thus their separation can be achieved taking advantage of their different chemical or physical properties by crystallisation, non-stereoselective chromatography or distillation. Then the pure enantiomeric substances are isolated *indirectly*, after release of the so-called chiral derivatizing agent (CDA). Indirect separation methods are frequently used, especially at the large-scale level, although the nature, enantiomeric purity, availability, costs and ease of cleavage of the chiral handle derived from the CDA, are sometimes limiting issues of this strategy. Alternatively, *direct* methodologies based on the formation of non-covalent diastereomeric pairs of molecule associates are most conveniently employed. Forces such as electrostatic, hydrogen bonding, repulsive/attractive van der Waals and π–π or dipolar interactions and inclusion phenomena contribute to the recognition process. Direct resolutions of enantiomers can be achieved by means of their interaction with a so-called chiral selector (SO), either being part of a chiral stationary phase (CSP), or as a chiral mobile phase (CMP) additive. Chiral selectors can be generated from natural or synthetic building blocks or can be obtained from natural sources (see Section 3.2). A huge number of CSPs have been developed over the years,[60] and a detailed description of such phases is beyond the scope of this paper. Almost the entire spectrum of separation techniques can be employed as potential tools for the resolution of enantiomers. The choice of the proper CSP will therefore be made in the light of the target molecule to be separated and the scale of the separation.

3.2 Chiral Stationary Phases (CSPs) Containing Glycopeptide Antibiotics

Naturally occurring macrocyclic antibiotics were introduced in 1994 as excellent selectors for the HPLC resolution of enantiomers both in organic and water-rich media,[61] among the vast repertoire of selectors for brush-type CSPs. Their molecular structure immediately appeared to be well suited for enantioselective recognition as well as for immobilization on a solid support.[62] The success of such molecules is due to the complex molecular architectures offered and to the presence of multiple stereogenic centres. Macrocyclic antibiotics represent a class of antibiotics used in therapy against infections caused by gram-positive and gram-negative bacteria and having ring structures with at least 10 members. Despite their common features, macrocyclic antibiotics differ in several of their physicochemical properties and in biological activity. There are hundreds of natural and semisynthetic macrocyclic antibiotics, which comprise a large variety of structural types. Among them, glycopeptide antibiotics of the teicoplanin family have a large, cyclic heptapeptide backbone rich in aromatic fragments, surrounded by polar and ionizable groups and carrying carbohydrate fragments at the macrocycle periphery (Fig. 3). The cyclic peptide backbone has a conformationally rigid, cup-shaped architecture, with the aromatic fragments rigidly interlocked in a well-defined stereochemical disposition. It is now well understood that all glycopeptide antibiotics exert antibiotic activity because they stereospecifically bind to the precursor peptidoglycan peptide terminus N-acyl-D-alanyl-D-alanine produced during bacterial cell wall biosynthesis, thereby inhibiting the action of bacterial enzymes that would otherwise use these termini to form new cross-links in peptidoglycan. Given such natural target of macrocyclic antibiotics, the early choice of suitable substrates for this kind of CSPs was that of amino acids. However, it turned out that the macrocyclic CSPs were very successful not only in amino acids enantioresolution, but also in the separation of a wide variety of different chiral structures.

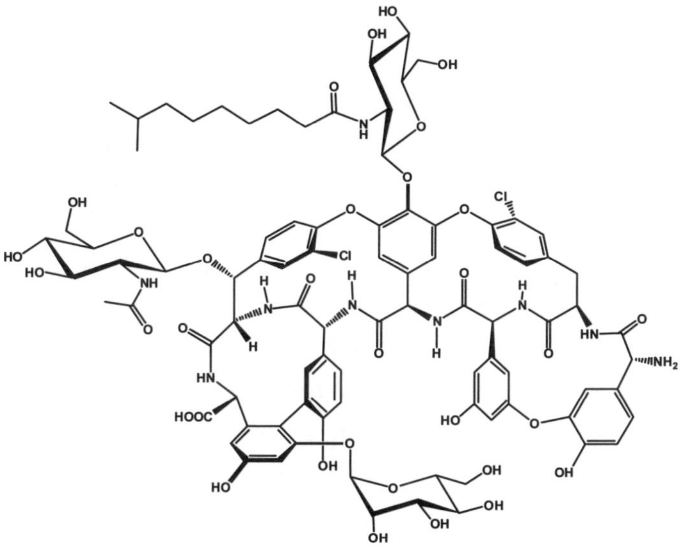

Fig. 3. Chemical Structure of the Main Component of the Teicoplanin Complex.

A detailed picture at molecular level of the events that govern the enantioselectivity of HPLC systems based on silica-bound glycopeptides is not yet available. However, indirect information on enantioselective complexation by silica-bound antibiotics in HPLC can be extracted from the analysis of retention data of several ligands whose structure is systematically varied to explore chemical diversity in terms of functional groups, stereogenic elements, molecular complexity, and rigidity-flexibility.

It can be concluded that molecular recognition by glycopeptide antibiotics is based on a range of cooperative interactions, accompanied by shape and size complementarities. Spectroscopic and computational analysis proved to be functional to uncover geometric and energetic details about the molecular mechanisms that regulate the complexation with small peptide molecules that mimic the natural ligand. It is expected that additional studies on a broader set of ligands will reveal new aspects and insights into the mode of actions of these naturally occurring chiral receptors, and will be useful to make predictions about the potential enantioselectivity of eventual separation systems that incorporate glycopeptide antibiotics or their derivatives.

3.3 Synthetic Strategies for the Preparation of Glycopeptide Antibiotics-based CSPs

Two conceptually different binding processes have been exploited in the literature for the grafting of glycopeptides antibiotics (Scheme 4):
(1) the stepwise assemblage of the target chiral selector on the silica surface;
(2) the direct attachment of a silyl derivative of the chiral selector on unmodified silica particles.
The first approach is rather simple, but, obviously, a more homogenous functionalization of the silica surface is achieved with the second approach.

Synthetic strategy (1) consists in the initial modification of the silica surface with organosilanes having suitable anchoring groups, which are either reactive themselves or can be additionally activated for the final attachment of the chiral selector. The most employed organosilanes were those terminated with carboxylic acid-, amino- and epoxy- functionalities. Carboxylic acid-terminated organosilanes were used in the early studies on chemically bonded glycopeptides to immobilize vancomycin and thiostrepton via their amino groups, leading to the formation of stable amide bonds between antibiotics and modified silica. Amino-terminated organosilanes were used to immobilize rifamycin B via its active carboxylic acid functionality leading, also in this case, to the formation of stable amide bonds between antibiotics and modified silica.

A number of glycopeptides-based CSPs have been prepared according to the aforementioned procedure, which is based on the use of 3-aminopropylated silica gel as starting material, but with a preliminary activation step with bifunctional spacers before the surface linking of the macrocycle. In particular, an innovative one-pot synthetic process was developed in 1999 to yield high loaded and stable glycopeptide containing CSPs.[63] Such a process is based on the activation of 3-aminopropylated silica gel with 1,6-diisocyanatohexane to give a monoureido–monoisocyanate intermediate in which the residual, pendant isocyanate groups are used to immobilize the macrocyclic receptors via their free amino groups (Scheme 4a). Additional linkage, in which carbamate groups are formed between the glycopeptide alcoholic or phenolic hydroxyls and the surface-linked isocyanate groups, may be present in the final material. In the resulting CSPs, the macrocycles are tethered to the aminopropylated silica via two embedded ureido functions and a six-carbon aliphatic spacer. This procedure afforded chiral phases featuring high chemical inertness and effective passivation of the underlying silica by the presence of the two ureido groups.

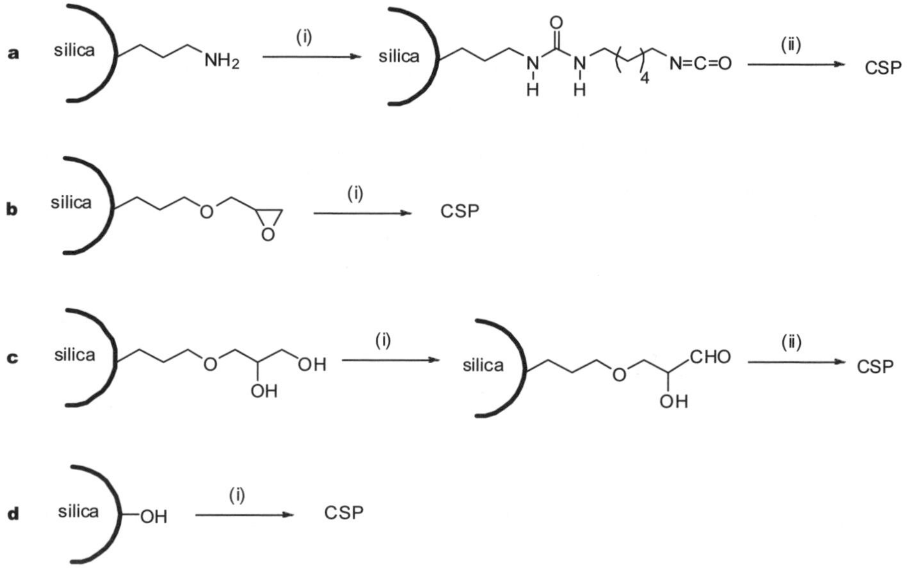

Scheme 4. General Synthetic Schemes for the Immobilization of Macrocyclic Antibiotics. (a) (*i*) 1,6-Diisocyanatohexane; (*ii*) Glycopeptide. (b) (*i*) Glycopeptide. (c) (*i*) NaIO$_4$; (*ii*) Glycopeptide. (d) (*i*) 3-Isocyanatopropyl-silyl Derivative of the Macrocyclic Antibiotic.[62]

This strategy, first developed to immobilize the main component of the teicoplanin complex,[63] was then extended to prepare a set of closely related CSPs containing just the aglycone portion of teicoplanin,[64] and a different glycopeptide of the teicoplanin family, namely A-40,926.[65]

The attachment of glycopeptides to silica gel via epoxy-terminated organosilanes (Scheme 4b) was first generally envisioned by Armstrong as a potential synthetic strategy for any macrocycle structure. In the final CSP structure, the glycopeptide is linked via a stable C–N bond to the silica, spaced-out by a dipropyl ether bridge.

Diol-derivatized silica gels were also employed in the preparation of glycopeptide-based CSPs due to the fact that the terminal diol functionality can be easily oxidized with sodium periodate to yield a silica surface with aldehyde functions (Scheme 4c). Chiral molecules having amino groups can be immobilized by reductive amination of the aldehyde-functionalized silica with sodium cyanoborohydride.

In synthetic strategy (2), the macrocyclic antibiotic is covalently bonded to the silica matrix in two steps: (a) chemical modification of the selector via reaction between suitable groups of the antibiotic and proper groups of the spacer, reacting also as a di- or trialkoxysilane; (b) immobilization of the functionalized selector on unmodified silica particles (Scheme 4d).

(3-Isocyanatopropyl)triethoxysilane is a suitable isocyanate-terminated organosilane, with different functionalities at the opposite terminals: on one end, it offers the highly reactive isocyanate group, and on the other one, it behaves as a triethoxysilane. This dual aspect made it a very helpful spacer in the immobilization of macrocycle compounds.

Irrespective of the synthetic strategy used for the immobilization of macrocyclic glycopeptides on the silica gel, surface coverage data obtained were in the 0.35–0.55 μmol m^{-2} range.

3.4 Enantioselective HPLC/MS Separations Using the Glycopeptides-based CSPs

Glycopeptide antibiotics have successfully been used as chiral selectors to resolve the enantiomers of a variety of chiral compounds by means of both liquid chromatography and electrophoresis. As outlined in Section 3.2, these natural compounds contain in their structure multiple stereogenic centers and a variety of functional groups. For these reasons, all glycopeptide-containing CSPs are multimodal CSPs, that is, they are capable of operating in three different elution systems, namely NP, RP, and the new POM (see Section 2.1). Moreover, as illustrated in Section 3.3, glycopeptide antibiotics are covalently bonded to silica gel through multiple linkages, therefore, there is no detrimental effect when switching from one mobile phase system to another. The NP mode can be applied without any irreversible change in enantioselectivity or denaturation of macrocycles, unlike protein-based CSPs. The RP mode was the first chromatographic system used in the early studies on glycopeptides containing

CSPs, owing to the polar nature of the eventual analytes; the new POM refers to the approach when methanol and/or acetonitrile are used as the mobile phases with small amounts of acid and/or base as the modifier. It offers the advantages of broad selectivity, high efficiency, low backpressure, short analysis time, high capacity, and excellent prospects for preparative-scale separations.

Although the majority of the HPLC chiral separations obtained with glycopeptides-based CSPs are achieved by UV/visible or fluorescence detection, a simple and efficient alternative to UV detection is nowadays represented by the evaporative light scattering (ELS) detector, which allows the direct chromatographic separation, with no need for preliminary derivatization. In the field of glycopeptides-based CSPs, it was applied for the first time in the chromatographic resolution of carnitine and O-acylcarnitine enantiomers on a teicoplanin-based CSP.[63]

In the last ten years, mass spectrometric detection has become a widely employed method for the analysis of chiral compounds on glycopeptide containing CSPs, without the need for preliminary derivatization. Various ionization techniques including ESI, APCI, and atmospheric pressure photoionization (APPI) interfaced with chiral liquid chromatographic methods were employed and compared in terms of their ionization efficiencies, matrix effects, and limitations. In the case of ESI, care must be taken to avoid situations where nonlinearity effects (such as those due to ion suppression at high concentration) occur; thus, diluted solutions must be prepared to achieve 1:1 areas ratio for a given racemate, especially in the case of high enantioselectivity values. Calibration curves in APCI are frequently linear over five orders of magnitude, whereas four are typical for ESI.

Several compounds of different classes have been analyzed by coupling existing HPLC methods with MS detection, or *ad hoc* developing new LC/MS hyphenated techniques, particularly in the monitoring of biological fluids: β-adrenoreceptors agonists and antagonists, calcium channel blockers, underivatized amino acids, native amino acids and peptides, antidepressants, local anesthetics, anticonvulsants, and amphetamine derivatives.

It must be noted that most existing HPLC methods developed using UV detection cannot be directly transferred to the MS detection, owing to the abovementioned mobile phase and additive incompatibilities (see Section 1.6). For example, NP solvents are highly incompatible when coupled with MS ionization sources such as ESI; however, the compatibility of the RP and POM modes with the MS interface makes them attractive direct approaches for the LC/MS analysis of chiral compounds. Some general rules of thumb when converting HPLC methods to MS amenable methodologies are gathered in a recent study,[43] and are as follows: (*i*) POM mobile phases are the most compatible and easily adaptable to the coupling to ESI-MS, yielding better limits of detection (as low as 100 pg ml^{-1}) and sensitivity over RP methods. (*ii*) NP methods are incompatible with direct coupling to ESI-MS; they can be used if post-column dilutions of a large excess of ESI-MS compatible solvents are acceptable in terms of sensitivity and band broadening. (*iii*) RP methods can be easily switched to APCI-MS with much greater sensitivity, while for ESI interfacing low water contents must be used as it tends to decrease the ionization efficiency. (*iv*) Ammonium trifluoroacetate enhances ionization for molecules with amine or amide functionalities. (*v*) Optimized concentrations of additives should be maintained, when compatible with the MS interface.

In summary, hyphenated techniques developed from the coupling of a separation device and an on-line spectroscopic detection technology have remarkably improved the analytical methodologies over the last two decades, significantly broadening their application fields in the analysis of biomaterials and of natural products.

REFERENCES

1. Norwood, D. L.; Mullis, J. O.; Feinberg, T. N. *Sep. Sci. Technol.* **2007**, *8*, 189.
2. Walker, G. S.; O'Connell, T. N. *Expert Opin. Drug Metab. Toxicol.* **2008**, *4*, 1295.
3. Heumann, K. G. *Anal. Bioanal. Chem.* **2002**, *373*, 323.
4. Gohlke, R. S. *Anal. Chem.* **1959**, *31*, 535.
5. Gohlke, R. S.; McLafferty, F. W. *J. Am. Soc. Mass Spectrom.* **1993**, *4*, 367.
6. Korhammer, S. A.; Bernreuther, A. *Fresenius J. Anal. Chem.* **1996**, *354*, 131.
7. Abian, J. *J. Mass Spectrom.* **1999**, *34*, 157.
8. Baldwin, M. A.; McLafferty, F. W. *Org. Mass Spectrom.* **1973**, *7*, 1111.
9. Arpino, P. J.; Dazwkins, B. G.; McLafferty, F. W. *J. Chromatogr. Sc.* **1974**, *12*, 574.
10. Blackey, C. R.; Carmody, J. J.; Vestal, M. L. *J. Am. Chem. Soc.* **1980**, *102*, 5931.
11. Vestal, M. L.; Fergusson, G. J. *Anal Chem.* **1985**, *57*, 2373.
12. Caprioli, R. M.; Fan, T.; Cotrell, J. S. *Anal. Chem.* **1986**, *58*, 2949.
13. Jones, P. R.; Yang, S. K. *Anal. Chem.* **1975**, *47*, 1000.
14. Takeuchi, T.; Hirata, Y.; Okumura, Y. *Anal. Chem.* **1978**, *50*, 659.
15. Willoughby, R. C.; Browner, R. F. *Anal. Chem.* **1984**, *56*, 2626.
16. Winkler, P. C.; Perkins, D. D.; Williams, D. K.; Browner, R. F. *Anal. Chem.* **1988**, *60*, 489.
17. Bier, M. E.; Winkler, P. C.; Herron, J. R. *J. Am. Soc. Mass Spectrom.* **1993**, *4*, 38.
18. Niessen, W. M. A.; Tinke, A. P. *J. Chromatogr. A* **1995**, *703*, 37.
19. Scott, R. P. W.; Scott, C. G.; Munroe, M.; Hess, J. *J. Chromatogr.* **1974**, *99*, 395.
20. McFadden, W. H.; Schwartz, H. L.; Evans, S. *J. Chromatogr.* **1976**, *122*, 389.
21. Alcock, N. H.; Eckers, C.; Games, D. E.; Games, M. P. L.; Lant, M. S.; McDowall, M. A.; Rossiter, M.; Smith, R. W.; Westwood, S. A.; Wong, H.-Y. *J .Chromatogr.* **1982**, *251*, 165.
22. Horning, E. C.; Horning, M. G.; Carroll, D. I.; Dzidic, I.; Stillwell, R. N. *Anal Chem.* **1973**, *45*, 936.
23. Horning, E. C.; Carroll, D. I.; Dzidic, I.; Haegele, K. D.; Horning, M. G.; Stillwell, R. N. *J. Chromatogr.* **1974**, *99*, 13.
24. Carroll, D. I.; Dzidic, I.; Stillwell, R. N.; Haegele, K. D.; Horning, E. C. *Anal. Chem.* **1975**, *47*, 2369.
25. Yamashita, M.; Fenn, J. B. *J. Phys. Chem.* **1984**, *88*, 4451.
26. Fenn, J. B.; Mann, M.; Meng, C. K.; Wong, S. F.; Whitehouse, C. M. *Science* **1989**, *246*, 64.
27. Fenn, J. B.; Mann, M.; Meng, C. K.; Wong, S. F.; Whitehouse, C. M. *Mass Spectrom. Rev.* **1990**, *9*, 37.
28. Huang, E. C.; Wachs, T.; Conboy, J. J.; Henion, J. D. *Anal. Chem.* **1990**, *62*, 713A.
29. Bruins, A. P. *Mass Spectrom. Rev.* **1991**, *10*, 53.
30. Bruins, A. P. *Trends Anal. Chem.* **1994**, *13*, 37.
31. Bruins, A. P. *Trends Anal. Chem.* **1994**, *13*, 81.
32. Niessen, W. M. A. *J. Chromatogr. A* **1998**, *794*, 407.
33. Snyder, L. R.; Kirkland, J. J.; Dolan, J. W. In *Introduction to Modern Liquid Chromatography*, John Wiley & Sons: New York, 2009.
34. Doyle, C. A.; Dorsey, J. G. *Chromatogr. Sci. Ser.* **1998**, *78*, 293.
35. Dorsey, J. G.; Cooper, W. T. *Anal. Chem.* **1994**, *66*, 857A.
36. Dole, M.; Mach, L. L.; Hines, R. L.; Mobley, R. C.; Ferguson, L. P.; Alice, M. B. *J. Chem. Phys.* **1968**, *49*, 2240.
37. Mach, L. L.; Kralik, P.; Rheude, A.; Dole, M. *J. Chem. Phys.* **1970**, *52*, 4977.
38. Clegg, G. A.; Dole, M. *Biopolymers* **1971**, *10*, 821.
39. Karas, M.; Hillenkamp, F. *Anal. Chem.* **1988**, *60*, 2299.
40. Iribarne, J. V.; Thomson, B. A. *J. Chem. Phys.* **1976**, *64*, 2287.
41. Mann, M.; Meng, C. K.; Fenn, J. B. *Anal. Chem.* **1989**, *61*, 1702.

42. Pearcy, J. O.; Lee, T. D. *J. Am. Soc. Mass Spectrom.* **2001**, *12*, 599 and references cited therein.

43. Desai, M. J.; Armstrong, D. W. *J. Chromatogr. A* **2004**, *1035*, 203.

44. Sproch, N.; Kruger, T. *J. Am. Soc. Mass Spectrom.* **1993**, *4*, 964.

45. Karas, M.; Bahr, U.; Dülcks, T. *Fresenius J. Anal. Chem.* **2000**, *366*, 669.

46. Naidong, W. *J. Chromatogr. B* **2003**, *796*, 209.

47. Rabel, R. M.; Caputo, A. G.; Butts, E. T. *J. Chromatogr.* **1976**, *126*, 731.

48. Alpert, A. J. *J. Chromatogr.* **1990**, *499*, 177.

49. Hemström, P.; Irgum, K. *J. Sep. Sci.* **2006**, *29*, 1784.

50. Wonnacott, D. M.; Patton, E. V. *J. Chromatogr.* **1987**, *389*, 103.

51. Ikegami, T.; Fujita, H.; Horie, K.; Hosoya K.; Tanaka, N. *Anal. Bioanal. Chem.* **2006**, *386*, 578.

52. Strege, M. A. *Anal. Chem.* **1998**, *70*, 2439.

53. Nguyen, H. P.; Schug, K. A. *J. Sep. Sci.* **2008**, *31*, 1465.

54. Langenhan, J. M.; Thorson, J. S. *Curr. Org. Synth.* **2005**, *2*, 59.

55. Kolb, H. C.; Finn, M. G.; Sharpless, K. B. *Angew. Chem.* **2001**, *113*, 2056.

56. Guo, Z.; Lei, A.; Zhang, Y.; Xu, Q.; Xue, X.; Zhang, F.; Liang, X. *Chem. Commun.* **2007**, 2491.

57. Moni, L.; Ciogli, A.; D'Acquarica, I.; Dondoni, A.; Gasparrini, F.; Marra, A. *Chem. Eur. J.*, **2010**, *16*, 5712.

58. Maier, N. M.; Franco, P.; Lindner, W. *J. Chromatogr. A* **2001**, *906*, 3.

59. FDA's Policy Statement on the Development of New Stereoisomeric Drugs (Stereoisomeric Drug Policy) Fed. Regist. (**1992**) 57 FR22249.

60. Armstrong, D. W.; Zhang, B. *J. Chromatogr. A* **2004**, *1053*, 89.

61. Armstrong, D. W.; Tang, Y.; Chen, S.; Zhou, Y.; Bagwill, C.; Chen, J.-R. *Anal. Chem.* **1994**, *66*, 1473.

62. D'Acquarica, I.; Gasparrini, F.; Misiti, D.; Pierini, M.; Villani, C. In *Advances in Chromatography*; Eds.; E. Grushka, N. Grinberg, CRC Press: Taylor & Francis Group, Boca Raton, FL, 2008; Vol. 46, p. 108-173.

63. D'Acquarica, I.; Gasparrini, F.; Misiti, D.; Villani, C.; Carotti, A.; Cellamare, S.; Muck, S. *J. Chromatogr. A*, **1999**, *875*, 145.

64. Berthod, A.; Chen, X.; Kullman, J.P.; Armstrong, D.W.; Gasparrini, F.; D'Acquarica, I.; Villani, C.; Carotti, A. *Anal. Chem.* **2000**, *72*, 1767.

65. D'Acquarica, I.; Gasparrini, F.; Misiti, D.; Zappia, G.; Cimarelli, C.; Palmieri, G.; Carotti, A.; Cellamare, S.; Villani, C. *Tetrahedron: Asymmetry*, **2000**, *11*, 2375.

Domenico Garozzo
Prof. of Biopolymers, Univerity of Catania

Address:
National Research Council, Institute of Chemistry and Technology of
Polymeric Materials CNR ICTP Via P. Gaifami 18 95125 Catania Italy
domenico.garozzo@cnr.it

Education:
1988, Ph. D. in Chemical Sciences, University of Catania;
1981, "Laurea" in Chemistry, University of Catania

Professional experience:
Dr. Garozzo's research centres on the characterization of natural and synthetic macromolecules. He developed Mass Spectrometry as analytical and structural tools for the analysis of polymers by the introduction of the SEC-MALDI technique, that is now universally used for the mass spectrometric analysis of polysaccharides and synthetic polymers. Afterwards, his studies explained why it is not possible to analyze polydisperse macromolecules, such as polysaccharides and most of the synthetic polymers, by MALDI MS without a pre-fractionation. In the field of natural macromolecules, his researches disclosed a sample preparation procedure able to obtain high quality MALDI mass spectra of R-type lipopolysaccharides. His most recent field of research is human serum glycomics.
Dr. Domenico Garozzo is in the editorial board of Journal of Proteomics (Elsevier), of the Open Proteomics Journal and of Open Glycoscience Journal.

From micro- to femto-moles, from small to giant molecules: the route of Modern Mass Spectrometry

*Angelo Palmigiano, Domenico Garozzo**

Consiglio Nazionale delle Ricerche Istituto di Chimica e Tecnologia di Polimeri Unità di Catania
Via P. Gaifami 18, 95126 Catania Italy

The mass spectrometer: a chemical lab

The definition of a mass spectrometer from the American Society for Mass Spectrometry (ASMS) is: "A mass spectrometer is an instrument that measures the masses of individual molecules that have been converted to ions; i.e., molecules that have been electrically charged"; using simply words: a machine used to weigh molecules. This definition is obviously correct, but the mass spectrometer is much more than an instrument. It is a complete laboratory for the investigation of molecules, clusters, and other species under the environment-free conditions of the gas phase. For this reason Mass Spectrometry (MS) is widely used today by almost all chemists and many researchers from neighbouring disciplines such as physics, medicine, or biology as a powerful analytical tool. Today the mass spectrometer is more and more present in the pharmaceutical industry, in the biological labs, in the hospitals and in many other bio-labs, while until few years ago it was present only in specialized chemical lab.

This revolution starts at the end of 1980s with the introduction of two different methods able to desorbs and ionize molecules with molecular weights in a range from few hundreds daltons to millions. Until then, a few techniques, fast atom bombardment (FAB), plasma desorption (PD) and desorption chemical ionization (DCI) were able to carry out in the gas phase and to ionize molecules with a molecular weight greater than one thousand, but they all required high concentrations of sample and they did not work at all for larger molecules such as proteins or polymers. Then, in 1988, Electrospray ionization (ESI) invented by John B. Fenn and Matrix assisted laser desorption (MALDI) introduced by Franz Hillenkamp and Michael Karas appeared almost simultaneously. These desorption-ionization methods revolutionized MS and are the main forms of ionization to this day. In the beginning the two techniques were not believed robust enough, but at the end of 1990s almost all MS instruments on the market were MALDI or ESI.

General Structure of a Mass Spectrometer

There are many different kinds of Mass Spectrometers, in order to describe all, it can be schematized as in figure 1

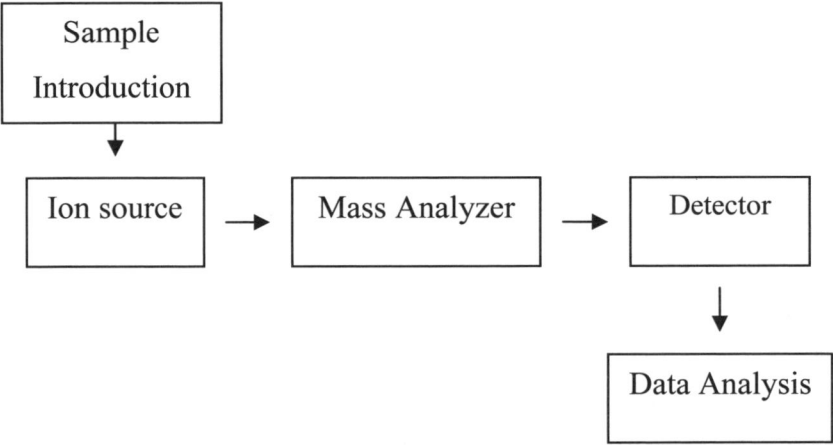

Sample Introduction

The selection of a sample inlet depends upon many factors: the sample, the matrix and the ion source are the most important.

A very important feature of MS is that it can be easily on-line interfaced with GC and LC. Off-line coupling are very significant such as, 2Dgel-MS (ESI or MALDI) off-line LC-MALDI, TLC MALDI. Recently, a new on-line TLC-ESI interface was presented.

Gas Chromatography. Gas chromatography (CG) is one of the most common technique for introducing samples into a mass spectrometer. Complex mixtures are separated by GC and MS is used to identify the individual component. The most common GG-MS interface now uses capillary GC column. Since the carrier gas flow rate is very small for these columns, the end of the capillary is inserted directly into the source region.

Liquid Chromatography Liquid chromatography inlets are used to introduce thermally labile compounds not easily separated by gas chromatography. These inlets have undergone considerable development and are now fairly routine. Normally Liquid Chromatography is coupled with ESI ionization even if some important LC-MALDI (off line) experiments were reported.

Ionization Sources

Prior to the 1980s, electron ionization (EI) was the primary ionization source for mass analysis. However, EI limited chemists small molecules well below the mass range of common bio-organic compounds. This limitation motivated scientists to develop the new generation of ionization techniques, including fast atom/ion bombardment (FAB), MALDI, and ESI (**Table 1**). These techniques have revolutionized biomolecular analyses, especially for large molecules.

Table 1

Ionization Source	Acronym	Event
Electrospray ionization	ESI	evaporation of charged droplets
Nanoelectrospray ionization	nanoESI	evaporation of charged droplets
Atmospheric pressure chemical ionization	APCI	corona discharge and proton transfer
Matrix-assisted laser desorption/ionization	MALDI	photon absorption/proton transfer
Desorption Ionization Mass Spectrometry	DESI	evaporation of charged droplets
Desorption/ionization on silicon	DIOS	photon absorption/proton transfer
Fast atom/ion bombardment	FAB	ion desorption/proton transfer
Electron ionization	EI	electron beam/electron transfer
Chemical ionization	CI	proton transfer

MALDI and ESI are now the most common ionization sources for biomolecular mass spectrometry, offering excellent mass range and sensitivity

Ionization Sources

Electron Ionization

Electron ionization (**EI**, formerly known as **electron impact**) is an ionization method in which energetic electrons interact with gas phase atoms or molecules to produce ions. This technique is widely used in mass spectrometry, particularly for gases and volatile organic molecules.

The following gas phase reaction describes the electron ionization process:

$$M + e^- \rightarrow M^{+\bullet} + 2e^- ,$$

where M is the analyte molecule being ionized, e^- is the electron and $M^{+\bullet}$ is the resulting ion.

In an EI ion source, electrons are produced through thermoionic emission by heating a wire filament that has electric current running through it. The electrons are accelerated to 70 eV in the region between the filament and the entrance to the ion source block. The accelerated electrons are then concentrated into a beam by being attracted to the trap electrode. The sample under investigation which contains the neutral molecules is introduced to the ion source in a perpendicular direction to the electron beam. Close passage of highly energetic electrons, referred to as a *hard* ionization source, causes large fluctuations in the electric field around the neutral molecules and induces ionization and fragmentation. The radical cation products are then directed towards the mass analyzer by a repeller electrode. The ionization process often follows predictable cleavage reactions that give rise to fragment ions which, following detection and signal processing, convey structural information about the analyte.

The ionization efficiency and production of fragment ions depends strongly on the chemistry of the analyte and the energy of the electrons. At low energies (around 20 eV), the interactions between the electrons and the analyte molecules do not transfer enough energy to cause ionization. At around 70 eV, the de Broglie wavelength of the electrons matches the length of typical bonds in organic molecules (about 0.14 nm) and energy transfer to organic analyte molecules is maximized, leading to the strongest possible ionization and fragmentation. Under these conditions, about 1 in 1000 analyte molecules in the source are ionized. At higher energies, the de Broglie wavelength of the electrons becomes smaller than the bond lengths in typical analytes; the molecules then become "transparent" to the electrons and ionization efficiency decreases.

Electrospray Ionization

Electrospray is a method by which ions, present in a solution, can be transferred to the gas phase by the application of an electric field to the tip of a capillary containing the solution of ions. The first electrospray experiments were carried out by Chapman in the late 1930s and the practical development of electrospray ionization for mass spectrometry was accomplished by Dole in the late 1960s. Dole also discovered the important phenomenon of multiple charging of molecules. It was Fenn's work that ultimately led to the modern day technique of electrospray ionization mass spectrometry and its application to biological macromolecules.

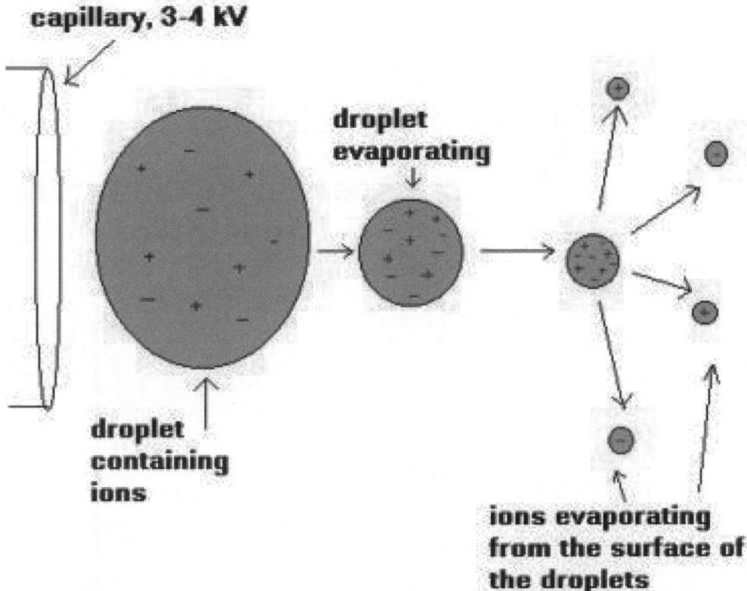

Electrospray ionization (ESI) is a method routinely used with peptides, proteins, carbohydrates, small oligonucleotides, synthetic polymers, and lipids. ESI produces gaseous ionized molecules directly from a liquid solution. It operates by creating a fine spray of highly charged droplets in the presence of an electric field. The sample solution is sprayed from a region of the strong electric field at the tip of a metal nozzle maintained at a potential of anywhere from 700 V to 5000 V. The nozzle (or needle) to which the potential is applied serves to disperse the solution into a fine spray of charged droplets. Either dry gas, heat, or both are applied to the droplets at atmospheric pressure thus causing the solvent to evaporate from each droplet. As the size of the charged droplet decreases, the charge density on its surface increases. The mutual Coulombic repulsion between like charges on this surface becomes so great that it exceeds the forces of surface tension, and ions are ejected from the droplet through a "Taylor cone" Another possibility is that the droplet explodes releasing the ions. In either case, the emerging ions are directed into an orifice through electrostatic lenses leading to the vacuum of the mass analyzer. Because ESI involves the continuous introduction of solution, it is suitable for using as an interface with HPLC or capillary electrophoresis.

Electrospray ionization is conducive to the formation of singly charged small molecules, but is also well-known for producing multiply charged species of larger molecules. This is an important phenomenon because the mass spectrometer measures the mass-to-charge ratio (m/z) and therefore multiple charging makes it possible to observe very large molecules with an instrument having a relatively small mass range. Fortunately, the software available with all electrospray mass spectrometers facilitates the molecular weight calculations necessary to determine the actual mass of the multiply-charged species.

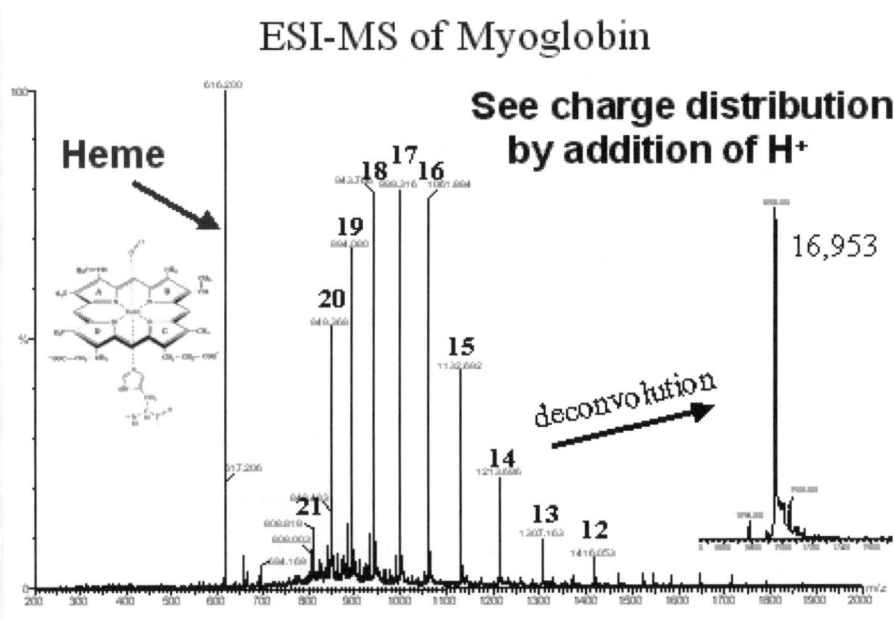

John Fenn told that the initial reaction of most mass spectrometrists to the peak multiplicity was one of horror! They were persuaded that the resulting spectral complexity would make interpretation very difficult or impossible. Matthias Mann, at that time a graduate student at the Fenn Lab, developed in few days a deconvolution algorithm that allowed to determine mw values up to 50,000 with accuracy of

0.01% (100 ppm) with a small, inexpensive quadrupole with a nominal mass limit of 2000 u! At that time (1988) mw values from other methods seldom had accuracies of about 10%. 14 years later, John Fenn that died December 2010, received the Nobel Prize for Chemistry for this discovery.

One very important feature of ESI is that it can be interfaced with almost all type of mass analyzer (magnetic sectors, quadrupoles, TOF, QTOF, ICR, ion trap, ICR, Orbitrap) and mass accuracy may be, today, better than 1 ppm!

Matrix-Assisted Laser Desorption/Ionization

The term matrix-assisted laser desorption ionization (MALDI) was coined in 1985 by Franz Hillenkamp and Michael Karas. MALDI is used successfully in biochemical areas for the analysis of proteins, peptides, glycoproteins, oligosaccharides, oligonucleotides, polysaccharides and synthetic polymers. It is relatively straightforward to use and reasonably tolerant to buffers and other additives. The mass accuracy depends on the type and performance of the analyser of the mass spectrometer, but most modern instruments should be capable of measuring masses to within 0.01% (100 ppm) of the molecular mass of the sample, at least up to ca. 40,000 Da.

MALDI is based on the bombardment of sample molecules with a laser light to bring about sample ionisation. The sample is pre-mixed with a highly absorbing matrix compound for the most consistent and reliable results, and a low concentration of sample to matrix works best. The matrix transforms the laser energy into excitation energy for the sample, which leads to sputtering of analyte and matrix ions from the surface of the mixture. In this way energy transfer is efficient and also the analyte molecules are spared excessive direct energy that may otherwise cause decomposition.

The sample to be analysed is dissolved in an appropriate volatile solvent, usually with a trace of trifluoroacetic acid if positive ionisation is being used, at a concentration of ca. 10 pmol/µL and an aliquot (1-2 µL) of this removed and mixed with an equal volume of a solution containing a vast excess of a matrix. A range of compounds is suitable for use as matrices: sinapinic acid is a common one for protein analysis while alpha-cyano-4-hydroxycinnamic acid is often used for peptide analysis. An aliquot (1-2 µL) of the final solution is applied to the sample target which is allowed to dry prior to insertion into the high vacuum of the mass spectrometer. The laser is fired, the energy arriving at the sample/matrix surface optimised, and data accumulated until a m/z spectrum of reasonable intensity has been amassed.

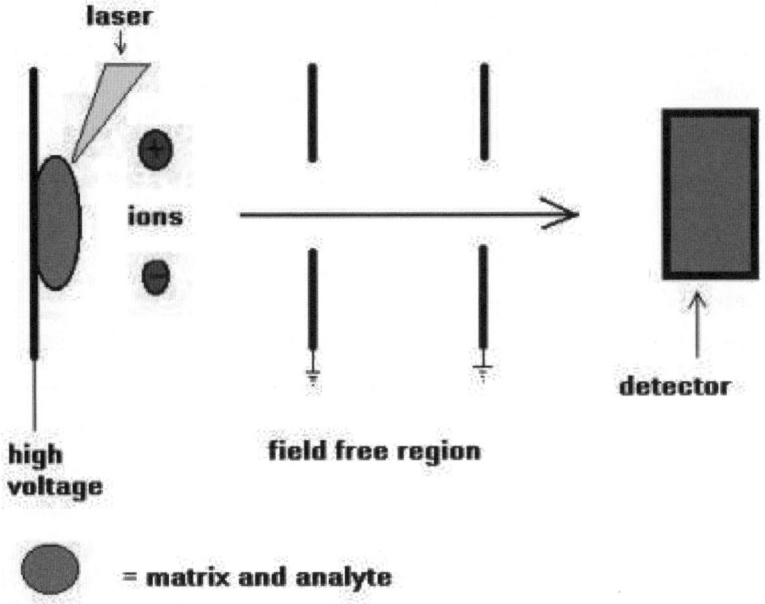

laser

ions

high voltage

field free region

detector

= matrix and analyte

The time-of-flight analyser separates ions according to their mass(m)-to-charge(z) (m/z) ratios by measuring the time it takes for ions to travel through a field free region known as the flight, or drift, tube. The heavier ions are slower than the lighter ones.

The m/z scale of the mass spectrometer is calibrated with a known sample that can either be analysed independently (external calibration) or pre-mixed with the sample and matrix (internal calibration).

MALDI is also a "soft" ionization method and so results predominantly in the generation of **singly charged molecular-related ions** regardless of the molecular mass, hence the spectra are relatively easy to interpret. Fragmentation of the sample ions does not usually occur.

FAB, DCI, Plasma Desorption, Laser Desorption

Now the importance of these desorption and ionization techniques is only historical, but for many years, late 70s to the beginning of 90s, they were the only way to obtain mass spectra of small peptides, oligosaccharides and very polar samples.

The mass analyzer

Sector instruments

A sector field mass analyzer uses an electric and/or magnetic field to affect the path and/or velocity of the charged particles in some way. As shown above, sector instruments bend the trajectories of the ions as they pass through the mass analyzer, according to their mass-to-charge ratios, deflecting the more charged and faster-moving, lighter ions more. The analyzer can be used to select a narrow range of *m/z* or to scan through a range of *m/z* to catalog the ions present.

Sector instruments are expensive and with a limited mass range (below 10.000). Sector instruments had huge commercial successes in the 1950's and 1960's as they were the only practical way of obtaining high-resolution data. In the last 20 years or so, with the development of high-resolution alternatives (for example TOF,/TOF Q-Tof, orbitrap) sector instruments are in decline. They still have their applications though, and are very well suited to EI and CI ionisation and consequently GC-MS. Single sector instruments are also used in the specialist area of isotope ratio mass spectrometry (IRMS).

Time-of-flight

The time-of-flight (TOF) analyzer uses an electric field to accelerate the ions through the same potential, and then measures the time they take to reach the detector. If the particles all have the same charge, the kinetic energies will be identical, and their velocities will depend only on their masses. Lighter ions will reach the detector first.

Quadrupole mass filter

Quadrupole mass analyzers use oscillating electrical fields to selectively stabilize or destabilize the paths of ions passing through a radio frequency (RF) quadrupole field created between 4 parallel rods. Only the ions in a certain range of mass/charge ratio are passed through the system at any time, but changes to the potentials on the rods allow a wide range of m/z values to be swept rapidly, either continuously or in a succession of discrete hops. A quadrupole mass analyzer acts as a mass-selective filter and is closely related to the quadrupole ion trap, particularly the linear quadrupole ion trap except that it is designed to pass the untrapped ions rather than collect the trapped ones, and is for that reason referred to as a transmission quadrupole. A common variation of the quadrupole is the triple quadrupole. Triple quadrupole mass spectrometers have three consecutive quadrupoles arranged in series to incoming ions. The first quadrupole acts as a mass filter. The second quadrupole acts as a collision cell where selected ions are broken into fragments. The resulting fragments can once again be filtered by the third quadrupole or all be allowed to pass though to the detector yielding an ms/ms fragmentation pattern.

Three-dimensional quadrupole ion trap

The quadrupole ion trap works on the same physical principles as the quadrupole mass analyzer, but the ions are trapped and sequentially ejected. Ions are trapped in a mainly quadrupole RF field, in a space defined by a ring electrode (usually connected to the main RF potential) between two endcap electrodes (typically connected to DC or auxiliary AC potentials). The sample is ionized either internally (e.g. with an electron or laser beam), or externally, in which case the ions are often introduced through an aperture in an endcap electrode.

There are many mass/charge separation and isolation methods but the most commonly used is the mass instability mode in which the RF potential is ramped so that the orbit of ions with a mass $a > b$ are stable while ions with mass b become unstable and are ejected on the z-axis onto a detector. There are also non-destructive analysis methods.

Ions may also be ejected by the resonance excitation method, whereby a supplemental oscillatory excitation voltage is applied to the endcap electrodes, and the trapping voltage amplitude and/or excitation voltage frequency is varied to bring ions into a resonance condition in order of their mass/charge ratio.

Linear quadrupole ion trap

A linear quadrupole ion trap is similar to a quadrupole ion trap, but it traps ions in a two dimensional quadrupole field, instead of a three-dimensional quadrupole field as in a 3D quadrupole ion trap. Thermo Fisher's LTQ ("linear trap quadrupole") is an example of the linear ion trap. A toroidal ion trap can be visualized as a linear quadrupole curved around and connected at the ends or as a cross section of a 3D ion trap rotated on edge to form the toroid, donut shaped trap. The trap can store large volumes of ions by distributing them throughout the ring-like trap structure. This toroidal shaped trap is a configuration that allows the increased miniaturization of an ion trap mass analyzer. Additionally all ions are stored in the same trapping field and ejected together simplifying detection that can be complicated with array configurations due to variations in detector alignment and machining of the arrays.

Orbitrap

Very similar nonmagnetic FTMS has been performed, where ions are electrostatically trapped in an orbit around a central, spindle shaped electrode. The electrode confines the ions so that they both orbit around the central electrode and oscillate back and forth along the central electrode's long axis. This oscillation generates an image current in the detector plates which is recorded by the instrument. The frequencies of these image currents depend on the mass to charge ratios of the ions. Mass spectra are obtained by Fourier transformation of the recorded image currents.

Similar to Fourier transform ion cyclotron resonance mass spectrometers, Orbitraps have a high mass accuracy, high sensitivity and a good dynamic range.

FT-ICR mass spectrometer

Fourier transform ion cyclotron resonance MS, measures mass by detecting the image current produced by ions cyclotroning in the presence of a magnetic field. Instead of measuring the deflection of ions with a detector such as an electron multiplier, the ions are injected into a Penning trap (a static electric/magnetic ion trap) where they effectively form part of a circuit. Detectors at fixed positions in space measure the electrical signal of ions which pass near them over time, producing a periodic signal. Since the frequency of an ion's cycling is determined by its mass to charge ratio, this can be deconvoluted by performing a Fourier transform on the signal. FTMS has the advantage of high sensitivity (since each ion is "counted" more than once) and much higher resolution and thus precision.

Few application examples

Pharmacokinetics

Pharmacokinetics is often studied using mass spectrometry because of the complex nature of the matrix (often blood or urine) and the need for high sensitivity to observe low dose and long time point data. The most common instrumentation used in this application is LC-MS with a triple quadrupole mass spectrometer. Tandem mass spectrometry is usually employed for added specificity. Standard curves and internal standards are used for quantitation of usually a single pharmaceutical in the samples. The samples represent different time points as a pharmaceutical is administered and then metabolized or cleared from the body. Blank or t=0 samples taken before administration are important in determining background and ensuring data integrity with such complex sample matrices. Much attention is paid to the linearity of the standard curve; however it is not uncommon to use curve fitting with more complex functions such as quadratics since the response of most mass spectrometers is less than linear across large concentration ranges. There is currently considerable interest in the use of very high sensitivity mass spectrometry for microdosing studies, which are seen as a promising alternative to animal experimentation.

Protein characterization

Mass spectrometry is an important emerging method for the characterization and sequencing of proteins. The two primary methods for ionization of whole proteins are electrospray ionization (ESI) and matrix-assisted laser desorption/ionization (MALDI). In keeping with the performance and mass range of available mass spectrometers, two approaches are used for characterizing proteins. In the first, intact proteins are ionized by either of the two techniques described above, and then introduced to a mass analyzer. This approach is referred to as "top-down" strategy of protein analysis. In the second, proteins are enzymatically digested into smaller peptides using proteases such as trypsin or pepsin, either in solution or in gel after electrophoretic separation. Other proteolytic agents are also used. The collection of peptide products are then introduced to the mass analyzer. When the characteristic pattern of peptides is used for the identification of the protein the method is called peptide mass fingerprinting (PMF), if the identification is performed using the sequence data determined in tandem MS analysis it is called de novo sequencing. These procedures of protein analysis are also referred to as the "bottom-up" approach.

Bottom-Up Proteomics

In bottom-up proteomics, the analytes introduced into the mass spectrometer are peptides generated by enzymatic cleavage of one or many proteins. The proteins can first be separated by 2Dgel, SDSGel or chromatography, in which case the sample will contain only one or a few proteins. Alternatively, a complex protein mixture initially can be digested to the peptide level, then separated by on-line chromatography coupled to electrospray mass spectrometry (ESI–MS). In the latter case, the digest can contain thousands to hundreds of thousands of peptides, and require separation in two or more chromatographic dimensions before MS analysis. The identity of the original protein is determined by comparison of the peptide mass spectra with theoretical peptide masses calculated from a proteomic or genomic database. There are two approaches for protein identification using the bottom-up approach, peptide mass fingerprinting and tandem MS (MS–MS).

Peptide mass fingerprinting: In peptide mass fingerprinting, peptide masses obtained from an MS scan are compared to calculated peptide masses generated by "in silico" cleavage of protein or gene sequences in the database using the same specificity as the enzyme that was employed in the experiment. One disadvantage of peptide mass fingerprinting is the requirement for pure proteins or simple mixtures of proteins. The purification steps therefore limit the throughput of the peptide mass fingerprinting approach. Another disadvantage is the requirement for several peptides to uniquely identify a protein. Peptide mass fingerprinting can be performed with ESI or MALDI instruments. Mass accuracy better then 100 ppm is mandatory.

Tandem MS: In MS-MS, peptide ions are isolated in the mass analyzer and subjected to dissociation to produce product ion fragments. The product ion spectra are compared with databases by cross-correlation analysis to identify the intact protein. (see below)

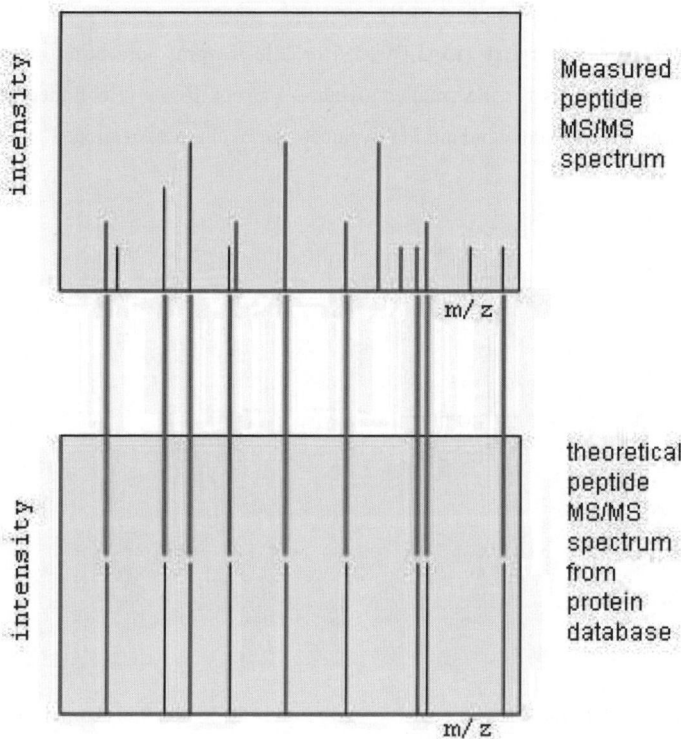

Measured peptide MS/MS spectrum

theoretical peptide MS/MS spectrum from protein database

Glycan analysis

Mass spectrometry (MS), with its low sample requirement and high sensitivity, has been the predominantly used in glycobiology for characterization and elucidation of glycan structures. Mass spectrometry provides a complementary method to HPLC for the analysis of glycans. Intact or permethylated glycans may be detected directly as singly charged ions by matrix-assisted laser desorption/ionization mass spectrometry (MALDI-MS) or, Electrospray ionization mass spectrometry (ESI-MS) also gives good signals for the smaller glycans. Various free and commercial software are now available which interpret MS data and aid in Glycan structure characterization

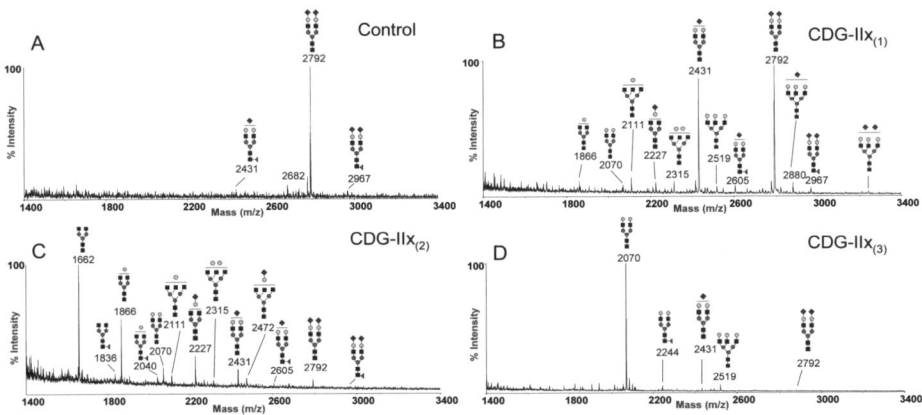

Differences between MALDI mass spectra of permethylated Tf N-glycans from three patients with unsolved CDG type II: (a) healthy control, (b) CDG-IIx(1) patient showing a remarkable hyposialylation and a less intense galactosylation defect, (c) CDG-IIx(2) patient with a severe sialylation and galctosylation deficiency and (d) CDG-IIx(3) patient presenting a pure sialylation defect.

Space exploration

As a standard method for analysis, mass spectrometers have reached other planets and moons. Two were taken to Mars by the Viking program. In early 2005 the Cassini-Huygens mission delivered a specialized GC-MS instrument aboard the Huygens probe through the atmosphere of Titan, the largest moon of the planet Saturn. This instrument analyzed atmospheric samples along its descent trajectory and was able to vaporize and analyze samples of Titan's frozen, hydrocarbon covered surface once the probe had landed. These measurements compare the abundance of isotope(s) of each particle comparatively to earth's natural abundance. Also onboard the Cassini-Huygens spacecraft is an ion and neutral mass spectrometer which has been taking measurements of Titan's atmospheric composition as well as the composition of Enceladus' plumes. A Thermal and Evolved Gas Analyzer mass spectrometer was carried by the Mars Phoenix Lander launched in 2007.

Mass spectrometers are also widely used in space missions to measure the composition of plasmas. For example, the Cassini spacecraft carries the Cassini Plasma Spectrometer (CAPS), which measures the mass of ions in Saturn's magnetosphere.

Further readings:

Interpretation of Mass Spectra by Fred W. McLafferty and Frantisek Turecek, University Science Books 1993.

MALDI MS: A Practical Guide to Instrumentation, Methods and Applications by Franz Hillenkamp and Jasna Peter-Katalinic (2007) Wiley-VCH

Electrospray and MALDI Mass Spectrometry: Fundamentals, Instrumentation, Practicalities, and Biological Applications by by Richard B. Cole (2010) Wiley

Electrospray Ionization Mass Spectrometry by Richard B. Cole (1997) Wiley

Internet resources:
http://masspec.scripps.edu/index.php
http://www.ionsource.com/

Rosangela Marchelli
Prof. of Organic Chemistry

Address:

Dipartimento di Chimica Organica e Industriale, Università degli Studi di Parma, Parco Area delle Scienze 17/A, 43124 Parma (Italy)

Education:

Diploma Classic Lyceum – Vigevano (PV)

Laurea in Chemistry – University of Pavia

Post-doctoral fellowship – National Research Council (NRC) of
Canada, Halifax, N.S. Canada

Professional experience:

Professor of Organic Chemistry at the University of Parma, first as lecturer, then as Associate Professor (1982-86), and Full Professor since 1986. She was Visiting Professor at the Department of Biology of the Dalhousie University, Halifax, N.S. Canada (1974), NATO Fellow at the University of California, Davis (1979) and Research Scientist at the Weizmann Institute of Rehovot (Israel) (1984). Delegate of the Italian Chemical Society (SCI) in the Division of Food Chemistry of EuCheMS (European Association for Chemical and Molecular Sciences). Dean of the Faculty of Agriculture of the University of Parma (1995-2009). Chair-woman of the Interdisciplinary Group of Food Chemistry of the Italian Chemical Society (2000-2005) and Scientific Coordinator of the Regional Laboratory for Food Safety and Quality (SIQUAL) (2006-2008). Since 2006 she is a Member of the Panel on Nutrition, Dietetics and Food Allergy of EFSA (European Food Safety Authority). Her scientific interests concern i) bioorganic chemistry, in particular synthesis and performance of chiral PNA (Peptide Nuclei Acids) in DNA and RNA recognition; ii) food chemistry, in particular, mycotoxins, food allergens and peptides. She is the author of more than 200 publications and of 3 Patents. She has received the Award Research 2010 from the Italian Chemical Society for her research on DNA recognition and biomedical and food applications.

MOLECULAR SECRETS OF GASTRONOMY

Rosangela Marchelli[1]*, Raffaele Sacchi[2], Roberto Corradini[1] and Stefano Sforza[1]

[1] Dipartimento di Chimica Organica e Industriale, Università di Parma, Parco Area delle Scienze 17/A, 43124 Parma, Italy
[2] Dipartimento di Scienza degli Alimenti, Università di Napoli Federico II, Reggia di Portici - Via Università 100 - 80055 - Portici (Na), Italy

1. Introduction

Gastronomy has been always considered as an art as well as an empirical technique based on experience and tradition. The word "chemistry" was generally avoided when speaking of food, on account of the tarnish identification with "unnatural" additives or sophistication. Indeed, the popular perception of food contamination was preferentially linked to chemicals, such as pesticides and dioxins despite the higher risk presented by microbial contaminants or natural products of moulds, such as mycotoxins. On the other side, recently there has been an enormous outburst of "health claims" related to foods, regardless of the existence of scientific demonstrations of cause-effect relationships.

In this lecture I will approach a fairly new subject, i.e. how chemistry can help to understand what makes a food pleasant and how food must be transformed during the technological process or the cooking procedure in order to be appreciated by the consumers and to maintain its nutritional value. I will unveil some molecular secrets of the "traditional gastronomy". I will give few hints about "molecular gastronomy",[1-2] the scientific study of phenomena occurring during culinary transformations and about the new born "note-by-note" cuisine, proposed by This,[3] who wants to use single components (the notes) of food to make a dish (the music). This said: "It's like a painter using primary colours or a musician composing note by note."

I will present this lecture under the form of a typical italian "*menu*", trying to show how gastronomy and science, in particular chemistry, can collaborate in order to rationalize what is happening during cooking or what it is that gives that particular colour or flavour or taste to our foods. Our menu will be:

Appetizer

Parma Ham

First course

Spaghetti with tomato sauce

Second course

Boiled or Roasted Meat

With French Fried Potatoes

Dessert

Chocolate Mousse

(according to a Molecular Gastronomy recipe)[1,4]

Champagne

2. Parma Ham

Parma ham is a well known dry-cured meat product, obtained from heavy pig thighs (12–14 kg), mildly salted and long aged (at least 12 months). The process of production of Parma ham has not changed through the centuries, and it is mainly based on the reduction of the water activity induced by sodium chloride and by removal of moisture along with the drying process which takes place during ageing.[5] In the manufacturing of Parma ham the only additive used is sodium chloride according to a protocol established by the Consorzio del Prosciutto di Parma. Despite the long ageing, Parma ham maintains its typical pink colour, which appears to be stable in time. What types of molecules are responsible for the colour?

The violet colour of meat is due mainly to myoglobin, the heme protein with the physiological function of oxygen storage in muscles. The bright red colour, instead, is due to a steady-state concentration of oxymyoglobin, in which oxygen is coordinated to the Fe(II) center of the Porphyrin. This gives rise to one-electron transfer, creating the superoxide radical anion and metmyoglobin, the brown and physiologically inactive Fe(III) form of myoglobin. Then the Fe(II) form of myoglobin is restored by a reductase in the presence of NADH.

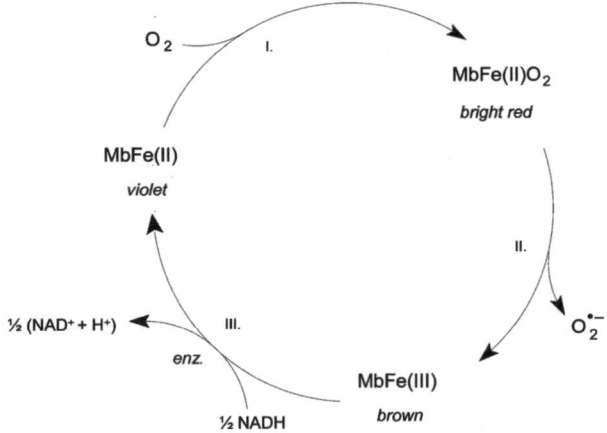

Fig. 1. Colour Cycle of Meat (reprinted with permission from ref. 2).

In Parma ham, as well as in the similar Serrano ham from Spain, oxymyoglobin is converted to Zinc-Protophorphyrin **1**, which is very stable to light and heat and is the principal colorant (pink) of this type of dry cured meat (Figure 2).

The structure of the Zn-Protoporphyrin **1** was elucidated by Wakamatsu et al.[6] by fluorescence, HPLC and further confirmed by Moller et al.[7] by Mass Spectrometry (Figure 3). The most interesting observation for the ESI(−)-MS data is the cluster of isotopologue ions of m/z 623.2 (the most intense). It was later reported that **1** is formed not from heme but from Protoporphyrin IX by means of a ferrochelatase, the same enzyme which inserts Fe(II) into the Protoporphyrin IX in heme biosynthesis, which catalyzes also its removal from heme.[8] However, the mechanism by which **1** is formed in meat products has not been fully elucidated.

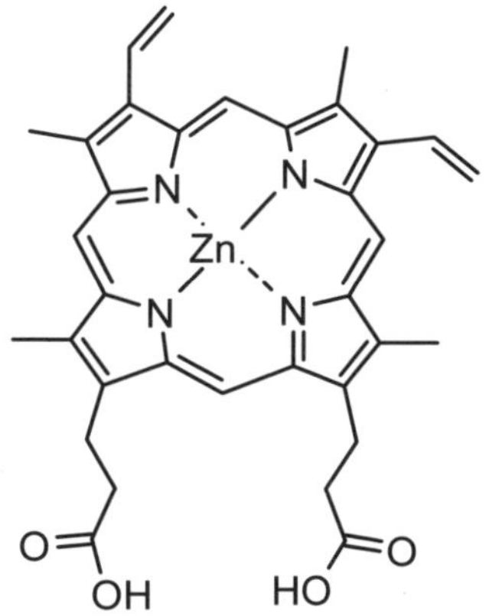

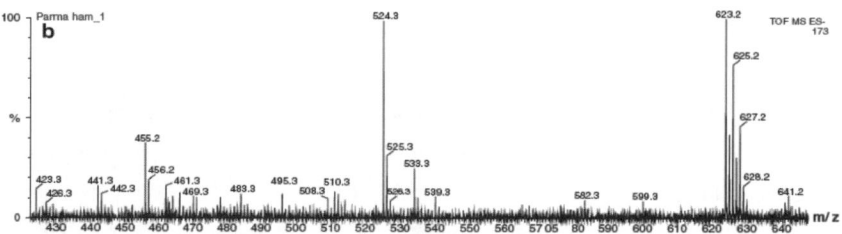

Fig. 2. The Structure of the Zn-Protoporphyrin Complex.

Fig. 3. Electrospray Ionization Tandem Mass Spectra of Negative Ions of m/z 623.2 from Dry-Cured Parma Ham Extracts in 75% Acetone/ Water Solution (reprinted with permission from ref. 7).

The flavour of dry-cured ham relies on a blend of volatile and non volatile molecules resulting from many reactions that take place during dry-cured ham processing. The positive role in the enhancement of aroma played by aldehydes and ketones, derived from lipid hydrolysis and autoxidation, and by ethyl and methyl esters, derived from the degradation of branched chain amino acids (BCAA), was demonstrated for several types of dry-cured hams.[9]

Non volatile molecules (peptides and free amino acids), mainly released by the activity of endogenous proteolytic enzymes, are also involved in the characteristic ham taste and flavour development. The formation of peptides during Parma ham processing was found to be highly correlated to the flavour formation, and in particular the hydrophobic peptides were associated with a bitterness perception.[10] A sharp rise of small peptides as compared to free amino acids has been reported in very aged hams.[11] Nevertheless, on average, sensory evaluation of flavour-related traits of dry cured ham improves by increasing the ham processing time.[12] In this respect, the role of bitter peptides may be counteracted by the combination with other taste-active compounds (amino acids, salt, other peptides) in adequate proportion, whereas an unbalanced accumulation could generate an unpleasant bitter taste. In our case, small peptides bearing hydrophobic residues are formed, some of them already found to be related to the bitter taste in aged hams, such as Gly-Phe and Leu-Leu.[13]

However, there are also molecules eventually able to counteract this undesired taste, for instance, the γ-glutamylamino acids, such as γ-glutamyl-phenylalanine **2**, reported in Figure 4, with their "umami taste".[14]

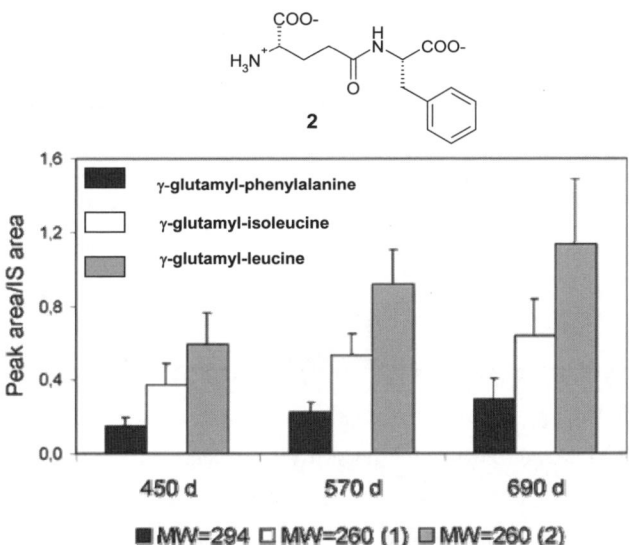

Fig. 4. Trend of γ-Glutamylamino Acids at Different Ageing Times (days) in Parma Ham (reprinted with permission from ref. 14).

Thus, the pleasant ham flavour may ultimately emerge from the correct balancing of different enzymatic activities, such as the proteolytic production of hydrophobic amino acids and small peptides and the generation of γ-glutamylamino acids masking their unpleasant bitterness.

3. Spaghetti with tomato sauce

3.1 Spaghetti

In order to meet consumer (at least Italian consumer) expectations and criteria, for example bright colour, firmness, springiness, lack of stickiness and tolerance to moderate overcooking , spaghetti must be cooked "*al dente*". Thus the right "pasta" must be produced with semolina from durum wheat, which contains a remarkable amount of gluten. Gluten proteins are composed of gliadins and glutenins, highly polymorphic polypeptides, ranging in MW from 30,000 to 90,000 kDa.[15,16] It is possible to distinguish flour/pasta made from durum wheat or soft wheat with several methods. In experiments of simulated digestion of the gliadin fractions, we have found a peptide (3909 MW), which is present only in the soft wheat (Figure 5).

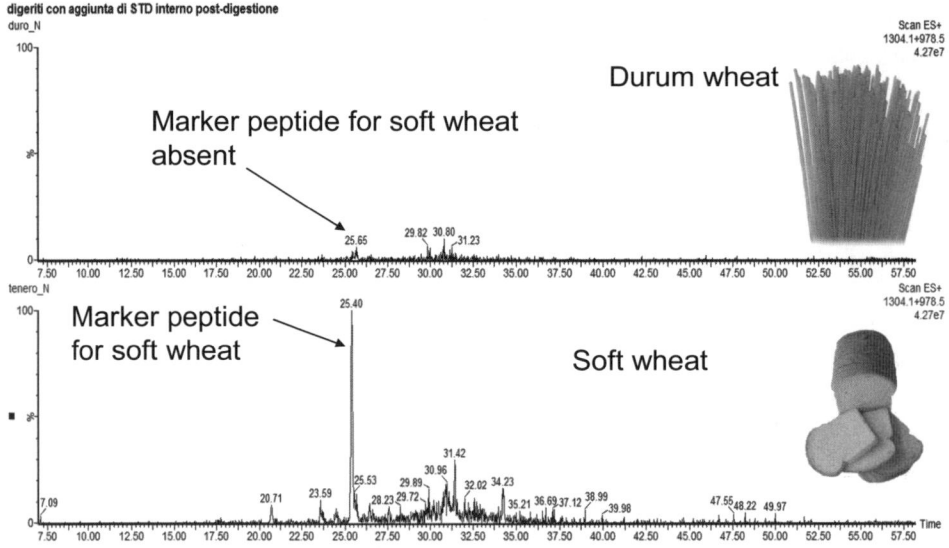

Fig 5. Marker Peptide for the Soft Wheat identified by LC/ESI-MS.

Gliadins and glutenins are responsible for gluten or dough extensibility (viscosity) and strength (elasticity), respectively. The term 'gluten strength' or 'dough strength' are generally used for evaluating and describing flour quality, as determined by mixing characteristics and baking performance. It has also been shown that the insoluble glutenins, mostly high molecular weight (HMWG) glutenin polymers, are the most significant components in dough strength.[17] It is clear that thermal stability of gluten depends on the relative composition of the gluten, that is, the relative amounts of gliadins, soluble glutenins and insoluble glutenins in the protein complex. The higher the amount of gliadins in the gluten, the higher the thermal stability and the longer it takes to reach maximum viscoelasticity. When the gluten is heat-denatured, it forms a network which resists water penetration and protects the starch granules from water uptake at temperatures where starch would begin to gelatinize.

The effect of cooking time on cooked gluten viscoelasticity, isolated with deionised water (A) and 2% NaCl solution (B), of several varieties of wheat is shown in Figure 6.[18] When deionised water was used for gluten washing (Fig. 6A), the relative recovery for the strong gluten type Glenlea was 40% at 0.5 min cooking time and increased steadily after 1.5 min cooking time for the first 5 min to 65%. The weak gluten types Fielder, Alpha 16 and 90W950, showed a limited recovery at 0.5 min cooking time, respectively, then increased rapidly at 5.0 min, similar to the relative recovery of Glenlea. When 2% salt solution was used (Fig. 6B) for gluten washing, the viscoelasticity of the strong gluten type Glenlea was already over 60% while for the other samples only a slight increase in recovery was evident in the first 5 min except for Fielder which had high soluble glutenin (SG) content. Upon heating, the increased viscoelasticity of the protein network between the starch granules limits water penetration, thus protecting the granules from excessive water uptake and disintegration, thereby, preventing pasta texture from becoming overly soft. Since starch gelatinizes in about 2–4 min, protein quantity and quality determine the firmness and chewiness of the cooked pasta during the first 5 min of cooking. Accordingly, with spaghetti prepared from varieties having high ratio of insoluble glutenin/monomeric proteins, the rate of the gelatinization of the starch plays an important role.

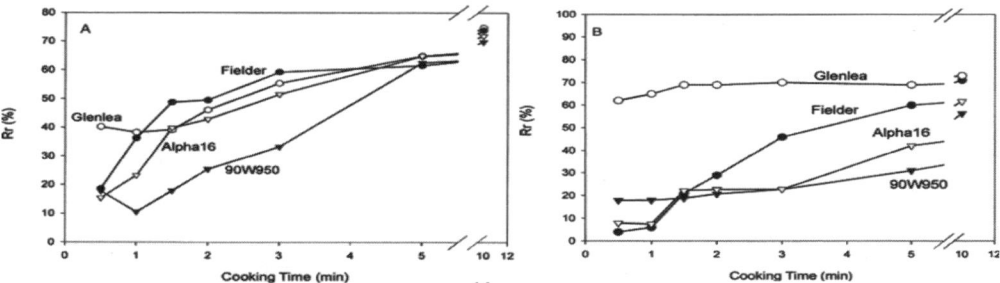

Fig 6. Effect of Cooking Time on Gluten Viscoelasticity from several wheat varieties. (A) Gluten was Extracted with Water without Salt or (B) with Water Containing 2% NaCl. (reprinted with permission from ref. 18)

To ensure that the spaghetti remain firm and chewy, quick denaturation of the protein around the starch granule is important to prevent the starch granules from excess water uptake and their consequent gelatinization and disintegration during cooking. If starch granules disintegrate prior to the denaturation of the protein, then the texture will be soft, especially if the flour has a low protein content. Gluten content, gluten composition, water content, starch swelling and drying temperature are important factors producing *"al dente"* spaghetti quality. Gluten with a high ratio of monomeric proteins to HMWG subunits delays the denaturation of the protein matrix. In contrast, the fast denaturation prevents the disintegration of the pasta or the pasta structure during cooking by delaying hydration and preventing extensive disintegration of starch granules. High temperature is important at the early stages of drying because the initial water content of pasta is essential to denature the protein, while at the same time, there is not enough water to hydrate and disintegrate (gelatinise) the starch granules. The viscoelasticity of heat denatured gluten is an important test to determine gluten quality.

The importance of moisture distribution in spaghetti is common knowledge among the pasta experts. Magnetic Resonance Imaging (MRI) was used to analyze moisture content and moisture distribution and also to evaluate the mobility of water molecules and to make a correlation with the mechanical properties of spaghetti. A correlation was made between the water proton T2 and the moisture content of the standard gel samples of pulverized durum semolina. The cooked samples of dried spaghetti showed the presence of a distinct low moisture region at the core representing the *"al dente"* state. Quantitative comparison of moisture distributions was possible using the average moisture profiles of sections segmented concentrically. The mechanical properties expressed on the force distance curves of the spaghetti samples reflected the differences in moisture distribution. Higher force was observed for dried spaghetti samples at the region corresponding to the low moisture core.[19]

3.2 The tomato sauce

To make a good tomato sauce, the following ingredients are needed: extra-vergin olive oil, sliced onions, garlic, tomatoes, basil, salt and pepper (and eventually chili pepper).

3.2.1 Extra-vergin olive oil

Extra-vergin olive oil is a traditional product of the Mediterranean area, it contains about 75% of monounsaturated fatty acids (mostly oleic acid) and a certain amount of polyunsaturated fatty acids (linoleic and linolenic acids) and is considered a healthy substitute of saturated fats. The product is well characterised also as far as its minor components, which provide the fruity flavours (volatile compounds) and a (bitter) taste (polyphenols). The main polyphenols present in extra-vergin olive oil are: tyrosol **3** and hydroxytyrosol **4**, secoiridoids (oleuropein **5a** and ligstroside **5b**) and lignanes (pinoresinol **6a** and 1-acetoxypinoresinol **6b**) (Figure 7). These products have antioxidant properties and contribute to the oxidative stability of the product.

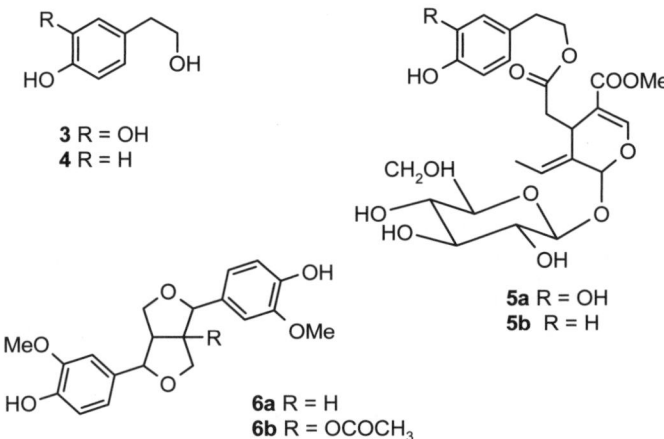

Fig. 7. Polyphenols in Extra Virgin Olive Oil: Tyrosol **3**, Hydroxytyrosol **4**, Oleuropein **5a**, Ligstroside **5b**, Pinoresinol **6a** and 1-Acetoxypinoresinol **6b**.

But what happens during cooking? The following reactions must be considered:

- lipid oxidation with loss of antioxidants and native flavour compounds;

- formation of new volatile compounds arising from the homolytic beta-scission of hydroperoxides from unsaturated fatty acids;

- hydrolysis of secoiridoid aglycons and loss of bitterness-pungency (mainly in water-acidenvironment);

- partitioning of compounds between oily and aqueous phases.

Many are the new volatiles formed, the main being hexanal, *trans*-2-hexenal, hexanol and *cis*-3-hexenol. The bitter and pungent taste of extra-vergin olive oil is mainly due to the dialdehydic forms of decarboxymethyleuropein **7** and ligstroside aglycons **8**, which by contact with the aqueous phase provided by tomatoes (see later), upon heating, undergo hydrolysis of the ester group giving raise to hydroxytyrosol **3** and tyrosol **4** respectively and to the dialdehydic form of decarboxymethylelenolic acid **9**, which are soluble in water and are not bitter (Figure 8).[20]

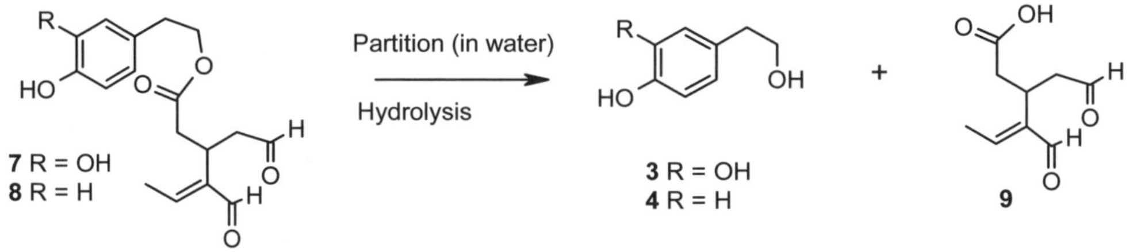

Fig. 8. Hydrolysis of the Dialdehydic Forms of Decarboxymethyloleuropein **7** and Ligstroside **8** Aglycons to Give Hydroxytyrosol **3** and, Tyrosol **4** Respectively.

3.2.2 Onion and garlic ("soffritto")

The first stage for making the "*soffritto*" is the cutting of onions and garlic. When the plant tissue is disrupted, the enzyme allinase is released , which breaks down odourless sulphur containing amino acids(S-alkenyl-L-cysteines) and their sulfoxides: S-(1-propenyl)-L-cysteine sulfoxide for onion (*Allium cepa*) **10a** and S-allyl-L-cysteine sulfoxide for garlic (*Allium sativum*) **10b**. These are cleaved to give pyruvic acid, ammonia and sulfur containing volatiles which provide the characteristic pungent and lacrimatory effects. In particular, a specific sulfenic acid, 1-propenesulfenic acid **11a**, formed from onions, is rapidly rearranged by a second enzyme, called the lachrymatory factor synthase or LFS, giving *syn*-propanethial *S*-oxide **12a**.[21] The gas diffusing through the air, on contact with the eye, stimulates sensory neurons creating a painful sensation. Tears are released from the tear glands to dilute and wash out the irritant. In the case of garlic, allylsulfenic acid **11b** by a coupling reaction forms allicin **13b**, a volatile compounds with the typical pungent flavour. If wishing a milder flavour, garlic may be used as an intact unpeeled clove (to be removed later) or marinated together with finely sliced onions at low pH (below 3), thus irreversibly inactivating alliinase and preventing these reactions to occur. During heating, pleasant flavours are formed, such as polysulfides **14** and thiosulfonates **15** (Figure 9), beside Maillard reaction products with the typical brown colour and aroma.[2]

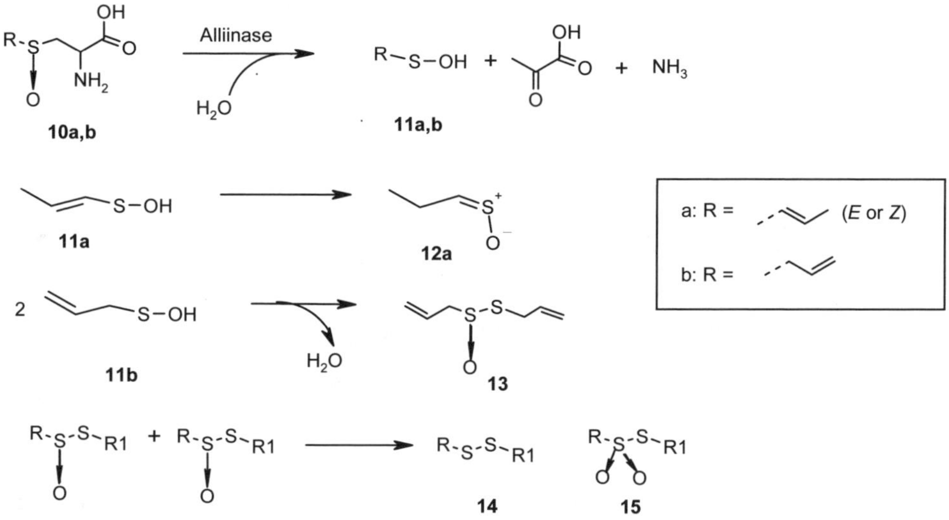

Fig. 9. Formation of Typical Flavour Compounds from Onions and Garlic: Sulfenic Acids **11a and b**, *syn*-Propanethial *S*-Oxide **12a** Allicin **13**, Polysulfides **14** and Thiosulfonates **15**

3.2.3. Tomatoes

Ripe fresh tomatoes or peeled canned tomatoes may be used. Tomatoes contain about 94% water, 3.4 carbohydrate, 0.9 fiber, 1% protein, vitamin C, niacin, retinol, and minerals (calcium and iron).[22] The colour is mainly due to lycopene **17** (although also ß-carotene **18** is present), which is mainly contained in the peel. Despite the many studies performed on the functional properties of lycopene, its potential cardioprotective effects on humans have not yet been substantiated. Hydroxycinnamic acids and flavonoids (rutine, naringenine) are present to different extent according to the variety of the fruit and to the agronomic and climatic conditions.[23]

Glycoalkaloids (tomatine **16**, dehydrotomatine, esculeosides,etc.) with weak toxicological properties are also present[24] (Figure 10).

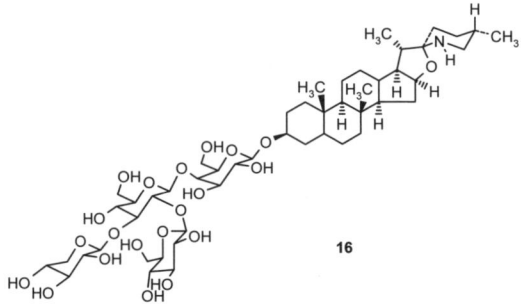

16

Fig. 10. Tomatine, a Glycoalkaloid Formed by Tomatidine (Aglycon) and Licotetraose, a Tetrasaccharide Constituted by Xylose, Galactose and two Glucose Moieties.

Actually tomatine (a glycoside) reaches its maximum amount when the tomato is in its veraison state and then, upon ripening, the glycosidic bond is cleaved to the aglycon tomatidine by the enzyme tomatinase. Tomatidine is not toxic. However, by heating to make the sauce, both polyphenols and glycoalkaloids practically disappear (Figure 11).[25] Isomerization of the all-*trans* to *cis* forms of lycopene occurs, and the *cis*-isomers increase with temperature and processing time.[26]

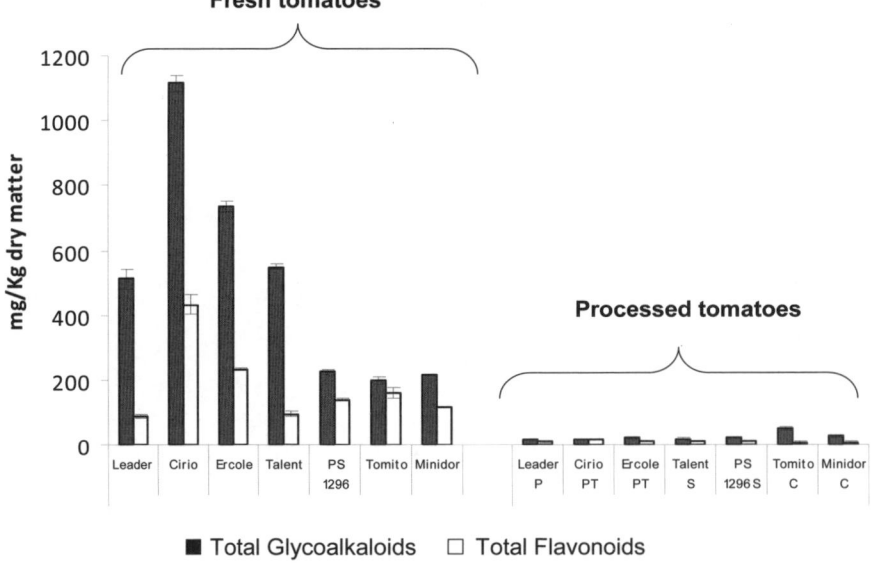

Fig. 11. Total Glycoalkaloids and Flavonoids in Fresh and Processed Tomatoes (reprinted with permission from ref. 25)

So, what happens during the interaction oil – tomato?

First, lycopene, being lipophylic, is partially solubilised in oil and becomes more bioavailable (Figure 12). The antioxidant capacity increases on account of the occurrence of the Maillard reaction, but also for the interchange of polyphenols from oil to tomatoes and from tomatoes to oil, resulting in a protective effect exerted by oil polyphenol on carotenoids.

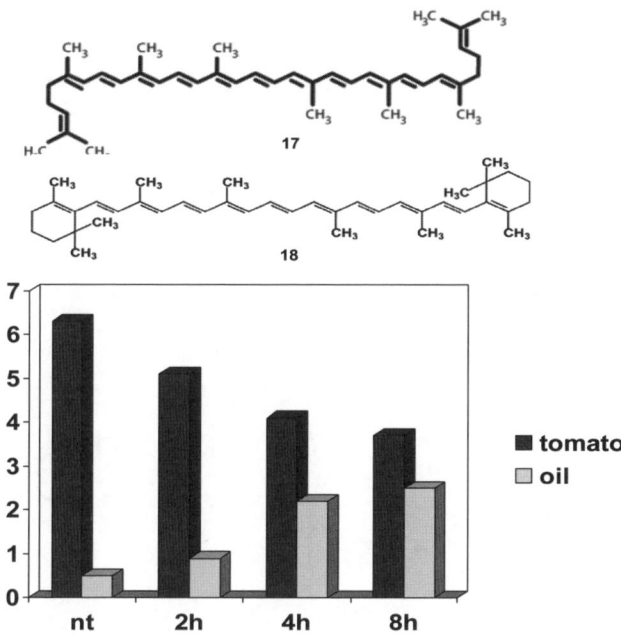

Fig. 12. Partitioning of Carotenoids from Tomatoes to the Oil Phase (Carotenoids mg/100g).

3.2.4. Basil

Basil must be added at the end of cooking to maintain its flavours, the main flavours being linalool, 1,8-cineol, estragol and eugenol **19**. Different varieties of *Ocimum basilicum* contain these compounds in different amounts: the Genovese basil practically does not contain estragol and this may be considered as a marker of authenticity. A lot has been said on the toxicity of basil, due to the presence of methyleugenol **20**, a carcinogenic for rats. However, there are neither known health effects in humans that result from typical dietary exposure to methyleugenol nor available clinical or epidemiology data. In any case, the amount of methyleugenol in the leaves of basil used for the tomato sauce is considered negligible. In a study performed in Italy[27] the aromatic composition of plants of *Ocimum basilicum cv. Genovese Gigante* at different growth stages was determined at 4 and 6 weeks after sowing and showed methyleugenol and eugenol as the main components. The content of these compounds was correlated with plant height rather than plant age. In particular, methyleugenol was found to be predominant in plants up to 10 cm in height, whereas eugenol was prevalent in taller plants (Figure 13). The enzyme performing methylation of eugenol is S-adenosylmethionine-O-methyl-transferase which is not active in taller plants.

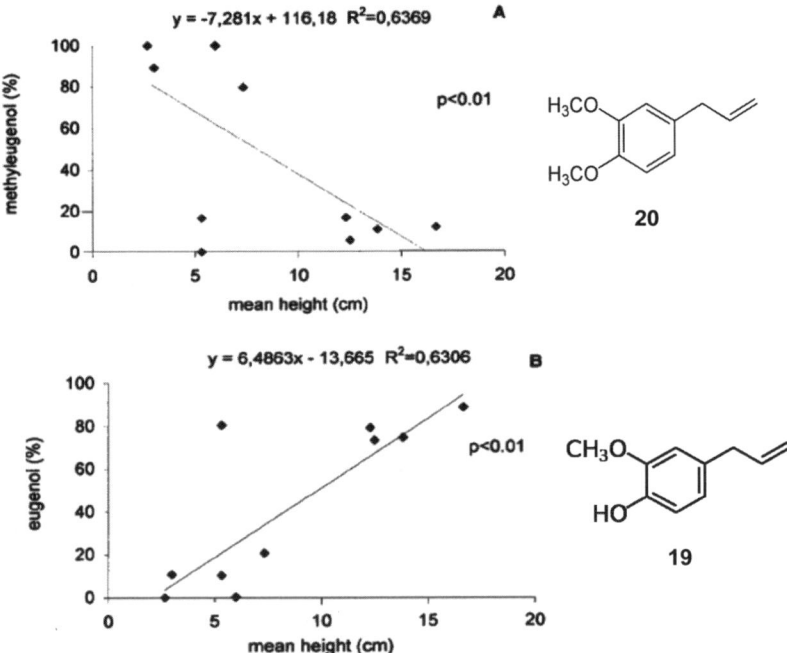

Fig 13. Correlation of the Amount of Eugenol **19** and Methyleugenol **20** with the Eight of the Plant: the Relative Percentage of **19** increases with the Height of the Plant (B) Whereas the Relative Percentage of **20** Decreases (A) (reprinted with permission from ref. 27).

3.2.5. Spices: pepper and chili pepper

Some people like it " hot". Although the traditional recipe does not require the use of other ingredients, it is quite common, specially in southern Italy to add pepper and chili pepper to the tomato sauce. We have mentioned so far the flavour of food, which is determined by a combination of taste and smell. These senses involve specialised chemoreceptors located in the mouth or in the nose, which give selective molecular interactions with the tastants or the odorants.

The sensation of oral pungency differs from the other tastes (sweet, bitter, salted, sour and umami), for the temporal nature (slow onset and prolonged persistence) and for the mechanism, which is the same as for the temperature sensing: it is chemesthesis.[2] In the mid-1970s, a systematic exploration of structure-activity relations for capsaicin-like activity was attempted. After extensive studies it was agreed that the perception of temperature on the tongue (as well as in other parts of the body) is mediated by receptors which are activated at different temperatures, transmitting signals to the brain, which are interpreted as different levels of temperature. A systematic review of these studies has been reported by Szallasi and Blumberg.[28] Thus, pepper, chili pepper or a thermic shock activate the nerves connected to an ionic channel, known as the vanilloid receptor of type 1 (TRPV1), opening it and allowing Na^+ and Ca^{2+} ions to flow across the membrane: the depolarisation of the membrane induces a neuronal stimuli to the brain. TRPV1 can be viewed as an integrator of painful chemical and physical stimuli. It has been suggested that is heat only that has the power to open TRPV1, capsaicin and low pH merely serving to reduce the heat threshold of the receptor. Consequently, even room temperature is able to open the channel under mildly acidic conditions.

A strict structure-activity relationship has been shown for pungent molecules: an aromatic ring bearing hydrogen donor and acceptor groups is linked through a spacer to a central H donor and acceptor group and hence, through another spacer to a lipophylic moiety. The main pungent molecules, capsaicin 21 in chili pepper and piperine 22 in pepper comply with these requirements (Figure 14).

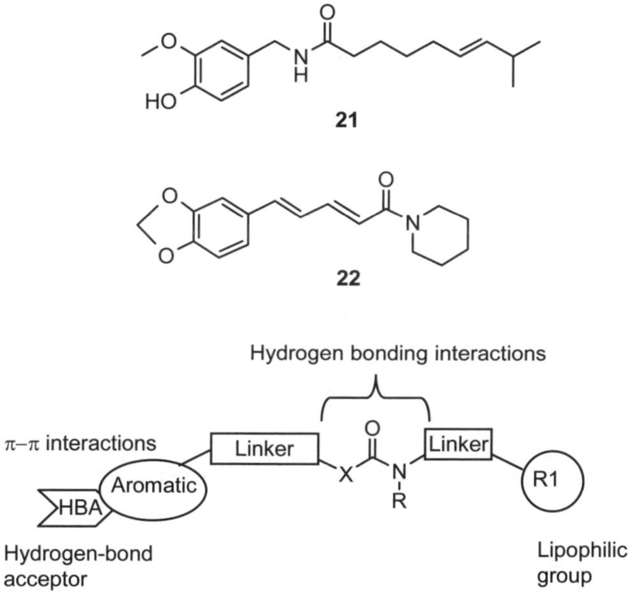

Fig. 14. General Structural Requirements for "Pungent" Molecules: Capsaicin 21 and Piperine 22 Comply with the Requirements.

3.3. Cheese

Once spaghetti have been dressed with the sauce, a generous spread of grated Parmigiano Reggiano cheese is welcomed. Parmigiano Reggiano is a hard type of cheese, aged from 12 to 36 or even more months. Free amino acids and small peptides are formed in cheese during ripening by the proteolytic activities arising from milk, rennet and lactic acid bacteria (LAB) proteases.[29] As a consequence, the free amino acids and, in particular, oligopeptides present in cheeses mainly derive from the proteolysis of caseins.[30] Beside their high nutritional value, these low molecular weight peptides play an important role in flavour development.[31] A correlation has been made between the microflora and the relative enzymes derived from milk, starter/non starter lactic acid bacteria, rennet and the evolution of peptides. This process has been extensively studied in our laboratories in collaboration with the Consorzio del Parmigiano Reggiano. The results are reported in Figure 15. After the first 1-2 months only peptides from the starter lactic acid bacteria enzymes and from rennet enzymes are present, the latter decreasing sharply up to five months. Peptides arising from non starter lactic acid bacteria enzymes increase reaching a maximum at 6-8 months.

New aminoacyl derivatives, of non-proteolytic origin but synthesised de novo in cheeses by so far unknown enzymatic activities were found, in particular, (L,L)-γ-glutamyl isopeptides **23** derived from L-glutamic acid: γ-glutamyl-phenylalanine (γ-Glu-Phe), γ-glutamyl-tyrosine (γ-Glu-Tyr) and γ-glutamyl-leucine (γ-Glu-Leu). Other so far unknown compounds were identified, such as (L,L)-pyroglutamyl-amino acids **24** and (L,L)- lactoyl-amino acids **25** (Figure 15).[32]

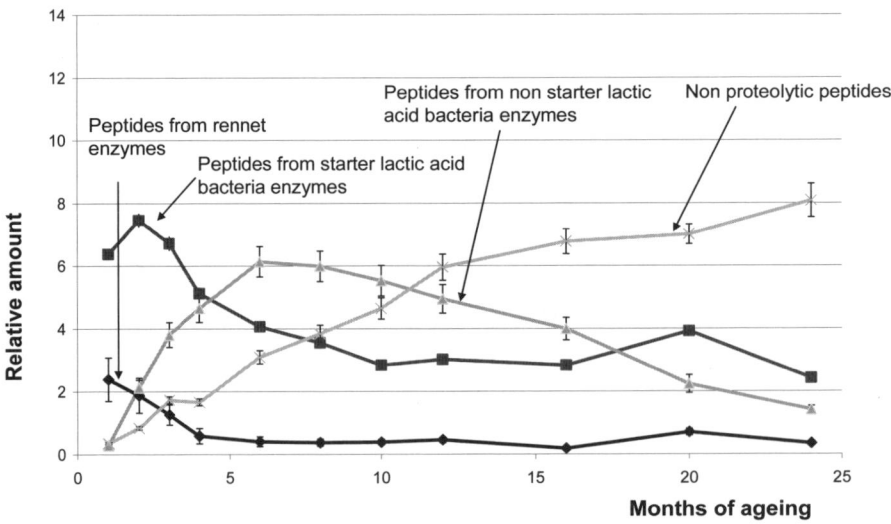

Fig. 15. Trends of Peptides Having Different Origins During the Ageing of Parmigiano-Reggiano Cheese.

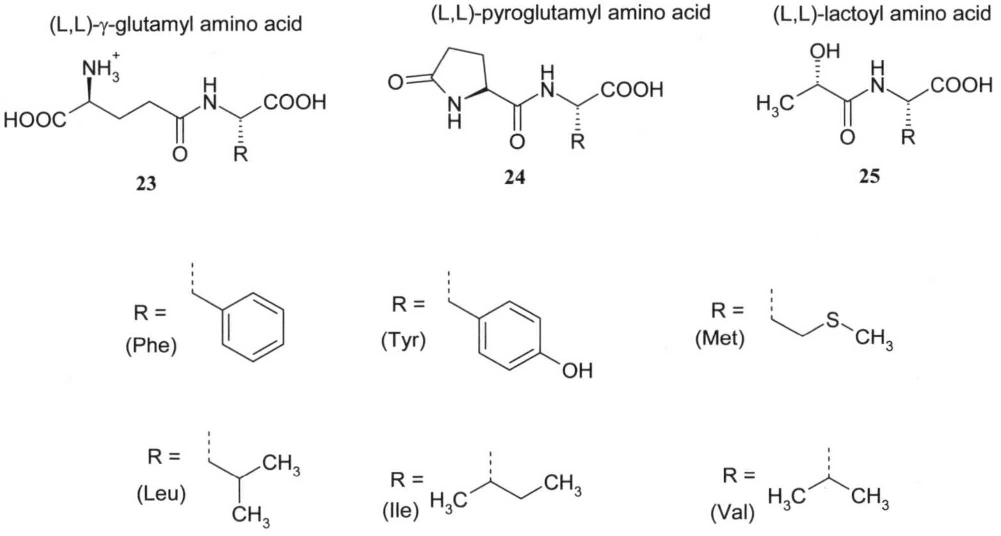

Fig. 16. Structures of (L,L)-γ-Glutamylamino Acids **23**, (L,L)-Pyroglutamylamino Acids **24** and (L,L)-Lactoylamino Acids **25**.

These molecules attracted attention since, unlike most of the other peptides, they seemed to increase regularly during ripening, being apparently unaffected by proteases. Thus, the amount of γ-glutamyl-, pyroglutamyl- and lactoyl-amino acids in Parmigiano-Reggiano cheeses could be usefully exploited to estimate the actual age of Parmigiano-Reggiano samples, a characteristic strongly linked to their quality and to their prices on the market.

Moreover, by formation of γ-glutamylamino acids, bitter amino acids (Phe, Leu, Ile, Val) are preferentially removed from the aminoacidic pool, being transformed in derivatives which have been demonstrated to have an "umami" taste. Umami is the fifth taste, it is a japanese word for defining a pleasant sensation, different from salty, sweet, bitter and sour, it is rather "savouriness". In the past few years the receptor proteins for bitter[33,34] sweet[35] and umami[36] have been identified. These receptors are a subclass of the super family of G-protein-coupled receptors (GPCRs) and have been classified as T1R1, T1R2, T1R3 and T2Rs. The activation of GPCRs by an external stimulus is the starting point of a succession of interactions between multiple proteins in the cell, leading to the release of chemical substances in the cell also called second messengers. Taste receptors share several structural homologies with the metabotropic glutamate receptors. Heterodimers T1R2-T1R3 are responsible for sweet sensing, T2Rs for bitter whereas T1R1-T1R3 are responsible for umami tasting. Receptors for salty and sour tastes are essentially ionic channels, although the former is still speculative and not completely characterised.

The D-amino acids D-alanine, D-aspartic and D-glutamic acids were also found in aged cheese, deriving from the lysis of the lactobacilli cell walls (Figure 17). They can also be considered as markers of the age of the cheese.[37]

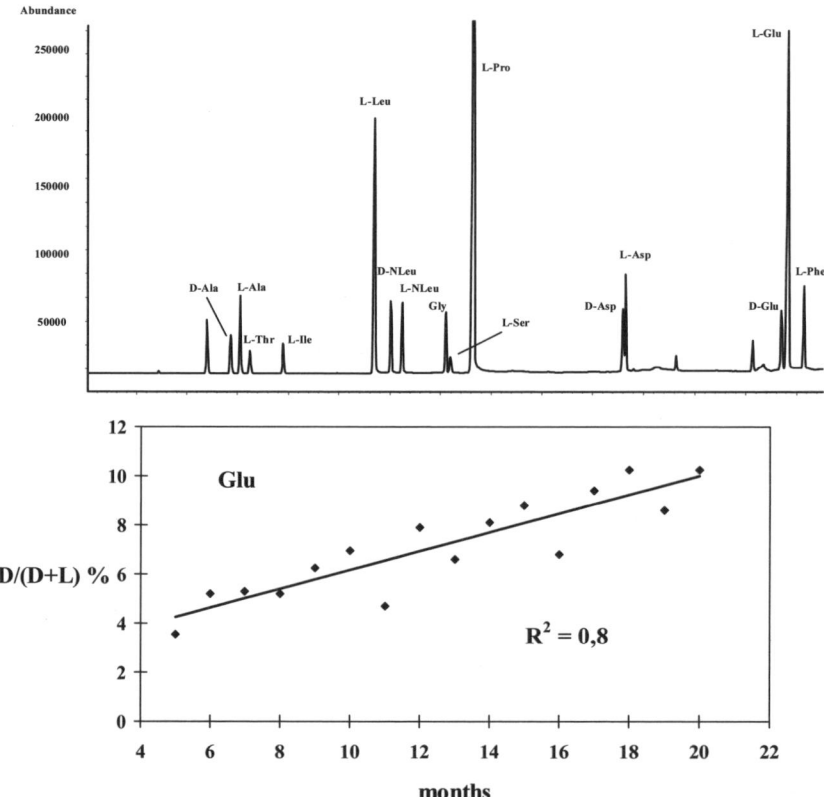

Fig. 17. GC Analysis of D,L-Amino Acids in Aged Parmigiano Reggiano Cheese (Chirasil-Val column (upper figure); Variation of the D/D+L % Ratio of Glutamic acid from 5 to 20 Months (lower figure).

4. Boiled or roast meat?

The principal proteins present in meat, actin and myosin, in the muscle fibers are outstretched, but above a temperature of 40°C they start to coagulate, thus inducing muscle contraction and hardening. As a consequence, there is a shrinkage in the muscle volume with a loss of fluids and development of rigidity. Thus, the longer the meat is cooked, the harder the meat is becoming. However, meat contains also connective tissues, cartilages, ligaments and tendons, which are constituted essentially by collagen. The *tropocollagen* or "collagen" is a subunit of larger collagen aggregates named "fibrils". It is constituted by three polypeptide strands (called alpha chains), each possessing the conformation of a left-handed helix. These three left-handed helices are twisted together into a right-handed triple helix, stabilized by numerous hydrogen bonds. Each triple-helix associates into a right-handed super-super-coil that is referred to as the "collagen microfibril".

When the meat is cooked in a water solution at about 70°C, the hydrogen bonds are broken and collagen starts to become irreversibly denatured, giving rise to protein aggregates with a reticular structure, which are able to include a number of water molecules forming a gel (gelatine) : thus, the meat becomes more tender. So, in order to define the optimum conditions for cooking of meat, we must know: i) the type of meat (more or less connective tissue); ii) the temperature at which the three main proteins of beef are denatured; iii) the duration of cooking. The effects of time and temperature on meat tenderness have been studied.[2] It was found that the best conditions to cook a meat rich in connective tissue should be to heat the meat at a temperature where the collagen is denatured but actin is still native, i.e. between 62 and 65°C. This is for the slow-cooking in water (boiled meat) or in stock (braised or stewed meat).

To make a roasted, barbecued, fried meat, instead, a meat poor in connective tissue (perhaps with the presence of some fats) is required and it must be heated at high temperature for a short time. Under these conditions heat conduction from the hot surface towards the center of the meat and transport of water toward the surface occur, where it evaporates. A crust is made at the surface of the meat and a temperature gradient is being established depending upon the thickness of the meat. The interior remain juicy and the proteins mostly maintain their native state. A number of chemical reactions occur at the surface, developing the typical flavour of roasted meat according to the Maillard and the Strecker reactions, as well as caramelisation, hydrolysis and oxidation. The main flavour compounds developed from boiled meats are sulphur containing compounds such as methyl-3-furanthiol **26** and furfurylthiol **27** and 2-bis(2-methyl-3-furyl)disulfide **28** whereas the flavours from roasted meats are mainly pyrazines such as **29, 30, 31**. (Figure 18).

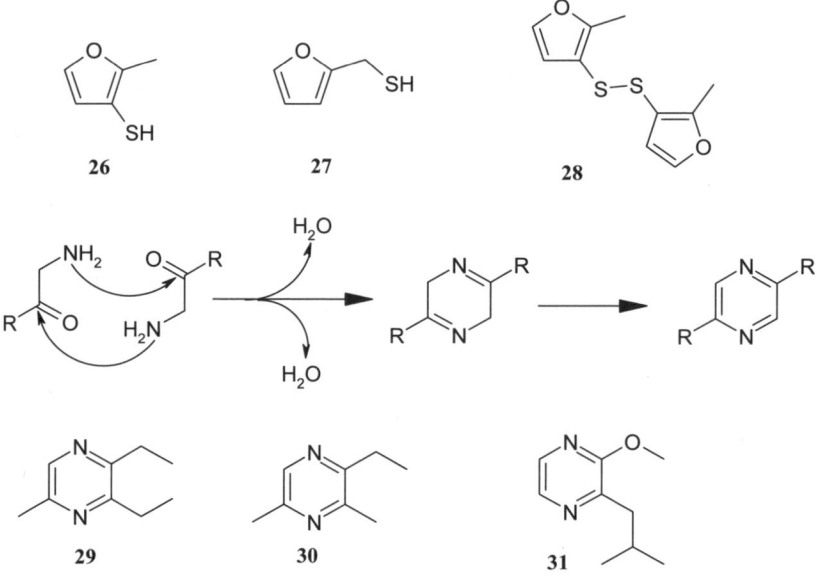

Figure 18. Flavour Compounds Developed from Boiled Meat (**26**, **27** and **28**) and from Roasted Meat (**29**, **30** and **31**)

5. Fried potatoes

The choice of which oil to use for frying is very important. First of all, the oil must be sufficiently stable to a prolonged heating without altering the colour and the flavour. This is achieved with a proper composition in fatty acids: max 20-25% of polyunsaturated fatty acids for an optimal stability/flavour balance; a higher content of monounsaturated and saturated fatty acids to grant a better stability to oxidation; a high smoke point ("punto di fumo"). The most suitable are olive oil and peanut oil.

In a study performed on the absorption of oil by sliced potatoes during frying,[38] oil does not seem to penetrate the potato slices during the actual frying, but to be taken up by the slice only when it is removed from the frying pan due to adhesion of oil to the surface of the slice. Therefore, oil uptake seems to be primarily a surface phenomena involving an equilibrium between adhesion and drainage of oil upon retrieval of the slice from the oil bath. On the other hand, raw and frozen potatoes have been compared and frozen potatoes were shown to absorb a much higher amount of oil (Figure 19).

Thin sections (0.5 cm x 0.5 cm x 10 μm) of potatoes were examined at the microscope before and after frying. Actually, potatoes are mainly "suspensions dispersed in gels", since amyloplasts (solid starch granules of less than 20 μm) are dispersed in the cytoplasm of cells, this phase being itself dispersed in the network of cell walls responsible for the "solid" behaviour of the whole potato. In frozen potatoes, due to the higher volume of ice, the cells are partially disrupted and consequently absorb a higher amount of oil.

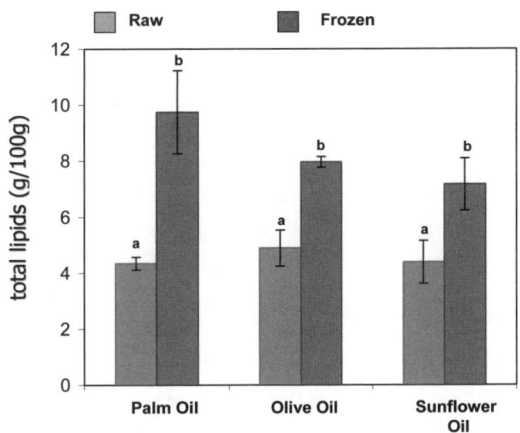

Figure 19. Lipid Uptake (g/100g) in French Fries Obtained from Raw Potatoes and Frozen Potatoes in Different Frying Oils (Palm, Olive, Sunflower).

6. The chocolate mousse

Chocolate mousse is a foam, a two-phase system constituted by a liquid and a gas phase (air bubbles). In the case of the chocolate mousse the liquid phase is an emulsion of fats in water. Is it possible to describe it in a more formal way? A formalism was introduced in 2002 by H. This[1] for the classification of the wide variety of possible structures in foods, by considering the physical nature rather than the chemical composition of foods. Symbols G, O, W, and S, respectively, were adopted for "gas", "oil", "water", and "solid".

Dark chocolate is a three phase system constituted by crystalline fat and sucrose and cocoa solids. Milk cream is primarily made of fat droplets dispersed in a water phase with micelles of caseins. Egg yolks are used for their content of lecithin, an emulsifier. However, it is possible to make a chocolate mousse by using only dark chocolate which commonly contain soy lecithin and water, without cream milk and egg yolks. The recipe has been proposed by H. This[39] as follows: "first make a chocolate emulsion, O/W, by heating chocolate in a water phase (the proportion of chocolate and water has to be chosen so that the final fat/water ratio is about the same as the fat/water ratio in ordinary cream). Then whip (+G) at room temperature while the emulsion is cooled: after some time (some minutes, depending of the efficiency of the cooling), a "chocolate mousse" [G + f(O,S)]/W is obtained. This mousse needs no eggs, contrary to traditional chocolate mousse, and the texture can be the same as the whipped cream:

$$f(O/S)/W + G \rightarrow [G+f(O,S)]/W$$

(1)

The recipe has been further developed by D. Bressanini[4]

7. Champagne

Finally, since there will be no time to discuss wine, on account of the amount of data concerning the huge varieties of wines, the colour, the flavours, the ageing of wines, let's finish by pouring a glass of champagne and toast to Chemistry!

Champagne is a multi-component hydro-alcoholic system supersaturated with CO_2 gas molecules formed during the second fermentation process. During this period the bottles are sealed, so that the CO_2 molecules cannot escape and progressively dissolve into the wine and establish an equilibrium with the CO_2 gaseous molecules in the space between the liquid and the cork. When a bottle of champagne is uncorked, the liquid becomes supersaturated and progressively releases the CO_2 gas molecules dissolved giving rise to bubble formation (the so-called effervescence process). In champagne tasting, the concentration of dissolved CO_2 is a parameter of great importance, since it has a direct impact at the sensory level on the mouth feel, i.e., the mechanical action of collapsing bubbles as well as the chemosensory excitation of the receptors in the oral cavity and on the *bouquet*.

But, what type of glasses shall we use and how will we serve champagne? Experiments were performed by using a flute and a coupe, respectively.[40] It was found that the loss of dissolved CO_2 concentration with time was significantly higher in the coupe than in the flute, proving analytically for the first time that the flute prolongs the drink chill and helps it to retain its effervescence in contrast with the coupe. Moreover, due to a much shallower liquid level in the coupe, bubbles collapsing at the free surface of the coupe were found to be significantly smaller than those collapsing at the free surface of the flute.

But is it better to serve champagne pouring it "with a tongue" from a certain height from the flute generating a certain turbulence or pouring in the slightly distorted glass in the same way as beer is poured in order to avoid excessive foam? It is to be considered that CO_2 molecules inevitably escape by invisible diffusion, through the free air/champagne interface. Experiments were performed pouring champagne in either ways in a flute to indirectly evaluate the respective losses of CO_2 as time proceeds.[41] The beer-like way of serving champagne is much softer than the champagne-like one. Turbulences in the flute are therefore expected to be much less important, thus reducing in turn the loss of dissolved CO_2 molecules by diffusion−convection in comparison with the champagne-like way of serving champagne. Thus, on account of the great importance that "bollicine" have in champagne tasting, should the way of serving champagne be reconsidered? Finally, a fundamental attention should be payed to the temperature of serving champagne, since the higher the champagne temperature is, the higher is the loss of dissolved CO_2 during the pouring process.

"Cin Cin!!!!"

REFERENCES

1. This, H. *Angew. Chem. Int. Ed.*, **2002**, *41*, 83.
2. Barham, P.; Skibsted, L.H.; Bredie, W.L.P.; Frost, M.B.; Moller, P.; Risbo, J.; Snitkjaer, P.; Mortensen, L.M. *Chem. Rev.*, **2010**, *110*, 2313 (and references therein).
3. White, M.; This H. *Nature*, **2010**, *464*, 355.
4. Bressanini, D. In *Le Scienze*, **2008**, Available from: http://lescienze.espresso.repubblica.it/edicola_mese/LE_SCIENZE/1331386.
5. Parolari, G. *Food Sci. Tech. Int.*, **1996**, *2*, 69.
6. Wakamatsu, J.; Nishimura, T.; Hattori, A. **2004**, *Meat Science*, *67*, 95.
7. Moller, J.K.S.; Adamsen, C.E.; Catharino, R.R.; Skipsted, L.H.; Eberlin, M.N. *Meat Science,* **2007**, *75*, 203.
8. Taketani, S.; Ishigaki, M.; Mizutani, A.; Uebayashi, M.; Numata, M.; Ohgari, Y.; Kitajima, S. *Biochemistry*, **2007**, *46*, 15054.
9. Hinrichsen, L.L.; Pedersen, S. B. *J. Agric. Food Chem.,* **1995**, *43*, 2932.
10. Hansen-Moller, J.; Hinrichsen, L.; Jacobsen, T. *J. Agric. Food Chem.*, **1997**, *45*, 3123.
11. Ruiz, J.; Garcia, C.; Diaz, M.; Cava, R.; Tejeda, J.F.; Ventanas, J. *Food Res. Int.*, **1999**, *32*, 643.
12. Ruiz, J.; Ventanas, J.; Cava, R.; Timon, M. L.; Garcıa, C. *Food Res. Int.*, **1998**, *31*, 53.
13. Sforza, S.; Pigazzani, A.; Motti, M.; Porta, C.; Virgili, R.; Galaverna, G.; Dossena, A.; Marchelli, R. *Food Chem.*, **2001**, *75*, 267.
14. Sforza, S.; Galaverna, G.; Schivazappa, C.; Marchelli, R.; Dossena, A.; Virgili, R. *J. Agric. Food Chem.*, **2006**, *54*, 9422.
15. Payne, P.I.; Seekings, J.A.; Worland, A.J.; Jarvis, M.G.; Holt, L.M. *J. Cereal Sci.*, **1987**, *6*, 103.
16. Shewry, P.M.; Hlaford, N.G.; Tatham, A.S. *J. Cereal Sci.*, **1992**, *15*, 105.
17. Orth, R.A.; Bushuk, W. *Cereal Chem.*, **1972**, *49*, 268.
18. Kovacs, M.I.P.; Fu, B.X.; Woods, S.M.; Khan, K. *J.Cereal Sci.*, **2004**, *39*, 9.
19. Irie, K.; Horigane, A.K.; Naito, S.; Motoi, H.; Yoshida, M. *Cereal Chem.*, **2004**, *81*, 350.
20. Sacchi, R.; Paduano, A.; Fiore, F.; Della Medaglia, D.; Ambrosino, M.L.; Medina, I. *J .Agric. Food Chem.,* **2002**, *50*, 2830.
21. Scott, T. Scientific American. http://www.sciam.com/article.cfm?id=what-is-the-chemical-proc. Retrieved 2007-04-28.
22. Hernandez Suarez, M.; Rodriguez Rodriguez, E.M.; Diaz Romero, C. *Food Chem.,***2008**,*106*,1046.
23. Valverdù-Queralt, A.; Jauregui, O.; Medina-Remon, A.; Andres-Lacueva, C.; Lamuela-Raventòs, R. M. *Rapid Commun.Mass Spectrom.*, **2010**,*24*,2986.
24. Friedman, M. *J.Agric.Food Chem.*, **2002**, *50*, 5751.
25. Dall'Asta, C.; Falavigna, C.; Galaverna, G.; Sforza, S.; Dossena, A.; Marchelli, R. *J. Sep. Sci.*, **2009**, *32*, 3664.
26. Shi, J. *Crit. Rev. Biotech.,* **2000**, *20*, 293.
27. Miele, M.; Dondero, R.; Ciarallo, G.; Mazzei, M. *J. Agric. Food Chem.*, **2001**, *49*, 517.
28. Szallasi, A.; Blumberg, P.M . *Pharmacol. Rev.,* **1999**, *51*, 159.
29. Sousa, M.J.; Ardo, Y.; McSweeney, P.L.H. *Int. Dairy J.*, **2001**, *11*, 327.

30. Piraino, P.; Upadhyay, V.K.; Rossano, R.; Riccio, P.; Parente, E.; Kelly, A.L.; McSweeney, P.L.H. *Food Chem.*, **2007**, *101*, 964.
31. McSweeney, P.L.H.; Sousa, M.J. *Lait*, **2000**, *80*, 293.
32. Sforza, S.; Cavatorta, V.; Galaverna, G.; Dossena, A.; Marchelli, R. *Int. Dairy J.*, **2009**, *19*, 582.
33. Adler, E.; Hoon, M. A.; Mueller, K. L.; Chandrashekar, J.; Ryba, N. J. P.; Zuker, C.S. *Cell,* **2000**, *100*, 693.
34. Chandrashekar, J.; Mueller, K. L.; Hoon, M. A.; Adler, E.; Feng, L.; Guo, W.; Zuker, C. S.; Ryba, N. J. P. *Cell,* **2000**, *100*, 703.
35. Nelson, G.; Chandrashekar, J.; Hoon, M. A.; Feng, L. X.; Zhao, G.;Ryba, N. J. P.; Zuker, C. S. *Nature* **2002**, *416*.
36. Beauchamp, G. *Am. J. Clin. Nutr.* **2009**, *90,* 723.
37. Marchelli, R.; Galaverna, G.; Dossena, A.; Palla, G.; Bobbio, A.; Santaguida, S.; Grozeva, K.; Corradini, R.; Sforza, S. *D-Amino Acids: A New Frontier in Amino Acid and Protein Research*; Nova Science Publishers: Hauppauge,N.Y. (USA); **2006**; 299.
38. Ufheil, G.; Escher, F. *Lebensmittel-Wissenschaft und Technologie*, **1996**, 29, 640.
39. This, H. *Acc. Chem. Res.*, **2009**, *42*, 575.
40. Liger-Belair, G.; Villaume, S.; Cilindre, C.; Polidori, G.; Jeandet, P. *J. Agric. Food Chem.*, **2009**, *57*, 4939.
41. Liger-Belair, G.; Bourget, M; Villaume, S.; Jeandet, P.; Pron, H.; Polidori, G. *J. Agric. Food Chem.,* **2010**, *58*, 8768.

Anna Barattucci
Assistant Professor of Organic Chemistry

Address:

Dipartimento di Chimica Organica e Biologica, Università di Messina, V.le Ferdinando Stagno d'Alcontres 31, 98166 Vill. S. AGATA. – MESSINA.

e-mail: abarattucci@unime.it

Education:

Year 1988: Scientific High School Diploma

Year 1993: Degree in Chemistry (Università di Messina)

Year 1998: PhD in Chemical Sciences (Università di Messina)

Professional experience:

Since her introduction in the list of the researchers of the University of Messina (2000), Dr. Barattucci has developed full-time research activity at Dipartimento di Chimica Organica e Biologica of the cited University. Her research regards organic synthesis involving chiral sulphur compounds, directed to the preparation of molecules having applicative interest. She is co-author of 35 articles on international referred journals and 47 communications to national and international congresses and schools.

The Faculty of Science MFN of the University of Messina has assigned to Dr Barattucci the teaching of the following Courses: 2003-2006: Laboratorio di Chimica Organica, CL3 in Scienze dell'Ambiente e della Natura; 2004-present: Laboratorio di Chimica Organica II, CL3 in Chimica; 2005-2008 Chimica Organica (L-Z), CL3 in Scienze Biologiche.

Critical Surveys Covering the Year 2010:
Introduction and Transformation of Functional Groups

Anna Barattucci

Dipartimento di Chimica organica e biologica, Università degli Studi di Messina
Viale F. Stagno D'Alcontres, 31 (vill. S. Agata), 98166 Messina

Sulfur is a very versatile element for organic chemists. This third-row soft element tends to create stable covalent bonds with carbon, nitrogen, oxygen and the same sulfur to give, in its different oxidation states, a large variety of functional groups which have found wide applications in organic chemistry: from stereoselective syntheses to C-C bond and heterocycle formation, organocatalysis, coordination, material and biological chemistry.

The aim of this lesson is to show the last innovation in the chemistry of those sulfur containing organic moieties, e.g. sulfoxides, sulfones, sulfinyl imines and derivatives, that have given a relevant contribution to organic synthesis in the last year, as well as to focus on the role played in the insertion and transformation of new functional groups, and their influence in C-C and C-heteroatom bond formation.

Papers published in 2010 dealing with these three arguments have been reviewed and selected from the following journals:

Acc. Chem. Res.; Angew. Chem., Int. Ed.; Appl. Catal., A; Appl. Catal., B; Arkivoc; Bioconjugate Chem.; Biomaterials; Bioorg. Med. Chem.; Bioorg. Med. Chem. Lett.; Catal. Today; Chem. Commun.; Chem. - Eur. J.; ChemMedChem; Chem. Rev.; Chem. Soc. Rev.; Chin. J. Chem.; Dalton Trans.; Eur. J. Inorg. Chem.; Eur. J. Org. Chem.; Green Chem.; Ind. Eng. Chem. Res.; J. Am. Chem. Soc.; J. Comb. Chem.; J. Fluorine Chem.; J. Mater. Chem.; J. Med. Chem.; J. Nat. Prod.; J. Org. Chem.; J. Organomet. Chem.; J. Sulfur Chem.; Org. Biomol. Chem.; Org. Lett.; Organometallics; Synlett; Synthesis; Tetrahedron; Tetrahedron: Asymmetry; Tetrahedron Lett.

All the paragraphs are divided into different sections: after a brief introduction about the past literature of each functional group, new synthetic methods, exploitation in organic syntheses and the application of the newly obtained structure will be commented.

SULFOXIDES

Sulfoxide is one of the most versatile sulfur functional groups. Its chemical features, joined with the stereogenicity of the unsymmetrically substituted sulfinyl group,[1] have made, in the last decades, sulfoxide become one of the most exploited chiral auxiliaries in stereoselective paths, from C-C bond formation by means of α-sulfinyl carbanions to pericyclic reactions and many other synthetic transformations. For what concerns the literature of 2010, despite other more recently exploited sulfurated groups (e.g. sulfimines), a critical survey of the recent literature will be able to show the still current applicability of this group in organic synthesis.

Besides new chemoselective clean methods that can find their application in industrial processes,[2] the relevant use of the enantiopure sulfinyl moiety in stereoselective transformations still keeps on prompting investigators to find new ways for the chemoselective and enantioselective oxidation of the sulfide group.[3] Reactions of organic sulfides with H_2O_2/Sharpless-like titanium(II) complexes[3c] and with Ru(NO) salen complexes in the presence of oxigen[3d] have given very high enantioselectivities, together with the almost total absence of sulfone overoxidation products; excellent stereochemical results have been obtained through the use of enzymatic catalysts,[3e,f] and, thanks to the active porous sites of Zn[3g] and Fe organic frameworks,[3h] highly efficient enantioselective heterogeneous oxidations have been carried out. A first efficient chemoselective oxidation through hypervalent iodine compounds has been reported,[3i] but no studies about the enantioselective oxidation have been done. Finally, there are few examples of organocatalyzed synthesis of sulfoxides.[3j,k]

The widespread use of sulfinyl group in asymmetric synthesis is due to its high optical stability, even at high temperatures; the stereochemical racemization at room temperature of α-sulfinyl carbanions of aryl dichloromethyl sulfoxides[4] has recently been reported but, in this case, the pyramidal inversion is helped by the chlorine atom assistance. Reduction of sulfoxide to sulfide catalyzed by oxo-complexes and redox photodeoxygenation of dibenzothiophene S-oxides have been recently reported.[5]

The main use of sulfoxides in the introduction and transformation of functional groups is directly connected to the easily accessible α-sulfinyl carbanions,[6] used in the last year in new stereocontrolled functional groups creation, like epoxides,[6b] but also to prepare key intermediates[6c,d] in synthesis. A nice application of this chemistry is given by Makosza and coll.[6e] reaction of 3-chloropropyl sulfoxides with non enolizable aldehydes, in the presence of a base, has given a first attack of the α-sulfinyl carbanion to the carbonyl group, followed by 1,5-intramolecular substitution of the suitably positioned chlorine of the resulting anion to give a tetrahydrofuran ring with good stereoselectivities (scheme n. 1).

Poli and coll.[6f] employ the most exploited α-sulfinyl carbanion, deriving from methyl p-tolyl sulfoxide, for the synthesis of a new sulfoxide whose palladium catalyzed intramolecular allylic alkylation leads to the stereoselective formation of naturally occurring γ-lactams.

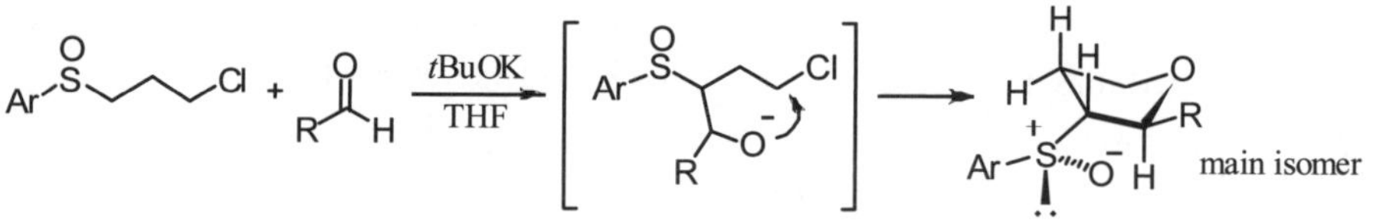

Scheme n. 1

The same (S_S)-anion has been added by Carreño and coll.[6g] to a chromone (scheme n. 2) to form a mixture of lactols, epimers at the newly formed stereocentre. After the formation of the corresponding methyl ethers, stereoselective (S_S)-sulfoxide directed Lewis acid promoted nucleophilic allylation to the upper face of the oxacarbenium intermediate, coming from the TiCl$_4$-promoted loss of the methoxy group, has led to the highly stereoselective formation of two allyl sulfinyl chromanes, then separated. The major diastereomer, with the resolved stereochemistry, is the key intermediate for the total synthesis of (S)-γ-CEHC, a natural γ-tocopherol metabolite.

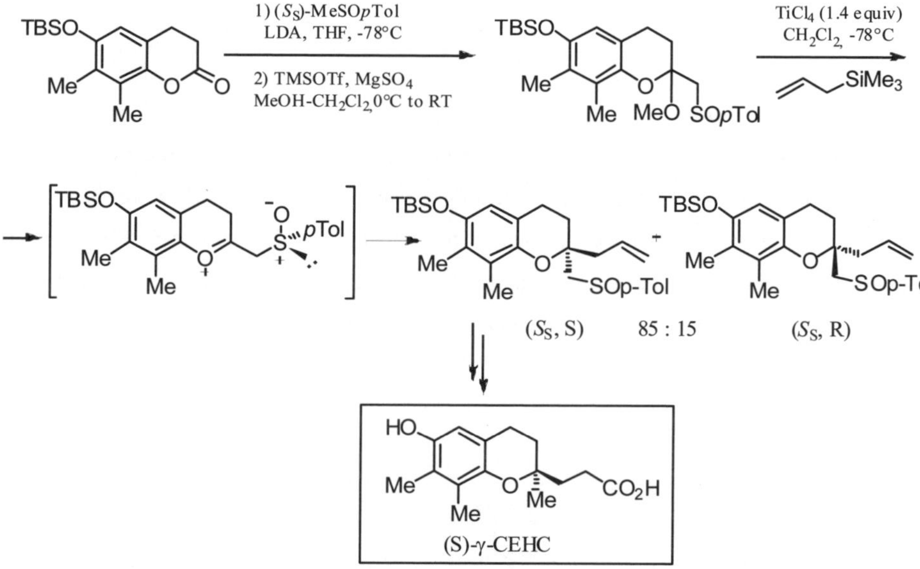

Scheme n. 2

In the total synthesis of (+)-Swainsonine,[6i] after an α-sulfinyl carbanion mediated cyclization and diastereoisomers separation, completely stereoselective reduction of the double bond, helped by the asymmetric sulfinyl moiety, is essential for the correct ring fusion formation (scheme n. 3).

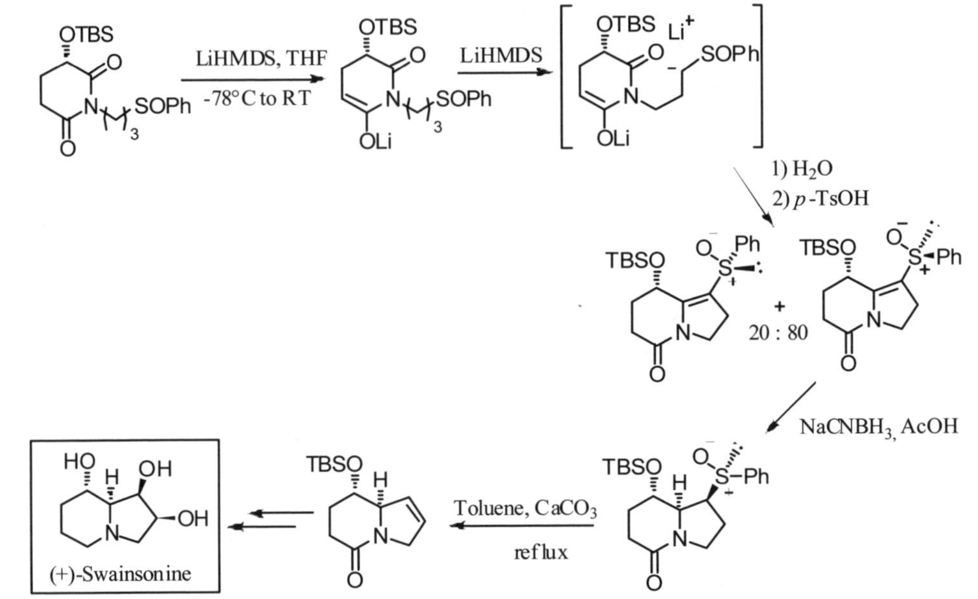

Scheme n. 3

The sulfinyl chiral auxiliary is then removed by an intramolecular β-elimination process that gives a cyclic alkene and a unstable sulfenic acid:[7] this thermolytic pattern has frequently been used to remove the sulfoxide moiety, once it has played its role, in other natural product synthesis[7c-e] but also for synthetic purposes to obtain allenyl β-lactams[7f] and target alkenes.[7g,h] The chemistry of sulfenic acids and derivatives, that doesn't stop at the above mentioned extrusion process, has been object of some publication in 2010.[7i-k]

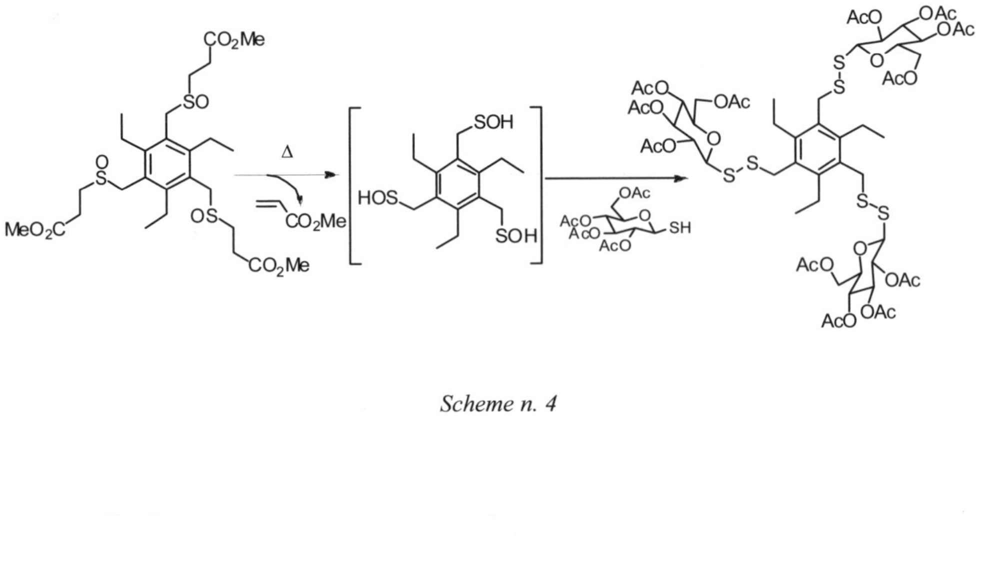

Scheme n. 4

One of their typical reactions, condensation with thiols, employed until a couple of years ago only to indirectly demonstrate the *in situ* generation of these unstable intermediates, has found its application in the synthesis of unsymmetrically substituted disulfides.[7l] Recently Aversa and coll. [7m] have reported the mild *in situ* thermolytic generation of trisulfenic acids and their exploitation in the efficient synthesis of tripodal tris(disulfides), containing carbohydrate and amino acidic functionalities, with potential application in biological processes (scheme n. 4).

In a work by García Ruano and coll.[8a] LDA treatment of resolved 2-*p*-tolylsulfinyl arenes gives γ-sulfinyl carbanions, whose nucleophilic addition to electronpoor vinyl sulfones leads to the almost complete diastereoselective formation of hydrocarbons containing two contiguous resolved stereocentres. A systematic study on differently substituted vinyl sulfones has been carried out to understand the electronic and steric effects on the stereochemical outcome of the addition. The remote control of the sulfinyl group on benzylation (scheme n. 5) has been proposed by the authors. Moving away from sulfinyl carbanions the same group, involved in a widespread use of the sulfinyl chiral auxiliary in asymmetric methods, reports a stereoselective synthesis of 2,3-epoxy alcohols, where the key step is the sulfinyl-directed addition of oxyranyl lithium salts to a (S_S)-2-*p*-(tolylsulfinyl)benzaldehyde,[8b] and the stereocontrolled formation of highly functionalized cyclopentenes and cyclopentanes precursors *via* a formal [3+2] cycloaddition of 3-sulfinyl-5-alkoxyfuran-2(5*H*)-ones with allenyl sulfones.[8c]

Scheme n. 5

For what concerns the recent use of sulfinyl derivatives in pericyclic reactions,[9] sigmatropic rearrangement of enantiopure allylic sulfoxides has found synthetic application in sulfenate anions generation with subsequent formation of aryl sulfoxides,[9a] while a DBU-catalyzed sigmatropic rearrangement, in the presence of triphenylphosphine, of enantiopure α-sulfinyl enones, leading to the γ-hydroxy-α-enones in high ee, has been reported by Miura and coll.[9b] A possible mechanism has been proposed to explain the almost complete enantioselection of the process (scheme n. 6).

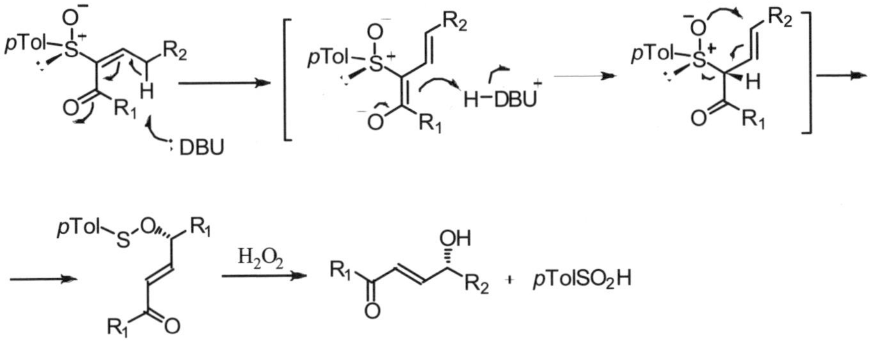

Scheme n. 6

On the other hand, while Maguire and coll.[9c] study the microwave-assisted Diels-Alder reaction with sulfinyl acrylamides as dienophiles, Fernández de la Pradilla and coll.[9d] exploit the highly diastereoselective intramolecular Diels-Alder cycloaddition of 2-sulfinyl dienes as the key step in the formation of carbo- and hetero-cycles with bicyclic or tricyclic structures, in generally good yields. As it is showed in the example in scheme n. 7, the cycloaddition leads to the diastereoselective formation of 1,4-cyclohexadienes, with two differently functionalized double bonds. Selective dihydroxylation of the first double bond of the major isomer, followed by oxidation to sulfone and subsequent removal of the sulfonyl group, leads to highly functionalized enantiopure cyclohexanes.

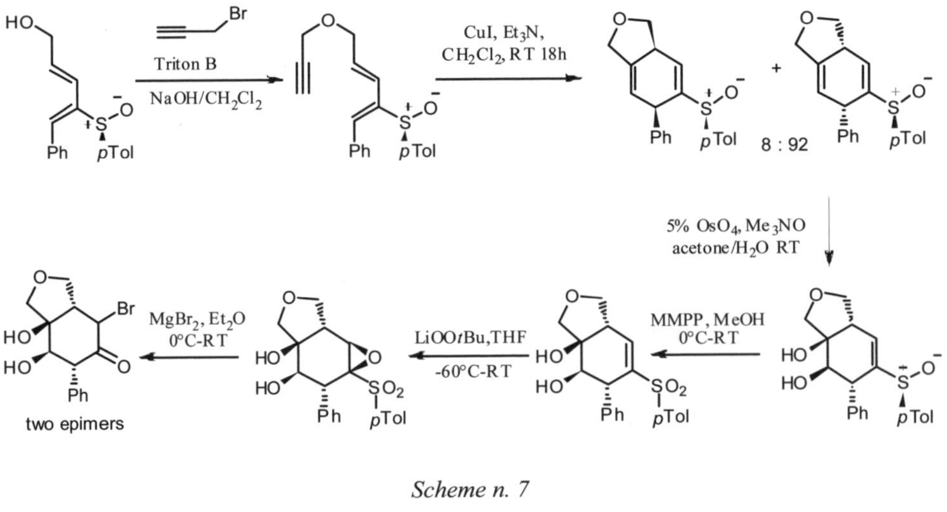

Scheme n. 7

Suitable vinyl sulfoxides[10] have also been employed as substrates for nucleophilic conjugated additions; recently, this kind of reactivity has been exploited as the first step in the synthesis of highly functionalized cyclopropane rings.[10b-d] In particular, a new stereocontrolled synthesis of (+)-2-aminobicyclo[3.1.0]hexane-2,6-dicarboxylic acid 6-ethyl ester (LY354740), a powerful and selective 2mGluR agonist, has been accomplished in four steps from the enantiopure (+)-(R)-2-(p-tolylsulfinyl)cyclopent-2-enone (scheme n. 8).[10b] The key step is represented by the sulfinyl-directed stereoselective cyclopropanation. After easy diastereomer separation, sulfinyl group is then removed by treatment with isopropylmagnesium bromide. Two more steps are required to join the formation of the target enantiopure molecule.

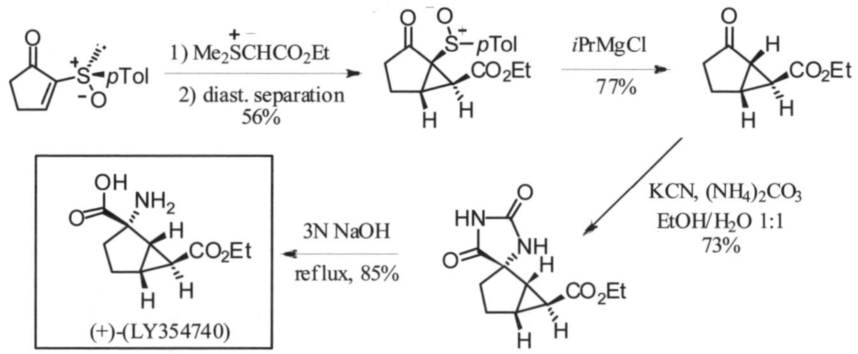

Scheme n. 8

Finally it has to be mentioned that sulfoxide, having a well known coordination chemistry,[11] still proves to be applied in efficient metal catalytic processes; recently there have been many examples of syntheses and application in catalysis of sulfoxide containing ligands.[11b-e] Despite its large use in catalysis, the number of papers concerning the use of the chiral sulfinyl moiety in organocatalyzed processes is poor. Few works have been found, and among these the group of Scettri employs aryl methyl sulfoxides[11f] and tetradentate[11g] imino sulfoxides, whose synthesis is reported in scheme n. 9, in the promotion of stereoselective allylation of aldehydes.

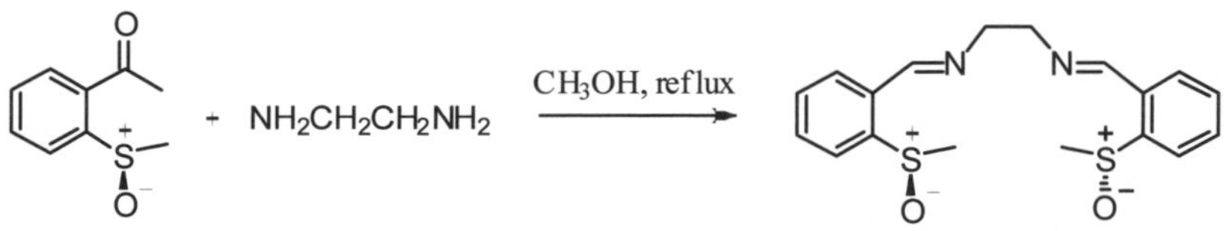

Scheme n. 9

SULFONES

The strong electronwithdrawing effect that it exercitates is directly responsible for the sulfonyl group's wide application in indirect C-C formation: in fact it favours the easy formation of relatively stable nucleophile α-sulfonyl carbanions, make vinyl sulfones optimal electrondeficient partners in pericyclic reactions and in conjugated additions of nucleophiles, without any risk of 1,2 addition.[12] Even if in this case no stereodifferentiation exists at the sulfonyl moiety, many recent reactions have been stereoselectively directed by the use of chiral catalysts. The great convenience in the use of this group lies also in the stable crystalline products that are always formed, and in the easy removal of the sulfonyl moiety from the synthetic product, with easy reductive methods or exploiting the leaving-group ability of sulfinate anion. Differently from the sulfinyl group, obtainable from thiothers by controlled oxidation, exhaustive oxidation of the corresponding thioether easily furnishes the sulfone; in 2010 clean methods of oxidation are present in the literature.[13] PhI(OAc)$_2$/KI mediated reaction of aryl sulfinates with alkenes, alkynes and α,β-unsaturated carbonyl compounds is a good method for the obtainement of vinyl sulfones;[14a] copper(I) catalyzed CH insertion reaction of α-diazosulfones that enantioselectively leads to cyclic sulfones[14b] and rhodium(I) promoted addition of arylboronic acids to arysulfonylacetonitriles, useful for the obtainment of β-ketosulfones,[14c] have also been recently reported.

The main application of α-sulfonyl carbanions is the Julia olefination, together with its modification (Julia-Kocienski), where their initial nucleophilic addition to aldehydes and ketones leads to unstable intermediates, that decompose giving the stereoselective formation of (E)-alkenes. Finding new precursors of α-sulfonyl anions for this purpose[15] and, among these, the α-fluorinated[15a-e] ones is object of the recent literature. Joined with the sulfonyl influence, the electronwithdrawing effect of fluorine helps the formation of very stable carbanions with the ultimate formation of target fluorinated alkenes (scheme n. 10).

Scheme n. 10

On the other hand, the use of a new strategy to mime the Julia reaction,[15f] developing a new methodology for the stereospecific conversion of epoxides into alkenes, have been object of the recent literature (scheme n. 11): in fact, starting from a nucleophilic attack of a thiolate to an epoxide (route a), and then oxidizing the corresponding thioether to sulfone, an alternative way to obtain the stable precursor of the key intermediate of Julia-Kociensky (route b) olefination has been found; the use of several bases and reaction conditions for its deprotonation and transformation into the target alkenes with optimal yields and good diastereoselection have been found.

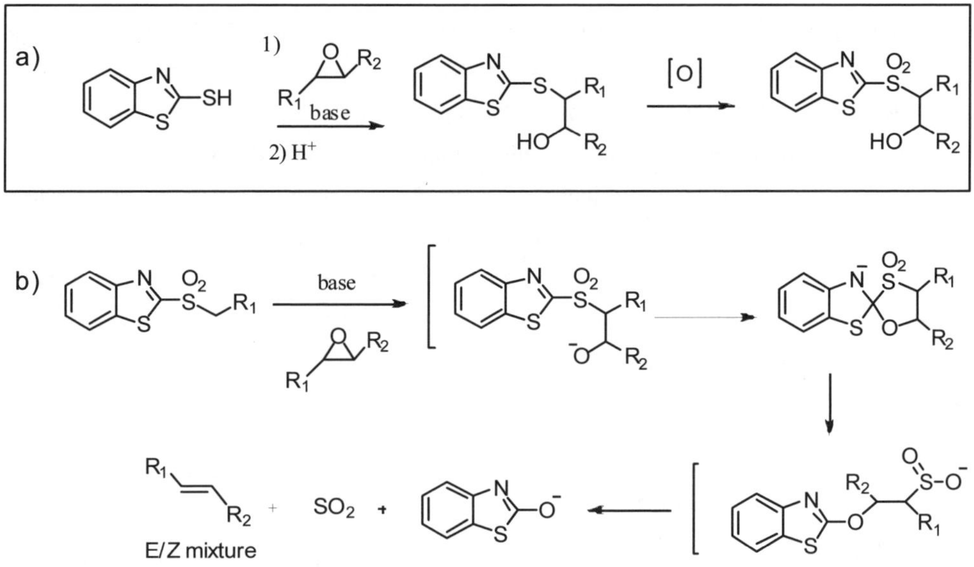

Scheme n. 11

Going on with sulfonyl carbanions, reaction of enantiopure terminal epoxides with the sodium salt of diethyl P-[(phenylsulfonyl)methyl]phosphonate[16] is a convenient way to convert epoxides into a variety of enantiopure *trans*-cyclopropyl sulfones (scheme n. 12). A phosphite group transfer and subsequent 3-*exo*-trig cyclopropanation by means of the α-sulfonyl carbanion moiety is the reaction pathway proposed by the authors.

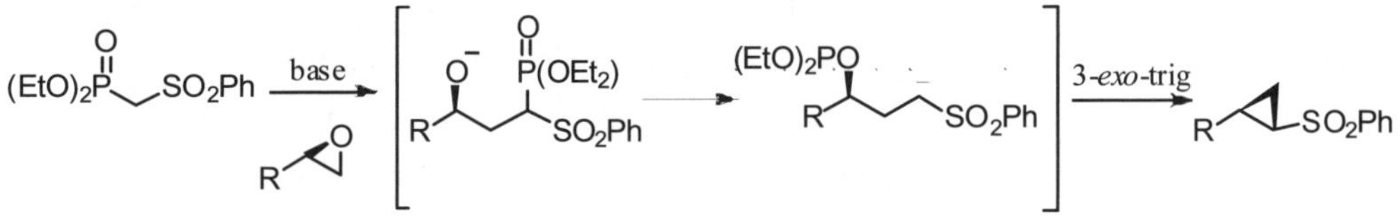

Scheme n. 12

315

A nice application of the fluoro-sulfone chemistry is given by Hu[17a] and Olah:[17b] base or Mg activated treatment of suitable perfluoroalkyl sulfones make them a source of transient perfluoroalkyl nucleophilic anions, as it is demonstrated by the isolation of their addition products to aldehydes or other electrophilic substrates.

Another useful application of sulfones in organic synthesis is their use as dienophiles and dipolarophiles in cycloaddition reactions.[18] A great contribution to this chemistry, leading to functionalized nitrogen heterocycles, has recently been given by Carretero and coll.[18a,b] Metal catalyzed regioselective 1,3-dipolar cycloaddition of α-iminoesters with bis(sulfonyl)ethene, β-sulfonylacrilates and β-sulfonylenones, followed by a base-promoted elimination of the sulfonyl group and subsequent aromatization, gave a library of 2,5-disubstituted, 2,3,5- and 2,4,5-trisubstituted pyrroles in satisfactory yields; one example is given in scheme n. 13. It has moreover been demonstrated that this method can be applied in an iterative and straightforward manner for the construction of oligopyrroles. Different types of acetylenes, bearing sulfone or thiol groups, have been employed as dipolarophiles in 1,3-cycloadditions with C-carboxymethyl-N-nitrilimines, giving the regioselective formation of thieno[2,3-c]pyrazoles.[18c]

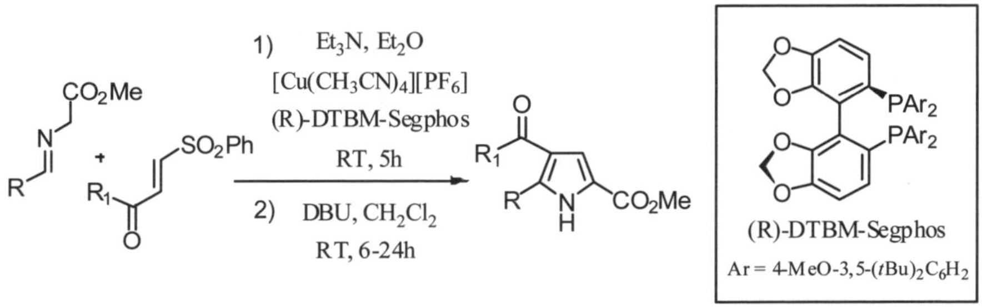

Scheme n. 13

The conjugated addition of organometallic reagents to α,β-unsaturated compounds, and among them vinyl sulfones, is one of the most versatile and widely used synthetic method for carbon-carbon bond formation. Thanks to the use of (S,R,R)-monodentate phosphoramidite ligand (scheme n. 14), the first enantioselective copper-catalyzed conjugated addition of diorganozinc reagents to a range of aromatic α,β-unsaturated sulfones have been reported by Feringa and coll.[19a] The enantioenriched sulfones obtained are potentially useful intermediates in the preparation of a wide variety of functionalized building blocks.

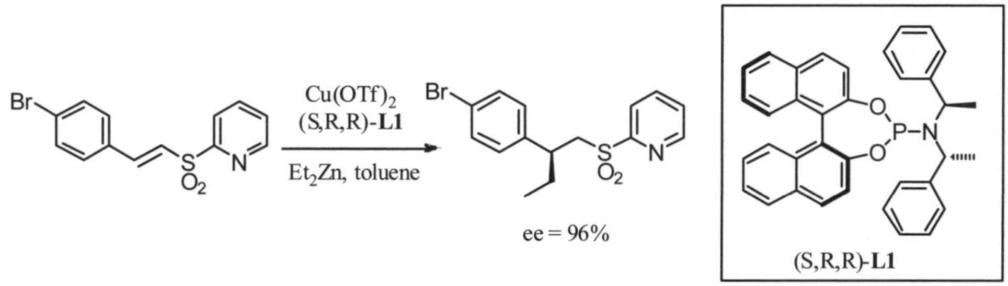

Scheme n. 14

On the other hand, chiral vinyl sulfone - modified carbohydrates having a suitably positioned leaving group have proven to be good substrates for diastereoselective two-steps formation of five membered carbo- and hetero-cycles.[19b]

Asymmetric dihydroxylation of vinyl sulfones under Sharpless conditions affords enantioenriched α-hydroxyaldehydes, involving the loss of a phenylsulfinate anion. This original method reported for the first time in 2003,[20a] has been employed by the same authors,[20b] starting from amino functionalized vinyl sulfones, as the first step in the stereoselective synthesis of piperidine containing building blocks in the synthesis of alkaloids. An initial cross-metathesis reaction gives the formation of the vinyl sulfonyl starting product and a subsequent modified dihydroxylation gives the first building block of the target molecules.[20c] The synthetic route to (+)-Febrifugine, reported in scheme n. 15, is an example of this chemistry.

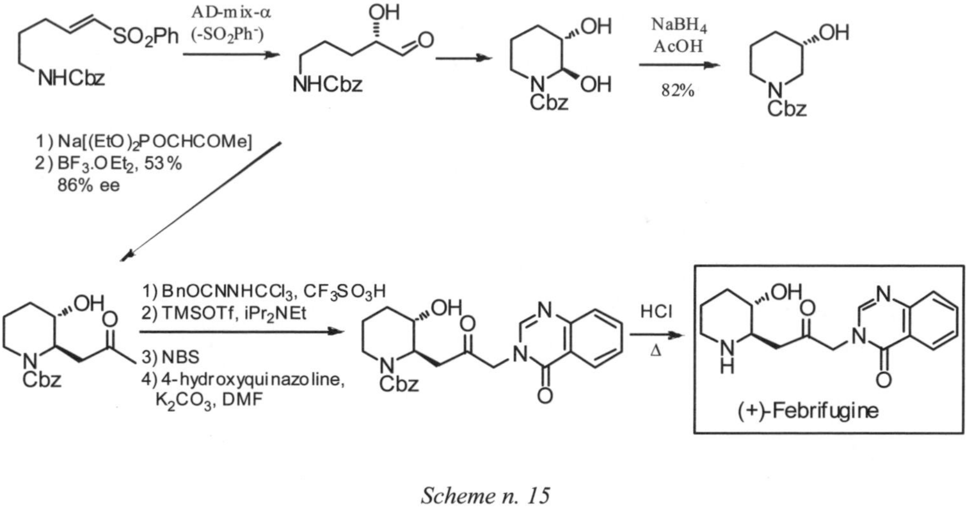

Scheme n. 15

Finally, a direct contribution to functional group transformation has been recently given by an oxidative base-mediated transformation of primary sulfones to carboxylic acids. Even if some oxidative desulfonations to carbonyl compounds were already reported, this appear to be a clean, versatile and efficient route to carboxylic acids, alternative to more toxic syntheses. The experimental results make the authors suggest the intermediation of a sulfonyl dianion, whose destiny to carboxylic acids is depicted in scheme n. 16. Furthermore, since ^{13}C-labelled methyl phenyl sulfone is commercially available, useful carbonyl-labelled acids can be accomplished using this chemistry.[21]

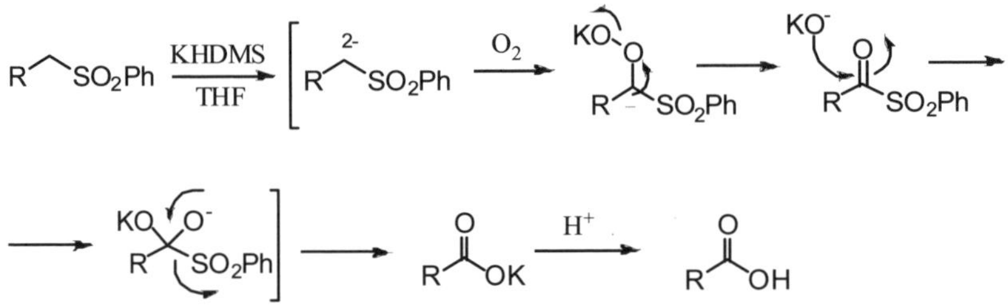

Scheme n. 16

Recently, the explosive application of sulfone chemistry in organocatalyzed transformations has become evident.[22] Anions derived from β-ketosulfones[23a-e] and bis-sulfones[23f] reacted as nucleophiles in inter- and intramolecular[23g] conjugated addition and nucleophilic substitution, giving excellent results in term of reaction rate acceleration and stereoselectivities. Organocatalysis is also present in intramolecular and intermolecular Michael addition of enoles, enolates and other carbanions[24] on different kind of vinyl sulfones, providing facile access to a wide range of enantiopure synthons after desulfonylation.

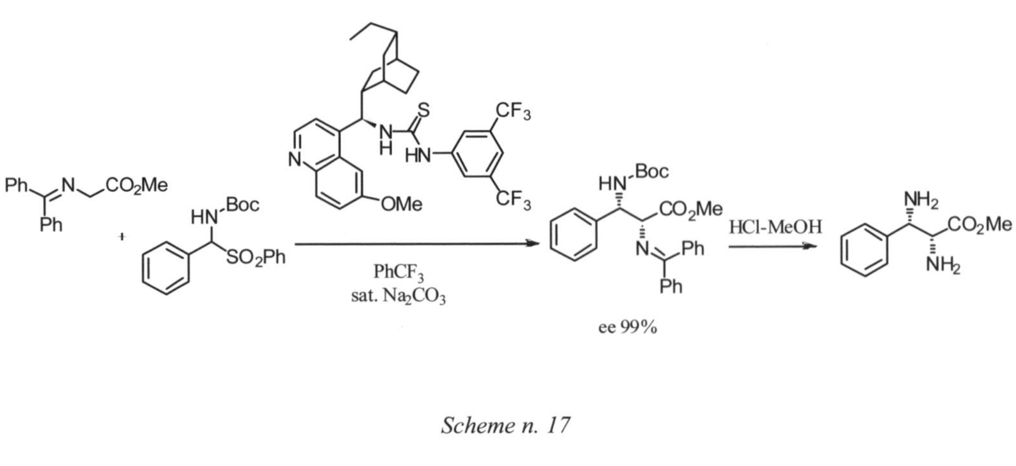

Scheme n. 17

Enantioselective reactions of stable α-amidosulfones, in particular Mannich type attack to their electronpoor electrophilic site with the loss of a sulfinate leaving group, is exploited in the stereoselective syntheses of β-cyano nitriles,[25a] β-amino aldehydes, β-amino acids,[25b] and α,β-diamino acids,[25c] these last ones formed by the highly enantio- and diastereoselective addition of glycine Schiff bases with N-Boc protected α-amidosulfones in the presence of a Cinchona alkaloid thiourea as organocatalyst (scheme n. 17).

The great sulfone ability as leaving group has also been exploited in the organocatalytic KF/alumina mediated α-alkylation diketopiperazines with arylsulfonylindoles.[25d]

Finally, new types of organocatalysts bearing a pyrrolidine and a sulfone functions have been prepared in few steps and in good yields from (S)-proline. They have been employed for the direct asymmetric Michael addition of cyclohexanone and nitro-olefins,[26] and have proven to have very high catalytic activity towards this reaction in water and favour high stereoselectivities.

SULFINIMINES, SULFAMINES AND DERIVATIVES

Since the pioneering work by Davis,[27] sulfinyl imines (sulfinimines) became a powerful group for the efficient synthesis of nitrogen containing organic molecules. The singularity of sulfinimines can be ascribed to the convergence of some factors: the electron-withdrawing effect of the sulfinyl group directly linked to the nitrogen makes the C=N bond a good substrate for conjugate nucleophilic additions that, on the other hand, frequently happen with very high degree of diastereoselection, thanks to the chiral sulfinyl sulfur influence. Furthermore, despite the harsh conditions that are frequently requested for many C-S bond cleavages in sulfoxides, in this case a simple acid-catalyzed hydrolysis of the reaction products leads to the formation of the desulfurized products. These kind of precursors found in the last decades a wide application in the synthesis of a large variety of enantiopure natural occurring and biologically active amines, not easily obtainable in other ways, just because of the difficulty of the α-carbon to the nitrogen to be functionalized by nucleophiles.

Two main research groups are the 'specialist' of this chemistry: the group of Davis, discoverer of the first sulfinimine, p-tolylsulfinyl immine and the group of Ellman, who enlarged the applicability of this chemistry with the introduction of a new sulfinimine, containing the t-butyl group instead of the p-tolyl one. The last two decades have witnessed the explosion of applications of various kind of nucleophiles addition to sulfinimines in organic syntesis remarked by the publication of many reviews:[28] the aim of this fascinating chemistry has lately been the finding of new and efficient carbon and heteronucleophiles that can give the stereoselective above mentioned addition, key step of longer routes to total syntheses. For what concerns sulfinimine synthesis, since their appearance in the literature an efficient method, the nucleophilic addition of an enantiopure sulfinamides to differently functionalized aldehydes and ketones, has basically remained the only one.

Even if the pioneers of this chemistry are Davis ad Ellman, different groups have joined this chemistry, as it can also be noticed by the literature of the last year. Many natural products syntheses employing these key intermediates have been reported exploiting the nucleophilic addition of many carbonucleophiles as α-nitro carbanions in aza-Henry reaction,[29a] α-sulfinyl carbanions,[29b,29c] Grignard reagents,[29d-29f] organolithium[29g] and organozinc compounds.[29h-k] Between these, the almost totally enantioselective LiCl and Zn-mediated asymmetric cynnamylation of *t*-butanesulfinyl imine have been reported by Lin group[29i] in the synthesis key intermediates for the preparation of α-methylene-γ-lactams (scheme n. 18).

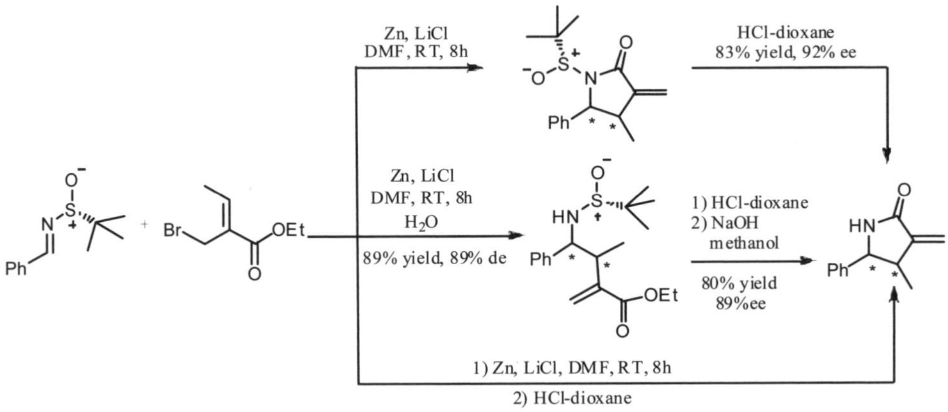

Scheme n. 18

An interesting new application of this chemistry comes from the finding of new nucleophiles and, among these, carbanions derived from the lithium treatment of alkyldiphenylsylanes added to sulfimine, to give the silylated precursors of silicium containing heterocycles or peptide mimics.[30]

The selective incorporation of fluorine atoms in organic molecules is a powerful methodology to modulate their biological properties. For this reason, in the last decades, many synthetic organic strategies have been directed toward finding new strategies to obtain fluorine containing compounds. In particular the presence of one or more fluorine atom in proximity of an amine, functional group that is often present in biologically occurring molecules, decreases its basicity, and consequently changes its chemical and biological activity. The chemistry of sulfinimines has been then applied to the synthesis of fluorinated compounds[31] too. One example is given by the enantioselective addition of CF$_3$ anion to both the enantiomers of suitable sulfinyl imines.[31a] Almost complete diastereosection have been obtained, leading, after acidic treatment of the obtained sulfinamide, to both the enantiomers of fluorine containing amines that found their application in the synthesis of trifluoromethylates analogues of Calcimimetic NPS R-568 (scheme n. 19).

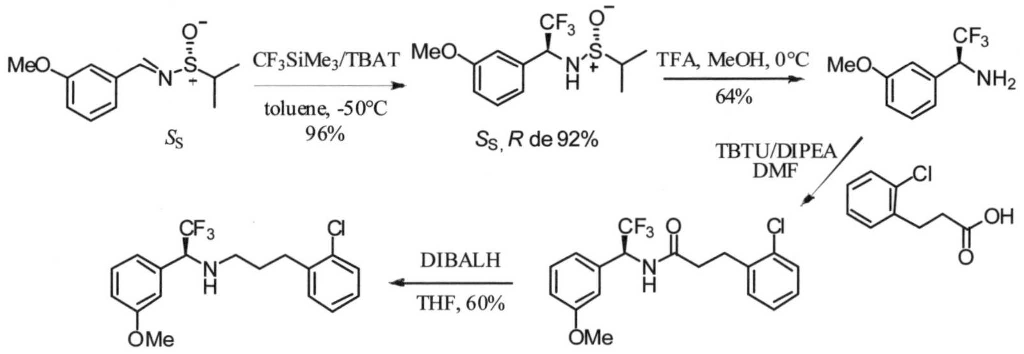

Scheme n. 19

The latest innovations reported by Davis concern the addition of Weinreb amide enolates to a suitably functionalized aldimine, with the final aim to synthesize alkaloids (+)-Euphococcine and (-)-Adaline (scheme n. 20).[32a,b] Even if the diastereomeric products of the addition are inseparable, they are obtained with very high diastereoselection, above of all in the precursor of Adaline; after treatment of the obtained N-sulfinyl-β-amido Weinreb amido ketal with 5 equivalents of methyl magnesium bromide and heating of the corresponding ketone with the buffer solution NH$_4$OAc:HOAc, a Mannich cyclization/cascade reaction gave the desidered homotropinone alkaloids in excellent yields.

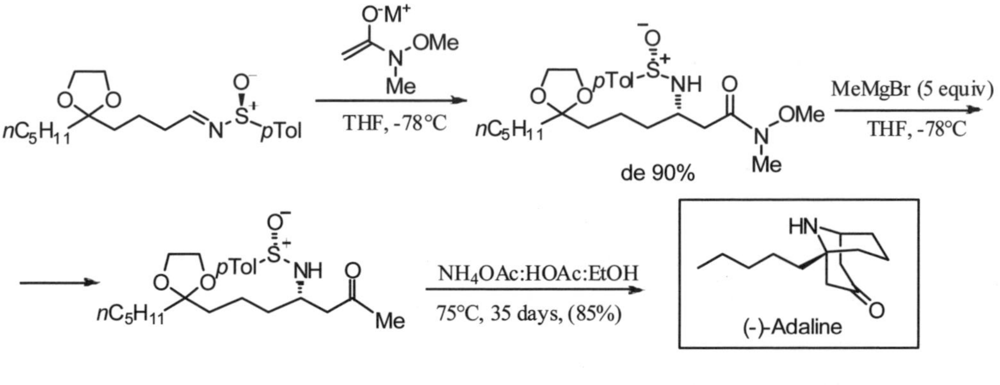

Scheme n. 20

Other enolate additions to these very useful synthetic intermediates are the key step for the final formation of Cocaine[32c] e N-tosyl aminoacids.[32d] A new method for desulfurizing N-sulfinamines and derivatives is worth mentioning: instead of the acid treatment, it exploits photolysis, without any racemisation.[32e]

The recent works by Ellman and coll. seem to refine this branch of organic chemistry too. If, from a synthetic point of view, a new mild synthetic strategy, to N-*t*-butanesulfinyl ketimmine has been reported,[33a] the new Rh(I) catalyzed alkenylation by means of MIDA boronates to Ellman sulfinimines have proven to happen with high yields and stereoselectivities.[33b] Starting from a commercially available aryl bromide, the synthetic route that leads to enantiopure (-)-Aurantioclavine, passing through the intermediation of a *t*-butane sulfinimine, is depicted in scheme n. 21. This method has been applied in the context of other natural products syntheses.[33c]

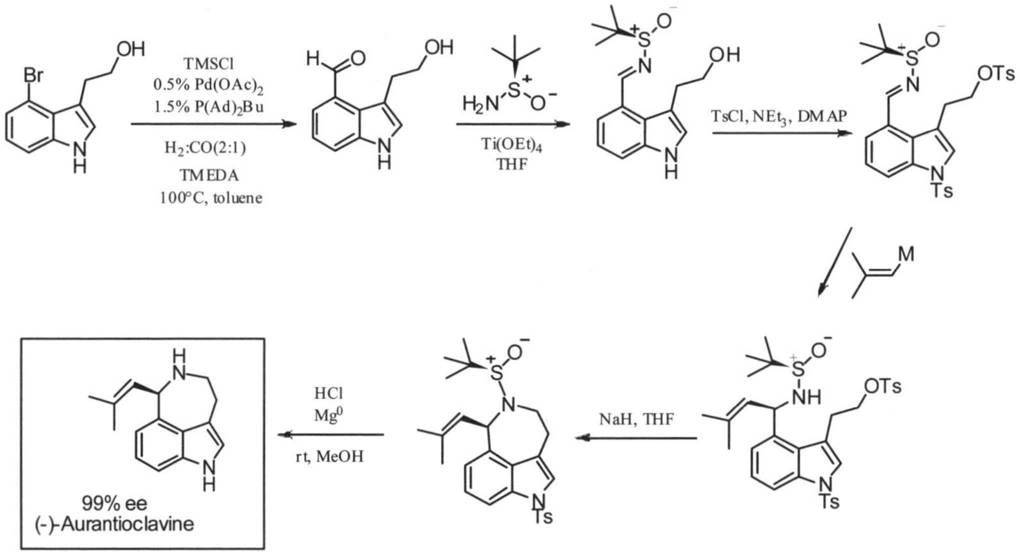

Scheme n. 21

In an interesting work recently reported by Da Kimpe and coll.[34] the different reactivity of N-sulfinyl imidates with respect to sulfinimine has been evidenced: the mobility of the α-hydrogen have been exploited for the formation of a carbanionic species that gave the *in situ* addition to sulfonylaldimines with the final formation, with good to excellent diastereomeric excesses, of new chiral β-sulfonylamino sulfinylimidates, new chiral β-aminoesters equivalents (scheme n. 22).

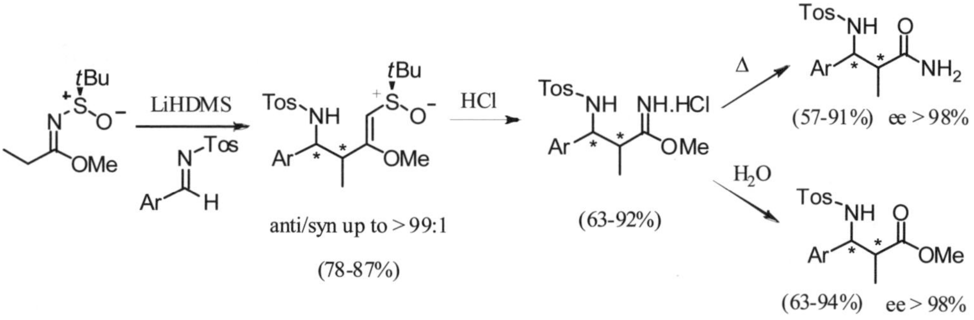

Scheme n. 22

An attempt of rationalization leading to the favored anti diastereomes is proposed (figure 1).

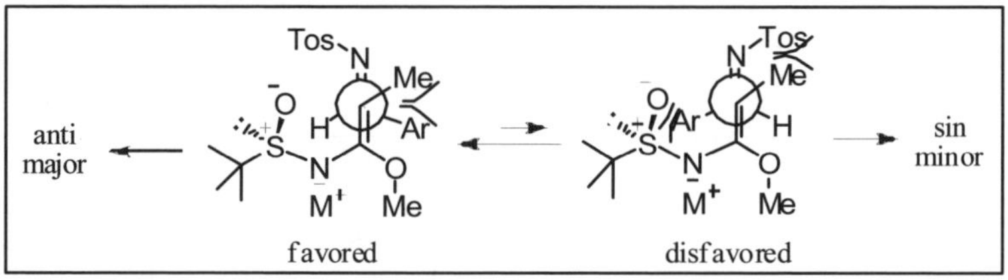

Figure n. 1

Selective reduction of C=N double bond of suitable sulfinimine[35] gives the formation of substituted nucleophilic N-substituted N-sulfinyl amines (or sulfinamides), whose intramolecular nucleophilic cyclization gives the stereoselective formation of naturally occurring five- and six-membered N-heterocycles.[35b-d]

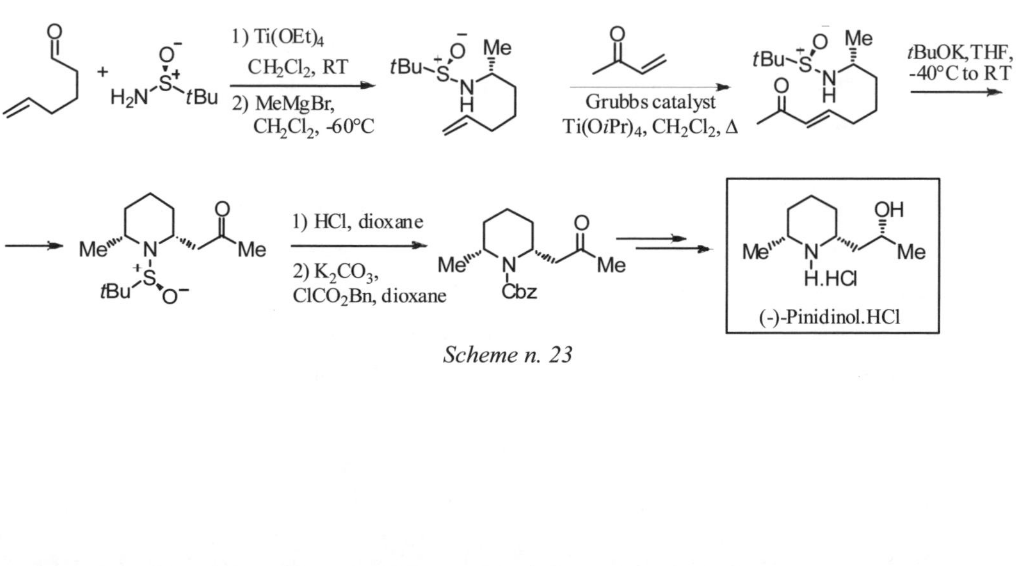

Scheme n. 23

A convenient application of intramolecular aza-Michael addition of this reaction intermediates is given by Fustero and coll.[35d] and shown in scheme n. 23; 2,6-disubstituted piperidines are formed with very high diastereocontrol, representing the key intermediates in the total syntesis of the piperidine alkaloids like (-)-Pinidinol.

A nice example of utilization of enantiopure C_2-symmetric bis-sulfinamides, whose syntheses consist of two simple steps including the above cited reduction, as organocatalysts in the allylation of acylhydrazones (scheme n. 24) is given by Khiar and coll.[35e] The preliminary studies done in order to rationalize the reaction mechanism confirm that a dicoordinate transition state seems to be responsible of the observed high enantioselectivity of the reactions.

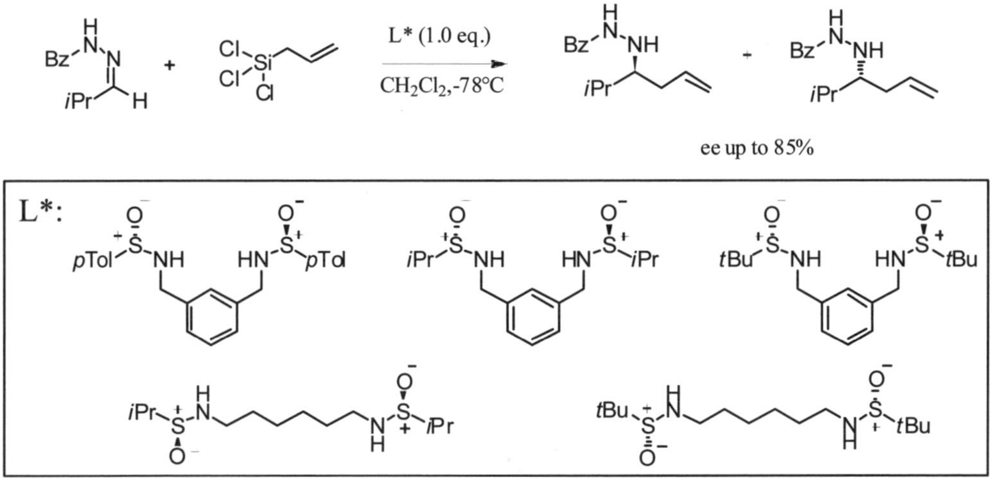

ee up to 85%

Scheme n. 24

Since chiral aziridines are key substrates in the synthesis of many naturally occurring nitrogen containing molecules, as alkaloids, amino acids etc, efficient synthesis of the *N*-sulfinyl ones and their use in organic routes have received increasing attention in the recent years. Starting from sulfimines, efficient synteses of this cyclic compounds have been realized.

Two nice examples come from De Kimpe[36] and García Ruano.[36b] While in the first case an efficient and almost completely diastereoselective method has been developed exploiting the 1,2 addition of a Grignard reagent across an α-chloro-N-*t*-butanesulfinyl ketimine and subsequent ring closure, the second method involves the stereoselective aziridation of N-sulfinylimines by means of 2-*p*-tolylsulfinyl benzyl iodide in the presence of hexamethyl disilazide (scheme n. 25), exploiting the double asymmetric induction exerted by both the sulfinyl groups. Treatment of the obtained sulfinylaziridine with 2.2 equivalents of *t*-BuLi gives the formation of totally desufurated aziridine without affecting the enantiopurity coming from the starting product.

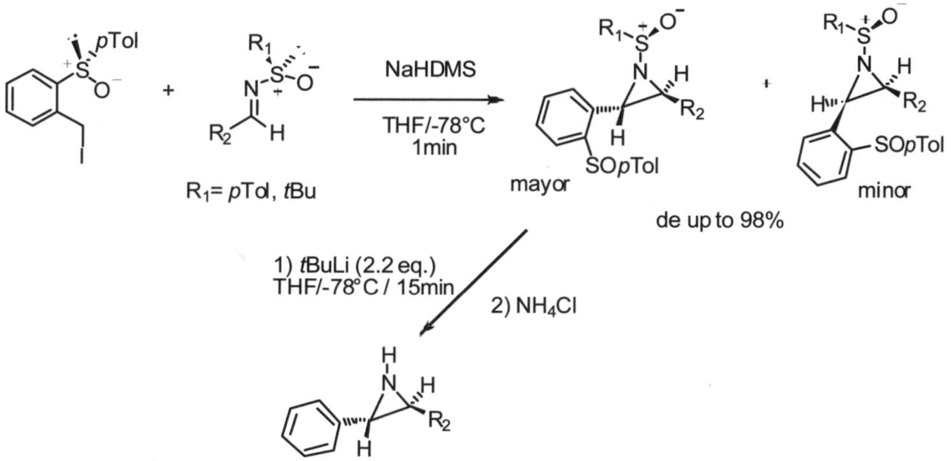

Scheme n. 25

Aziridation of cyclic α-bromoenones has been achieved by the use of lithium salt of (S_S)-(+)-*p*-toluenesulfinamide, with moderate to good diastereomeric excesses.[36c] The chemistry of sulfinimines and sulfinyl aziridine have finally been employed in a straightforward route to 3-aminooxaethanes, opening the way to the synthesis of target molecules containing this four-membered ring, useful in drug discovery.[37]

REFERENCES

1. a) Carreño, M. C.; Hernández-Torres, G.; Ribagorda, M.; Urbano, A. *Chem. Commun.* **2009**, 6129; b) Fernández, I.; Khiar, N. *Chem. Rev.* **2003**, *103*, 3651 and ref. cited therein.
2. a) Liu, F.; Fu, Z.; Liu, Y.; Lu, C.; Wu, Y.; Xie, F.; Ye, Z.; Zhou, X.; Yin, D. *Ind. Eng. Chem. Res.* **2010**, *49*, 2533; b) Cojocariu, A. M.; Mutin, P. H.; Dumitriu, E.; Fajula, F.; Vioux, A.; Hulea, V. *Appl. Cat., B* **2010**, *97*, 407; c) Cojocariu, A. M.; Mutin, P. H.; Dumitriu, E.; Aboulaich, A.; Vioux, A.; Fajula, F.; Hulea, V. *Catal. Today* **2010**, *157*, 270.
3. a) Wojaczyńska, E; Wojaczyński, J. *Chem. Rev.* **2010**, *110*, 4303; b) Stingl, K. A.; Tsogoeva, S. B. *Tetrahedron: Asymmetry* **2010**, *21*, 1055; c) Panda, M. K.; Shaikh, M. M.; Ghosh, P. *Dalton Trans.* **2010**, *39*, 2428; d) Tanaka, H.; Nishikawa, H.; Uchida, T.; Katsuki, T. *J. Am. Chem. Soc.* **2010**, *132*, 12034; e) Rioz-Martínez, A.; de Gonzalo, G.; Torres Pazmiño, D. E.; Fraaije, M. W.; Gotor, V. *Eur. J. Org. Chem.* **2010**, 6409; f) Rioz-Martínez, A.; Bisogno, F. R.; Rodríguez, C.; de Gonzalo, G.; Lavandera, I.; Torres Pazmiño, D. E.; Fraaije, M. W.; Gotor *Org. Biomol. Chem.* **2010**, *8*, 1431; g) Dybtsev, D. N.; Yutkin M. P.; Samsonenko, D. G.; Fedin, V. P.;

Nuzhdin, A. L.; Bezrukov, A. A.; Bryliakov, K. B.; Talsi, E. P.; Belosludov, R. V.; Mizuseki, H.; Kawazoe, Y.; Subbotin, O. S.; Belosludov, V. R. *Chem. - Eur. J.* **2010**, *16*, 10348; h) Chen, L.; Yang, Y.; Jiang, D. *J. Am. Chem. Soc.* **2010**, *132*, 9138; i) Moorthy, J. N.; Senapati, K.; Parida, K. N. *J. Org. Chem.* **2010**, *75*, 8416; j) Wei, S.; Stingl, K. A.; Weiß, K. H.; Tsogoeva, S. B. *Synthesis* **2010**, 707; k) Marsh, B. J.; Carbery, D. R. *Tetrahedron Lett.* **2010**, *51*, 2362.

4.   Satoh, T.; Momochi, H.; Noguchi, T. *Tetrahedron: Asymmetry* **2010**, *21*, 382.

5.   a) Cabrita, I.; Sousa, S. C. A.; Fernandes, A. C. *Tetrahedron Lett.* **2010**, *51*, 6132; b) Korang, J.; Grither, W. R.; McCulla, R. D. *J. Am. Chem. Soc.* **2010**, *132*, 4466.

6.   a) Appel, R.; Mayr, H. *Chem. - Eur. J.* **2010**, *16*, 8610; b) Midura, W. H.; Cypryk, M. *Tetrahedron: Asymmetry* **2010**, *21*, 177; c) Ohmori, K.; Yano, T.; Suzuki, K. *Org. Biomol. Chem.* **2010**, *8*, 2693; d) Caron, P.-Y.; Deslongchamps, P. *Org. Lett.* **2010**, *12*, 508; e) Komsta, Z.; Barbasiewicz, M.; Mąkoska, M. *J. Org. Chem.* **2010**, *75*, 3251; f) Vogel, S.; Bantreil, X.; Maitro, G.; Prestat, G.; Madec, D.; Poli, G. *Tetrahedron Lett.* **2010**, *51*, 1459; g) Lecea, M.; Hernández-Torres, G.; Urbano, A.; Carreño, M. C.; Colobert, F. *Org. Lett.* **2010**, *12*, 580; h) Rodrigues, A.; Wladislaw, B.; De Vitta, C.; Pandini Cardoso Filho, J. E.; Marzorati, L.; Alves Bueno, M.; Olivato, P. R. *Tetrahedron Lett.* **2010**, *51*, 5344; i) Chooprayoon, S.; Kuhakarn, C.; Tuchinda, P.; Reutrakul, V.; Pohmakotr, M. *Org. Biomol. Chem.* **2011**, *9*, 531.

7.   a) *The Chemistry of sulfenic acids and derivatives*, Patai S.; *Ed. John Wiley and Sons*, Chichester, **1990**; b) Aversa, M. C.; Barattucci, A.; Bonaccorsi, P.; Giannetto, P. *Curr. Org. Chem.* **2007**, *11*, 1034; c) Zhang, Y; Wu, Y. *Chin. J. Chem.* **2010**, *28*, 1635; d) Kambutong, S.; Kuhakarm, C.; Tuchinda, P.; Pohmakotr, M. S. *Synthesis* **2010**, 143; e) Trost, B. M.; Nguyen, H. M.; Koadin, C. *Tetrahedron Lett.* **2010**, *51*, 6232; f) Bari, S. S.; Arora, R.; Balla, A.; Venugopalan, P. *Tetrahedron Lett.* **2010**, *51*, 1719; g) Lu, X.; Long. T. E. *J. Org. Chem.* **2010**, *75*, 249; h) O'Byrne, A.; Murray, C.; Keegan, D.; Palacio, C.; Evans, P.; Morgan, B. S. *Org. Biomol. Chem.* **2010**, *8*, 539; i) Maitro, G.; Prestat, G.; Madec, D.; Poli, G. *Tetrahedron: Asymmetry* **2010**, *21*, 1075; j) McGrath, A. J.; Garrett, G. E.; Valgimigli, L.; Pratt, D. A. *J. Am. Chem. Soc.* **2010**, *132*, 16759; k) Singh, S. P.; O'Donnell, J. S.; Schwan, A. L. *Org. Biomol. Chem.* **2010**, *8*, 1712; l) Aversa, M. C.; Barattucci, A.; Bonaccorsi, P. *Eur. J. Org. Chem.* **2009**, 6355; m) Aversa, M. C.; Barattucci, A.; Bonaccorsi, P. *Synlett* **2011**, 254.

8.   a) García Ruano, J. L.; Schöpping, C.; Alvarado, C.; Aleman, J. *Chem. - Eur. J.* **2010**, *16*, 8968; b) García Ruano, J. L.; Martín-Castro, A. M.; Tato, F.; Torrente, E.; Tocco, M. G.; Florio, S.; Capriati, V. *Tetrahedron* **2010**, *66*, 1581; c) Nuñez, A. jr.; Martin, M. R.; Fraile, A.; García Ruano, J. L. *Chem. - Eur. J.* **2010**, *16*, 5443.

9.   a) Bernoud, E.; Le Duc, G.; Bantreil, X.; Prestat, G.; Madec, D.; Poli, G. *Org. Lett.* **2010**, *12*, 320; b) Miura, M.; Toriyama, M.; Kawakubo, T.; Yasukawa, K.; Takido, T.; Motohashi, S. *Org. Lett.* **2010**, *12*, 3882; c) Kissane, M.; Lynch, D.; Chopra, J.; Lawrence, S. E.; Maguire, A. R. *Org. Biomol. Chem.* **2010**, *8*, 5602; d) Fernández de la Pradilla, R.; Tortosa, M.; Castellanos, E.; Viso, A.; Baile, R. *J. Org. Chem.* **2010**, *75*, 1517.

10. a) Gamba-Sanchez, D.; Prunet, J. *J. Org. Chem.* **2010**, *75*, 3129; b) Krysiak, J.; Midura, W. H.; Wieczorek, W.; Sierón, L.; Mikołajczyk, M. *Tetrahedron: Asymmetry* **2010**, *21*, 1486; c) Satoh, T.; Kuramoto, T.; Ogata, S.; Watanabe, H.; Saitou, T.; Tadokoro, M. *Tetrahedron: Aymmetry* **2010**, *21*, 1; d) Watanabe, H.; Ogata, S.; Satoh, T. *Tetrahedron* **2010**, *66*, 5675; e) Mancha, G.; Cuenca, A. B.; Rodríguez, N.; Medio-Simón M.; Asensio, G. *Tetrahedron* **2010**, *66*, 6901; f) Sokolenko, L. V.; Maletina, I. I.; Yagupolskii, Y. L. *Synlett* **2010**, 2075.

11. a) Calligaris, M.; Carugo, O. *Coord. Chem. Rev.* **1996**, *153*, 83; b) Chen, J.; Chen, J.; Lang, F.; Zhang, X.; Cun, L.; Zhu, J.; Deng, J.; Liao, J. *J. Am. Chem. Soc.* **2010**, *132*, 4552; c) Chen, Q.-A.; Dong, X.; Chen, M.-W.; Wang, D.-S.; Zhou, Y.-G.; Li, Y.-X. *Org. Lett.* **2010**, *12*, 1928; d) Qi, X.; Rice, G. T.; Lall, M. S.; Plummer, M. S.; White, M. C. *Tetrahedron* **2010**, *66*, 4816; e) Martinez Mallorquin, R.; Chelli, S.; Brebion, F.; Fensterbank, L.; Goddard, J.-P.; Malacria, M. *Tetrahedron: Asymmetry* **2010**, *21*, 1695; f) De Sio, V.; Massa, A.; Scettri, A. *Org. Biomol.*

Chem. **2010**, *8*, 3055; g) Massa, A.; Acocella, M. R.; De Sio, V.; Villano, R.; Scettri, A. *Tetrahedron: Asymmetry* **2009**, *20*, 202.

12. El-Awa, A.; Noshi, M. N.; Mollat du Jourdin, X.; Fuchs, P. L. *Chem. Rev.* **2009,** *109,* 2315 and ref. cit. therein.

13. a) Karmee, S. K.; Greiner, L.; Kraynov, A.; Müller, T. E.; Niemeijer, B.; Leitner, W. *Chem. Commun.* **2010**, *46*, 6705; b) Jain, S. M.; Rana, B. R.; Singh, B.; Sinha, A. K.; Bhaumik, A.; Nandi, M.; Sain, B. *Green Chem.* **2010**, *12*, 374.

14. a) Katrun, P.; Chiampanichayakul, S.; Korworapan, K.; Pohmakotr, M.; Reutrakul, V.; Jaipetch, T., Kuhakarn, C. *Eur. J. Org. Chem.* **2010**, 5633; b) Flynn, C. J.; Elcoate, C. J.; Lawrence. S. E.; Maguire, A. R. *J. Am. Chem. Soc.* **2010**, *132*, 1184; c) Tsui, G. C.; Glenadel, Q.; Lau, C.; Lautens, M. *Org. Lett.* **2011**, *13*, 208.

15. a) Zajc, B.; Kumar, R. *Synthesis* **2010**, 1822; b) Zhao, Y.; Huang, W.; Zhu, L.; Hu, J. *Org. Lett.* **2010**, *12*, 1444; c) Calata, C.; Pfund, E.; Lequeux, T. *Tetrahedron* **2011**, *67*, 1398; d) Allendörfer, N.; Es-Sayed, M.; Nieger, M.; Bräse, S. *Synthesis* **2010**, 3439; e) Puget, B.; Jahn, U. *Synthesis* **2010**, 2579; f) Wu, F.-L.; Ross, B. P.; McGeary, R. P. *Eur. J. Org. Chem.* **2010**, 1989.

16. Bray, C. D.; de Faveri, G. *J. Org. Chem.* **2010**, *75*, 4652.

17. a) Zhao, Y.; Zhu, J.; Ni, C.; Hu, J. *Synthesis* **2010**, 1899; b) Prakash, G. K. S.; Wang, Y.; Mogi, R.; Hu, J.; Mathew, T.; Olah, G. A. *Org. Lett.* **2010**, *12*, 2932.

18. a) Robles-Machín R.; González-Esguevillas, M.; Adrio, J.; Carretero, J. C. *J. Org. Chem.* **2010**, *75*, 233; b) Robles-Machín R.; Lopéz-Peréz, A.; González-Esguevillas, M.; Adrio, J.; Carretero, J. C. *Chem. - Eur. J.* **2010**, *16*, 9864; c) Chandanshvile, J. Z.; Bonini, B. F.; Gentili, D.; Fochi, M. F.; Bernardi, L.; Comes Franchini, M. *Eur. J. Org. Chem.* **2010**, 6440.

19. a) Bos, P. H.; Macía, B.; Fernández-Ibańez, M. A.; Minnaard, A. J.; Feringa, B. L. *Org. Biomol. Chem.* **2010**, *8*, 47; b) Atta, A. K.; Pathak, T. *Eur. J. Org. Chem.* **2010**, 872.

20. a) Evans, P.; Leffray, M. *Tetrahedron* **2003**, *40*, 7973; b) McLaughlin, N. P.; Evans, P. *J. Org. Chem.* **2010**, *75*, 518; c) Au, C. W. G.; Nash, R. J.; Pyne, S. G. *Chem. Commun.* **2010**, *46*, 713.

21. Bonaparte, A. C.; Betush, M. P.; Panseri, B. M.; Mastarone, D. J.; Murphy, R. K.; Murphee, S. S. *Org. Lett.* **2011**, *6*, 1447.

22. a) Nielsen, M.; Borch Jacobsen, C.; Holub, N.; Weber Paixão, M.; Jørgensen, K. A. *Angew. Chem., Int. Ed.* **2010**, *49*, 2668; b) Alba, A.-N. R.; Companyó, X.; Rios, R. *Chem. Soc. Rev.* **2010**, *39*, 2018.

23 a) Weber Paixão, M.; Holub, N.; Vila, C.; Nielsen, M.; Jørgensen, K. A. *Angew. Chem., Int. Ed.* **2009**, *48*, 7338; b) Holub, N.; Jang, H.; Weber Paixão, M.; Tiberi, C.; Jørgensen, K. A. *Chem. Eur. J.* **2010**, *16*, 4337; c) Zweifel, T.; Nielsen, M.; Overgaard, J.; Borch Jacobsen, C.; Jørgensen, K. A. *Eur. J. Org. Chem.* **2011**, 47; d) Indumathi, S.; Peramul, S.; Menendez, J. C. *J. Org. Chem.* **2010**, 75, 472; e) Alemán, J.; Marcos, V.; Marzo, L.; García Ruano, J. L. *Eur. J. Org. Chem.* **2010**, 4482; f) Garcìa Ruano, J. L.; Marcos, V.; Alemán, J. *Chem. Commun.* **2009**, 4435; g) Yang, H.; Carter, R. G. *J. Org. Chem.* **2010**, *75*, 4929.

24. a) Bournaud, C.; Marchal, E.; Quintard, A.; Sulzer-Mossé, S.; Alexakis, A. *Tetrahedron: Asymmetry* **2010**, *21*, 1666; b) Xiao, J.; Liu, Y.-L.; Loh, T.-P. *Synlett* **2010**, 2029; c) Xiao, J.; Lu, Y.-P.; Liu, Y.-L.; Wong, P.-S.; Loh, T.-P. *Org. Lett.* **2011**, *13*, 876; d) Quintard, A.; Belot, S.; Marchal, E.; Alexakis, A. *Eur. J. Org. Chem.* **2010**, 927; e) Zhu, Q.; Lu, Y. *Chem. Commun.* **2010**, *46*, 2235; f) Zhu, Q.; Lu, Y. *Angew. Chem., Int. Ed.* **2010**, *49*, 7753; g) Moteki, S. A.; Xu, S.; Arimitsu, S.; Maruoka, K. *J. Am. Chem. Soc.* **2010**, *132*, 17074; h) Quintar, A.; Alexakis, A. *Chem. Commun.* **2010**, *46*, 4085.

25. a) González, P. B.; Lopez, R.; Palomo, C. *J. Org. Chem.* **2010**, *75*, 3920; b) Deiana, L.; Zhao, G.-L.; Dziedzic, P.; Rios, R.; Vesely, J.; Ekström, J.; Córdova, A. *Tetrahedron Lett.* **2010**, *51*, 234; c) Zhang, H.; Syed, S.; Barbas, C. F. III *Org. Lett.* **2010**, *12*, 708; d) Dubey, R.; Olenyuk, B. *Tetrahedron Lett.* **2010**, *51*, 608.

26. Syu, S.-E.; Kao, T.-T.; Lin, W. *Tetrahedron* **2010**, *66*, 891.

27. Davis, F. A.; Friedman, A. J.; Kluger, E. W. *J. Am. Chem. Soc.* **1974**, *96*, 5000.

28. a) Davis, F. A. *J. Org. Chem.* **2006**, 71, 8993; b) Ferreira, F.; Botuha, C.; Chelma, F.; Pérez-Luna, A. *Chem. Soc. Rev.* **2009**, *38*, 1162; c) Robak, M. A.; Herbage, M. A.; Ellmann, J. A. *Chem. Rev.* **2010**, *110*, 3600 and ref. cit. therein.

29. a) Weng, J.; Li, Y.-B.; Wang, R.-B.; Li, F.-Q.; Liu, C.; Chan, A. S. C.; Lu, G. *J. Org. Chem.* **2010**, *75*, 3125; b) Raghavan, S.; Krishnaiah, V. *J. Org. Chem.* **2010**, *75*, 748; c) Raghavan, S.; Krishnaiah, V.; Sridhar, B. *J. Org. Chem.* **2010**, *75*, 498; d) Ishikawa, S.; Noguchi, F.; Kaminura, A. *J. Org. Chem.* **2010**, *75*, 3578; e) Reddy, L. R.; Prashad, M. *Chem. Commun.* **2010**, *46*, 222; f) Chen, B.-L.; Wang, B.; Lin, G.-Q. *J. Org. Chem.* **2010**, *75*, 941; g) García, D.; Foubelo, F.; Yus, M. *Eur. J. Org. Chem.* **2010**, 2893; h) Almansa, R.; Collados, J. F.; Guijarro, D.; Yus, M. *Tetrahedron: Asymmetry* **2010**, *21*, 1421; i) Shen, A.; Liu, M.; Jia, Z.-S.; Xu, M.-H.; Lin, G.-Q. *Org. Lett.* **2010**, *22*, 5154; j) Liu, M.; Shen, A.; Sun, X.-W.; Deng, F.; Xu, M.-H.; Lin, G.-H. *Chem. Commun.* **2010**, 8460; k) Fandrick, D. R.; Johnson, C. S.; Fandrick, K. R.; Reeves, J. T.; Tan, Z.; Lee, H.; Song, J. J.; Yee, N. K.; Senayake, C. H. *Org. Lett.* **2010**, *12*, 748.

30. a) Hernández, D.; Nielsen, L.; Lindsay, K. B.; López-García, M. A.; Bjerglund, K.; Skrydstrup, T. *Org. Lett.* **2010**, *12*, 3528; b) Hernández, D.; Lindsay, K. B.; Nielsen, L.; Mittag, T.; Bjerglund, K.; Friis, S.; Mose, R.; Skrydstrup, T. J. *J. Org. Chem.* **2010**, *75*, 3283.

31. a) Fernández, I.; Valdivia, V.; Alcudia, A.; Chelouan, A.; Khiar, N. *Eur. J. Org. Chem.* **2010**, 1502; b) Fustero, S.; Moscardó, J.; Sánchez-Roselló, M.; Rodríguez, E.; Barrio, P. *Org. Lett.* **2010**, *12*, 5494; c) Mimura, H.; Kawada, K.; Yamashita, T.; Sakamoto, T.; Kikugawa, Y. *J. Fluorine Chem.* **2010**, *131*, 477; d) Zhang, F.; Liu, Z.-J.; Liu, J.-T. *Tetrahedron* **2010**, *66*, 6864.

32. a) Davis, F. A.; Edupuganti, R. *Org. Lett.* **2010**, *12*, 848; b) Davis, F. A.; Theddu, N. *J. Org. Chem.* **2010**, *75*, 3814; c) Davis, F. A.; Theddu, N. J.; Edupuganti, R. *Org. Lett.* **2010**, *12*, 4118; d) Davis, F. A.; Ramachandar, T.; Chai, J.; Qiu, H. *Arkivoc* **2010**, *8*, 17; e) Davis, F. A.; Ramachandar, T.; Zhang, Y.; Chai, J.; Qiu, H.; Deng, J.; Velvadapu, V. *Tetrahedron Lett.* **2010**, *51*, 4042.

33. a) Datta., G. K.; Ellman, J. A. *J .Org. Chem.* **2010**, *75*, 6283; b) Brak, K.; Ellman, J. A. *J. Org. Chem.* **2010**, *75*, 3147; c) Brak, K.; Ellman, J. A. *Org. Lett.* **2010**, *12*, 2004.

34. Colpaert, F.; Mangelinckx, S.; De Kimpe, N. *Org. Lett.* **2010**, *12*, 1904.

35. a) Guijarro, D.; Pablo, O.; Yus, M. *J. Org. Chem.* **2010**, *75*, 5265; b) Leemans, E.; Mangelinckx, S.; De Kimpe, N. *Chem. Commun.* **2010**, 3122; c) Reddy, L. R.; Das, S. G.; Liu, Y.; Prashad, M. *J. Org. Chem.* **2010**, *75*, 2236; d) Fustero, S.; Monteagudo, S.; Sánchez-Roselló, M.; Flores, S.; Barrio, P.; del Pozo, C. *Chem. - Eur. J.* **2010**, *16*, 9835; e) Fernández, I., Alcudia, A.; Gori, B.; Valdivia, V.; Recio, R.; García, M. V.; Khiar, N. *Org. Biomol. Chem.* **2010**, *8*, 4388.

36. a) Colpaert, F.; Mangelinckx, S.; Leemans, E.; Denolf, B.; De Kimpe, N. *Org. Biomol. Chem.* **2010**, *8*, 3251; b) Arroyo, Y.; Meana, A.; Sanz-Tejedor, M. A.; Alonso, I.; García-Ruano, J. L. *Chem. - Eur. J.* **2010**, *16*, 9874; c) Bonifácio, V. D. B.; González-Bello, C.; Rzepa, H. S.; Prabhakar, S.; Lobo, A. M. *Synlett* **2010**, 145.

37. Hamzik, P. J.; Brubaker, J. S. *Org. Lett.* **2010**, *12*, 1116.

Luca Bernardi

Assistant Professor in Organic Chemistry

Address:

Department of Organic Chemistry "A. Mangini", Faculty of Industrial Chemistry, University of Bologna.

Education:

Luca Bernardi studied Industrial Chemistry at the University of Bologna (I) and at the University of Kent (UK), receiving his BSc (Distinction) and MSc (*cum laude*) in 1998 and 2000, respectively. He then carried out his PhD studies working in the groups of Prof. A. Ricci at the University of Bologna and Prof. K. A. Jørgensen, at the University of Aarhus (DK), which were completed in 2004. After a post-doctoral stay in Aarhus as fellow of the Danish National Research Foundation, he was back to Bologna, where he was appointed as "Ricercatore" (Assistant Professor) at the end of 2010.

Professional experience:

Luca Bernardi has been involved in a number of academic research projects, mainly focussed on the development of sustainable chemical processes. His research activity resulted in more than 40 publications in international chemistry journals, and was recognised by the C.I.N.M.P.I.S. award in 2004, the Thieme Journal Prize in 2009, and the Ciamician medal of the Italian Chemical Society in 2010.

Organocatalysis in the Asymmetric Synthesis of Chiral Compounds

Luca Bernardi

Department of Organic Chemistry "A. Mangini", Faculty of Industrial Chemistry, University of Bologna, Viale Risorgimento 4, 40136 Bologna (Italy); e-mail: luca.bernardi2@unibo.it

1. INTRODUCTION

Asymmetric organocatalysis is the use of organic molecules as catalysts in enantioselective transformations. The first example of organocatalytic (non-asymmetric) reaction stems from the middle of the XIX century, when Justus von Liebig used acetaldehyde to catalyse cyanamide hydrolysis.[1] Several groundbreaking reports describing asymmetric organocatalytic transformations appeared in last decades of the XX century. However, only in the last ten years has this synthetic technique surged to a recognised scientific field, often complementary to the more traditional catalytic asymmetric syntheses which make use of chiral organometallic complexes or enzymes/microorganisms. The sudden interest and development in asymmetric organocatalysis is generally ascribed to different reasons.[2] First, it was only in 2000 that MacMillan conceptualised and defined the field of organocatalysis, recognising that the use of organic molecules as catalysts for asymmetric synthesis is not an oddity but a promising and general approach.[3] This approach features several advantages such as moisture/oxygen tolerance and ready availability of catalysts and substrates. Second, several new modes of substrate activation by organic catalysts have been discovered since 2000, and, more importantly, generalised in a number of transformations. In this context, the (re)discovery by List, Lerner and Barbas III that secondary amines like proline combine reversibly with ketones/aldehydes to give nucleophilic enamines,[4] together with Jacobsen's disclosure of hydrogen-bond donor catalysts (like thioureas) for asymmetric nucleophilic additions to carbonyl compounds,[5] have been certainly major turning points.

The aim of this presentation is to give a focussed overview of some of the major progresses in the area of asymmetric organocatalysis, which appeared in the literature in the year 2010. Literature reports have been selected on the following rational:

i)   future directions (*i.e.* emerging activation modes and catalytic strategies).

ii)   outstanding achievements or realisation of long sought synthetic goals.

Significantly, the development of some of the most challenging and hitherto unrealised asymmetric transformations (*i.e.* point *ii)*), has often taken advantage of new emerging catalytic strategies (*i.e.* point *i*)). Being the two points closely intertwined, literature reports would not fit in a rigorous classification. Examples will be thus treated singularly, divided according to the chemical transformation object of study.

As a general consideration and as outlined in many of the examples described below, *cooperativity* has strongly emerged during 2010 as an essential concept in the development of new organocatalytic transformations. This powerful word features different facets: *i)* cooperativity is the use of a network of weak interactions between catalyst and substrates operating in concert, analogous with the operating mode of many natural enzymes. The deriving rigid geometrical organisation might result in outstanding stereochemical control even using weak interactions between catalyst and substrate, such as hydrogen bonds; *ii)* cooperativity is the merger of two (or more) catalytic cycles in the same transformation. Organocatalysis can be productively combined not only with a second organocatalytic system, but also with transition metal or Lewis acid catalysis, providing new opportunities for new reactivity; *iii)* cooperativity is the formation of a new catalytic species by the combination of different catalytic modules, which assemble by weak interactions. The result is a more rapid screening of the variation in the catalyst structure during the identification of the optimal catalyst system, as well as generation of new catalytic species avoiding tedious synthetic steps.

2. COOPERATIVE CATALYSIS. STRONG BRØNSTED ACIDS CAN BE TUNED BY COORDINATION WITH A WEAK ACID.

Activation of electrophilic substrates (*i.e.* imines) by coordination (or protonation) with a chiral Brønsted acid has become a broadly used and general approach in asymmetric catalysis.[6] More recently, counterion coordination of carbocation intermediates by (thio)urea catalysts, well known anion recognising motives, has also demonstrated great potential.[7] The combination of both concepts by Jacobsen has lead to an outstanding example of *cooperative catalysis* in the Povarov cycloaddition reaction (Scheme 1), wherein *a fast but non selective catalyst (an achiral strong Brønsted acid) is tuned by coordination with a weakly acidic chiral urea system*.[8] This tuning of an undesired fast but unselected activation (given by the strong acid forming an iminium ion) is highly remarkable. In fact, it is *not* based on the usual *increase* of substrate reactivity given by the chiral catalysts, but rather on the *sequestration* of a highly reactive species by effective coordination, preventing unwanted reactivity.

In particular, a screening of the combination of different strong Brønsted acids with chiral (thio)ureas, in the reaction between an *N*-phenyl imine and dihydrofuran DHF, identified *ortho*-nitrobenzenesulfonic acid NBSA and catalyst **1** as the most competent combination system (Scheme 1).

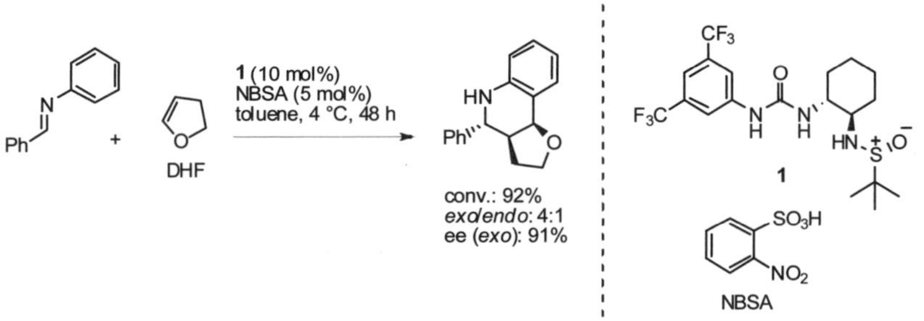

Scheme 1.

After having demonstrated the efficiency of the catalyst system for the Povarov reaction between a range of *N*-aryl imines and dienophiles, a detailed mechanistic study was undertaken. Kinetic studies, combined with ^{1}H NMR experiments, indicated quantitative protonation of imine by a strong Brønsted acid, such as TfOH (Scheme 2, *(a)*). This sparingly soluble (in apolar solvents such as C_6D_6) iminium ion undergoes the Povarov cycloaddition reaction with a rate constant k_{rac} of about 0.407 M^{-1} s^{-1}. In the presence of the chiral catalyst **1** a sharp decrease in reaction rate was surprisingly observed. High enantioselectivity cannot thus be explained by rate acceleration given by *e.g.* increase in charge separation by coordination of the urea **1** to the triflate. As a rate law given by two contributions (the racemic and the asymmetric reaction) needs to be considered (Scheme 2, *(b)*), high enantioselectivity can be observed only if the racemic contribution is suppressed. The concentration of the complex [imine·TfOH] must thus be negligible compared to [imine·**1**·TfOH]. ***This is the only way to suppress the racemic contribution.*** The prevalent formation of the enantiomer depicted in Scheme 2 can therefore be rationalised considering a very high efficient binding of the iminium ion by the urea catalyst **1**. Indeed, a high equilibrium constant (ca. 9000 M^{-1}) could be derived for the formation of this complex. Theoretical calculations as well as ^{1}H NMR experiments furnished additional support to its formation, wherein interactions with the formyl hydrogen seem to play a pivotal role. Calculations suggested a highly asynchronous, yet concerted, cycloaddition reaction with DHF occurring in the transition state depicted in Scheme 2 *(b)*. A ***highly organised hydrogen bond network operating cooperatively*** guarantees high stereoselectivity, as well as complex stability. A rapid deprotonation-rearomatisation step then follows, giving the tetrahydroquinoline product.

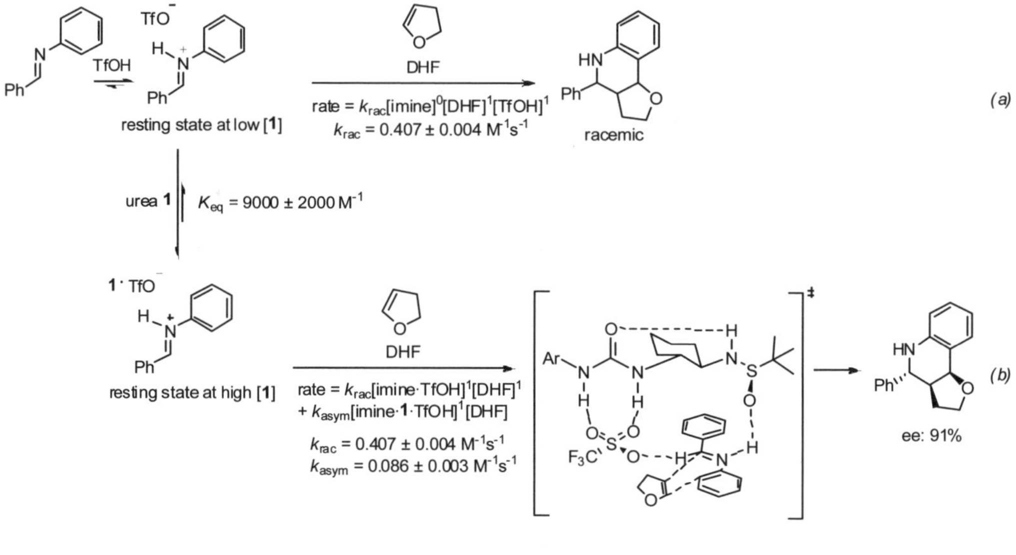

Scheme 2.

2. COOPERATIVE CATALYSIS. COMBINING LEWIS BASE AND LEWIS ACID FOR THE FIRST CATALYTIC ENANTIOSELECTIVE FLUORIDE ANION ADDITION.

The unique properties given by the incorporation of fluorine atom in organic molecules have spurred the invention of a number of methods leading to fluorine containing compounds. Most of these methods, which in many instances have a catalytic enantioselective counterpart, are based on the reaction of a nucleophilic substrate with an electrophilic "F$^+$" source. Less studied is the nucleophilic addition of fluoride anions. Fluoride anions are in fact poorly reactive in protic solvents, and have a very pronounced Brønsted base reactivity in non-polar reaction media, which coupled with low nucleophilicity promotes side reactions. The identification of a suitable fluoride anion source, giving the *first catalytic enantioselective fluoride anion addition*, was recently challenged by Kalow and Doyle. Their work capitalised the efficient generation of fluoride anion for silicon activation in the presence of Lewis bases,[9] as well as the epoxide opening with pyridine·9HF.[10] On these precedents, Kalow and Doyle considered the possibility of using chiral Lewis bases, in combination with benzoyl fluoride, for generating a fluoride anion able to open epoxides. Preliminary experiments indicated the feasibility of the proposed process for cyclohexene epoxides. However, application of chiral Lewis bases such as **2** did not result in measurable enantioselectivity, until some chiral Lewis acids complexes, ((salen)Co(II) **3** and (salen)Co(III)OTs, typically used in enantioselective epoxide ring openings), were applied to the system (Scheme 3).[11] Interestingly, a strong matched/mismatched effect between the stereochemistry of the Lewis acid and the Lewis base was observed, thus giving an exquisite example of *cooperative* *catalysis involving two chiral catalytic species*. The nature of the catalyst species involved and operative mechanism is still not clear. Several hypotheses accounting for the observed results can be considered. The generation of a (salen)Co(III) fluoride seems plausible, whereas the Lewis base might participate either as an axial ligand, either during epoxide activation, but also during fluoride delivery.

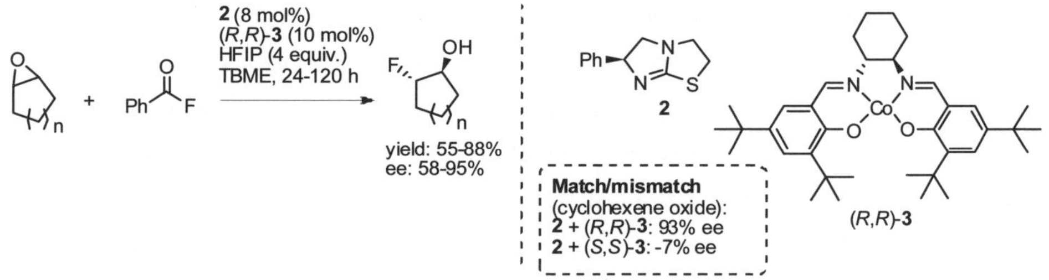

Scheme 3.

3. COOPERATIVE CATALYSIS. COMBINING ENAMINE CATALYSIS WITH ORGANO- AND METAL-CATALYSED PATHWAYS FOR α- AND γ-ALKYLATION OF ALDEHYDES WITH ELECTROPHILIC ALKYLATING AGENTS.

Since the (re)discovery by List, Lerner and Barbas III that secondary amines can be productively combined in a catalytic cycle with carbonyl compounds (aldehydes and ketones), giving nucleophilic enamine intermediates,[4] this catalytic strategy has received immense attention directed at encompassing different electrophiles.[12] Direct alkylation reactions of these catalytic enamine intermediates, with net substitution of a leaving group at a carbon atom, proved to be however a formidable challenge. The major obstacle to this transformation is certainly a *catalyst deactivation pathway, through irreversible N-alkylation*. Low reactivity of the alkylating agent towards the rather soft enamine nucleophiles, neutralisation of the acid resulting from the substitution reaction, as well as product racemisation, are also possible issues. The inherent unfeasibility of a direct catalytic asymmetric α-alkylation of aldehydes using most common alkylating agents, such as benzyl bromide, had already been circumvented before 2010 using a number of alternative strategies. Intramolecular examples, merging enamine and palladium catalytic cycles, combination of radical or cationic pathways have been the methods of choice to effect this powerful transformation.[13] In 2010, several reports appeared, implementing considerably previous methodologies. These reports can be classified according to the strategy employed.

3.1 Photoredox organocatalysis.

The catalytic enantioselective α-alkylation of aldehydes with electron deficient benzyl bromides[14] was developed in 2010 by MacMillan merging two catalytic cycles, the first involving the formation of radicals from the benzyl bromides through a ***photoredox process***, the second providing enamine intermediates from aldehydes and amine catalyst (Scheme 4). The perfect intertwine of the reduction potentials related to the different species involved in the two catalytic cycles is essential for the success of this approach. The photocatalytic cycle depicted in the scheme involves the formation of electron deficient intermediates through reduction. Light irradiation of the photoredox catalyst commercially available *fac*-Ir(III)(ppy)₃ (**4**) provides the necessary strong reductant species *fac*-*Ir(III)(ppy)₃ (**5**), featuring a suitable potential for the reduction of electron poor benzyl bromides (nitroaryl, pyridyl etc.). The resulting electron poor radicals can then be intercepted by the nucleophilic enamine **8**, formed through the concurrent organocatalytic cycle. The resulting radical amine **9** can then be oxidised to a hydrolysable iminium ion **10** by the reduced form **6** of the photocatalyst (*fac*-Ir(IV)(ppy)₃$^+$), acting as a powerful oxidant.

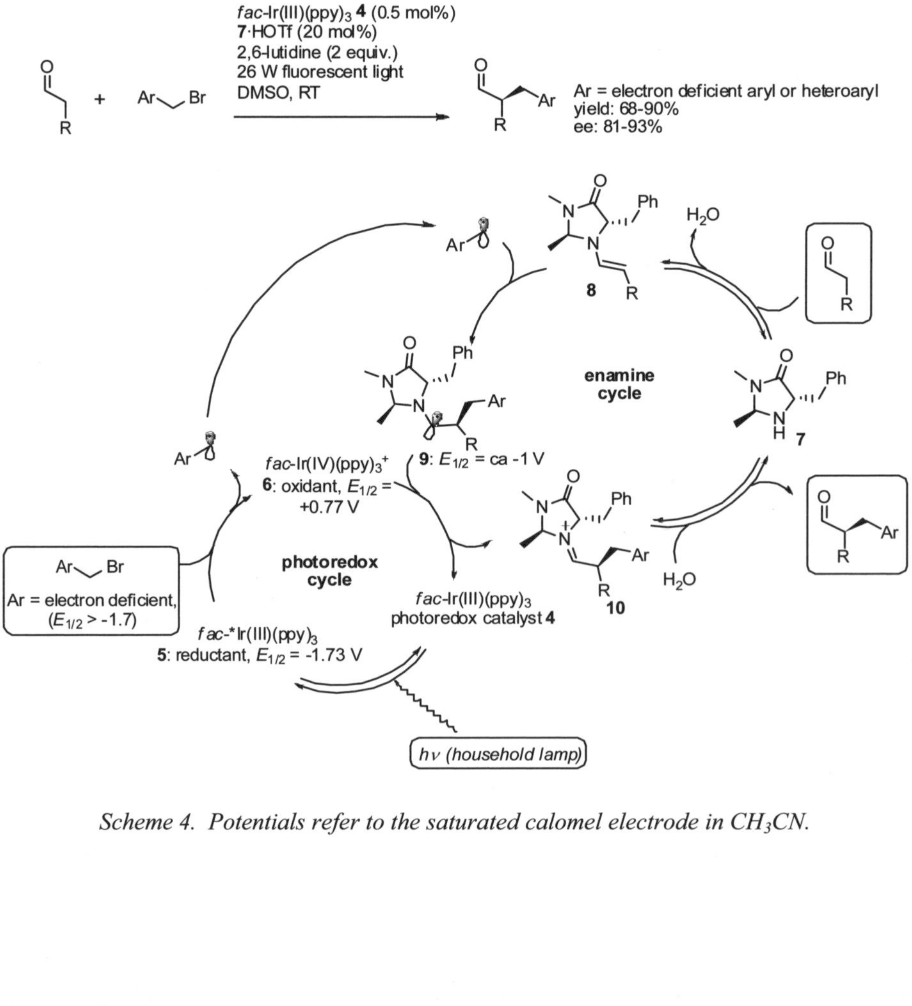

Scheme 4. Potentials refer to the saturated calomel electrode in CH₃CN.

The first realisation of the powerful merge between enamine and photoredox catalyses was reported by Nicewicz and MacMillan in 2008.[15] In that seminal report, the system was based on $Ru(bpy)_3^{2+}$ as photocatalyst, and allowed alkyation with only rather electron poor halides such as bromomalonates, due to the lower reduction potential of the oxidised form of the photocatalyst employed ($Ru(bpy)_3^+$, $E_{1/2}$ = -1.33 V). Later, a related alkylation with trifluoromethyliodide was developed,[16] wherein it was demonstrated that using a more strongly reductant photocatalyst ($Ir(ppy)_2(dtb-bpy)^+$) it was possible to extend these alkylation manifolds to a less reactive alkylating agent such as CF_3I. The report herein described extends very significantly the limit of the oxidation potential of the alkylating agents which can be used. Besides the use of a stronger reductant, key to success was the identification of catalyst **7**, furnishing a less hindered enamine compared to previously employed secondary amines. A less hindered enamine seemed to be necessary for also accommodating heteroarylmethyl radicals as reaction partners.

More recently, Zeitler demonstrated that readily available *organic dyes* can be used in this and in related chemistry,[17] in place of the more expensive and toxic ruthenium or iridium photocatalysts. Based on stability, reduction potential, as well as λ_{max} considerations, as series of red and orange organic dyes were tested, eventually leading to the identification of eosin Y as a very competent photoredox catalyst, able to substitute the more expensive $Ru(bpy)_3$ and $Ir(ppy)_2(dtb-bpy)^+$ catalysts also in MacMillan alkylation reactions, despite the lower reduction potential of its excited form (Figure 1).

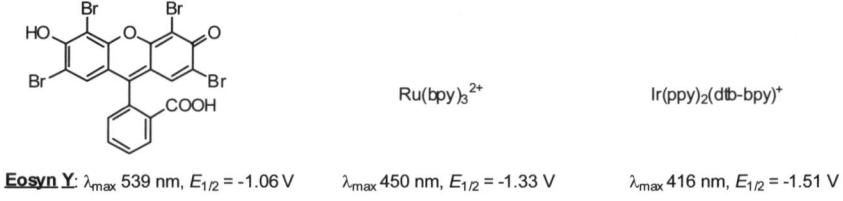

Eosyn Y: λ_{max} 539 nm, $E_{1/2}$ = -1.06 V λ_{max} 450 nm, $E_{1/2}$ = -1.33 V λ_{max} 416 nm, $E_{1/2}$ = -1.51 V

Figure 1. Potentials refer to the saturated calomel electrode in CH_3CN.

3.2 Alkylation via S_N1 pathways.

As predicted by Mayr's reactivity scale,[18] enamines should readily react with *stabilised carbocations*, such as benzhydryl. Furthermore, catalyst deactivation by N-alkylation might be reversible with these alkylating agents, due to the stability of the carbocation. In fact, few methods had been already developed in the last few years providing suitable procedures for the alkylation of enamines with stabilised carbocations.[13] Advances reported in 2010 are herein highlighted (Scheme 5).

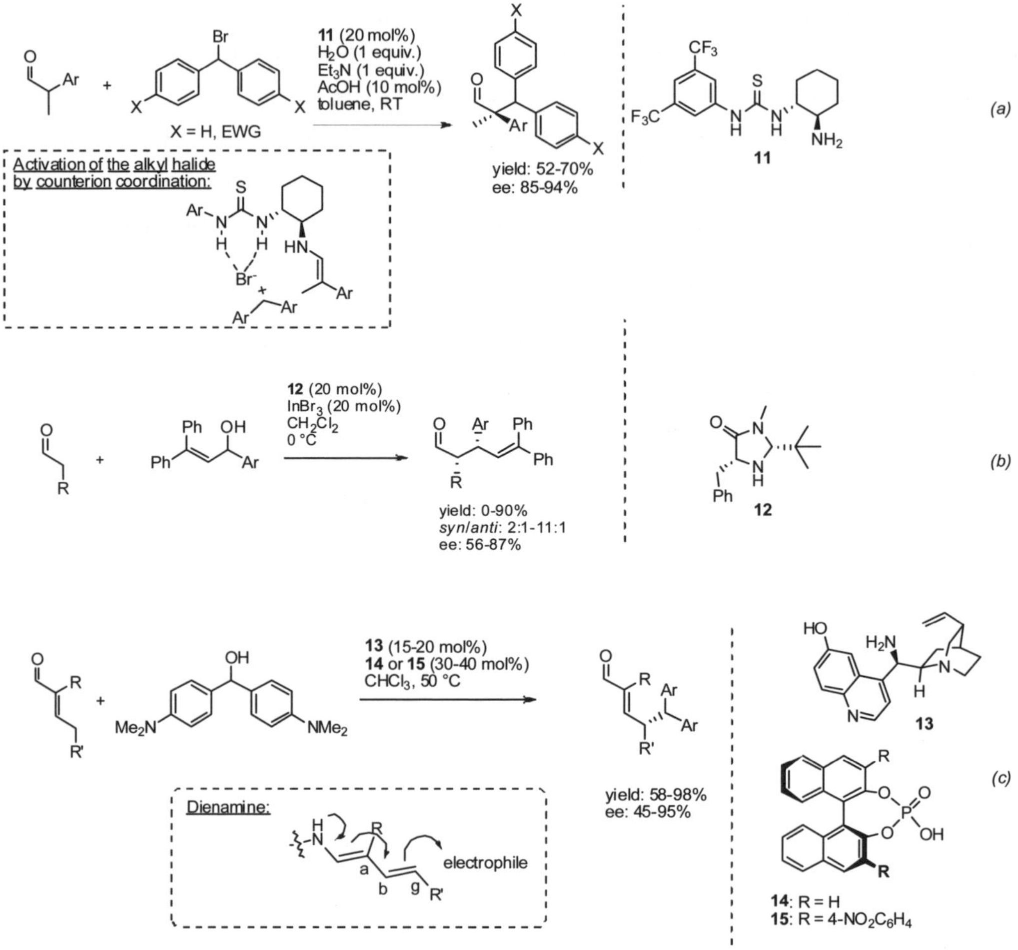

Scheme 5.

Jacobsen employed halodiphenylmethane as alkylating agents, in the reaction with α-aryl-propionaldehydes (Scheme 5, *(a)*).[19] The primary amine catalyst **11**, bearing a thiourea moiety, was found to be able to catalyse the reaction with very good efficiency. Kinetic isotope effects, as well as Hammett studies, provided evidence for a S_N1 reaction pathway, wherein the thiourea moiety assists in halide extraction by coordination. This report is thus an additional example of the ability of (thio)ureas in the coordination of the counterion of carbocation, which had been limited to α-oxo and α-amino stabilised carbocations so far. Cozzi used instead the combination of a Lewis acid (InBr$_3$),[20] for increasing the scope of their previously reported alkylation of stabilised carbocations with enamines

(Scheme 5, *(b)*). This Lewis acid can in fact assist carbocation formation even in the case of the less reactive allylic alcohols, unreactive under their previously reported conditions. Melchiorre's report on the γ-alkylation of α-branched enals (Scheme 5, *(c)*) features different interesting aspects.[21] It represents one of the few example of reactions with dienamine intermediates, providing new opportunities for the γ-functionalisation of carbonyl compounds. Besides, it gives a very good example of the utility of *combining two different chiral species in a cooperative manner*. Using a similar catalytic combination as the one developed by List for the epoxidation of the same α-branched enal substrates (a *Cinchona* alkaloid derived primary amine and a BINOL-derived phosphoric acid),[22] matched-mismatched effects depending on the stereochemistry of the two catalysts **13** and **14/15** were in fact observed. Alkylation of ketones using similar alkylating agents but a different catalytic system was also reported.[23]

3.3 *Propargylation with a ruthenium complex catalytic system:*

Nishibayashi described the **cooperative combination of an enamine catalytic cycle with a ruthenium complex catalytic system**, for the enantioselective propargylation of aldehydes.[24] Propargylic alcohols were used as alkylating agents, able to form a ruthenium allenylidene intermediate undergoing attack by the enamine (formed from Jørgensen's catalyst **16** and aldehyde) at its least hindered face (Scheme 6). Experiments carried out under stoichiometric conditions, as well as the lack of reactivity of internal propargylic alcohols, were taken as evidences for this allenylidene intermediate. This report considerably expands the utility of the cooperative combination of enamine catalysis with transition metal catalysed alkylation reactions, restricted to palladium catalysed allylations so far.[13]

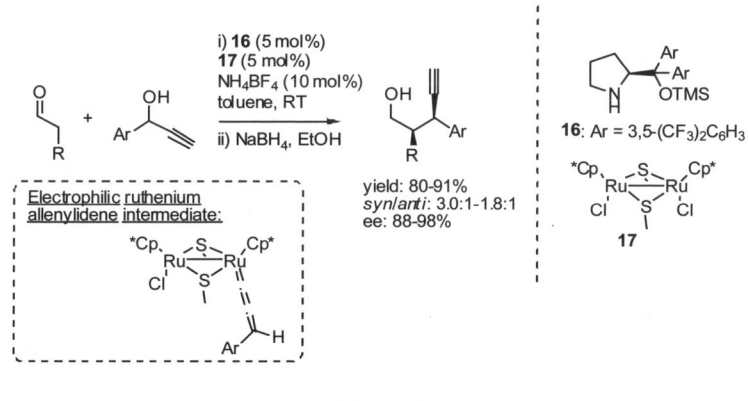

Scheme 6.

3.4 *Trifluoromethylation of aldehydes using hypervalent iodine reagent:*

Besides the trifluoromethylation of aldehydes using the merger of photocatalysis with enamine catalysis, Allen and MacMillan reported on the use of *Togni's trifluoromethylating reagent* for the same transformation. In this case the reaction proceeds through a close shell reaction pathway (Scheme 7).[25] Use of Lewis acids for the activation of the hypervalent iodine species was found to be critical, with CuCl outperforming other salts. Considering the usual reaction pathways operative in hypervalent iodine chemistry, the reaction is thought to occur through attack of the enamine at the positive iodine atom of the Lewis acid activated Togni's reagent, followed by reductive elimination with retention of stereochemistry. Addition of *tert*-amylalcohol was necessary in some cases, to prevent, by formation of the product hemiacetal, post-addition racemisation induced by the Lewis acid.

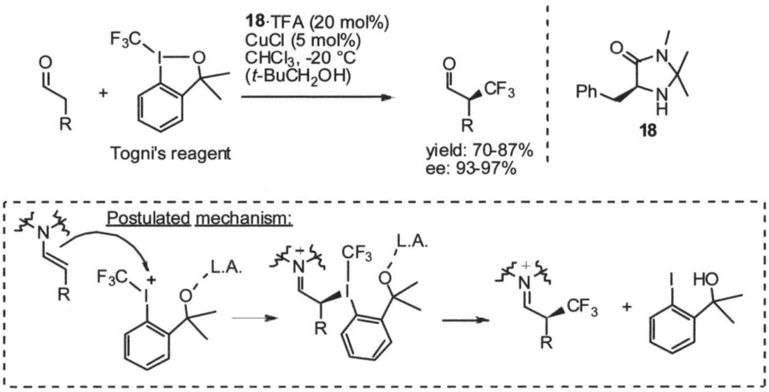

Scheme 7.

3.5 *Alkylation through S$_N$2' addition-elimination pathway:*

Reasoning that a **soft alkylating agent** should be able to alkylate enamines in preference of amine catalyst, Palomo employed α-halomethylacrylates in the reaction (Scheme 8).[26] The reaction between aldehydes and this alkylating reagent, in the presence of secondary amine catalysts, did not proceed. However, using DABCO or DMAP as the base, it was possible to obtain the desired alkylated aldehydes with excellent results. The lack of reactivity of these allylic halides in the absence of Lewis bases indicates the reaction to proceed through the first formation of an ammonium salt. DFT calcula-

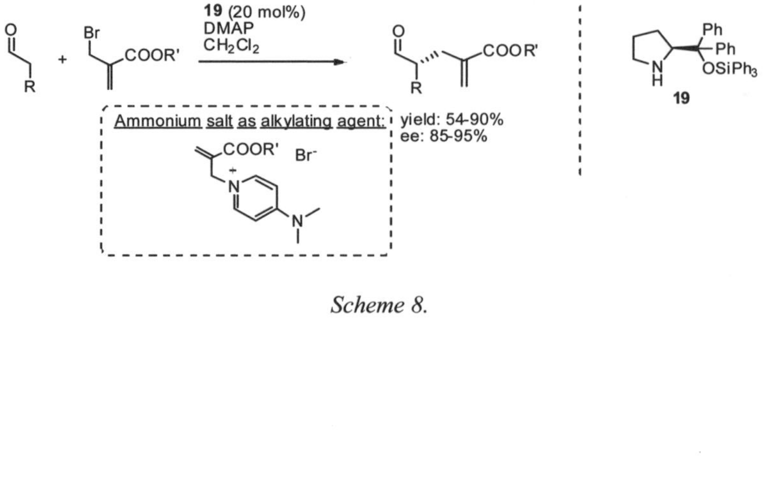

Scheme 8.

tions as well as control experiment on pre-formed ammonium salts confirmed this hypothesis. Attack of the enamine on this species in a S_N2' fashion follows, giving elimination of the Lewis base.

4. ENANTIOSELECTIVE HALOCYCLISATION REACTIONS. NEW SOLUTIONS TO AN OLD PROBLEM.

Intramolecular nucleophilic addition to an halonium activated double bond by an alcohol or a carboxylic acid (halolactonisation and haloetherification), or a nucleophile in general, is a direct and straightforward method for the preparation of (hetero)cyclic systems. Despite their tremendous synthetic utility, catalytic asymmetric versions of these transformations have proven to be extremely challenging. Only two examples (a iodocyclisation with malonate as nucleophile,[27] and a iodoetherification[28]) providing acceptable levels of enantioselectivities were known in the literature at the beginning of 2010, both dealing with chiral Lewis acids. The elegant policyclisation reported by Ishihara,[29] promoted by *stoichiometric* amounts of chiral phosphoramidite bases, also deserves to be mentioned. It has long been recognised that Lewis bases are able to catalyse these transformations through a halide-shuttle process,[30] besides the more classical Lewis acid activation of the halonium source. However, mechanistic studies by Brown[31] and Denmark[32] pointed out some major obstacles in the realisation of catalytic asymmetric halocyclisation reaction. Besides the need of fine tuning the reactivity of the halide source, to suppress background reactivity, bromonium or iodonium ion (but not chloronium) scrambling between two olefins, leading to racemisation, was found to occur at rates comparable to the irreversible intramolecular nucleophilic addition. Five reports finally appeared in 2010 demonstrating that careful choice of organocatalyst, halide source and reaction conditions could successfully lead to efficient organocatalytic halolactonisation reactions. These contributions represent also the first entries to efficient catalytic asymmetric *halolactonisation* reactions, as previous examples were related to different nucleophilic systems. It is worth noting that in these five contributions four distinct working models justifying reactivity and enantioselectivity were proposed, as described below.

4.1 *Brønsted acid activation of halonium source*:

The first report described a chlorolactonisation of 4-aryl-4-alkylidenebutyric acids, catalysed by the Cinchona alkaloid derivative (DHQD)$_2$PHAL **20**, and was reported by Borhan (Scheme 9 *(a)*).[33] Optimisation of the structure of chlorine source was found to be crucial to achieve excellent levels of enantiocontrol. The 1,3-dichlorohydantoin derivative depicted in the Scheme was identified as the optimum source, with intermediate reactivity between NBS (fast but unselective reaction) and NCS (not reactive at the low temperatures necessary to achieve high stereoselectivity). Having proved that it is the N-3 chlorine of the hydantoin which is transferred, the role of the second chlorine at N-1 was later demonstrated to be as an electronic activator, decreasing the electron density of N-3 chlorine by inductive effects.[34] Two possible associative complexes between catalyst and chorine source were considered, the first involving **hydrogen bond activation of the 1,3-dichlorohydantoin by the protonated catalyst** (the proton derives from acid substrate or additive), the second based on a base interaction between the halide and the catalyst (halide shuttle by base catalysis). The former was favoured by the authors. Shortly after, a bromolactonisation of some conjugated (Z)-enynes giving optically active bromoallenes was reported by Tang (Scheme 9, *(b)*).[35] NBS was used as the bromonium source, and a tosylurea derived from quinine **21** was the most competent catalytic species. Based on low enantioselectivity observed when NBS and catalyst were premixed, the authors favoured also in this case a working model where the protonated catalyst activates NBS by hydrogen bond.

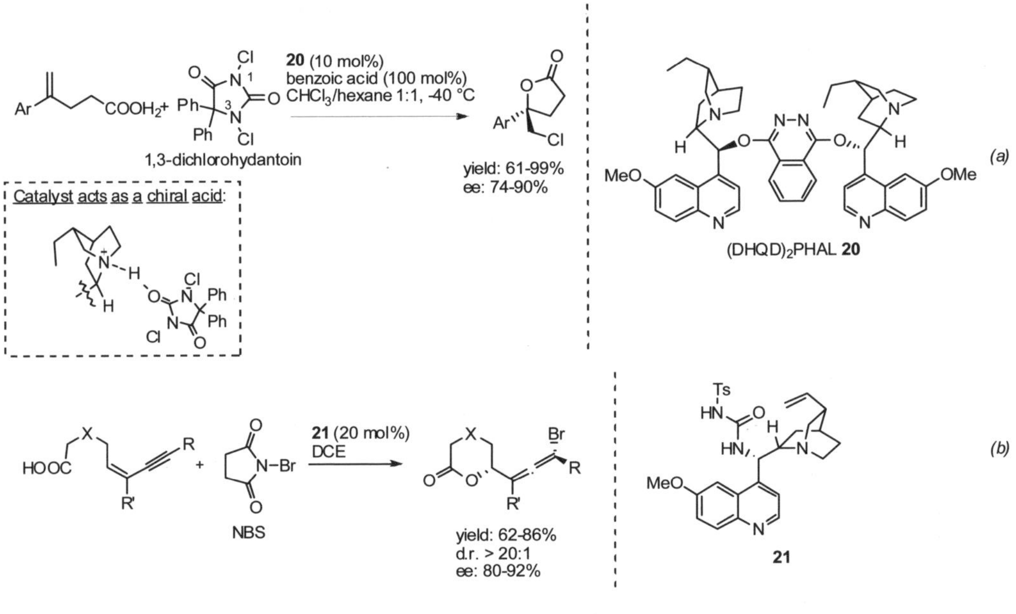

Scheme 9.

4.2 *Halide shuttle by base catalysis*:

A halide shuttle system was instead exploited by Yeung for the realisation of a bromolactonisation reaction (Scheme 10).[36] First, achiral NH-thiocarbamates were found to have an exceptional reactivity in catalysing bromolactonisation reactions with NBS. Thiocarbamate based chiral systems were thus prepared and screened, demonstrating that the presence of a basic moiety (a tertiary amine) in the catalyst structure was crucial for reactivity and enantioselectivity. This screening lead to the identification of the cinchonine derived catalyst **22** as the optimum structure. A number of factors (based on results of structurally modified catalysts) pointed out to the formation of a ***tight ternary complex between catalyst and substrates***, preventing bromonium exchange between olefins. In this complex, the thiocarbamate should be able to bicoordinate NBS, while tertiary amine deprotonates and coordinates the carboxylic moiety. $NsNH_2$ was found to have a positive effect on the enantioselectivity of the reaction. This additive was thought to act as a bromonium sink, quickly reacting with NBS forming NsNHBr at the beginning of the reaction, and then regenerating NBS slowly, thus limiting unselective background reactivity.

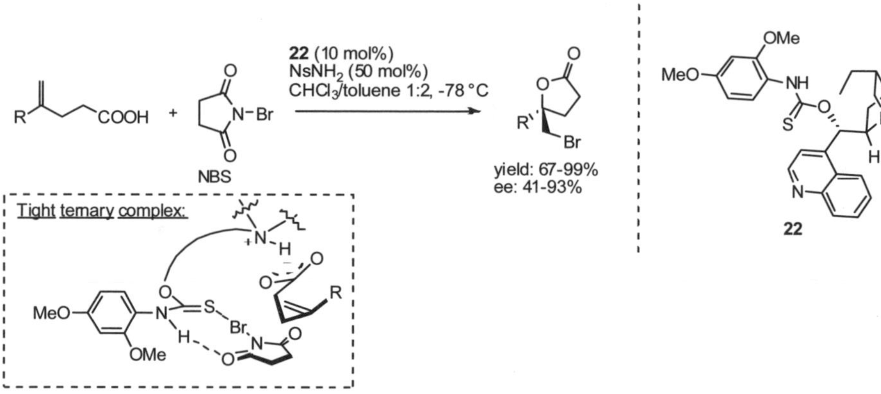

Scheme 10.

4.3 *Halide shuttle with base catalyst and counterion coordination:*

The careful choice and optimisation of the halonium source was the key in the development of the only catalytic enantioselective ***iodolactonisation*** reported so far.[37] In the work of Veitch and Jacobsen, bifunctional catalysts and different iodonium sources were screened in the iodolactonisation of 5-methylidene-5-phenyl hexanoic acid. Addition of catalytic amounts of molecular iodine proved to be essential to achieve useful reactivity and enantioselectivity. As enantioselectivity was also dependent on iodonium source structure, the working model proposed involves the catalytic activation of the iodosuccinimide by molecular iodine, which then reacts with catalyst **23** giving the complex depicted in Scheme 11, as observed by ^{1}H NMR. Succinimidate is then responsible for carboxylic acid deprotonation, whereas the iodonium coordinated by the amine activates the olefin. Importantly, preliminary DFT calculations seems to indicate that ***amine coordination is still present in the intermediate, thus rigidifying the complex and facilitating chirality transfer***. Surprisingly, γ-lactones derived from pentanoic acids were found to have opposite absolute configuration than δ-lactones derived from hexanoic acids.

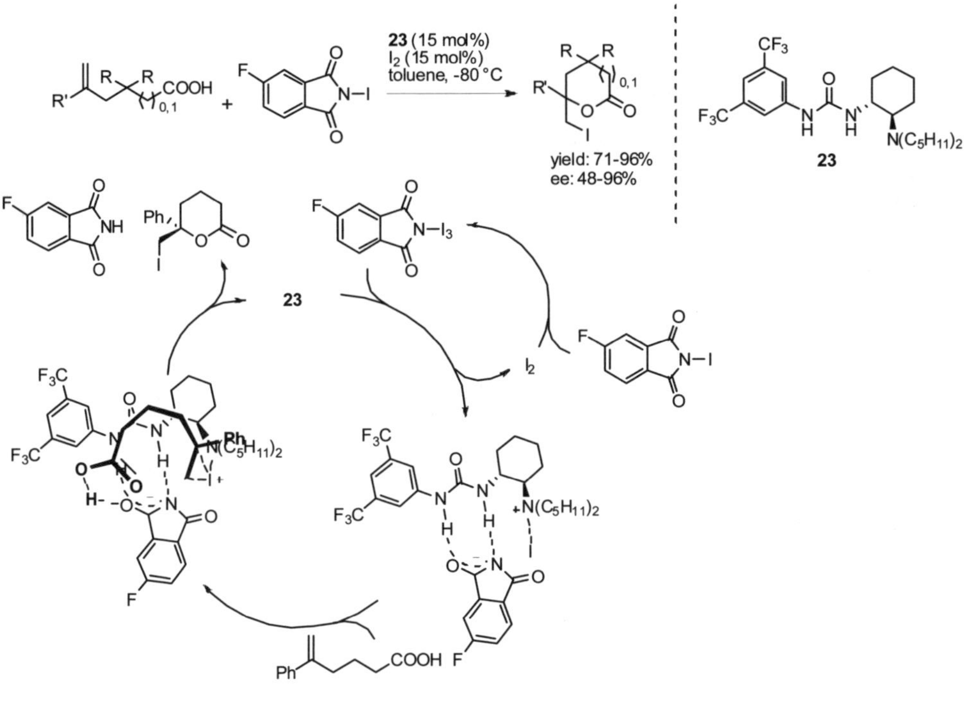

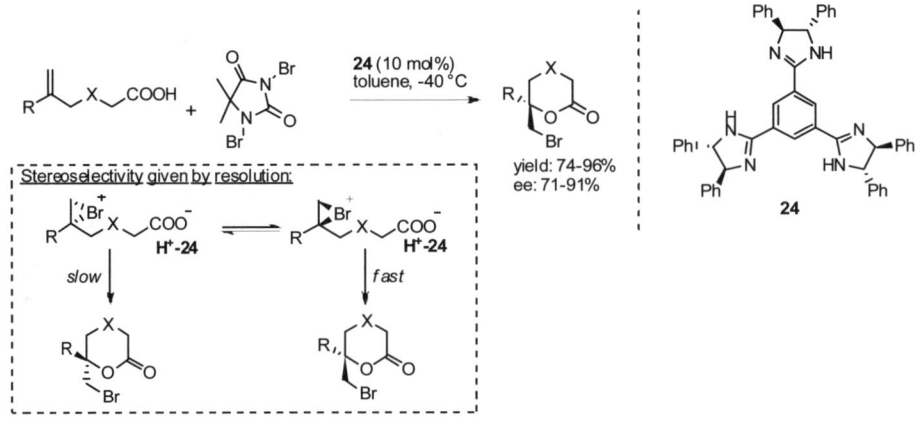

Scheme 11.

4.4 *Carboxylate coordination*:

The C_3-symmetric trisimidazoline **24** was used as a very efficient catalyst by Fujioka for the bromolactonisation reaction of 5-alkylidene hexanoic acids (Scheme 12).[38] Based on ^{1}H NMR

Scheme 12.

experiments, as well as previous observation of the molecular recognition of carboxylic acids by these imidazoline systems, the authors proposed a first deprotonation of the carboxylic acid by the catalyst, followed by uncoordinated activation of the double bond by the dibromohydantoin, used as bromonium source. Being the carboxylate in a chiral environment, it will eventually react selectively with one of the two bromonium ion formed, which possibly exist in rapid equilibrium. Instead of being based on the stereocontrol in the formation of the bromonium ion, as in the previously described systems, stereoselectivity is given in this case by a *dynamic kinetic resolution on the two stereomeric bromonium intermediates, taking advantage of their rapid equilibration*.

5. A CATALYTIC ASYMMETRIC OXIDATION WITH A NEW HYPERVALENT IODINE SPECIES. AN ENVIRONMENTALLY FRIENDLY AND FLEXIBLE ALTERNATIVE TO CLASSIC HYPERVALENT IODINE REAGENTS.

Hypervalent iodine reagents are widely used as oxidants in organic synthesis.[39] Recent progress directed at the use of aryl-λ^3- or aryl-λ^5-iodanes in sub-stoichiometric amount, in combination with stoichiometric oxidants (typically *m*-CPBA) have been reported.[40] Scattered examples of enantioselective variants of these transformations have also appeared in the literature. A virtually unexplored alternative to these useful catalytic processes is instead the use of iodine derived oxo-acids (*i.e.* **hypoiodites [IO]⁻, I^I** or **iodites [IO₂]⁻, I^{III}**).

Ishihara has recently reported that these negative oxidant species can be used very efficiently in asymmetric catalysis, in combination with a chiral quaternary ammonium salt able to impart stereoselectivity in the reaction (Scheme 13).[41] Significant advantages over the use of more popular aryl-λ^3-iodanes such as ArI(OAc)₂ in asymmetric catalysis are: *i) environmentally more benign and milder nature of the process*, as *hydrogen peroxide instead of m-CPBA is used as terminal oxidant. ii) easier variation of catalyst structure*. Hypervalent iodine reactions based on aryl-λ^3-iodanes have traditionally been very reluctant to asymmetric catalysis. One of the reasons is the troublesome preparation of rather complex chiral enantiopure aryliodine structures, preventing rapid catalyst screening. The new approach reported by Ishihara seems very promising in this respect, as chirality information is given by a chiral quaternary ammonium salt. A large variety of these salts, featuring significant structural diversity, is readily available as they have been broadly used for other phase-transfer catalytic reactions.

The non-asymmetric cycloetherification of ketophenols was studied first. Significantly, employment of tetra-*n*-butyl ammonium iodide in combination with hydrogen peroxide was very advantageous in terms of yield, compared to more classic aryl-λ^3-iodanes (in combination with *m*-CPBA), which gave very messy reaction mixtures. As the reaction performed well even in solvents nonmiscible with water, application of chiral quaternary ammonium iodine salts as phase-transfer catalysts was straightforward, ultimately leading to a highly enantioselective process when the Maruoka-type spiroammonium salt **25** was used. Control experiment excluded the involvement of iodate(V) or periodate(VII) species in the reaction. Attempted reaction with molecular iodine or NIS gave a complex reaction mixture. The reaction performed relatively well instead using a combination of tetra-*n*-butylammonium hydroxide and molecular iodine, which is known to generate a hypoiodite ([IO]⁻) species. Considering the tendency of hypoiodite ([IO]⁻) to disproportionate to iodite ([IO₂]⁻), both species were considered possible as active oxidants in this asymmetric process.

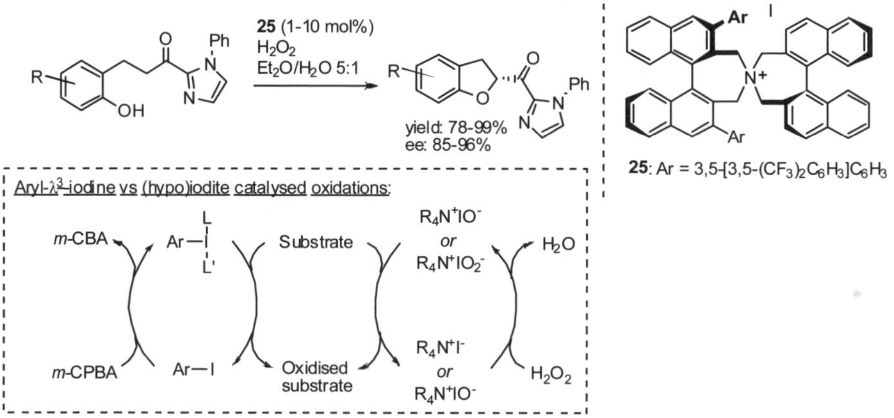

Scheme 13.

6. ASYMMETRIC BRØNSTED ACID CATALYSIS IN WATER. COOPERATIVITY, HYDROPHOBIC EFFECTS AND COULOMBIC FORCES, OVERRIDE WATER HYDROGEN BONDS.

Water is abundant, cheap, non-toxic, safe, and has a high heat capacity. It is thus one of the most convenient reaction solvents, for reactions on both small and large scale. Water as reaction medium renders also possible the use of hydrophilic substrates, insoluble in non-polar organic solvents. Finally, the reactivity profile of many organic substrates, (*i.e.* acidity and nucleophilicity), is often different in water compared to organic solvents, thus providing distinct opportunities for reactivity.

Asymmetric organocatalysis in aqueous medium (both **on water**, wherein substrates and catalysts are not water soluble, and **in water**, wherein the system is homogeneous) has been the subject of intense studies in the last few years.[42] However, nearly all examples involved the use of Lewis basic primary and secondary amines as catalysts, which form covalently bound enamines and iminium ions with aldehyde/ketone substrates. In these cases, stereodiscrimination is based very often on steric factors. It was thus envisioned that water could be a suitable medium, not interfering to a great extent with the substrates-catalyst tight covalent assemble. Brønsted acid catalysis is based instead on intermolecular hydrogen bond interactions between catalyst and substrates.[6] Being water itself an excellent hydrogen bond acceptor-donor, it was not considered as suitable medium for Brønsted acid catalysis, despite scattered examples indicated the tolerance of some catalytic systems to relevant amounts of water. However, in ***non-asymmetric*** Brønsted acid catalysis Kleiner and Schreiner demonstrated in 2006 that activation by hydrogen bond is possible in water.[43] Hydrophobic effects[44] were considered crucial for the activation of an epoxide by a thiourea catalyst, giving hydrolysis. Non-covalent interactions resulting in epoxide activation by the catalyst were given thus not only by direct attractive forces between the organic components, but also by the tendency of ***limiting the interactions between water molecules and lipophilic organic substrates, minimising contact surface and "squeezing" catalyst and substrate together (hydrophobic effect).***

Rueping and Theissmann have recently reported a catalytic asymmetric Brønsted acid catalysed reaction in water.[45] Instead of using a simple hydrogen bond interaction between catalyst and substrate, which can hardly compete with the medium, they envisioned that some catalysts can in fact bind the substrate in their "protected" hydrophobic pocket ***taking advantage of both Coulombic forces and hydrophobic effects.*** Addition of inorganic salts was also assumed to increase polarity of the medium thus maximising these effects. All this considered, a well defined tight ion pair might result, giving a highly enantioselective reaction even in the strong hydrogen-bond donor water, as interactions with water are overridden.

The biomimetic reduction of quinolines with Hantzsch esters catalysed by BINOL-derived phosphoric acids, previously developed by the same group,[46] was taken as a case study (Scheme 14). The high organisation of the reaction transition state, wherein calculations support bifunctionality of these types of catalysts with bicoordination to both electrophile and Hantzsch ester,[47] should further assist in the formation of a highly organised transition state (***cooperativity***). The optimal catalyst **26** previously used showed greatly reduced enantioselectivity in water media (water-dioxane or pure water, 41-44% ee) compared to the reaction carried out in dry toluene or dioxane (97% ee). Salt addition improved only slightly the results. However, the sterically more demanding phosphoric acid catalyst **27** gave the product with greatly improved enantioselectivity, matching the enantioselectivity obtained in dry organic solvents, especially when a hindered Hantzsch ester bearing *tert*-butyl groups was used. These results were rationalised considering the maximisation of hydrophobic effects with more lipophilic substrates, as well as the more closed conformation of catalyst **27** compared to **26** (as observed in the solid state of both catalysts by X-ray), which might prevent water from entering the active site in the case of **27**, thus "protecting" the substrates from interactions with water.

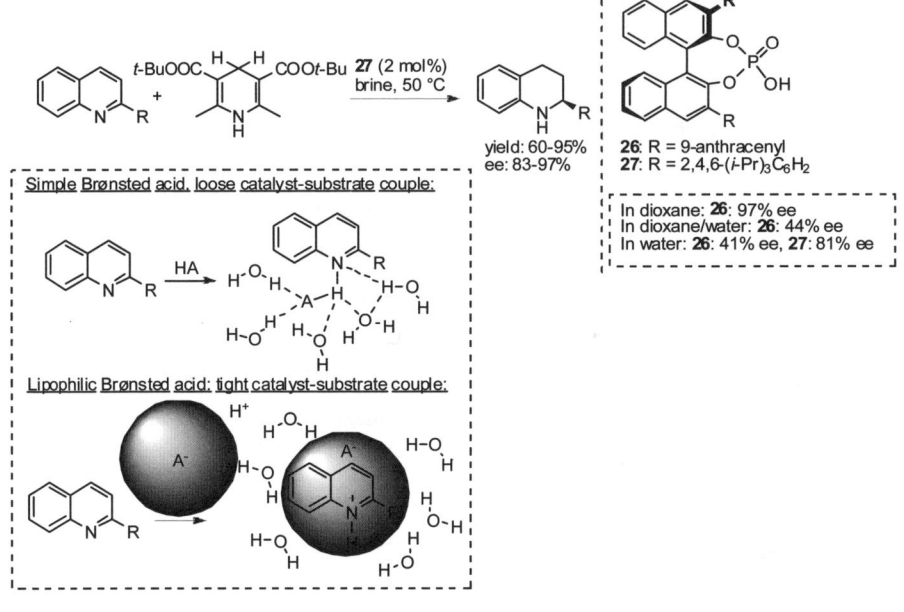

Scheme 14.

7. SILICON CHEMISTRY. ASYMMETRIC SUPER BRØNSTED ACID CATALYSTS AND PHASE-TRANSFER OF A SILICON PROMOTER.

In the quest for new chiral strong Brønsted acid catalysts, great efforts have been devoted at expanding the range of substrates which can be effectively activated by protonation. Whereas basic imines can be protonated by relatively weak acids as the popular BINOL-derived phosphoric acids disclosed by Terada and Akiyama,[6] activation of less basic substrates such as aldehydes and ketones is still a largely unsolved challenge. Triflimide, Tf_2NH, is a very useful catalyst for non-asymmetric reactions. Its usefulness goes beyond its Brønsted acidity. In fact, when combined with silicon based nucleophiles, a strong silicon Lewis acid (Tf_2NSiR_3) is readily formed. The Lewis acidity of Tf_2NSiR_3 is even higher than $TfOSiR_3$,[48] despite the lower Brønsted acidity of Tf_2NH compared to TfOH in water.[49] The very low coordinative properties of the triflimide anion, which make it a good counterion for metal-based Lewis acid catalysts, are considered responsible for this striking Lewis acidity. In addition, a self-repair mechanism is able to restore the active silicon Lewis acid catalytic species in the presence of traces of water. As a result, extremely low catalyst loadings are often possible.[50]

In the development of chiral analogues of Tf_2NH, Yamamoto's BINOL derived phosphoric triflimide is probably the most prominent example.[51] Originally developed for Diels-Alder reactions on ketones, this highly acidic catalytic species has found great use in a number of transformations, including reactions involving ketones.[52] In an effort to further enhance activity as well as selectivity, List designed and prepared in 2009 chiral disulfonamides like **28**.[53] C_2-symmetry, sterically important aryl substituents at the 3,3' position of the binaphthyl core, together with the good encasing of the active proton (or silicon) in the chiral pocket of the catalyst, should guarantee a good transmission of the stereochemical information from catalyst to substrates. Indeed, these catalysts were very successfully applied to a catalytic asymmetric Mukaiyama aldol reaction. The corresponding silicon Lewis acid, and not the parent Brønsted acid, was found to be responsible for the catalytic activity, allowing very low catalyst loading (0.1 mol%) due to the mentioned self-repair mechanism.

More recently, the same group used these super-acid catalysts for a remarkable bisvinylogous (and more common vinylogous) Mukaiyama aldol reaction (Scheme 15).[54] Vinylogy is the propagation of the electronic effects (and reactivity) through a π-system. Vinylogous Mukaiyama aldol and related reactions, wherein extended enolates react at the γ-position of a carbonyl group, have become a rather common practice, thanks to the use of stabilised silicon enolates. In contrast, propagation of the nucleophilic reactivity through an additional double bond, thus giving addition at the ε-position, had never been reported. Preliminary calculations suggested that this reactivity could be possible, as electron density was higher at the ε-carbon, compared to the other nucleophilic α- and γ-positions. Application of these enolates to the Mukaiyama aldol reaction in the presence of catalyst **29**, furnished indeed the corresponding α,β-γ,δ-unsaturated esters with good enantioselectivities and moderate yields. In most

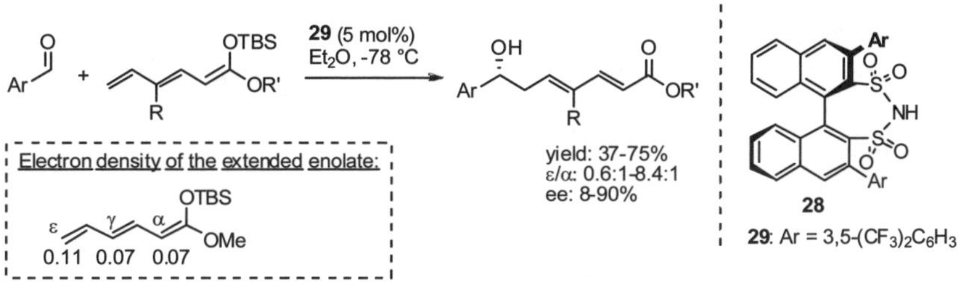

Scheme 15.

cases, competing α-addition was also observed, as predicted from calculations from the similar (albeit lower) electron density at the α-carbon. The utility of catalyst **29** thus now encompasses Mukaiyama, vinylogous Mukaiyama, and bisvinylogous Mukaiyama reactions.

Activation of the silicon reagent with a Lewis base is an alternative strategy for promoting additions of silicon nucleophiles to carbonyl compounds. This approach is extremely useful especially when low-reactive nucleophiles, such as trimethyl(trifluoromethyl)silane, are employed. These reactions proceed through an autocatalytic cycle, triggered by the Lewis base, wherein the anionic adduct generated in the initiation step serves as the actual promoter. However, whereas alkoxides formed upon addition to aldehydes and ketones are able to carry out the autocatalytic cycle, the corresponding amide adducts derived from imines do not always feature sufficient silicon affinity to activate the silicon reagent. A stoichiometric amount of Lewis base is thus required.

We have recently developed a new a new approach to the addition of silicon-based nucleophiles to imines relying on phase-transfer catalysis (Scheme 16).[55] This reaction manifold was initially developed and optimised for the trifluoromethylation of imines, but is also amenable to various silicon nucleophile additions. Our approach is based on the ***phase-transfer extraction of a stoichiometric promoter, an insoluble metal phenoxide, by a quaternary ammonium salt (eventually chiral). The catalyst ammonium phenoxide is regenerated in the organic layer (toluene) after each catalytic cycle through a phase-transfer process, with concomitant release of the amide adduct.*** The inability of the anionic amide adduct in promoting the autocatalytic cycle is thus overcome, rendering possible the use of ammonium salts in substoichiometric amounts. The choice of phenoxides as stoichiometric promoters was dictated by their higher polarisability and lipophilicity, compared to other popular anionic Lewis bases such as fluorides and hydroxides. These features secure a good affinity with lipophilic ammonium salts, facilitating the phase-transfer process. Although the method has been initially

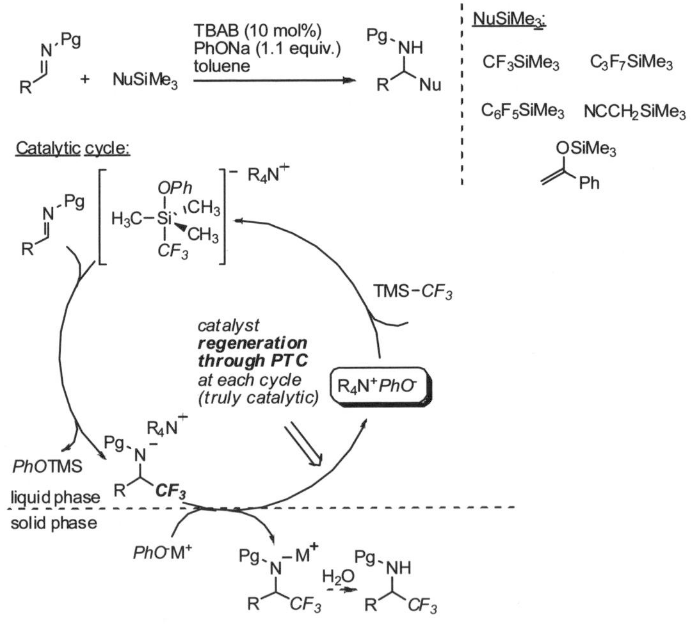

Scheme 16.

studied using an achiral phase-transfer catalyst (tetra-*n*-butylammonium bromide), a moderately enantioselective example demonstrates its possible implementation for asymmetric transformations.

8. CONCLUDING REMARKS.

As shown in these examples, the field of organocatalysis has nowadays become an indispensable tool for the realisation of challenging catalytic transformations. Development of new activation modes, merger of different catalytic systems operating in concert, implementation of biomimetic concepts, catalyst development taking advantage of the assembly of different modules by weak interactions, are the fields were most innovations are most likely to be found in the near future. Increasing applications of organocatalysis in total synthesis, medicinal chemistry, material science, can also be expected.

9. ACKNOWLEDGEMENTS.

I am grateful to Prof. Alfredo Ricci, Prof. Bianca F. Bonini, Dr. Marco Bella, and Claudio Gioia, for proof reading the manuscript.

10. REFERENCES AND NOTES.

1. von Liebig, J. *Annalen der Chemie und Pharmazie* **1860**, *113*, 246.
2. MacMillan, D. W. C. *Nature* **2008**, *455*, 304.
3. Ahrendt, K. A.; Borths, C. J.; MacMillan, D. W. C. *J. Am. Chem. Soc.* **2000**, *122*, 4243.
4. List, B.; Lerner, R. A.; Barbas III, C. F. *J. Am. Chem. Soc.* **2000**, *122*, 2395.
5. Sigman, M. S.; Jacobsen, E. N. *J. Am. Chem. Soc.* **1998**, *120*, 4901.
6. Reviews: (a) Doyle, A. G.; Jacobsen, E. N. *Chem. Rev.* **2007**, *107*, 5713; (b) Akiyama, T. *Chem. Rev.* **2007**, *107*, 5744; (c) Terada, M. *Synthesis* **2010**, 1929.
7. Examples: (a) Taylor, M. S.; Jacobsen, E. N. *J. Am. Chem. Soc.* **2004**, *126*, 10558; (b) Raheem, I. T.; Thiara, P. S.; Peterson, E. A.; Jacobsen, E. N. *J. Am. Chem. Soc.* **2007**, *129*, 13404; (c) De, C. K.; Klauber, E. G.; Seidel, D. *J. Am. Chem. Soc.* **2009**, *131*, 17060. For a review: (d) Zhang, Z.; Schreiner, P. R. *Chem. Soc. Rev.* **2009**, *38*, 1187.
8. (a) Xu, H.; Zuend, S. J.; Woll, M. G.; Tao, Y.; Jacobsen, E. N. *Science* **2010**, *327*, 986. Other examples of organocatalytic asymmetric Povarov reactions: (b) Akiyama, T.; Morita, H.; Fuchibe, K. *J. Am. Chem. Soc.* **2006**, *128*, 13070; (c) Liu, H.; Dagousset, G.; Masson, G.; Retailleau, P.; Zhu, J. *J. Am. Chem. Soc.* **2009**, *131*, 4598. (d) Bergonzini, G.; Gramigna, L.; Mazzanti, A.; Fochi, M.; Bernardi, L.; Ricci, A. *Chem. Commun.* **2010**, *46*, 327.
9. Fluoride anion generation for silicon activation from benzoyl fluorides and Lewis bases: (a) Bappert, E.; Muller, P.; Fu, G. C. *Chem. Commun.* **2006**, 2604; (b) Poisson, T.; Dalla, V.; Marsais, F.; Dupas, G.; Oudeyer, S.; Levacher, V. *Angew. Chem. Int. Ed.* **2007**, *47*, 7090.
10. Olah, G. A.; Meidar, D. *Isr. J. Chem.* **1978**, *17*, 148.
11. Kalow, J. A.; Doyle, A. G. *J. Am. Chem. Soc.* **2010**, *132*, 3268.
12. Reviews: a) Mukherjee, S.; Yang, J. W.; Hoffmann, S.; List, B. *Chem. Rev.* **2007**, *107*, 5471; b) Melchiorre, P.; Marigo, M.; Carlone, A.; Bartoli, G. *Angew. Chem. Int. Ed.* **2008**, *47*, 6138; c) Bertelsen, S.; Jørgensen, K. A. *Chem. Soc. Rev.* **2009**, *38*, 2178.
13. Overview: Alba, A. N.; Viciano, M.; Rios, R. *ChemCatChem* **2009**, *1*, 437.
14. Shih, H.-W.; Vander Wal, M. N.; Grange, R. L.; MacMillan, D. W. C. *J. Am. Chem. Soc.* **2010**, *132*, 13600.
15. Nicewicz, D. A.; MacMillan, D. W. C. *Science* **2008**, *322*, 77.
16. Nagib, D. A.; Scott, E. M.; MacMillan, D. W. C. *J. Am. Chem. Soc.* **2009**, *131*, 10875.
17. Neumann, M.; Füldner, S.; König, B.; Zeitler, K. *Angew. Chem. Int. Ed.* **2011**, *50*, 951.

18. Kempf, B.; Hampel, N.; Ofial, A. R.; Mayr, H. *Chem. Eur. J.* **2003**, *9*, 2209.

19. Brown, A. R.; Kuo, W.-H.; Jacobsen, E. N. *J. Am. Chem. Soc.* **2010**, *132*, 9286.

20. Capdevila, M. G.; Benfatti, F.; Zoli, L.; Stenta, M.; Cozzi, P. G. *Chem. Asian J.* **2010**, *16*, 11237.

21. Bergonzini, G.; Vera, S.; Melchiorre, P. *Angew. Chem. Int. Ed.* **2010**, *49*, 9685.

22. Lifchits, O.; Reisinger, C. M.; List, B. *J. Am. Chem. Soc.* **2010**, *132*, 10227.

23. Zhang, L.; Cui, L.; Li, X.; Li, J.; Luo, S.; Cheng, J.-P. *Chem. Eur. J.* **2010**, *16*, 2045.

24. Ikeda, M.; Miyake, Y.; Nishibayashi, Y. *Angew. Chem. Int. Ed.* **2010**, *49*, 7289.

25. Allen, A. E.; MacMillan, D. W. C. *J. Am. Chem. Soc.* **2010**, *132*, 4986.

26. Gómez-Bengoa, E.; Landa, A.; Lizarraga, A.; Mielgo, A.; Oiarbide, M.; Palomo, C. *Chem. Sci.* **2011**, *2*, 353.

27. Inoue, T.; Kitagawa, O.; Ochiai, O.; Shiro, M.; Taguchi, T. *Tetrahedron Lett.* **1995**, *36*, 9333.

28. Kang, S. H.; Lee, S. B.; Park, C. M. *J. Am. Chem. Soc.* **2003**, *125*, 15748.

29. Sakakura, A.; Ukai, A.; Ishihara, K. *Nature* **2007**, *445*, 900.

30. Denmark, S. E.; Burk, M. T. *Proc. Natl. Acad. Sci.* **2010**, *107*, 20655, and references therein.

31. Brown, R. S. *Acc. Chem. Res.* **1997**, *30*, 131.

32. Denmark, S. E.; Burk, M. T.; Hoover, A. J. *J. Am. Chem. Soc.* **2010**, *132*, 1232.

33. Whitehead, D. C.; Yousefi, R.; Jaganathan, A.; Borhan, B. *J. Am. Chem. Soc.* **2010**, *132*, 3298.

34. Yousefi, R.; Whitehead, D. C.; Mueller, J. M.; Staples, R. J.; Borhan, B. *Org. Lett.* **2011**, *13*, 608.

35. Zhang, W.; Zheng, S.; Liu, N.; werness, J. B.; Guzei, I. A.; Tang, W. *J. Am. Chem. Soc.* **2010**, *132*, 3664.

36. Zhou, L.; Tan, C. K.; Jiang, X.; Chen, F.; Yeung, Y.-Y. *J. Am. Chem. Soc.* **2010**, *132*, 15474.

37. Veitch, G. E.; Jacobsen, E. N. *Angew. Chem. Int. Ed.* **2010**, *49*, 7332.

38. Murai, K.; Matsushita, T.; Nakamura, A.; Fukushima, S.; Shimura, M.; Fujioka, H. *Angew. Chem. Int. Ed.* **2010**, *49*, 9174.

39. Zhdankin, V. V.; Stang, P. J. *Chem. Rev.* **2008**, *108*, 5299.

40. Richardson, R. D.; Wirth, T. *Angew. Chem. Int. Ed.* **2006**, *45*, 4402.

41. Uyanik, M.; Okamoto, H.; Yasui, T.; Ishihara, K. *Science* **2010**, *328*, 1376.

42. For a review, see: Gruttadauria, M.; Giacalone, F.; Noto, R. *Adv. Synth. Catal.* **2009**, *351*, 33.

43. Kleiner, C. M.; Schreiner, P. R. *Chem. Commun.* **2006**, 4315.

44. Engberts, J. B. F. N. *Pure Appl. Chem.* **1995**, *67*, 823.

45. Rueping, M.; Theissmann, T. *Chem. Sci.* **2010**, *1*, 473.

46. Rueping, M.; Antonchick, A. P.; Theissmann, T. *Angew. Chem. Int. Ed.* **2006**, *45*, 6751.

47. (a) Simón, L.; Goodman, J. M. *J. Am. Chem. Soc.* **2008**, *130*, 8741; (b) Marcelli, T.; Hammar, P. Fahmi, L. *Chem. Eur. J.* **2008**, *14*, 8562.

48. Mathieu, B.; Ghosez, L. *Tetrahedron Lett.* **1997**, *38*, 5497.

49. Tf$_2$NH: pK_a = 1.7 in H$_2$O; TfOH: pK_a = -5.9 in H$_2$O.

50. Boxer, M. B.; Alber, B. J.; Yamamoto, H. *Aldrichimica Acta* **2009**, *42*, 3.

51. Nakashima, D.; Yamamoto, H. *J. Am. Chem. Soc.* **2006**, *128*, 9626.

52. Rueping, M.; Nachtstein, B. J.; Koenigs, R. M.; Ieawsuwan, W. *Chem. Eur. J.* **2010**, *16*, 13116.

53. García-García, P.; Lay, F.; García-García, P.; Rabalakos, C.; List, B. *Angew. Chem. Int. Ed.* **2009**, *48*, 4363.

54. Ratjen, L.; García-García, P.; Lay, F.; Beck, M. E.; List, B. *Angew. Chem. Int. Ed.* **2011**, *50*, 754.

55. Bernardi, L.; Indrigo, E.; Pollicino, S.; Ricci, A. *Chem. Commun.* in press, DOI: 10.1039/C0CC05777K.

Alessandra Silvani
Assistant Professor in Organic Chemistry

Address:

     Dipartimento di Chimica Organica e Industriale, Università degli Studi di Milano, via Venezian 21, 20133 Milano, Italy

Education:

1967, born in Milan. 1992, degree in Chemistry at the University of Milan. 1996, Ph.D. degree, with a thesis on the asymmetric synthesis of natural products. From 1996 to 1999, postdoctorial fellow at the University of Milan. 2000, awarded with a "Young Researchers" Grant for a project entitled "New chiral ligands for asymmetric catalysis". 2001, visiting researcher at the University of Barcelona, in the frame of the Italy-Spain Cooperation Project (Azione Integrata). From 2001 to now, Permanent Researcher at the Department of Organic and Industrial Chemistry of the University of Milan.

Professional experience:

Her work includes the total synthesis of enantiopure bioactive natural compounds, and the development of new methodologies for the asymmetric synthesis of nitrogen containing compounds. Her current research interests are also focused on the design and synthesis of new reverse turn inducers as peptidomimetics for medicinal chemistry. Until now, she is co-author of 62 scientific publications on peer-reviewed international journals.

Critical Survey Covering the Year 2010:
Total Synthesis of Natural Products

Alessandro Sacchetti[a] and Alessandra Silvani[b]*

[a] Politecnico di Milano, Dipartimento di Chimica, Materiali ed Ingegneria Chimica 'Giulio Natta', via Mancinelli 7, 20131 Milano, Italy

[b] Dipartimento di Chimica Organica e Industriale, Università degli Studi di Milano, via G. Venezian 21, 20133 Milano, Italy.

E-mail: alessandra.silvani@unimi.it

It would be hard to imagine how organic chemistry would have developed without exciting inputs from the world of natural products. The engagement of the two fields started with the challenge of isolating pure products from complex, naturally derived mixtures. When the process of structure determination, including stereochemistry, was greatly accelerated with major advances in spectroscopy and crystallography-based elucidations, a whole new world of fascinating molecules, isolated from plants, bacteria, fungi and marine sources, insinuated itself into the minds of organic chemists, providing an invaluable source for searching new lead agents of medicinal import.[1]

The close link of organic chemistry with the field of natural products offers entry into the drug discovery process in a number of ways. In the most direct case, a natural product may itself possess the potency, selectivity, and pharmacokinetic traits required to render it a clinically useful drug agent. More frequent are the cases where the natural products themselves serve as lead agents, providing the chemist with a structural platform which can be elaborated to yield a therapeutically valuable pharmaceutical.

The impact of natural products on drug development can be felt across virtually every major therapeutic area. To what can be ascribed this remarkable record of connectivity between natural products and agents of interest as potential pharmaceuticals? In most cases we don't yet know the reason why natural products are being produced in their ecosystems. So, while the purpose of their biosynthesis is far from clear and presumably varies from case to case, it is altogether a matter of fact that natural products are primarily built and evolutionarily optimized to interact with proteins, such as enzymes or receptors. However, although natural products benefit from the wisdom of lessons which nature has learned, it would be also recognized that their activities in humans are serendipitous and not optimized, but go on with providing a rich source of information and inspiration, often through the challenge of the total synthesis.

Nowadays, the total synthesis of natural products has reached an extraordinary level of sophistication.[2] Indeed, the development of new types of chemical reactions over the past few decades has enabled synthetic chemists to assemble almost every discovered natural product. Lately, literature reports on total synthesis of natural products appears daily, as a result of the many advances in reaction methods. Through remarkable advances in tools, theories and protocols, synthesis has provided a reliable supply of many compounds which are available only in trace amounts from natural sources. Moreover, the field of total synthesis has served and continues to serve as the ultimate testing ground for new reactions, methods, catalysts and strategies.

The question of whether a molecule from nature could be made is increasingly moving to whether it could be made in efficient, practical and environmentally friendly ways. Now, the challenge in total synthesis is therefore to employ reactions that are more selective (minimizing the number of side-products), efficient (providing good yields) and not too long (reducing costs). In recent years several new criteria have been brought forward to face this demand and to improve total synthesis: atom,[3] step[4] and redox economy,[5] protecting-group-free[6] and biomimetic synthesis.[7] Besides, over the past decade the research area of organocatalysis[8] has rapidly grown to become a third pillar of asymmetric catalysis together with metal and biocatalysis, paving the way for new and powerful strategies in total synthesis.

The idea of step economy is not new, but it is important in considering the current state and future of synthesis. Syntheses based on reactions that provide only small increases in target-relevant complexity generally require many steps. A way of reducing the number of steps is to perform a series of reactions in one vessel so that, as the products of each reaction form, they immediately go on to take part in the next reaction of the sequence, ultimately ending up in a final isolable product. Such single-operation, multiple-step processes — described variously as cascade, tandem, or domino reactions[9] — often coupled with avoidance of protecting-group operations, are an increasingly important focus of many research groups, and are certainly having an impact on the ability of synthesis to produce complex targets in a step-economical fashion. This strategy draws inspiration from biosynthetic pathways, such as the route to terpenes and steroids or to polyethers, that often proceed through such serial processes.

A further approach to step economy is to invent new reactions able to generate great increases in target-relevant complexity per step. These reactions can suggest to chemists new strategic disconnections and often change the way they think about a synthesis. The unique impact of the olefin metathesis reactions and the innovative synthetic strategies they enable, is a recent noteworthy example of this approach.

In a *JOC Perspective* of June 2010 entitled "Aiming for the Ideal Synthesis",[10] Professor Phil S. Baran, the recipient of the 2010 ACS Award in Pure Chemistry, states: "The stage is now set for organic chemists to aim for "ideality" in the way molecules are synthesized." The "ideal synthesis" was defined by Hendrickson in 1975[11] as one which: "...creates a complex molecule...in a sequence of only construction reactions involving no intermediary refunctionalizations, and leading directly to the target, not only its skeleton but also its correctly placed functionality." In an attempt to quantify the ideality percentage in terms of number of construction reactions and strategic redox reactions with respect to the total number of steps, Baran doesn't forget to highlight that the definition of ideality should be restricted to the comparison of different routes leading to the same target structure. Finally, it should be noted that "ideality" in synthesis is just one variable for the consideration of a synthetic route, since other strategic factors have to be evaluated, also depending on the purpose and urgency of the synthesis.

In this review, some total syntheses appeared on high impact factor journals in 2010 will be examined, with the aim of illustrating the exciting advances achieved in the area. The chosen examples represent a variety of natural products subtypes, ranging from alkaloids to peptides, to macrolides and compounds of mixed biosynthetic origin. The purpose of the syntheses is related to promising pharmacological activities in most cases. The synthetic sequences will be examined through the critical lens of ideality, taking also into account, besides the overall yield and step count, molecular complexity, comparison of different routes, and chemo- and stereoselectivity challenges. Of course, the selection is guided by personal criteria, and it may be not the best one. In spite of the fact that a 100% ideal total synthesis is for the present unattainable, it should only suggest that such a pursuit must serve as a constant source of inspiration.

Total Synthesis of Palau'amine[12]

Introduction. Palau'amine is a marine natural product belonging to the pyrrole-imidazole alkaloid family possessing a very high degree of complexity. Its rare structural feature is just an example of the extraordinary biodiversity found in the island nation of Palau and it posed an inspiring synthetic challenge to chemists since its first isolation in 1993[13] from the sponge *Stylotella aurantium* and the definitive structural elucidation in 2007.[14] It possesses daunting structural and physical attributes, including nine nitrogen atoms, eight contiguous stereogenic centres, reactive (hemi)aminal moieties, an oxidation-prone pyrrole ring, and a highly polar, non-crystalline morphology. The first total synthesis of racemic Palau'amine is now described by P.S. Baran and co-workers of the Scripps Research Institute of California, by means of a successful route featuring transformations on unprotected intermediates, cascade reactions, and a remarkable finale involving a transannular cyclization to forge the critical bond.

Synthesis. The hexacyclic core of Palau'amine **1**, containing a highly strained *trans*-azabyciclo[3.3.0]octane substructure, is the result of a complex biosynthetic pathway[15] starting from histidine and proline amino acids and involving the key ambivalent reactivity of 2-aminoimidazole as the crucial element for molecular diversity in the class of polycyclic pyrrole-imidazole alkaloids (Figure 1).

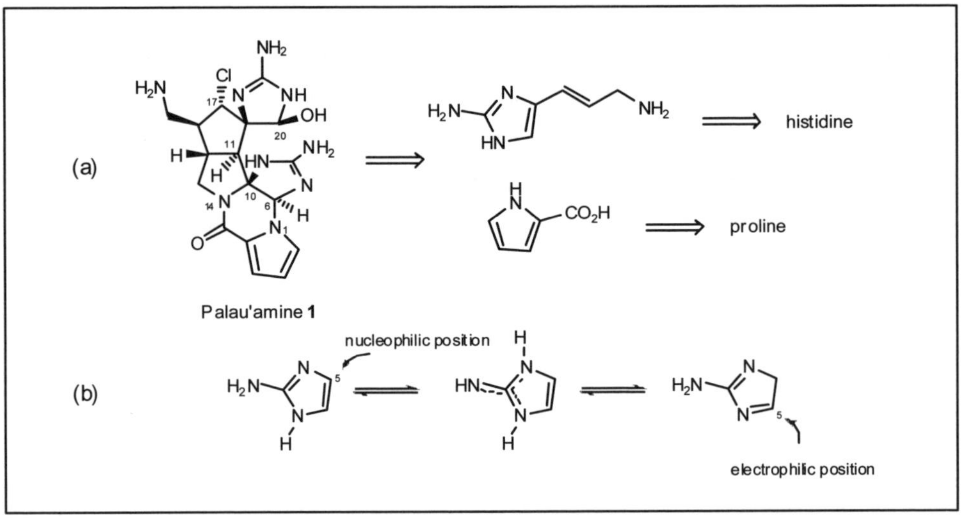

Figure 1. (a) Biogenetic Precursors of Palau'amine. (b) Tautomerism and Ambivalent Reactivity of 2-Aminoimidazole.

Taking inspiration from some elements of the biosynthetic hypothesis and relying on previous efforts towards the total synthesis of closely related less strained compounds,[16] Baran focused the retrosynthetic analysis depicted in Scheme 1. It exploits the idea that an irreversible transannular ring-chain tautomerization would convert macrocycle **2** into the constitutional isomer **1**, bypassing all the difficulties previously encountered in the construction of the strained azabyciclo[3.3.0]octane core. Macro Palau'amine **2** was believed to be accessible through macrolactamization of the diamine derived from diazide **3**. This intermediate was envisioned to arise from the S$_N$Ar of a pyrrole derivative to the bromo-aminoimidazole **4**.

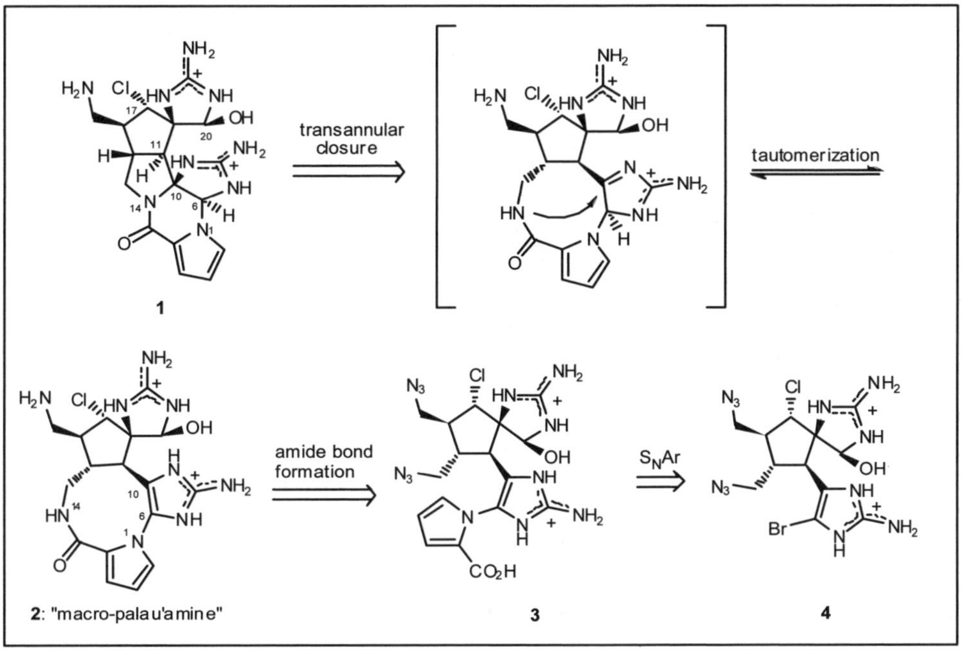

Scheme 1. Retrosynthetic Analysis of Palau'amine.

The reaction sequence commenced with the single diastereoisomer **5** (Scheme 2), which was readily obtained in multi-gram amounts from a Diels-Alder reaction.[17] The cyclohexene diazide **6** (obtained after a five-step sequence) was subsequently contracted to the cyclopentane **7**: ozonolytic scission of the olefin in **6** to a bis(methyl ketone) was followed by dibromination and cyclization by an intramolecular aldol reaction; the more reactive bromo group was then exchanged for a chloro group, thus attenuating the halogen reactivity for later stage conversions; cleavage of the protecting group then furnished diol **7**. At this stage, exposure of **7** to SO_2Cl_2 afforded chloro-enone **8**, which was submitted to Luche reduction and selective displacement of the allylic bromide affording allylic alcohol **9**. Compound **9** was spirocyclized by intramolecular aza-Michael reaction, after oxidation with IBX. Final elaboration with sodium diformylamide gave rise to the desired diastereoisomer *rac*-**10** in slight excess.

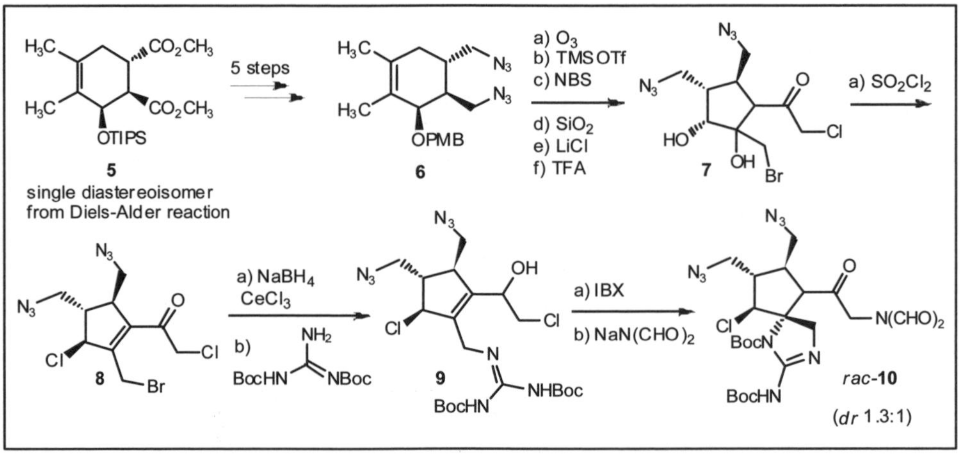

Scheme 2. Preparation of the Cyclopentane Core *rac*-**10** of Palau'amine.

Spirocycle **10** was converted into Palau'amine **1** in an impressively small number of steps, combining a highly chemoselective oxidation, cascade reactions, and by utilizing the intrinsic ambivalent reactivity of 2-aminoimidazole (Scheme 3).

After deprotection of *rac*-**10** under acidic conditions, Baran developed a protocol using silver(II)-picolinate to obtain hemiaminal **11** regio- and stereoselectively in a one-pot procedure. This hemiaminal functionality remains untouched in subsequent steps, thus underlining the detailed and subtle knowledge of the inherent reactivity of functional groups leading to highly efficient processes. Two further steps converted the α-amino ketone **11** into a 2-amino-4-bromoimidazole derivative, of which the amidine tautomer **4'** is shown in Scheme 3. As transition-metal-catalyzed reactions failed to give the desired product **3**, Baran switched to an unconventional method of introducing the pyrrole moiety. Exploiting the ambiphilicity of the 2-aminoimidazole, he developed a cascade reaction which starts with an uncatalyzed nucleophilic displacement using the pyrrole surrogate **12**, and combines acid-catalyzed methanol eliminations with a cyclization and a deprotection reaction. At least five chemical transformations have been tied up in a one-pot reaction with a yield of 44% from **4'**. The final elaboration was carried out by another cascade sequence: reduction of both azido groups followed by EDC mediated coupling gave the nine-membered macrolactam **2**. Again, the inherent ambiphilic nature of the imidazole–amidine system was successfully exploited: heating of the crude reaction mixture in TFA elicited the crucial transannular cyclization (presumably proceeding through amidine tautomer **2'**) that fastened the remaining two stereocenters and cemented the *trans*-5,5 ring system to deliver racemic Palau'amine **1** in 17% overall yield from **3**.

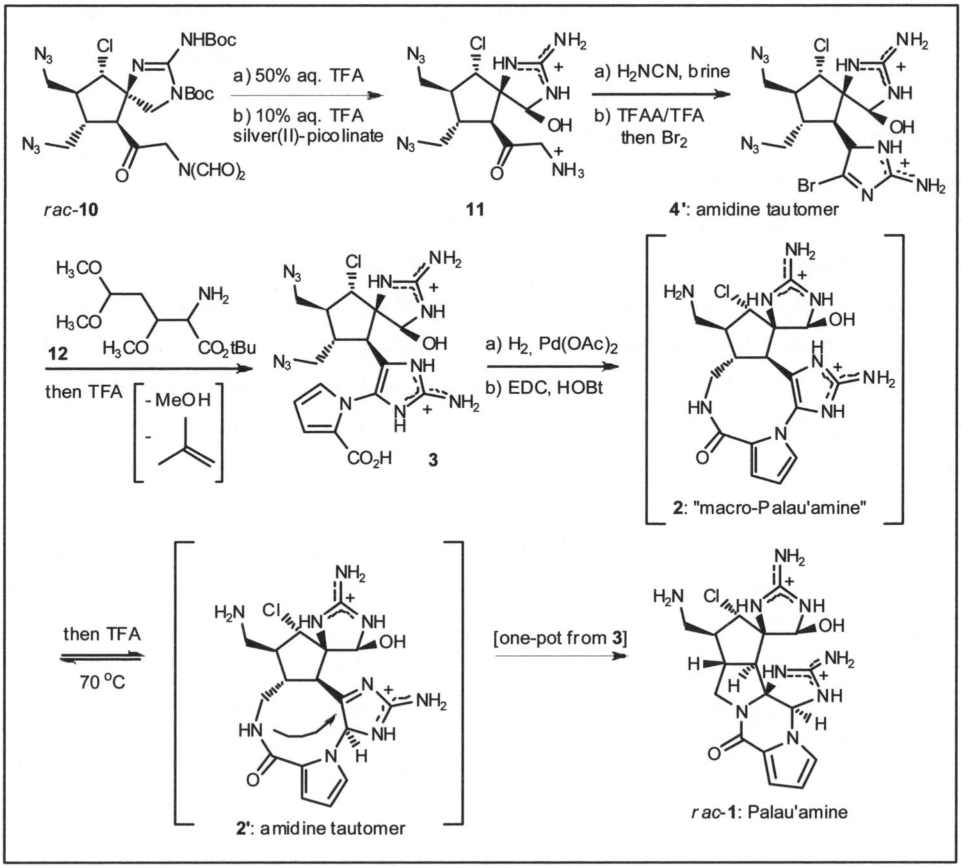

Scheme 3. Total Synthesis of *rac*-Palau'amine **1**.

With synthetic Palau' amine now in hand, the remaining questions about its bioactivity and biogenesis can be addressed. For instance, it will certainly be interesting to confirm the hypothesis that the ring closure of macro-Palau'amine **2** to the target **1** would be a synthetic reflection of the inherent reactivity hidden in the natural product or in its biogenetic precursors.

Total Synthesis of Enigmazole A[18]

Introduction. Enigmazole A (**1**), a novel phosphate-containing macrolide, was isolated in 2010 from the marine sponge *Cinachyrella enigmatica* of Papua New Guinea.[19] It was tested for antiproliferative activity in the NCI 60-cell antitumor assay and it was shown to be active against the 60 human tumor cell lines, with a mean GI50 value of 1.7 μM, but there was no particular pattern of tumor panel specificity or differential activity among the cell lines. On the other hand, some chromatographic fractions from the *C. enigmatica* extract, containing additional enigmazole analogues, selectively target aberrant receptor tyrosine kinase c-Kit. Since mutations causing constitutive activation of c-Kit have been implicated in a variety of cancers, the core enigmazole structure could provide a potential scaffold for the discovery and development of selective target-oriented antitumors.

Synthesis. The synthesis was accomplished at the University of California by Molinski and co-workers. Compound **1**, including eight stereogenic centers, is comprised of an 18-membered phosphomacrolide that contains an embedded exomethylene-substituted tetrahydropyran ring and an acyclic portion, spanning an oxazole moiety. Retrosynthetic analysis (Scheme 1) shows disconnession at the macrolide ester bond and C12-C13 bond leading to the "Eastern" fragment **2** that could be united with the remainder of the molecule by Wittig olefination. The central pyran ring was envisioned to arise in turn from a diastereoselective hetero-Diels-Alder (HDA) cycloaddition between aldehyde **3** and diene **4**. For construction of the Eastern fragment **2**, Molinski planned a key asymmetric allylation on a preformed oxazole synthon, in turn obtained from oxazol-2-ylzinc reagent **5**, by C20-C21 bond formation through Negishi coupling with **6**.[20]

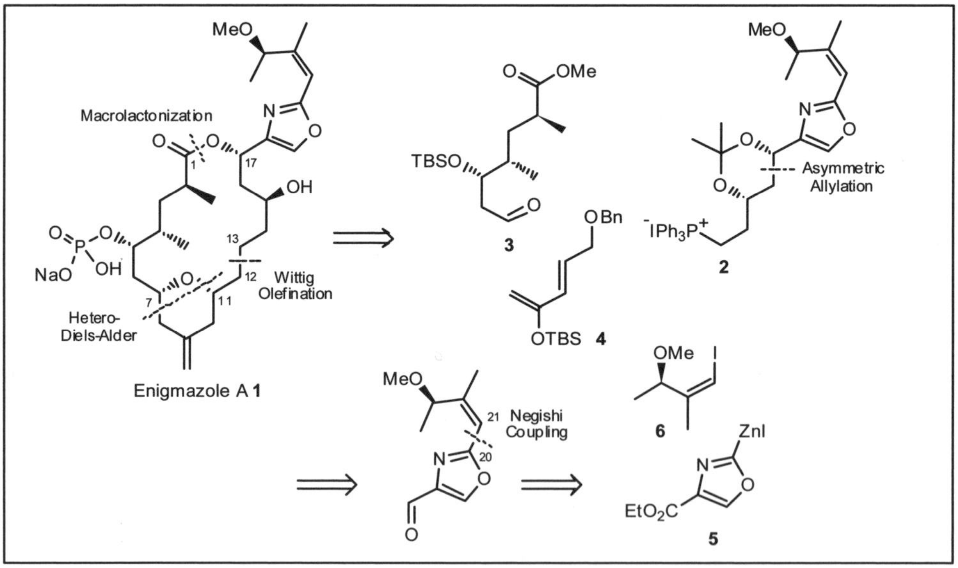

Scheme 1. Retrosynthetic Analysis of Enigmazole A.

Treatement of oxazol-2-yl zincate **5** with vinyl iodide **6** in the presence of Pd(0) resulted in smooth Negishi cross-coupling to give **7** (Scheme 2). Reduction to aldehyde **8** and subsequent asymmetric allylation with **9** gave alcohol **10** with excellent diastereoselectivity. Oxidative cleavage of the vinylidene double bond gave ketone **11**, which was subsequently reduced under Narasaka's conditions[21] to give the *syn*-1,3-diol and immediately protected as the acetonide **12**. Reductive removal of the benzoyl protecting group and elaboration of the resulting alcohol through the terminal iodide **13** afforded the eastern hemisphere phosphonium salt **2**.

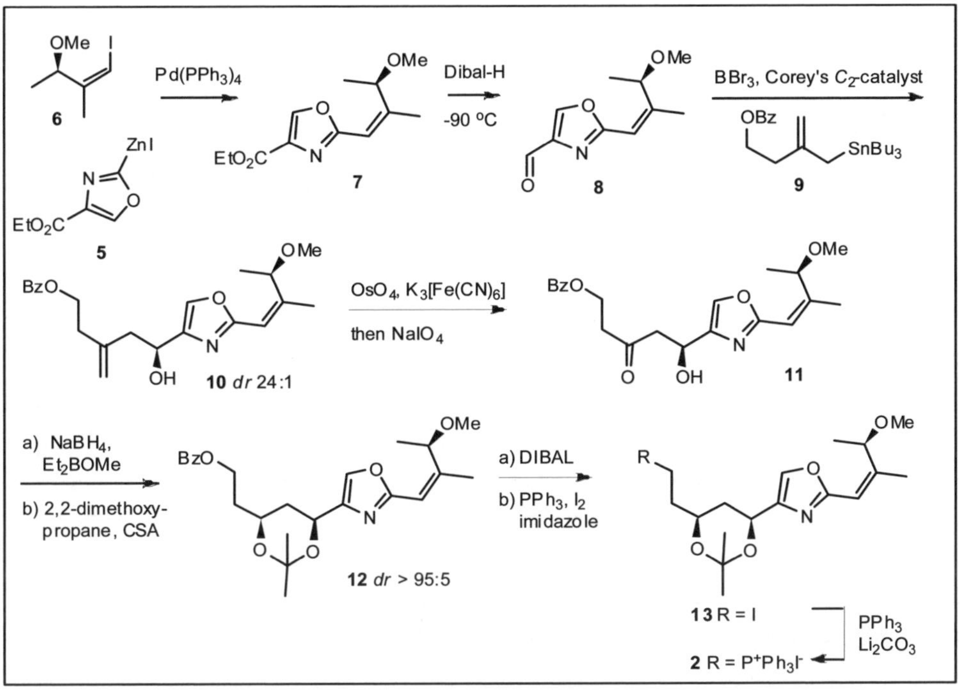

Scheme 2. Synthesis of Eastern Hemisphere Phosphonium Salt **2**.

Synthesis of aldehyde **3** proceeded from **14** (prepared by resolution of the racemic acid as the phenethylamine salt).[22] Key step was the Roush allylation[23] of **15**, providing **16** in good yield and diastereoisomeric ratio. Protection of the alcohol as its TBS ether followed by ozonolysis gave the aldehyde **3** (Scheme 3). Unification of "Eastern" and "Western" hemispheres of Enigmazole A began with HDA cycloaddition between aldehyde **3** and diene **4**. Treating a mixture of **3** and **4** with catalytic BF$_3$·OEt$_2$ led to a diastereoselective *substrate*-controlled HDA reaction. Optimization of the reaction included *in situ*, one-pot conversion to dimethylketal **17**. Hydrogenolysis of **17**, followed by Swern oxidation, gave aldehyde **18** that was immediately subjected to Wittig reaction with the ylide derived from **2**, affording intermediate **19**.

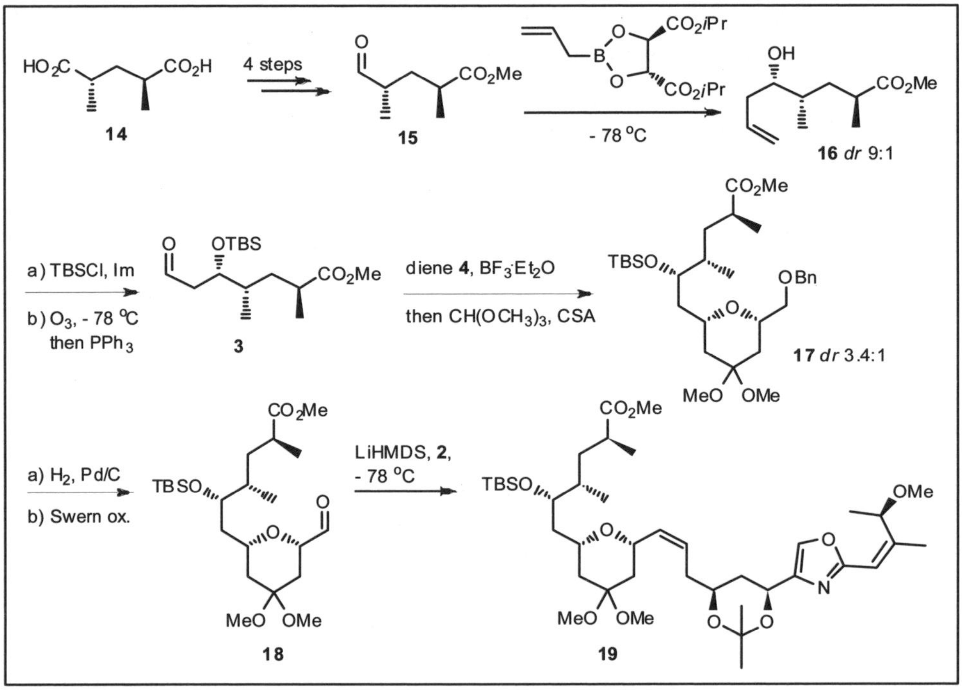

Scheme 3. Synthesis of Aldehyde **3** and Unification of Eastern and Western Hemispheres.

A deep investigation was necessary in order to obtain the 18-membered macrolide ring of Enigmazole A, instead of the competing 16-membered one. At the end, Molinski found that the order of C12-C13 hydrogenation-macrolactonization steps was crucial and carried out the sequence depicted in Scheme 4. Deprotection of **19** gave the corresponding dihydroxy acid, on which Keck macrolactonization furnished the desired macrolide **20** cleanly and exclusively. Selective hydrogenation using Wilkinson's catalyst, followed by acetal hydrolysis, provided ketone **21**. While standard Wittig olefination gave poor conversions, the nonbasic Lombardo's reagent[24] led to smooth olefination of ketone, affording **22**, after desilylation. The phosphate was introduced to the C5 secondary hydroxyl group of **22** as a protected phosphoramidite to give the fully protected natural product **23**. Treatment of **23** with K_2CO_3 smoothly cleaved the C15 acetate and both 9-fluorenylmethyl groups of the phosphate ester. Ion exchange provided finally Enigmazole A (**1**) as the Na^+ salt, identical in all respects to the natural product.

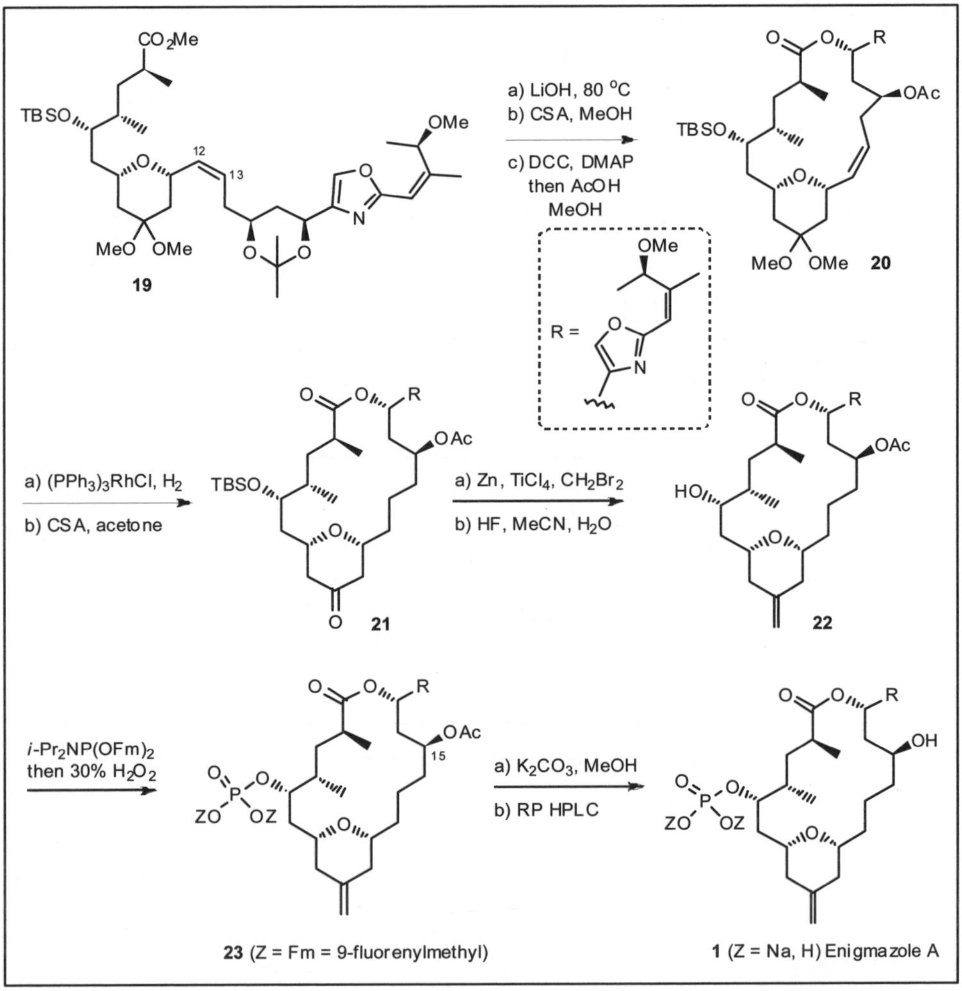

Scheme 4. Macrolactonization and Completion of the Synthesis of Enigmazole A **1**.

Total Synthesis of Syringolin A[25,26]

Introduction. The natural product Syringolin A (SylA, **1**) was isolated from strains of the bacterial plant pathogen *Pseudomonas syringae pv. Syringae* (*Pss*) in 1998.[27] It is biosynthesized under infection conditions by a mixed nonribosomal peptide synthetase (NRPS)/polyketide synthetase (PKS) cluster. SylA exerts potent biological activities such as inhibition of proliferation of neuroblastoma and ovarian cancer cells, resulting from potent, selective and, under physiological conditions, irreversible proteasome inhibition. Its potential as a promising lead structure for drug discovery efforts goes hand in hand with the fact that even slight structural variations of the natural product SylA have a significant impact on proteasome inhibition and subsite selectivity. Consequently, derivatization of SylA holds promise to lead to proteasome inhibitors with enhanced inhibitory and pharmacokinetic properties. To this end, the establishment of a synthetic route that allows the facile and rapid generation of SylA derivatives would be highly desirable.

Synthesis. Two total syntheses of SylA appeared in literature in 2010, at a distance of just a month and a half the former[25] from the latter.[26] In both cases the authors envisaged a convergent assembly strategy, utilizing commercially available amino acids to introduce all of the stereocenters and relying on the synthesis of a fully functionalized SylA macrocycle **2**, followed by introduction of the exocyclic side chain **3** at a late stage. This strategy would expedite the synthesis of other analogues (Scheme 1).

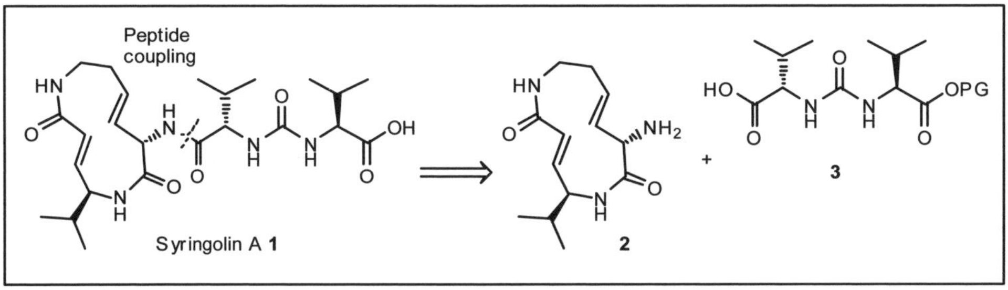

Scheme 1. Common Retrosynthetic Step.

Synthesis A. The first work was accomplished by Kaiser and co-workers of the Max-Planck Institute of Dortmund.[25] After that a macrolactamization approach proved to be not a suitable strategy for the synthesis of **2**, Kaiser revised his original synthetic approach to include ring-closing metathesis (RCM) of a conformationally preorganized precursor as a key reaction for ring closure.

The corresponding retrosynthesis (Scheme 2) from protected **2a** (PG = Boc) leads to the RCM tripeptide precursor **4**, that features a vinylglycine residue at its *N*-terminus. As vinylglycines are known to easily undergo unwanted isomerization to the α,β-conjugated system under peptide coupling conditions, the use of a previously reported phenyl selenyl derivative that can be transformed into a vinylglycine just prior to RCM was envisaged instead. Further disconnections at the peptide bonds result in three major building blocks that are all rapidly accessible by known methods. By means of standard peptide coupling procedures, RCM precursor **4** was readily achieved. However, all attempts to cyclise it to macrocycle **2a** failed again. Although different Grubbs catalysts and reaction conditions were screened, the desired product could not be isolated. It was clear that, to achieve ring closure, an alteration of the conformation prior to ring closure was necessary. The authors imagined that this could be achieved through simple modification of the RCM precursor by replacing the central double bond by a spatially different five-membered ring system, arising from dihydroxylation and subsequent acetonide formation. Reinstallation of the double bond could then be achieved by a Corey-Winter elimination[28] (Scheme 3).

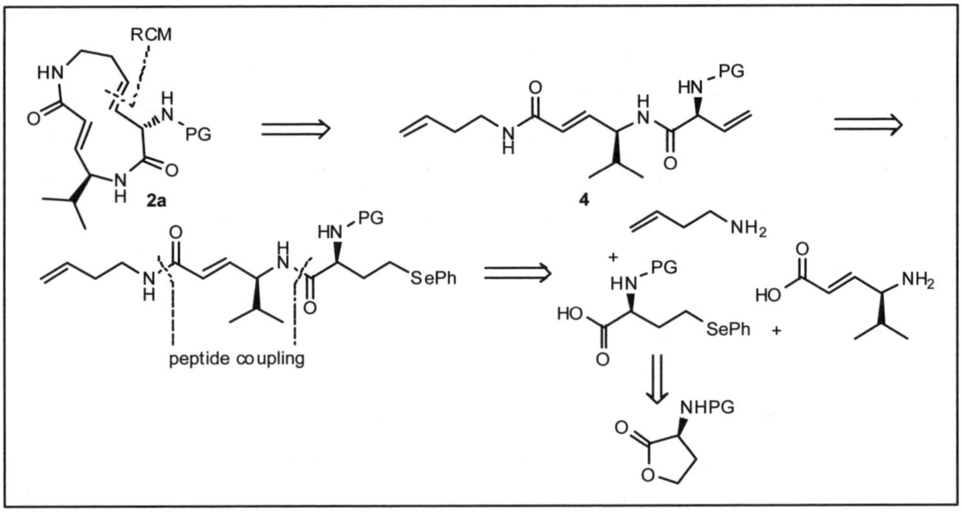

Scheme 2. Retrosynthesis for the Generation of the Core Macrocycle by Using RCM as a Key Step.

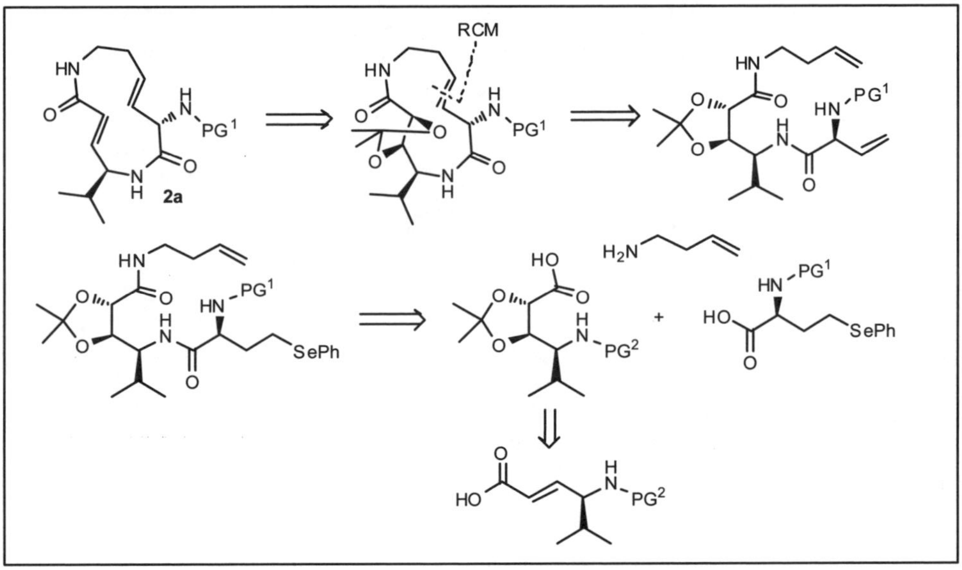

Scheme 3. Retrosynthesis for the Generation of the Core Macrocycle by Using RCM on a Conformationally Pre-Oriented Precursor.

Thus far, α,β-unsaturated methyl ester **5** was dihydroxylated with a high diastereomeric ratio (96% *dr*) by using osmium tetroxide and 4-methyl morpholine *N*-oxide (Scheme 4). Protection of diol as an acetonide, followed by methyl ester saponification and derivatization by peptide coupling with 3-butenylamine yielded **6**. Its conversion into **7** was achieved by chemoselective Boc deprotection with TMSOTf and 2,6-lutidine, followed by coupling with phenylselenyl building block **8**. One-pot oxidation and elimination yielded the vinylglycinyl intermediate and RCM precursor **9**. After screening of various catalysts and reaction conditions, Grubbs II catalyst allowed the generation of **10** with 2.6:1 *E/Z* diastereoselectivity and 49% yield of the desired (*E*) isomer. As the next step, in order to chemoselectively deprotect the acetonide group, a protecting group exchange was performed, yielding Troc-protected amine **11**. Microwave-assisted acidic cleavage of the acetonide to the dihydroxy intermediate and further conversion into thiocarbonate with 1,1'-thiocarbonyldiimidazole and DMAP, provided the necessary precursor for the Corey-Winter elimination. Elimination due to the desired (*E*)-configured double bond of SylA core macrocycle **12** was then achieved by heating the thiocarbonate in trimethyl phosphite for 2.5 h at 130 °C.

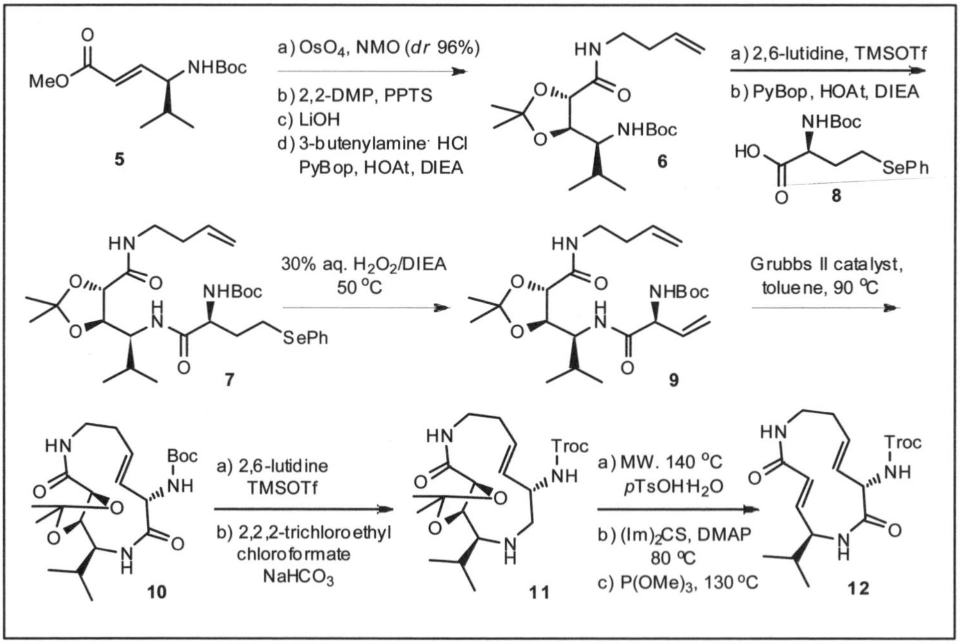

Scheme 4. Synthesis of the SylA Macrocycle Core Structure.

To complete the synthesis of SylA, the urea building block **3** (PG = Me) for attachment to the macrocyclic core was also required and was prepared by standard chemistry (Scheme 5). Then, Zn-mediated cleavage of the Troc-protecting group of **12** to free amine, followed by peptide coupling of **3** and methyl ester cleavage, led to the desired natural product Syringolin A **1**. Exploiting the convergent nature of the synthesis, all four stereoisomers of the urea dipeptide unit could be synthesised and coupled to the macrocycle core, in order to further biological assays and to study the influence of the side-chain stereochemistry on the overall structure of SylA.

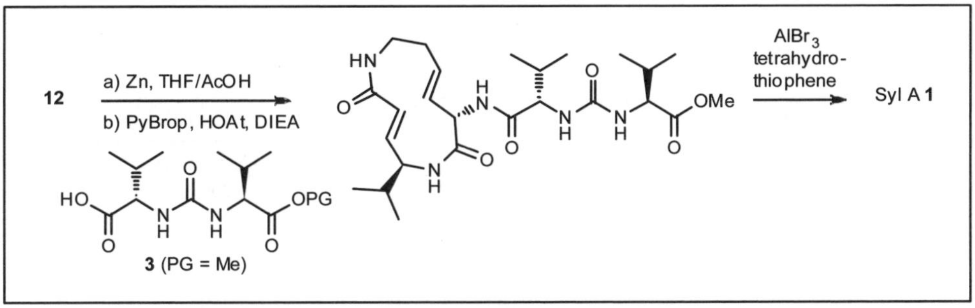

Scheme 5. Completion of the Synthesis of Syringolin A **1**.

Synthesis B. The second work was accomplished by Stephenson and co-workers of the Boston University of Massachusetts.[26] Also this total synthesis features a convergent approach, relying on 13 steps from commercially available materials, Garner's aldehyde and L-valine. The key 12-membered macrocycle core **2a** would be formed through an intramolecular peptide coupling at the less hindered position. Horner-Wadsworth-Emmons olefination would provide the α, β-unsaturated ester **13**. The (*E*)-olefin of **14** could be furnished by Johnson-Claisen rearrangement, and its precursor could be generated from easily accessible Garner's aldehyde **15**[29] (Scheme 6).

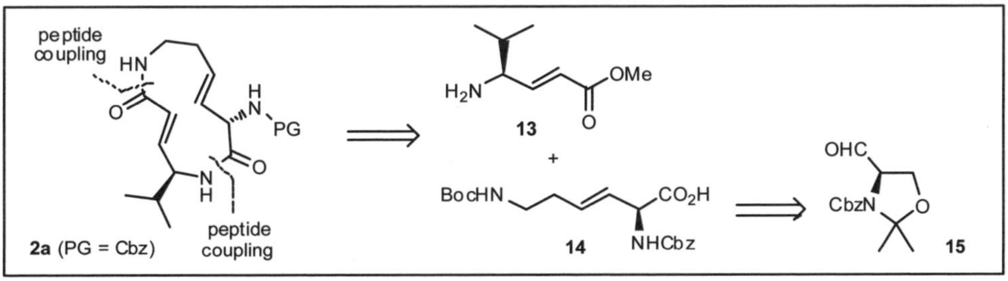

Scheme 6. Retrosynthetic Analysis.

Garner's aldehyde **15** was treated with vinyl magnesium bromide to afford vinyl alcohol **16** (Scheme 7). Johnson-Claisen rearrangement employing excess $CH_3C(OMe)_3$ in xylene at reflux provided the desired α, β-unsaturated ester **17** in high yield and excellent stereoselectivity ($E/Z > 95: 5$). Hydrolysis led to the carboxylic acid, which was subjected to Curtius rearrangement using DPPA, Et_3N, and *tert*-butanol, affording N-Boc protected amine **18**. Selective deprotection using *p*-TsOH and subsequent one-step oxidation with the Jones reagent, efficiently provided the desired amino acid **14**, with complete preservation of the enantiomeric purity.

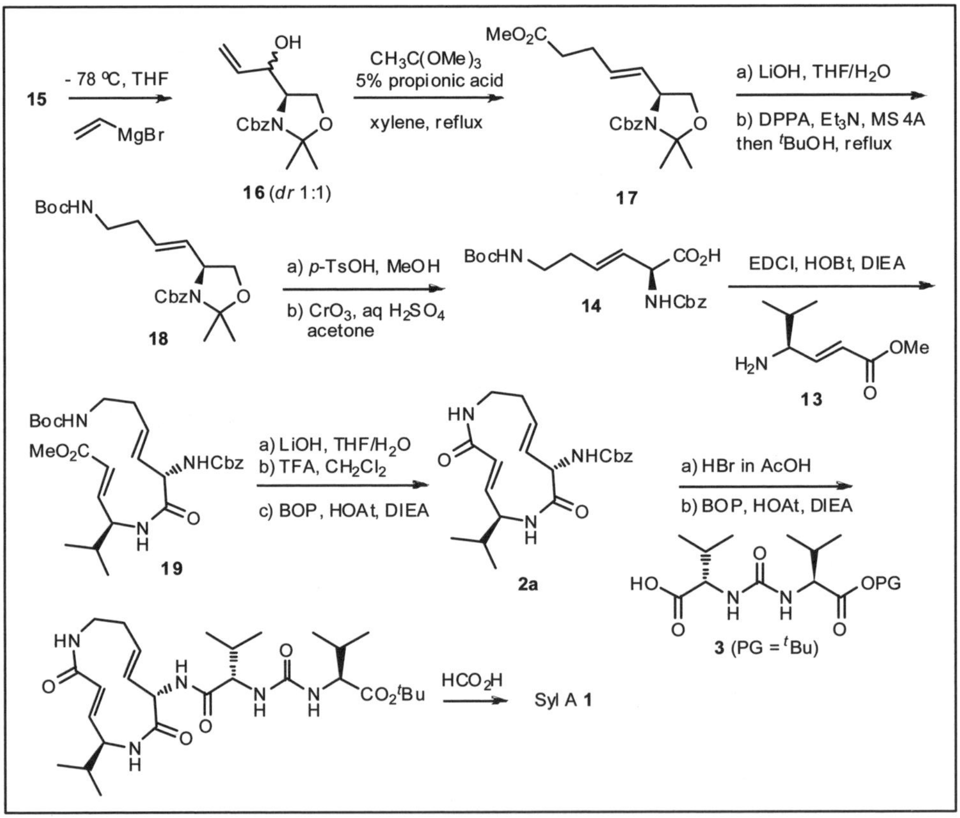

Scheme 7. Synthesis of Syl A **1**.

With both coupling fragments **13** and **14** in hand, the first peptide bond formation was accomplished using EDCI, HOBt, and *i*-Pr$_2$NEt to afford **19** in 70% yield. Removal of the C- and N-terminal protecting groups, followed by treatment with peptide coupling reagents BOP and HOAt, provided the macrocycle **2a** in 15% yield. Although the yield of this macrolactamization is low, this approach can be considered the most straightforward to Syringolin A and its analogues. The final stage of the synthesis required introduction of the urea side chain **3**, which was obtained by standard protecting group and peptide coupling chemistry.

Total Synthesis of Englerin A[30, 31, 32]

Introduction. Englerin A (**1**) is a guaiane sesquiterpene that was recently isolated[33] from the stem bark of *Phyllanthus engleri* collected in Tanzania. Biological evaluation revealed that Englerin A has potent and selective ability to inhibit renal cancer cell growth at the nanomolar level, showing to be 1–2 orders of magnitude more potent than taxol, in some cases. Its unique structure includes a tricyclic motif carrying two esters, one to a cinnamic acid (C6, Englerin A numbering) and the other to a glycolic acid (C9) residue, the latter being apparently crucial for potency and selectivity. Structure-activity relationship investigations of Englerin A would be helpful in identifying analogues for further development as drug targets. These studies are of vital importance because kidney cancer is a major cause of morbidity and mortality in adults, and until now no satisfactory drugs are available for its treatment.

Synthesis. Given this background, it is not surprising that synthetic interest in this target has been considerable in 2010. Two research groups, Ma and co-workers,[30] from the Chinese Academy of Science, and Echavarren and co-workers,[31] from the University of Tarragona, envisaged a common synthetic strategy. It relies on a gold-catalyzed cyclization of a proper functionalized enyne as the key cascade step to give the oxotricyclic core of Englerin A. On the other hand, Nicolaou and co-workers,[32] from the Chemical Synthesis Laboratory of Singapore, adopted a strategy featuring a [5 + 2] cycloaddition reaction to cast the seven-membered oxabicyclic key intermediate, in both racemic and optically active forms. While only the similar and impressive gold-catalyzed cyclization key step is here related for the syntheses set up by Echavarren and Ma, the approach by Nicolaou is reported in details.

Syntheses A[30] and B.[31] Both the syntheses start from inexpensive chiral pool-derived compounds, namely (*R*)-citronellal or geraniol. The key step is the gold(I)-catalyzed [2+2+2] alkyne/alkene/carbonyl cycloaddition of similar 1,6-enynes bearing a carbonyl group, in which two C-C and one C-O bonds are formed in a domino process.[34] In both cases, the propargylic stereocenter of the substrate exerts an exquisite stereocontrol in the cyclization process, yielding a single diastereoisomer of the desired oxatricyclic derivative. Both the total syntheses are completed in a concise, efficient and easily scalable way and can provide access to intermediates potentially useful for the preparation of a variety of analogues. In the work of Ma, the final route to (-)-Englerin A doesn't need of protective groups and provides another example of protecting-group free synthesis of a natural product.

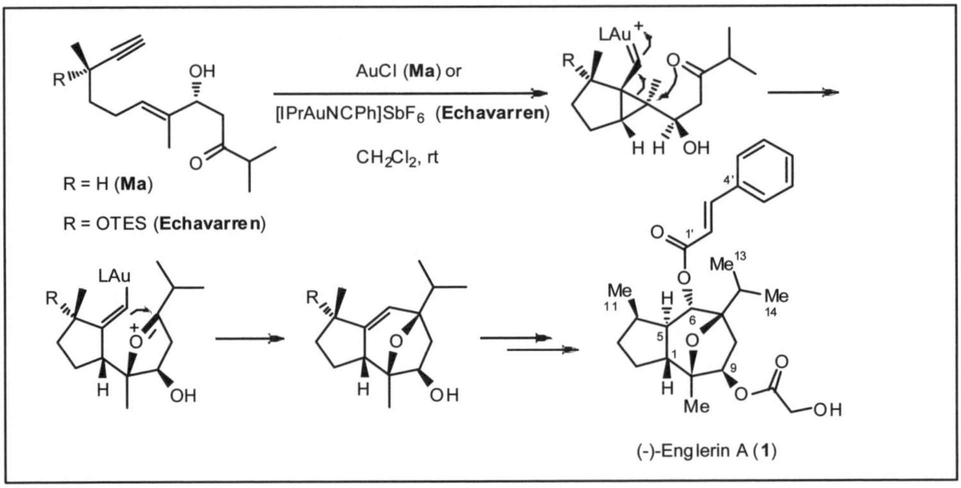

Scheme 1. Asymmetric Synthesis of (-)-Englerin A (**1**): Mechanistic Rationale for the Key Gold(I)-Catalyzed Cyclization.

Synthesis C.[32] Scheme 2 shows, in retrosynthetic format, the key carbon-carbon bond disconnections employed to devise the synthetic strategy of Nicolaou toward Englerin A (**1**), in which a [5 + 2] cycloaddition reaction played the key role in the formation of the bicyclo [3.2.1] ring framework of the molecule. The envisioned [5 + 2] cycloaddition reaction between oxopyrilium species **2** and ethyl acrylate (**3**) was expected to lead to oxabicyclic enone **4** in its racemic form, whereas an asymmetric synthesis of either enantiomer of the natural product could, in principle, arise through the use of this reaction engaging chiral sulfonamide acrylate derivative **5** or its enantiomer.

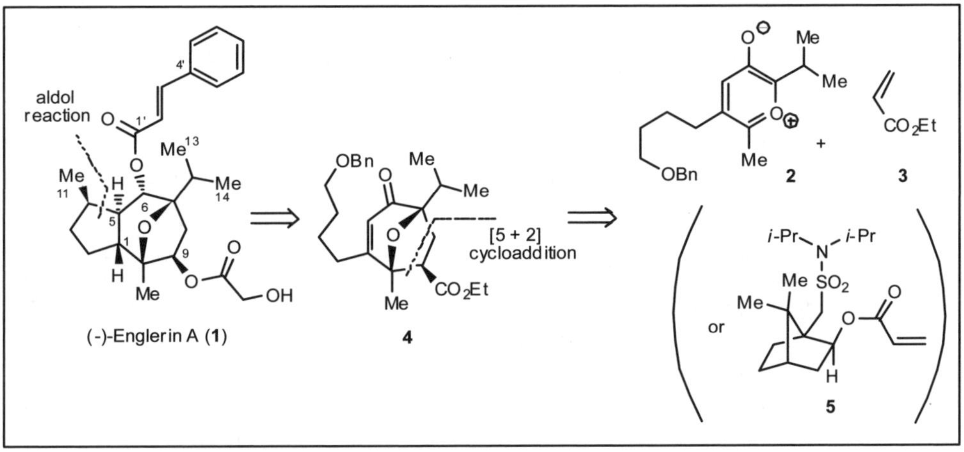

Scheme 2. Retrosynthetic Disconnection of Englerin A.

The synthesis of oxabicyclic enone **4** commenced from the readily available propargylic alcohol **6**, whose regioselective iodination led to vinyl iodide **7** (Scheme 3). Coupling of the latter intermediate with trimethylsilyl (TMS) acetylene in the presence of cat. Pd(PPh$_3$)$_4$, followed by TMS group removal, afforded hydroxyl enyne **8**. Gold-catalyzed ring closure of **8** (Ph$_3$PAuCl, AgOTf)[35] then generated furan system **9**. Formylation of the latter compound, followed by reaction with i-PrMgCl afforded hydroxy furan **10**. The latter intermediate underwent Achmatowicz rearrangement[36] upon exposure to m-CPBA to give ring-expanded lactol **11**. With the stage set for the crucial [5 + 2] cycloaddition reaction, generation of oxopyrilium species **2** (see Scheme 2) from **11** (MsCl, i-Pr$_2$NEt) and its reaction with ethyl acrylate (**3**) were investigated.

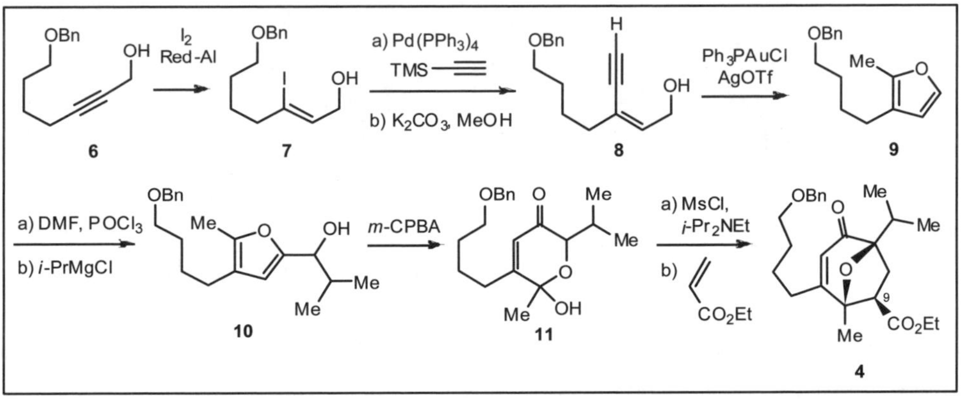

Scheme 3. Synthesis of Oxabicyclic Enone **4**.

The optimized conditions were established by employing substoichiometric amounts of i-Pr$_2$NEt under high dilution to furnish bicycle **4** together with its 9α isomer (C$_{9\beta}$:C$_{9\alpha}$ ca. 8:1) in 46% yield. Sequential exposure of pure **4** to catalytic hydrogenation conditions resulted in reduction of the olefinic bond and cleavage of the benzyl ether to afford, stereoselectively, hydroxy ketoester **12** (Scheme 4).

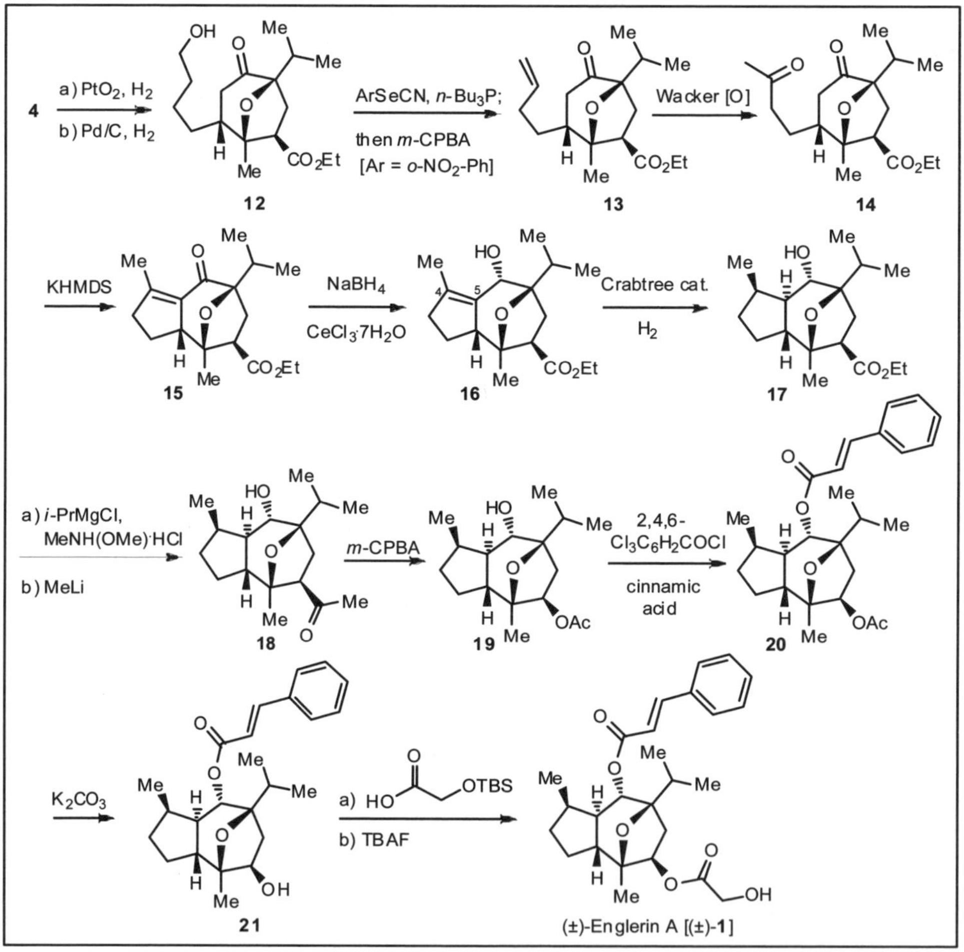

Scheme 4. Completion of the Total Synthesis of (±)-Englerin A [(±)-**1**].

Dehydration of **12** through its selenide derivative, followed by subjecting the resulting terminal olefin **13** to Wacker oxidation[37] conditions (O$_2$, PdCl$_2$ cat., CuCl cat.), afforded diketone **14**. Treatment of compound **14** with KHMDS led, through an intramolecular aldol/dehydration sequence, to enone **15**.

Stereoselective reduction of ketone **15** under Luche conditions[38] furnished allylic alcohol **16**, which underwent stereoselective hydrogenation (C4-C5 olefinic bond) in the presence of Crabtree's catalyst[39] to afford tricyclic system **17** as a single diastereoisomer. The desired oxygenation at C9 was achieved through a Baeyer-Villiger oxidation of methyl ketone **18**, obtained, through intermediate Weinreb amide, from ethyl ester **17**. In order to attach the proper ester side chains on its cyclic framework, **19** was first coupled with cinnamic acid through the Yamaguchi protocol[40] to afford the natural analogue (±)-Englerin B acetate [(±)-**20**] from which the acetyl group was removed to give (±)-Englerin B [(±)-**21**]. Installation of the desired glycolic acid residue on the latter compound proceeded smoothly under Yamaguchi conditions to afford, upon desilylation, (±)-Englerin A [(±)-**1**].

Having secured a racemic entry to Englerin A, an asymmetric synthesis of oxabicyclic enone **4** through the [5 + 2] cycloaddition reaction between oxopyrilium species **2** and chiral sulfonamide acrylate derivative **5**[41] was pursued, as shown in Scheme 5. Thus, generation of oxopyrilium species **2** from **11** in the presence of **5** delivered oxabicyclic enones **22** and **23** as a chromatographically separable mixture (**22:23** ca. 1:2).

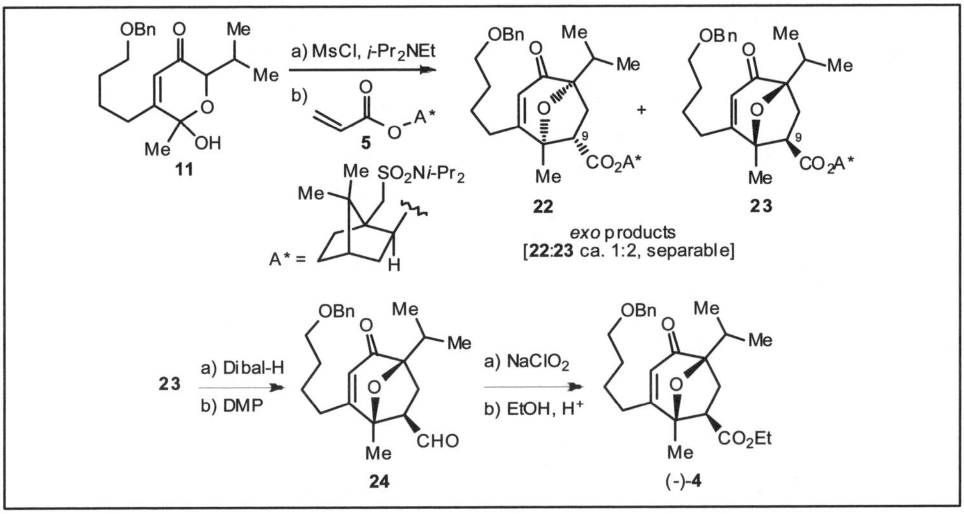

Scheme 5. Asymmetric Synthesis of Oxabicyclic Enone **4**.

Caution was exercised at this stage for possible epimerization at C9 under basic saponification conditions for **23**. Therefore, enone ester **23** was subjected to a two-step reduction/oxidation sequence to afford aldehyde **24**. Oxidation of aldehyde **24** to the corresponding acid, followed by esterification of the latter under acidic conditions, proceeded smoothly to furnish optically active enone ethyl ester (-)-**4**.

The arrival at optically active **4**, therefore, constitutes a formal asymmetric synthesis of (-)-Englerin A [(-)-**1**].

It should be recognized that, even though rather far from the so called "ideal synthesis", the classical "step by step" approach of Nicolaou (for 2010, see also Nicolaou, K. C. et al. in The Total Synthesis and Structural Revision of Vannusals A and B[42]) is highly instructive, always disclosing a richness of fitting procedures, not all quite recent, but all exhibiting extraordinary performances on substrates possessing a variety of sensitive functional groups.

REFERENCES

1.   For lead references on the role of natural products in drug development, see: (a) Wilson, R. M.; Danishefsky, S. J. *J. Org. Chem.* **2006**, *71*, 8329-8351. (b) Cragg, G. M.; Grothaus, P. G.; Newman, D. J. *Chem. Rev.* **2009,** *109,* 3012-3043. (c) Butler, M. S. *J. Nat. Prod.* **2004,** *67,* 2141-2153. (d) Kinghorn, A. D. *J. Med. Chem.* (Book Review) **2010**, *53*, 2329-2329.

2.   Wender, P.A., Miller, B. L. *Nature* **2009**, *460*, 197-201.

3.   (a) Trost, B. M. *Angew. Chem. Int. Ed. Engl.* **1995**, *34*, 259-281. (b) Trost, B. M. *Science*, **1991**, *254*, 1471-1477.

4.   Wender, P. A.; Verma, V. A.; Paxton, T. J.; Pillow, T. H. *Acc. Chem. Res.* **2008**, *41*, 40-49.

5.   Burns, N. Z.; Baran, P. S.; Hoffmann, R. W. *Angew. Chem.Int. Ed.* **2009**, *48*, 2854-2867.

6.   (a) Young, I. S.; Baran, P. S. *Nature Chem.* **2009**, *1*, 193-205. (b) Baran, P. S.; Maimone, T. J.; Richter, J. M. *Nature* **2007**, *446*, 404-408.

7.   (a) Kim, J. ; Movassaghi, M. *Chem. Soc. Rev.* **2009**, *38*, 3035-3050. (b) Bulger, P. G.; Bagal, S. K.; Marquez, R. *Nat. Prod. Rep.* **2008**, *25*, 254-297.

8.   Grondal, C.; Jeanty, M.; Enders, D. *Nature Chem.* **2010**, *2*, 167-178.

9.   (a) Davies, H. M. L.; Sorensen, E. J. *Chem. Soc. Rev.* **2009**, *38*, 2981-2982. (b) Nicolau, K. C.; Chen, J. S. *Chem. Soc. Rev.* **2009**, *38*, 2993-3009. (c) Nicolau, K. C.; Edmonds, D. J.; Bulger, P. G. *Angew. Chem. Int. Ed.* **2006**, *45*, 7134-7186.

10.   Gaich, T., Baran, P. S. *J. Org. Chem.* **2010**, *75*, 4657-4673.

11.   Hendrickson, J. B. *J. Am. Chem. Soc.* **1975**, *97*, 5784.

12.   Seiple, I. B.; Su, S.; Young, I. S.; Lewis, C. A.; Yamaguchi, J.; Baran, P. S. *Angew. Chem. Int. Ed.* **2010**, *49*, 1095-1098.

13.   (a) Kinnel, R. B.; Gehrken, H. -P.; Scheuer, P. J. *J. Am. Chem. Soc.* **1993**, *115*, 3376-3377. (b) Kinnel, R. B.; Gehrken, H. -P.; Swali, R.; Skoropowski, G.; Scheuer, P. J. *J. Org. Chem.* **1998**, *63*, 3281-3286.

14.   a) Grube, A.; Köck, M. *Angew. Chem. Int. Ed.* **2007**, *46*, 2320-2324. b) Kobayashi, H.; Kitamura, K.; Nagai, K.; Nakao, Y.; Fusetani, N.; van Soest, R. W. M.; Matsunaga, S. *Tetrahedron Lett.* **2007**, *48*, 2127-2129, c) Buchanan, M. S.; Carroll, A. R.; Addepalli, R.; Avery, V. M.; Hooper, J. N. A.; Quinn, R. J. *J. Org. Chem.* **2007**, *72*, 2309-2317.

15.   Al Mourabit, A.; Potier, P. *Eur. J. Org. Chem.* **2001**, 237-243.

16.   Köck, M.; Grube, A.; Seiple, I. B.; Baran, P. S. *Angew. Chem. Int. Ed.* **2007**, *46*, 6586-6594.

17.   Yamaguchi, J.; Seiple, I. B.; Young, I. S.; O'Malley, D. P.; Maue, M.; Baran, P. S. *Angew. Chem. Int. Ed.* **2008**, *47*, 3578-3580.

18.   Skepper, C. K.; Quach, T.; Molinski, T. F. *J. Am. Chem. Soc.* **2010**, *132*, 10286-10292.

19.   Oku, N.; Takada, K.; Fuller, R. W.; Wilson, J. A.; Peach, M. L.; Pannell, L. K.; McMahon, J. B.; Gustafson, K. R. *J. Am Chem. Soc.* **2010**, *132*, 10278-10285.

20.   Negishi, E.; Van Horn, D. E.; King, A. O.; Okukado, N. *Synthesis* **1979**, 501-502.

21.   Narasaka, K.; Pai, F.-C. *Tetrahedron* **1984**, *40*, 2233-2238.

22.   Gruenfeld, N.; Stanton, J. L.; Yuan, A. M.; Ebetino, F. H.; Browne, L. J.; Gude, C.; Huebner, C. F. *J. Med. Chem.* **1983**, *26*, 1277-1282.

23.   (a) Roush, W. R.; Walts, A. E.; Hoong, L. K. *J. Am. Chem. Soc.* **1985**, *107*, 8186-8190. (b) Roush, W. R.; Palkowitz, A. D.; Ando, K. *J. Am.Chem. Soc.* **1990**, *112*, 6348-6359.

24.   Lombardo, L. *Org. Synth.* **1987**, *65*, 81-85.

25.   Clerc, J.; Schellenberg, B.; Groll, M.; Bachmann, A. S.; Huber, R.; Dudler, R.; Kaiser, M. *Eur. J. Org. Chem.* **2010**, 3991-4003.

26.   Dai, C.; Stephenson, C. R. J. *Org. Letters* **2010**, *12*, 3453-3455.

27.   (a) Wäspi, U.; Blanc, D.; Winkler, T.; Ruedi, P.; Dudler, R. *Mol. Plant-Microbe Interact.* **1998**, *11*, 727. (b) Wäspi, U.; Hassa, P.; Staempfli, A. A.; Molleyres, L. P.; Winker, T.; Dudler, R. *Microbiol. Res.* **1999**, *154*, 89.

28.   Corey, E. J.; Winter, R. A. E. *J. Am. Chem. Soc.* **1963**, *85*, 2677-2678.

29.   Garner, P.; Park, J. M. *J. Org. Chem.* **1987**, *52*, 2361.

30.   Zhou, Q.; Chen, X.; Ma, D. *Angew. Chem. Int. Ed.* **2010**, *49*, 3513-3516.

31.   Molawi, K.; Delpont, N.; Echavarren, A. M. *Angew. Chem. Int. Ed.* **2010**, *49*, 3517-3519.

32.   Nicolaou, K. C., Kang, Q.; Ng, S. Y.; Chen, D. Y.-K. *J. Am. Chem. Soc.* **2010**, *132*, 8219-8222.

33.   (a) Ratnayake, R.; Covell, D.; Ransom, T. T.; Gustafson, K. R.; Beutler, J. A. *Org. Lett.* **2009**, *11*, 57-60. (b) Beutler, J. A.; Ratnayake, R.; Covell, D.; Johnson, T. R. WO2009/088854, 2009.

34.   a) Jiménez-Núñez, E.; Claverie, K. C.; Nieto-Oberhuber, C.; Echavarren, A. M. *Angew. Chem.* **2006**, *118*, 5578; *Angew. Chem. Int. Ed.* **2006**, *45*, 5452. b) Jiménez-Núñez, E.; Echavarren, A. M. *Chem. Commun.* **2009**, *45*, 7327. For reviews, see: c) Fürstner, A.; Davies, P. W. *Angew. Chem.* **2007**, *119*, 3478; *Angew. Chem. Int. Ed.* **2007**, *46*, 3410. d) Michelet, V.; Toullec, P. Y.; Genêt, J.-P. *Angew. Chem.* **2008**, *120*, 4338; *Angew. Chem. Int. Ed.* **2008**, *47*, 4268. e) Jiménez-Núñez, E.; Echavarren, A. M. *Chem. Rev.* **2008**, *108*, 3326. f) Gorin, D. J.; Sherry, B. D.; Toste, F. D. *Chem. Rev.* **2008**, *108*, 3351.

35.   Du, X.; Song, F.; Lu, Y.; Chen, H.; Liu, Y. *Tetrahedron* **2009**, *65*, 1839-1845.

36.   Achmatowicz, O.; Bielski, R. *Carbohydr. Res.* **1977**, *55*, 165-176.

37.   Clement, W. H.; Selwitz, C. M. *J. Org. Chem.* **1964**, *29*, 241-243.

38.   (a) Molander, G. A. *Chem. Rev.* **1992**, *92*, 29-68. (b) Luche, J. L. *J. Am. Chem. Soc.* **1978**, *100*, 2226-2227.

39.   Crabtree, R. H.; Davis, M. W. *J. Org. Chem.* **1986**, *51*, 2655-2661.

40.   Inanaga, J.; Hirata, K.; Saeki, H.; Katsuki, T.; Yamaguchi, M. *Bull.Chem. Soc. Jpn.* **1979**, *52*, 1989-1993.

41.   Oppolzer, W.; Chapuis, C.; Bernardinelli, G. *Tetrahedron Lett.* **1984**, *25*, 5885-5888.

42.   (a) Nicolaou, K. C.; Ortiz, A.; Zhang, H.; Dagneau, P.; Lanver, A.; Jennings, M. P.; Arseniyadis, S.; Faraoni, R.; Lizos, D. E. *J. Am. Chem. Soc.* **2010**, *132*, 7138-7152. (b) Nicolaou, K. C.; Ortiz, A.; Zhang, H.; Guella, G. *J. Am. Chem. Soc.* **2010**, *132*, 7153-7176.

Davide Tessaro
Prof. of Fundamentals of Chemistry

Address:

Dipartimento di Chimica, Materiali ed Ingegneria Chimica "G. Natta"
Politecnico di Milano
Via Mancinelli, 7
20131 Milano
davide.tessaro@polimi.it

Education:
-       October 2001: Master Degree in Chemistry at Università degli Studi di Milano. Thesis title: "Cyanuration of polycyclic and di- aldehydes catalysed by oxynitrilase". Relator: prof. Bruno Danieli; co-relator: dr. Sergio Riva.
-       May 2005: PhD in Industrial Chemistry and Chemical Engineering, at Politecnico of Milano. Thesis title: "Use of L-Hydantoinase as a new approach to the synthesis of β-amino acids". Tutor: prof. Stefano Servi. Relator: prof. Claudio Fuganti.
-       February 2007: Assistant Professor at Politecnico of Milano, Department of Chemistry, Materials and Chemical Engineering "G. Natta"

Professional experience:
-       November-December 2001: stage in Praha (Czech Republic) at the Microbiology Institute working on sugar acylation catalysed by subtilisin. Tutors: prof. Vladimir Křen, dr. Sergio Riva.
-       October-December 2002 and March-May 2003: stage in Stuttgart (Germany) at the Biotechnology Institute of the Stuttgart University, working for the PhD thesis on dihydrouracyl hydrolysis catalysed by hydantoinases. Tutors: prof. Crystoph Syldatk, prof. Stefano Servi

Dr. Tessaro has been working for a decade on the biocatalysis field. In particular, in the latest period he focused on chemo-enzymatic strategies for deracemization of amino acid derivatives. He is currently working on exploitation of thioesters for carrying out a Dynamic Kinetic Resolution based on simultaneous base-catalysed substrate racemization and enantioselective enzymatic hydrolysis, thus obtaining a product high both in yield and in enantiomeric excess.
So far, he teaches Fundamentals of Chemistry at the Politecnico di Milano, in the Industrial Engineering School.

Biocatalysis in Organic Synthesis:
New Items In The Chemist's Toolbox
Critical Survey covering the years 2009-2011

Davide Tessaro[*]
Politecnico di Milano
Dip. CMIC "G. Natta"
Via Mancinelli 7, 20131 Milano
davide.tessaro@polimi.it

The following journals have been abstracted:

Advanced Synthesis & Catalysis (22 references)

Analytical chemistry (13 references)

Angewandte Chemie (International ed. in English) (40 references)

Applied Biochemistry and Biotechnology (62 references)

Biocatalysis and Biotransformation (27 references)

Biotechnology and Bioengineering (32 references)

Chemical Communications (Cambridge, United Kingdom) (17 references)

Chemical reviews (5 references)

Chemical Society Reviews (9 references)

Chemistry A European Journal (40 references)

Current Opinion in Biotechnology (8 references)

Dalton transactions (Cambridge, England : 2003) (9 references)

European Journal of Organic Chemistry (11 references)

Green Chemistry (15 references)

Journal of Agricultural and Food Chemistry (12 references)

Journal of Biotechnology (30 references)

Journal of Industrial Microbiology & Biotechnology (12 references)

Journal of Molecular Catalysis B: Enzymatic (119 references)

Journal of Organic Chemistry (20 references)

Journal of the American Chemical Society (102 references)

Nature (47 references)

Nature Chemical Biology (16 references)

Nature Chemistry (4 references)

Nature structural & molecular biology (4 references)

Organic Letters (9 references)

Organic Process Research & Development (20 references)

Proceedings of the National Academy of Sciences of the United States of America (66 references)

Process Biochemistry (49 references)

Science (New York, N.Y.) (19 references)

Tetrahedron (5 references)

Tetrahedron Letters (15 references)

Tetrahedron: Asymmetry (30 references)

Trends in Biotechnology (11 references)

1. INTRODUCTION

Enzymes, though pertaining most appropriately to the biological world, are catalyst whose action is fully explainable in terms of rational chemical and physical principles. Consequently, their *ex vivo* action can be fully exploited by the chemist who is searching for alternative to the more traditional chemocatalysis. Progress in the biotechnology field continuously increases their commercial availability, their stability, their flexibility, their "chemist-friendliness": several biocatalysts are accessible as on-the-shelf reagents and can be used in organic chemistry with no need to be highly accomplished enzymologists.

Moreover, their peculiar characteristics of activity, specificity, mildness, low toxicity, environmental compatibility obviously deserve an important role on the stage.

So far, enzymes are known to catalyse almost any reaction of organic chemistry, and can constitute a very valuable tool when aiming for new synthetic strategies. Scope of the present survey is to focus on the most recent developments in biocatalysis for utilization in organic synthesis, giving examples of successful methodologies as well as new promising approaches, and showing how enzymes can be easily integrated in the more known chemical procedures to supply the desired product with excellent quality features.

2. OXIDATIONS

Oxidations are a central transformation in organic chemistry. Unfortunately, they often require transition metal or otherwise noxious reagents raising environmental issues together with safety concerns about storage and handling of the oxidizing agents. In addition, it is not always easy to achieve good results in terms of chemo- or regioselectivity, so reaction yields can be disappointing, purification of the target molecule from similar byproducts may be challenging and it is sometimes crucial to control thoroughly the reaction conditions all over the conversion. Together with newly developed inorganic or metallorganic reagents, biocatalysis can provide a "greener" alternative to traditional reagents. A recent review[1] effectively describes the present state of the art in the use of enzyme and microorganisms for catalytic oxidation and oxifunctionalization chemistry. Here, a great variety of biocatalytic oxidation reactions is discussed, including their general usefulness, their respective outcome and, if necessary, the possible cofactor regeneration strategies.

A significant example of the biocatalysis potential is the one-pot tandem oxidation of methylene groups[2] illustrated in *Scheme 1*.

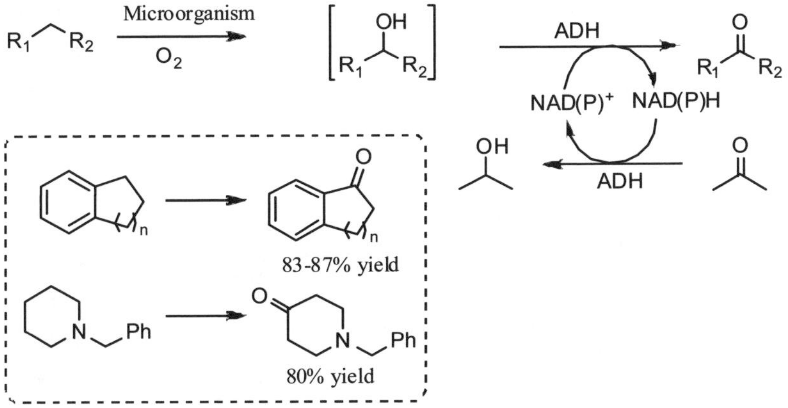

Scheme 1

In this case, a whole-cell biocatalyst introduces selectively an hydroxyl group on the target carbon; then an isolated alcohol dehydrogenase carries out a second oxidation leading to the final ketone and employing acetone as a cosubstrate for NADH regeneration. On the studied substrates, the regioselectivity was absolute, the yields were satisfactory and high turnover numbers for the cofactor recycling were achieved. Most interestingly, even non activated methylene groups can be converted to the respective ketones.

Laccases are copper metalloenzymes whose importance has been growing for the latest years.[3] Working in mild conditions, they are able to oxidatively couple phenol derivatives effectively and selectively: for instance they have been employed for the synthesis of phenoxazine derivatives[4] for therapeutic and bioanalytical applications (*Scheme 2*).

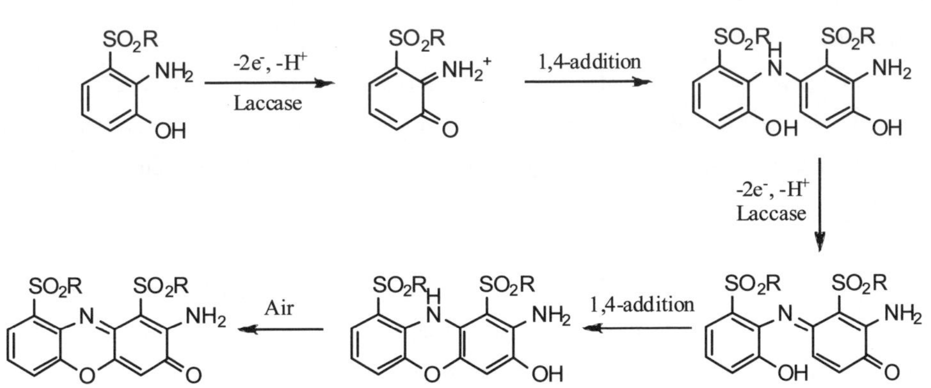

Scheme 2

Monoaminooxidases (MAO) convert primary and secondary amines into the respective imine. A novel application consists in the desimmetrization of *cis*-disubstituted (*meso*) pyrrolidines which can undergo a selective nucleophilic addition[5] (*Scheme 3*) or be directly employed in an Ugi-type multi-component reaction.[6]

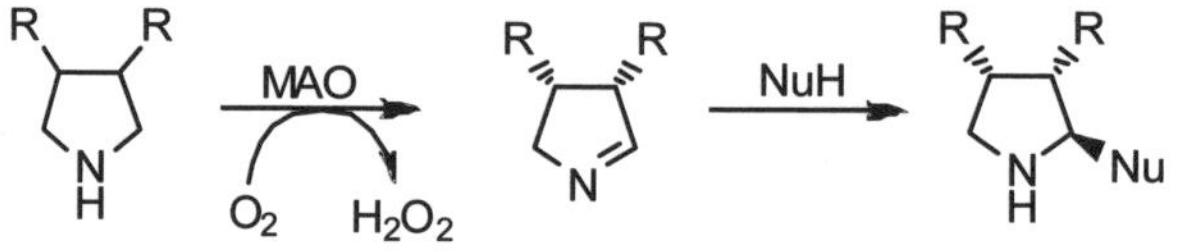

Scheme 3

Baeyer-Villiger monoxygenases (BVMO) can be exploited for the enantioselective synthesis of β-amino acids or β-amino alcohols, as shown in *Scheme 4*.[7]

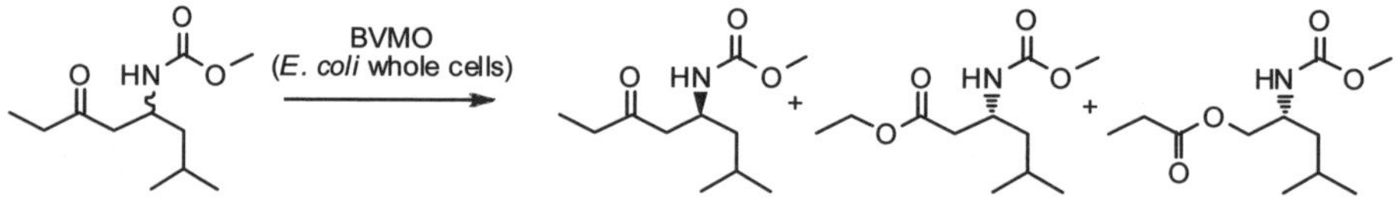

Scheme 4

3. REDUCTIONS

Old Yellow Enzyme is the first flavin-dependent enzyme identified in yeast.[8] Lately, there is a renewed interest in the catalytic performances of this protein, particularly concerning the selective reduction of an activated carbon-carbon double bond, where the enzymatic reaction is often a very convenient alternative compared to the metal-catalyzed hydrogen transfer, in terms of selectivity, catalyst cost, environmental impact and downstream processes.

For example, both isomers of citronellal can be produced through the selective reduction of the unsaturated precursor with excellent e.e. and absolute chemoselectivity,[9] and the so-called "Roche-Ester" can be synthesized with good yield in enantiopure fashion (*Scheme 5*).[10]

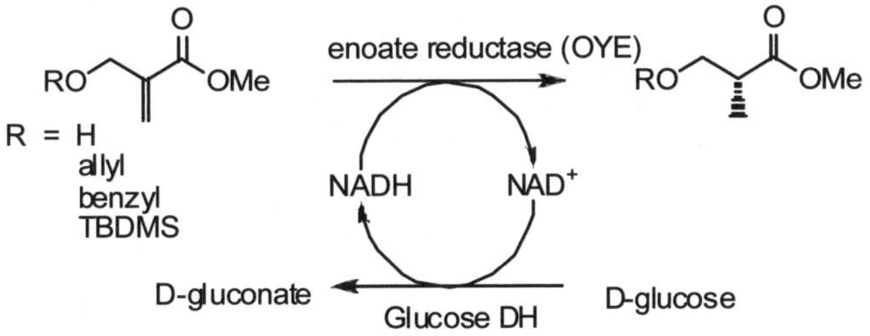

Scheme 5

The cofactor regeneration issue has been recently approached from an unusual point of view: the addition of a proper amount of a photocatalyst into the reaction mixture promotes the light-driven reduction of the flavin, thus allowing the photoenzymatic reduction of the double bond (*Scheme 6*).[11]

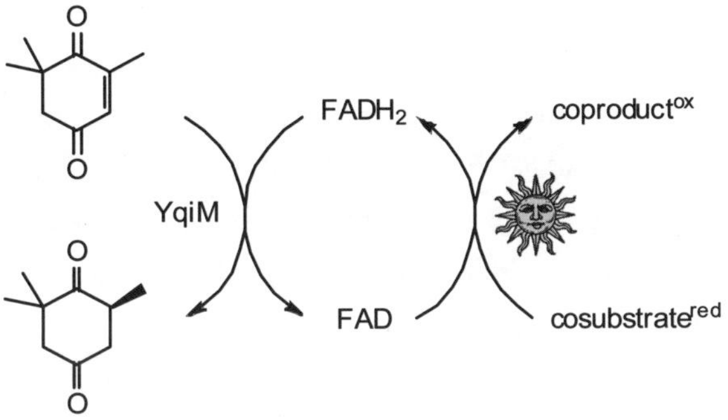

Scheme 6

Usually, enzyme can't catalyze the direct hydrogen addition to double bond, even if there is a very interesting example of a metal substitution in the zinc-dependent carbonic anhydrase leading to a new semi-artificial protein capable of reducing or isomerising the model substrate *cis*-stilbene (*Scheme 7*).[12]

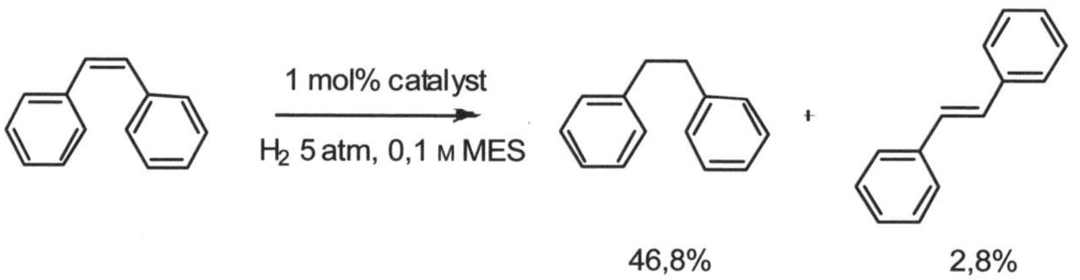

1 mol% catalyst

H$_2$ 5 atm, 0,1 M MES

46,8%   2,8%

Scheme 7

4. RACEMISATION AND DERACEMISATION

Racemisation is usually an undesired event along the course of a chemical sequence. Normally, the reaction conditions of every step are optimized in order to avoid any kind of isomerisation which would spoil the yield and introduce separation issues or downstream complications. Nevertheless, racemisation would be useful to the chemist when performing a resolution where the undesired enantiomer accumulates representing the major byproduct at the end of the reaction. If, in fact, a continuous substrate racemisation occurs in the reaction vessel, the constant interconversion will lead to the complete consumption of both the isomers of the starting compound. In the view of maximum efficiency, both reactions (racemisation and resolution) should occur independently, with comparable rate and in the same pot, this meaning that the harsh conditions (strong acids or bases, high temperature or pressure) usually associated to an efficient racemisation are to be avoided. Enzymes, working with high efficiency in mild conditions, can be very useful for this purpose.

Enantiopure amines are very important as building blocks in fine chemistry, can be resolved by a variety of enzymatic and chemical agents, but are not prone to racemisation, thus constituting a challenging substrate. ω-Transaminases efficiently transfer an amino group from a donor to an acceptor and are quite enantioselective. Coupling two transaminases (TA) with opposite selectivity it is possible to attain the complete racemisation of a number of chiral amines making their resolution feasible in an one-pot process.[13] (*Scheme 8*)

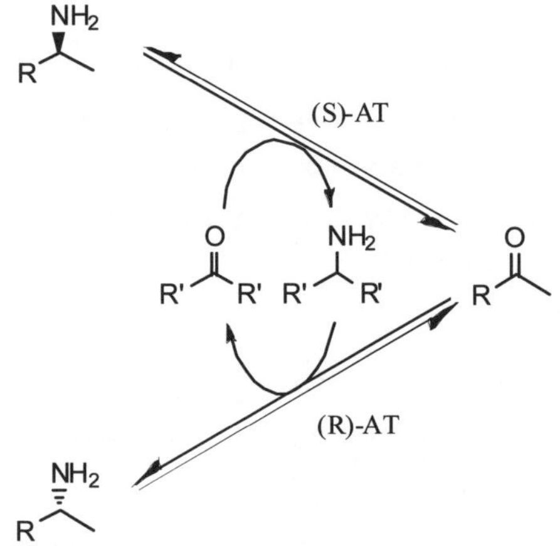

Scheme 8

For instance, both enantiomers of the therapeutically relevant amine mexiletine have been separately obtained from the racemate through a combination of two ω-transaminases with opposite selectivity whose respective reactions have been coupled with two irreversible processes (*Scheme 9*).[13]

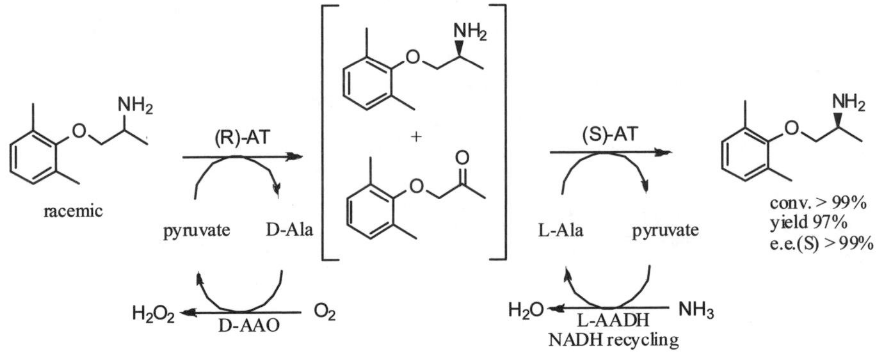

Scheme 9

5. CARBON-CARBON BONDS

The creation of new carbon-carbon bonds is crucial in increasing the complexity of a molecule, therefore a regio- or stereoselective method for coupling carbon fragments is particularly valuable.

It is possible, for instance, to perform a regioselective biocatalytic Friedel-Crafts alkylation on aminocoumarins using modified cofactors as alkyl donors (*Scheme 10*).[14]

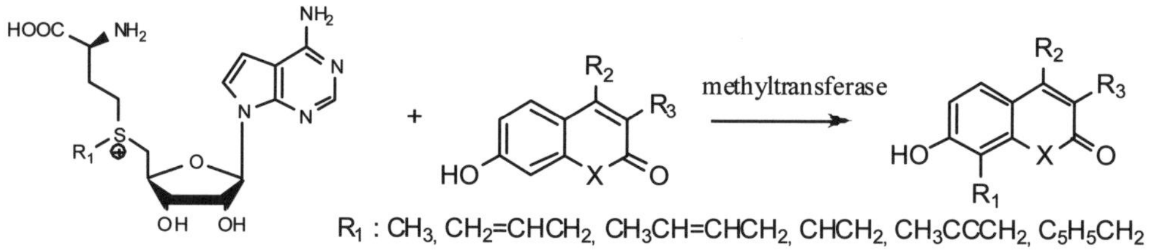
methyltransferase
R1 : CH3, CH2=CHCH2, CH3CH=CHCH2, CHCH2, CH3CCCH2, C5H5CH2

Scheme 10

D-Fructose-6-phosphate aldolase (FSA) from *E. coli* is a protein exploitable for carrying out a cascade chemo-enzymatic synthesis of complex sugar-related polyhydroxylated compounds starting from simple precursors (*Scheme 11*).[15]

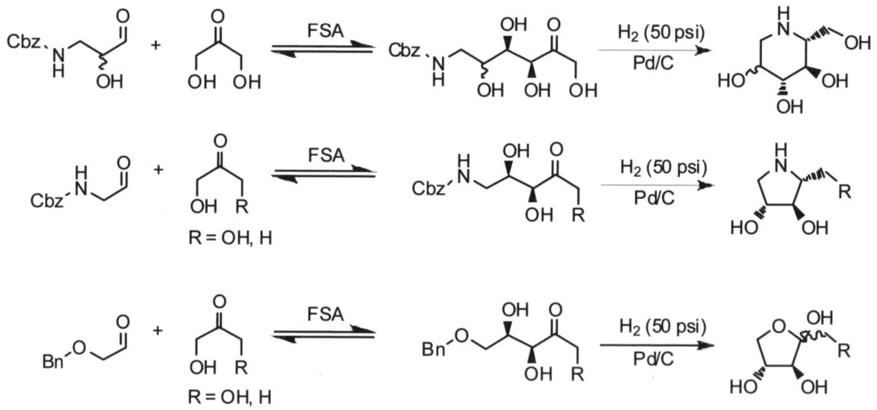
FSA
H2 (50 psi)
Pd/C
R = OH, H

Scheme 11

6. ENZYME PROMISCUITY

The enzymes are usually very specific, being the product of a long evolution directed towards a definite transformation. Nevertheless, it has been sometimes shown that a number of biocatalysts (usually hydrolytic enzymes) can promote a second, albeit unexpected, reaction.[16]

For instance, as illustrated in *Scheme 12*, the lipase from *Candida anctartica* is able to activate hydrogen peroxide towards the oxidation of secondary aryl alcohols to ketones, for a novel metal-free reaction taking advantage of ionic liquids.[17]

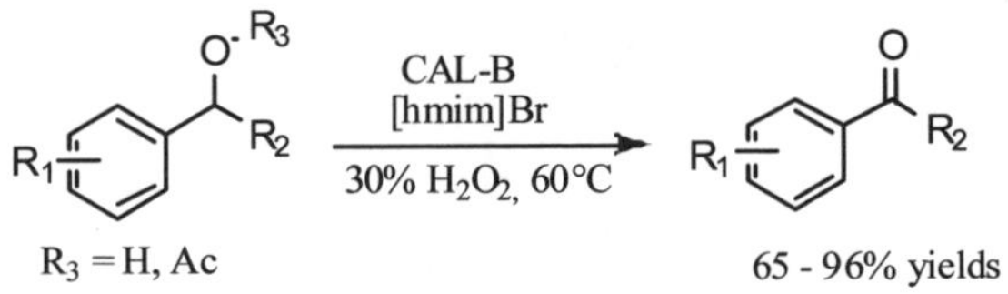

Scheme 12

Lipases can also catalyse the formation of new carbon-carbon bonds through an aza-Michael addition[18] or a direct Mannich reaction.[19] In the former case, a β-amino ester is obtained, whereas in the latter a β-amino ketone is the product of a three-component reaction (*Scheme 13*).

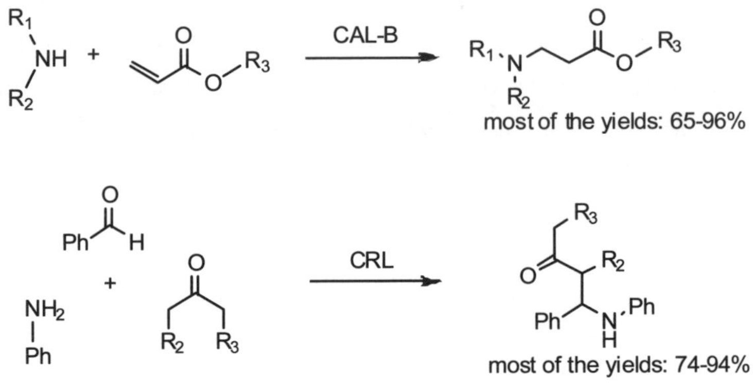

Scheme 13

In a recent work, the lipase from porcine pancreas (PPL) has been shown to catalyse (in a multicomponent system) a Knoevenagel condensation, a Michael addition and a cyclization at the same time; the combination of these three reactions allows to build-up a variety of spirooxiindole derivatives in very good yields, as illustrated in *Scheme 14*.[20]

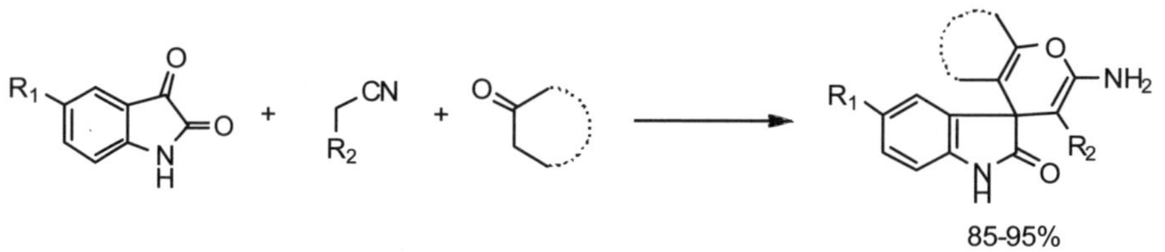

85-95%

Scheme 14

7.   SMART SYSTEMS

The mildness of conditions in which enzymes usually work permits to the chemist to design and set up one-pot systems or sequential cascade process who lead straight to desired product, thus diminishing the need for protective group chemistry, product isolation methodologies and special reactor design. Here we summarize some examples of cascade reactions in which enzymes work in tandem with other catalyst or different biocatalysts.

Example 1: Chiral 1,3-diols are widely used as building block in the synthesis of pharmaceutically active compounds and are precursor of chiral ligands for organometallic catalysis. A number of chiral diols have been prepared in all the four isomers exploiting a modular synthesis in which organocatalysis and biocatalysis work independently on separate stereogenic centers, as shown in *Scheme 15*.

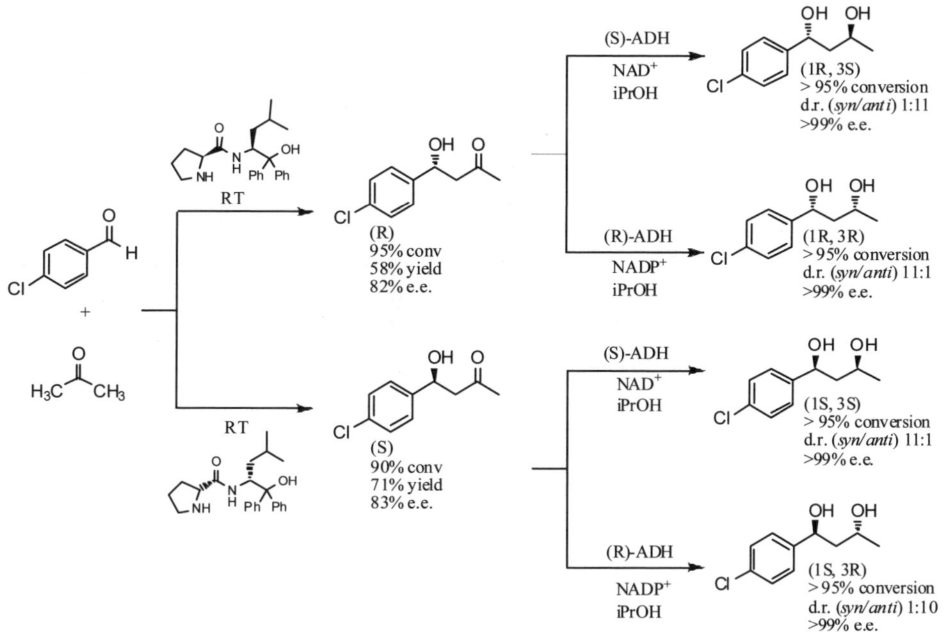

Scheme 15

Example 2: Working in similar fashion, a sequential synthesis of chiral 1,3-aminoalcohols has been reported, in which the first step is a Mannich asymmetric reaction, catalysed by D- or L-proline, leading to a chiral β-aminoketone. The latter is subjected to a one-pot reduction/resolution step employing a metal catalyst together with a selective lipase, eventually obtaining the desired product with excellent optical purity (see *Scheme 16*).[21]

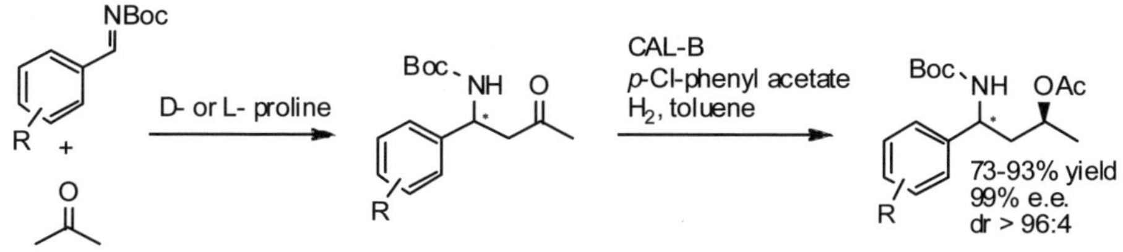

Scheme 16

Example 3: With a single enzyme it is possible to catalyse two concurrent processes and to obtain, from a ketone and a racemic alcohol, two enantiopure alcohols at the same time. (*Scheme 17*) Changing the enzyme, the opposite enantiomers are obtained in the same conditions.[22]

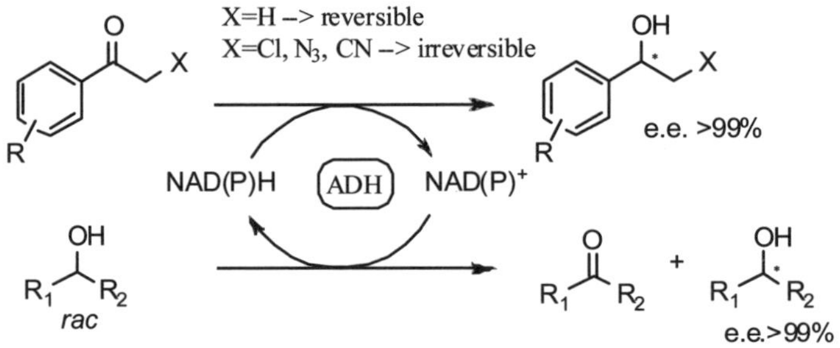

Scheme 17

Example 4: Starting from α-bromo ketones, enantiopure bromohydrins or terminal epoxides have been obtained employing an alcohol dehydrogenase of the convenient stereoselectivity; in the former case, if an azide moiety is present in the substrate, it is also possible to subsequently perform in the same pot a Staudinger reduction leading to optically pure five- or six-membered N-heterocycles, as illustrated in *Scheme 18*.[23]

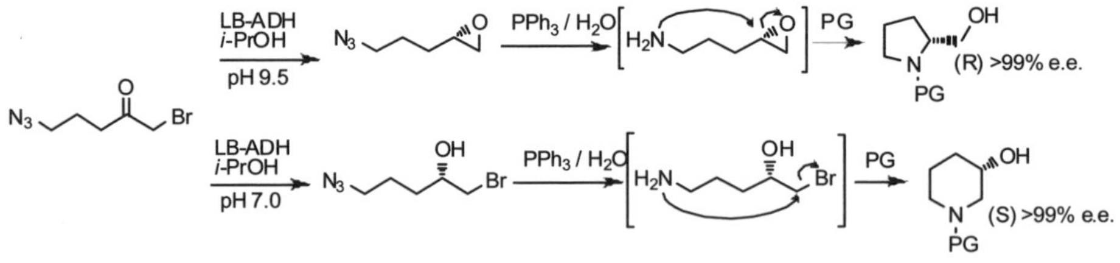

Scheme 18

Example 5: In a cascade process, racemic β-phenylalanine is resolved using a selective phenylalanine aminomutase (PAM) which selectively converts in the α-isomer the (R)-enantiomer. This is a reversible reaction which is shifted to the desired side by using a lyase (PAL) capable of converting irreversibly the α-amino acid into the corresponding cinnamic acid (*Scheme 19*). The (S)-β-phenylalanine is left unreacted in the reaction mixture and could be recovered with very good optical purity.[24]

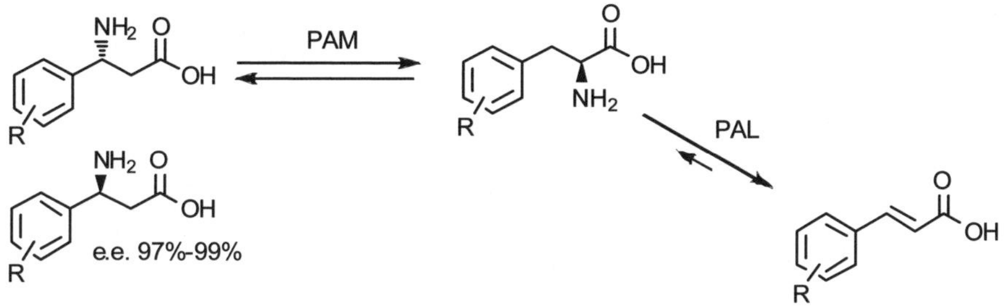

Scheme 19

Example 6: Halohydrin dehalogenase can perform enantioselective azidolysis of aromatic epoxides to 1,2-azido alcohols which are in turn ligated to alkynes through click chemistry producing chiral hydroxyl triazoles in a one-pot procedure with enantiomeric excess ranging from fair to excellent, as shown in *Scheme 20*.[25]

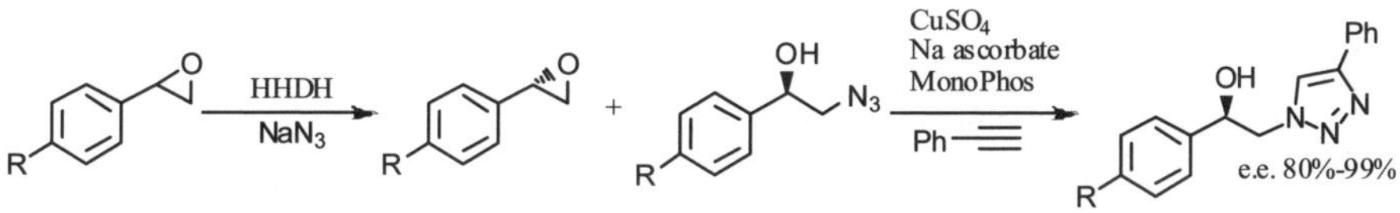

Scheme 20

Example 7: When a metallorganic catalyst and an enzyme catalyse the same reaction with opposite selectivity, it is possible to couple them in a single-pot procedure provided that they are able to work in the same conditions and that it is possible to shift the equilibrium towards the desired side of the reaction. In a recent work, a titanium chiral complex is used to produce an enantiomerically enriched mixture of acylated cyanohydrins; at the same time, a lipase selectively hydrolyses the minor enantiomer converting it back to the starting substrate and thus increasing cycle after cycle the optical purity of the product (*Scheme 21*).[26]

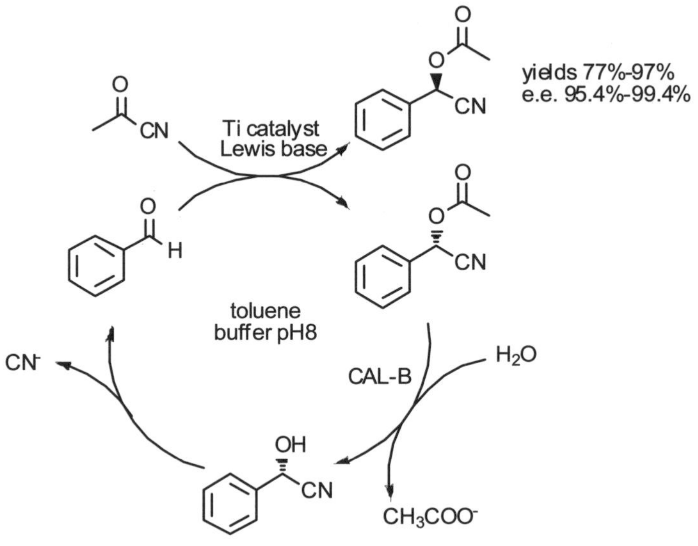

Scheme 21

Example 8: 12-Ketoursodeoxycholic acid has been produced starting from cholic acid using five enzymes working in a row (*Scheme 22*). Here the reaction is performed in three different setups allowing to carry the reaction to completion exploiting the cofactor specificities.[27]

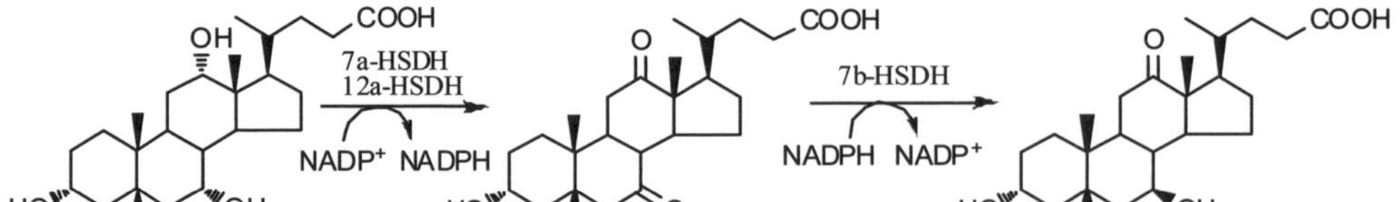

Scheme 22

Example 9: Novel iminocyclitols consisting of polihydroxylated pyrrolidine derivatives with different stereochemical configurations and different substitutions at position C5 have been synthesized by means of a sequence aldolase-phosphatase-chemical reduction from a number of N-protected aminoaldehydes, as illustrated in *Scheme 23*.[28] Aldolases are in general quite interesting because they introduce two new stereogenic centres at the same time, thus dramatically increasing the molecular complexity.

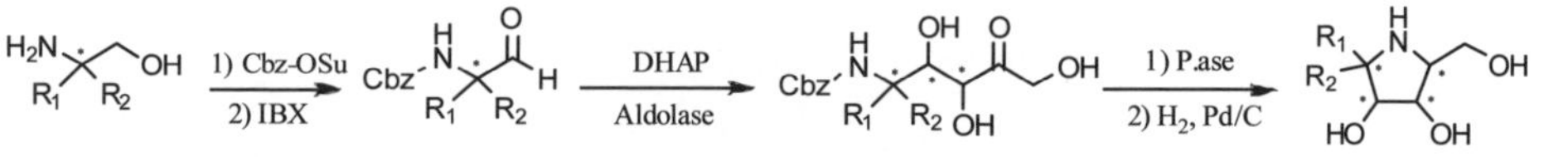

Scheme 23

In a comprehensive review[29], the role of enzymes in generating downstream molecular complexity has been elucidated in order to show the convenience of introducing biocatalytic procedures in the synthesis of a target compound, especially if in the early steps.

8. PRODUCT-FOCUSED SYNTHESIS

Due to the great number of synthetically useful enzyme-catalyzed reactions, it is not unusual to find a biocatalytic reaction to be part of the synthesis of a given target compound. Here we present a summary of sequences in which a biocatalytic step plays a fundamental role.

Sertraline and norsertraline are important active molecules: the former is currently marketed for the treatment of depression, the latter is so far under clinical trials for treatment of CNS disorders.

As explained in *Scheme 24*, both molecules can be synthesized through a sequence in which the obtainment of an enantiopure chiral compound, from a combination of metal-catalyzed racemisation and enzymatic resolution, is the key step for what concerns the source of chirality.[30-31]

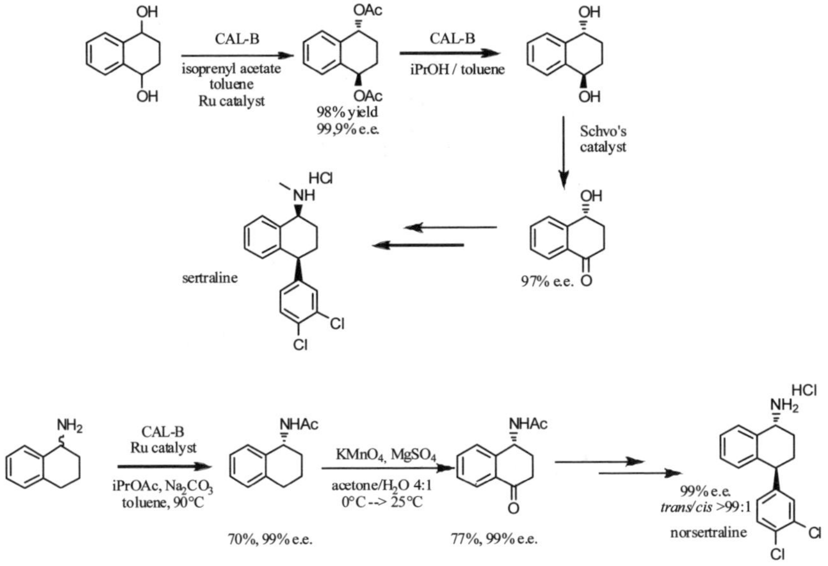

Scheme 24

Atorvastatin calcium is the active component for the best selling drug in the world (Lipitor®). Its synthesis involves a chiral hydroxynitrile, whose preparation is in any case achieved by addition of an alkaline cyanide to a chiral precursor at elevated temperatures. This operation leads to the formation of several byproducts which are difficult to separate and, of course, has obvious safety issues. An engineered halohydrin dehalogenase accepts cyanide ion as non-natural nucleophile and is able to catalyse the addition reaction (see *Scheme 25*). It is thus possible to design a new process whose "green properties" are much superior to the traditional process,[32] especially concerning waste prevention, safety and energy efficiency.

On the same topic, in a recent review[33] the various biocatalytical methodologies carrying out the synthesis of the key intermediates for atorvastatine have been discussed in detail.

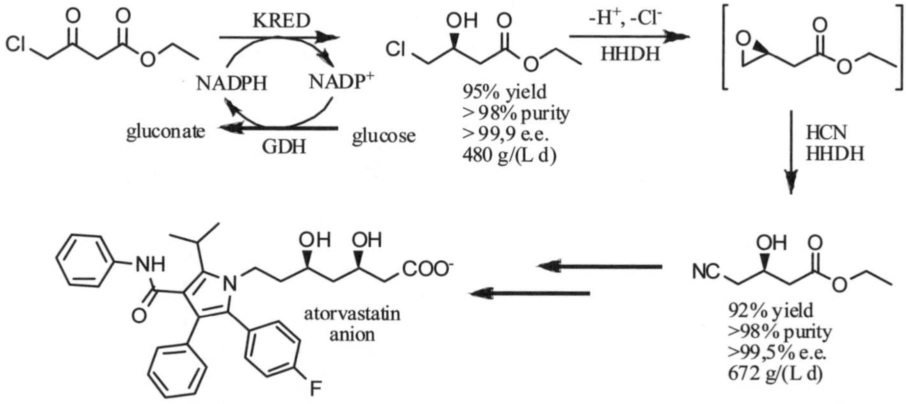

Scheme 25

Chiral 2-amino-1,3-diols are an important class of pharmaceutically relevant compound. In a recent work a new strategy has been devised and demonstrated in which the succession of two enzymatic steps converts two achiral precursors (propanal and hydroxypyruvate) to the enantiopure (2S,3S)-2-aminopentane-1,3-diol.[34] (see *Scheme 26*)

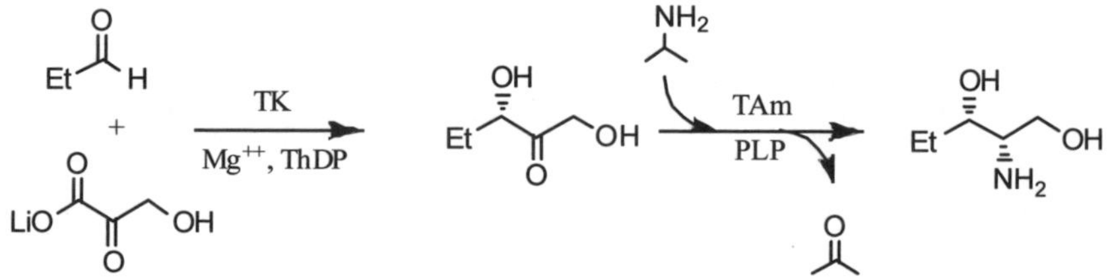

Scheme 26

(S)-allysine ethylene acetal is a key intermediate in several angiotensin converting enzymes (ACE) and neutral endopeptidase (NEP) inhibitors currently in clinical trials. A new strategy shown in *Scheme*

27, including an enzymatic resolution step, has been designed for its manufacture, and it has been transferred to a production plant, yielding multihundred kilograms of this valuable chiral intermediate.[35]

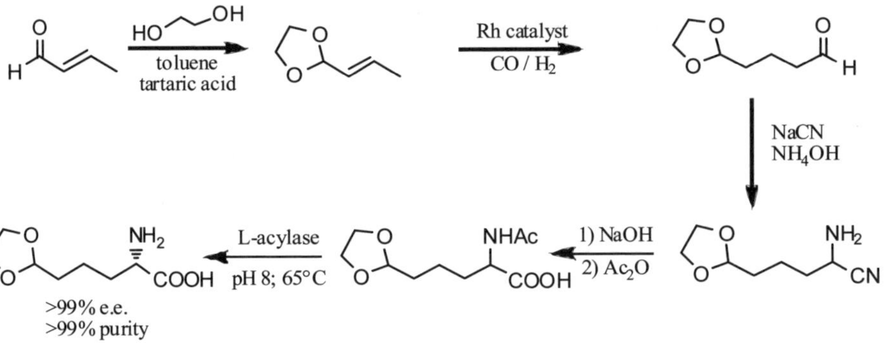

Scheme 27

Biocatalytic steps can also provide valuable chiral material for exploiting a successive multicomponent reaction (MCR). For example, an enantiopure imine generated from the desymmetrization of a disubstituted *meso*-pyrrolidine through the action of a monoamine oxidase has been employed in a Ugi MCR for synthesizing prolyl peptides,[6] the hepatitis C virus NS3 protease inhibitor Telaprevir,[36] and a number of synthetic alkaloids,[37] as illustrated in *Scheme 28*.

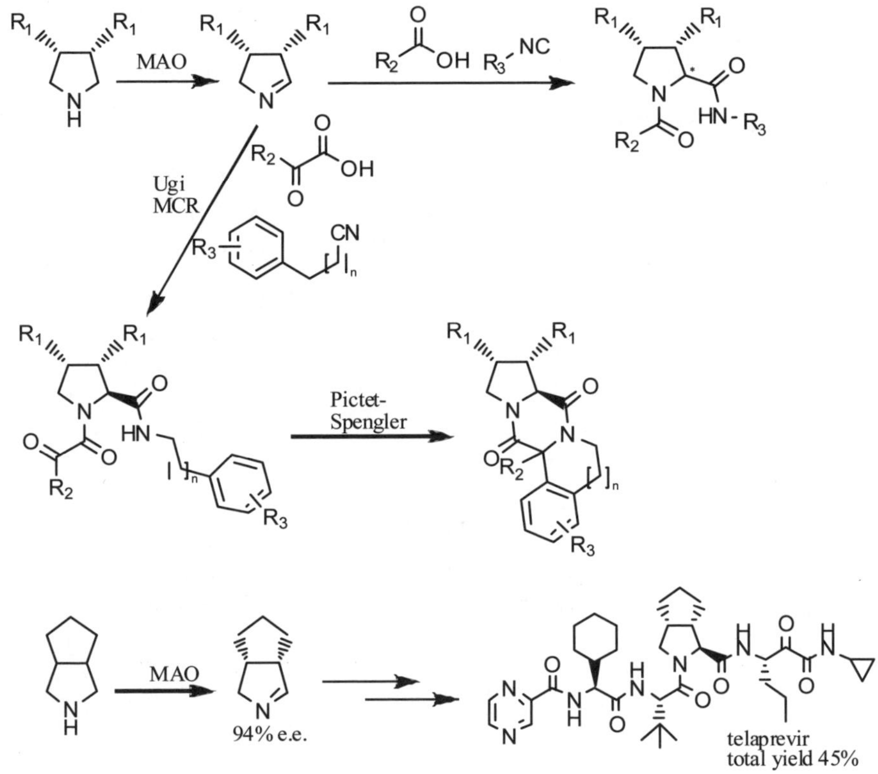

Scheme 28

β-Thimidine is a nucleoside which is a fundamental precursor to the anti-AIDS drugs stavudine (d4T) and zidovudine (AZT). This nucleoside can be efficiently obtained from 5-methyl uridine, which is in turn derived by an enzymatic transglycosilation converting an inexpensive substrate,[38] as shown in *Scheme 29*.

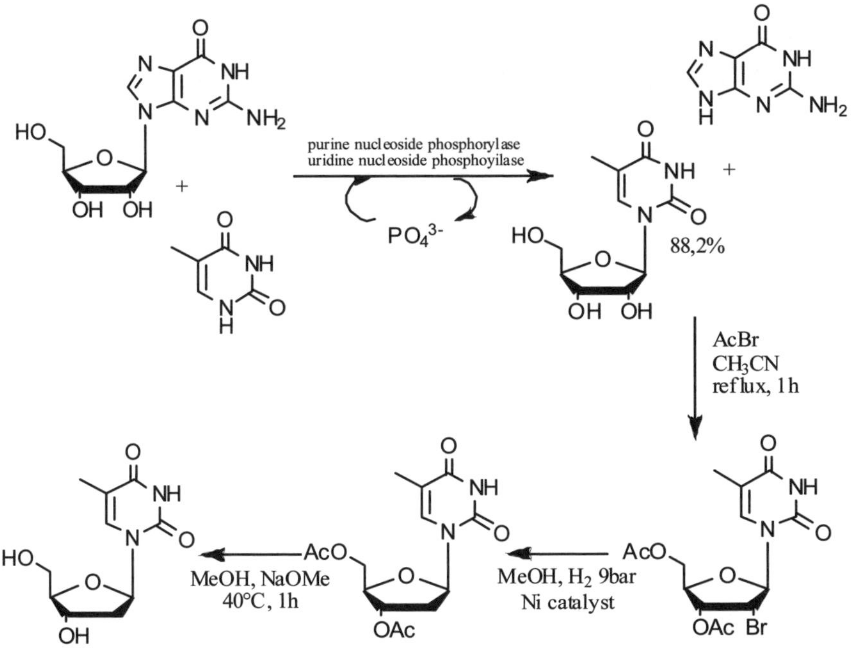

Scheme 29

Pancrastatin analogs have been lately synthesized by a combination of Biocatalysis and click chemistry: the cyclitol ring has been prepared by action of a toluene dioxygenase (TDO) followed by an epoxidation and a nucleophilic substitution. Further elaboration concerning the insertion of a triazole group have led to the target product (*Scheme 30*).[39]

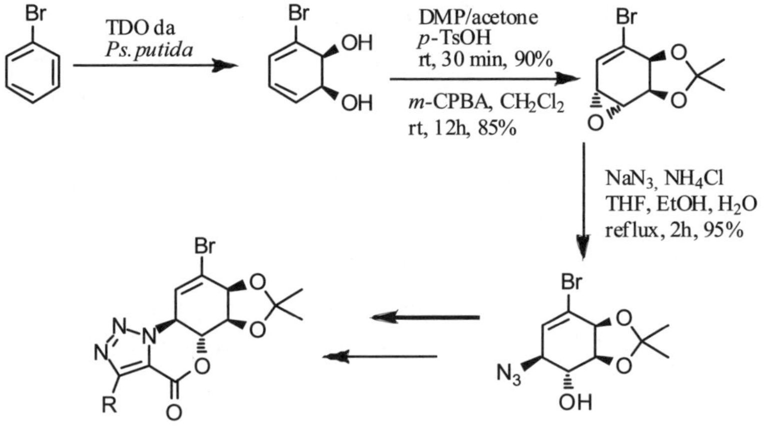

Scheme 30

A novel ketoreductase (KRED) engineered via directed evolution technologies have been developed for an economical and simple process for the production of a key intermediate for the synthesis of montelukest sodium (Singulair), an important leukotriene receptor antagonist (*Scheme 31*).[40] The evolved enzyme is robust and efficient, and can stand the presence of 70% organic solvent at 45°C, permitting to the process to be scaled up to >200 kg scale.

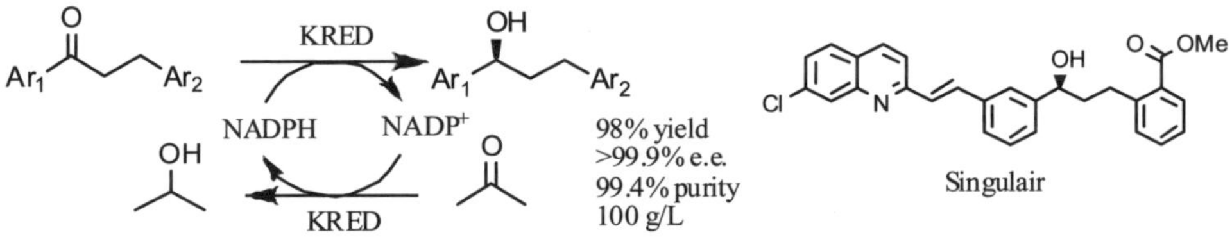

98% yield
>99.9% e.e.
99.4% purity
100 g/L

Singulair

Scheme 31

9. CONCLUSION AND FUTURE PERSPECTIVES

The recent literature confirms how enzymes can easily play an important role in organic synthesis. Their compatibility with other catalysts, their specificity, their green features can be exploited in designing convenient multi-step processes.

Moreover, in the latest years the ever-growing molecular biology tools has permitted to tailor enzymes with novel catalytic activities.[41] The great variety of available biocatalysts permits even to imagine multi-step pure *in-pot* enzymatic synthesis thus mimicking *in vivo* processes.[42]

Aspects like catalyst stability or commercial availability have been greatly improved, and successful immobilization is often the key for a profitable employment in industrial processes.[43]

10. REFERENCES

1. Hollmann, F.; Arends, I. W. C. E.; Buehler, K.; Schallmey, A.; Buehler, B. *Green Chem.* **2011,** *13*, 226

2. Zhang, W.; Tang, W. L.; Wang, D. I. C.; Li, Z. *Chem. Commun. (Cambridge, U. K.)* **2011,** *47*, 3284

3. Witayakran, S.; Ragauskas, A. J. *Adv. Synth. Catal.* **2009,** *351*, 1187

4. Bruyneel, F.; Payen, O.; Rescigno, A.; Tinant, B.; Marchand-Brynaert, J. *Chemistry* **2009,** *15*, 8283

5. Koehler, V.; Bailey, K. R.; Znabet, A.; Raftery, J.; Helliwell, M.; Turner, N. J. *Angew. Chem., Int. Ed.* **2010,** *49*, 2182

6. Znabet, A.; Ruijter, E.; de Kanter, F. J. J.; Koehler, V.; Helliwell, M.; Turner, N. J.; Orru, R. V. A. *Angew. Chem., Int. Ed.* **2010,** *49*, 5289

7. Rehdorf, J.; Mihovilovic Marko, D.; Bornscheuer Uwe, T. *Angew Chem Int Ed Engl* **2010,** *49*, 4506

8. Warburg, O.; Christian, W. *Naturwissenschaften* **1932,** *20*, 688

9. Bougioukou, D. J.; Walton, A. Z.; Stewart, J. D. *Chem. Commun. (Cambridge, U. K.)* **2010,** *46*, 8558

10. Stueckler, C.; Winkler, C. K.; Bonnekessel, M.; Faber, K. *Adv. Synth. Catal.* **2010,** *352*, 2663

11. Grau, M. M.; van der Toorn, J. C.; Otten, L. G.; Macheroux, P.; Taglieber, A.; Zilly, F. E.; Arends, I. W. C. E.; Hollmann, F. *Adv. Synth. Catal.* **2009,** *351*, 3279

12. Jing, Q.; Okrasa, K.; Kazlauskas Romas, J. *Chemistry* **2009,** *15*, 1370

13. Koszelewski, D.; Grischek, B.; Glueck Silvia, M.; Kroutil, W.; Faber, K. *Chemistry* **2011,** *17*, 378

14. Stecher, H.; Tengg, M.; Ueberbacher, B. J.; Remler, P.; Schwab, H.; Griengl, H.; Gruber-Khadjawi, M. *Angew. Chem., Int. Ed.* **2009,** *48*, 9546

15. Concia Alda, L.; Lozano, C.; Castillo Jose, A.; Parella, T.; Joglar, J.; Clapes, P. *Chemistry* **2009,** *15*, 3808

16. Busto, E.; Gotor-Fernandez, V.; Gotor, V. *Chem. Soc. Rev.* **2010,** *39*, 4504

17. Sharma, U. K.; Sharma, N.; Kumar, R.; Kumar, R.; Sinha, A. K. *Org. Lett.* **2009,** *11*, 4846

18. Dhake, K. P.; Tambade, P. J.; Singhal, R. S.; Bhanage, B. M. *Tetrahedron Lett.* **2010,** *51*, 4455

19. He, T.; Li, K.; Wu, M.-Y.; Feng, X.-W.; Wang, N.; Wang, H.-Y.; Li, C.; Yu, X.-Q. *J. Mol. Catal. B Enzym.* **2010,** *67*, 189

20. Chai, S.-J.; Lai, Y.-F.; Xu, J.-C.; Zheng, H.; Zhu, Q.; Zhang, P.-F. *Adv. Synth. Catal.* **2011,** *353*, 371

21. Millet, R.; Traff Annika, M.; Petrus Michiel, L.; Backvall Jan, E. *J Am Chem Soc* **2010,** *132*, 15182

22. Bisogno Fabricio, R.; Lavandera, I.; Kroutil, W.; Gotor, V. *J Org Chem* **2009,** *74*, 1730

23. Bisogno, F. R.; Cuetos, A.; Orden, A. A.; Kurina-Sanz, M.; Lavandera, I.; Gotor, V. *Adv. Synth. Catal.* **2010,** *352*, 1657

24. Wu, B.; Szymanski, W.; de Wildeman, S.; Poelarends, G. J.; Feringa, B. L.; Janssen, D. B. *Adv. Synth. Catal.* **2010,** *352*, 1409

25. Campbell-Verduyn, L. S.; Szymanski, W.; Postema, C. P.; Dierckx, R. A.; Elsinga, P. H.; Janssen, D. B.; Feringa, B. L. *Chem. Commun. (Cambridge, U. K.)* **2010,** *46*, 898

26. Wingstrand, E.; Laurell, A.; Fransson, L.; Hult, K.; Moberg, C. *Chemistry* **2009,** *15*, 12107

27. Monti, D.; Ferrandi, E. E.; Zanellato, I.; Hua, L.; Polentini, F.; Carrea, G.; Riva, S. *Adv. Synth. Catal.* **2009,** *351*, 1303

28. Calveras, J.; Egido-Gabas, M.; Gomez, L.; Casas, J.; Parella, T.; Joglar, J.; Bujons, J.; Clapes, P. *Chemistry* **2009,** *15*, 7310

29. Hudlicky, T.; Reed, J. W. *Chem. Soc. Rev.* **2009,** *38*, 3117

30. Krumlinde, P.; Bogar, K.; Backvall Jan, E. *Chemistry* **2010,** *16,* 4031

31. Thalen Lisa, K.; Zhao, D.; Sortais, J.-B.; Paetzold, J.; Hoben, C.; Backvall Jan, E. *Chemistry* **2009,** *15,* 3403

32. Ma, S. K.; Gruber, J.; Davis, C.; Newman, L.; Gray, D.; Wang, A.; Grate, J.; Huisman, G. W.; Sheldon, R. A. *Green Chem.* **2010,** *12,* 81

33. Patel, J. M. *J. Mol. Catal. B Enzym.* **2009,** *61,* 123

34. Smith, M. E. B.; Chen, B. H.; Hibbert, E. G.; Kaulmann, U.; Smithies, K.; Galman, J. L.; Baganz, F.; Dalby, P. A.; Hailes, H. C.; Lye, G. J.; Ward, J. M.; Woodley, J. M.; Micheletti, M. *Org. Process Res. Dev.* **2010,** *14,* 99

35. Cobley, C. J.; Hanson, C. H.; Lloyd, M. C.; Simmonds, S.; Peng, W.-J. *Org. Process Res. Dev.* **2011,** *15,* 284-

36. Znabet, A.; Polak, M. M.; Janssen, E.; de Kanter, F. J. J.; Turner, N. J.; Orru, R. V. A.; Ruijter, E. *Chem. Commun. (Cambridge, U. K.)* **2010,** *46,* 7918

37. Znabet, A.; Zonneveld, J.; Janssen, E.; De Kanter, F. J. J.; Helliwell, M.; Turner, N. J.; Ruijter, E.; Orru, R. V. A. *Chem. Commun. (Cambridge, U. K.)* **2010,** *46,* 7706

38. Gordon, G. E. R.; Bode, M. L.; Visser, D. F.; Lepuru, M. J.; Zeevaart, J. G.; Ragubeer, N.; Ratsaka, M.; Walwyn, D. R.; Brady, D. *Org. Process Res. Dev.* **2011,** *15,* 258

39. de la Sovera, V.; Bellomo, A.; Gonzalez, D. *Tetrahedron Lett.* **2011,** *52,* 430

40. Liang, J.; Lalonde, J.; Borup, B.; Mitchell, V.; Mundorff, E.; Trinh, N.; Kochrekar, D. A.; Nair Cherat, R.; Pai, G. G. *Org. Process Res. Dev.* **2010,** *14,* 193

41. Lutz, S. *Science* **2010,** *329,* 285

42. Santacoloma, P. A.; Sin, G.; Gernaey, K. V.; Woodley, J. M. *Org. Process Res. Dev.* **2011,** *15,* 203

43. Sheldon, R. A. *Org. Process Res. Dev.* **2011,** *15,* 213

Finito di stampare nel mese di maggio 2011
presso **ARTELITO** S.p.A. - industria grafica - Camerino (MC)